IEE TELECOMMUNICATIONS SERIES 23

Series Editors: Professor J. E. Flood
Professor C. J. Hughes
Professor J. D. Parsons

RADIO SPECTRUM MANAGEMENT

Other volumes in this series:

Volume 1 **Telecommunications networks** J. E. Flood (Editor)
Volume 2 **Principles of telecommunication-traffic engineering** D. Bear
Volume 3 **Programming electronic switching systems** M. T. Hills and S. Kano
Volume 4 **Digital transmission systems** P. Bylanski and D. G. W. Ingram
Volume 5 **Angle modulation: the theory of system assessment** J. H. Roberts
Volume 6 **Signalling in telecommunications networks** S. Welch
Volume 7 **Elements of telecommunications economics** S. C. Littlechild
Volume 8 **Software design for electronic switching systems** S. Takamura, H. Kawashima, N. Nakajima
Volume 9 **Phase noise in signal sources** W. P. Robins
Volume 10 **Local telecommunications** J. M. Griffiths (Editor)
Volume 11 **Principles and practices of multi-frequency telegraphy** J. D. Ralphs
Volume 12 **Spread spectrum in communications** R. Skaug and J. F. Hjelmstad
Volume 13 **Advanced signal processing** D. J. Creasey (Editor)
Volume 14 **Land mobile radio systems** R. J. Holbeche (Editor)
Volume 15 **Radio receivers** W. Gosling (Editor)
Volume 16 **Data communications and networks** R. L. Brewster (Editor)
Volume 17 **Local telecommunications 2** J. M. Griffiths (Editor)
Volume 18 **Satellite communication systems** B. G. Evans (Editor)
Volume 19 **Telecommunications traffic, tariffs and costs** R. E. Farr
Volume 20 **An introduction to satellite communications** D. I. Dalgleish
Volume 21 **SPC digital telephone exchanges** F. J. Redmill and A. R. Valdar
Volume 22 **Data communications and networks II** R. L. Brewster (Editor)
Volume 23 **Radio spectrum management** D. J. Withers
Volume 24 **Satellite communication systems II** B. G. Evans (Editor)
Volume 25 **Personal mobile radio systems** R. C. V. Macario (Editor)

RADIO SPECTRUM MANAGEMENT

D J Withers

Peter Peregrinus Ltd. on behalf of the Institution of Electrical Engineers

Published by: Peter Peregrinus Ltd., London, United Kingdom

Peter Peregrinus Ltd.,
Michael Faraday House,
Six Hills Way, Stevenage,
Herts. SG1 2AY, United Kingdom

British Library Cataloguing in Publication Data
Withers, D. J.
Radio spectrum regulation and management.—
(Telecommunications).
1. Broadcasting. Radio frequency wavebands. Allocation
I. Title II. Series
384.5452

ISBN 0 86341 177 0

Printed in England by Redwood Press Ltd., Wiltshire

Contents

Preface

A government has authority over the use of the radio spectrum within its jurisdiction and with that authority goes a responsibility to ensure that the spectrum is managed so that its citizens can use radio effectively. Spectrum management within a single isolated society would not be difficult. However, radio signals do not respect political frontiers, and the international dimension which is inescapable in spectrum management brings complexities, mitigated by collaboration. This collaboration is disciplined, mainly, by a code of practice which has been developed under the aegis of the International Telecommunication Union (ITU). The development of this code, the ITU Radio Regulations, and its implementation, are backed by extensive international technical studies, done through the International Radio Consultative Committee (CCIR), an organ of the ITU.

The use of radio is growing rapidly and new applications are emerging all the time. Growth and innovation require frequent modification of the Radio Regulations. However, the needs of different countries evolve in different ways and at different rates. It is seldom possible to sweep away old regulations to make way for new ones; instead, the new must usually be grafted onto the old. This process of accretion, continued over decades, has produced extensive and complex regulations, set down in a text which few find user-friendly.

Moreover, the time is passing when, in most countries, the government spectrum managers and a few engineers in major telecommunications and broadcasting organisations are the only people interested in the management of the spectrum. Liberalisation of service provision is creating a much wider concern for the way in which the use of the spectrum is regulated and may more engineers could benefit from access to the results of the CCIR technical studies. The CCIR radio propagation data, for example, are uniquely valuable for radio system planning.

The broad survey which this book contains is intended as an introduction to radio regulation and as a directory to sources of relevant technical information. I am hopeful that it will give spectrum managers an introduction to areas of radio regulation with which they are not familiar and that it will help system engineers to make fuller use of the system planning material that is available. For others with less specialised concerns, this survey may include all that they need to know of the subject.

The information on matters concerning the ITU is up to date to the end of 1989. However, the CCIR held its XVII Plenary Assembly in May 1990. New and amended Recommendations and Reports were agreed and the texts will, no doubt, be published during the course of 1991. Changes were made to the study group structure. Former Study Groups 2 and 7 were combined to form a new Study Group 7, responsible for 'science services'. Former Study Groups 3 and 9 were combined to form a new Study Group 9, responsible for studies of fixed service systems at all frequencies. The newly formed Study Group 12 will study specific problems of inter-service sharing. And new arrangements were made for coordinating terminology within the CCIR, in liaison with the CCITT, to replace Study Group CMV. Finally, substantial changes were planned for the working methods used in the CCIR, designed to facilitate a faster response to urgent problems.

Acknowledgements

In writing this book I have been drawing on the stored wisdom of generations of administrators and engineers. The kind advice of friends and ex-colleagues in British Telecommunications plc and the Radiocommunications Division of the United Kingdom Department of Trade and Industry has steered me away from a number of errors. I take responsibility for any errors that remain and for any opinions expressed. But above all, in acknowledging my debts, I must give my heartfelt thanks for the education I have received from delegates of a hundred nations at many international conferences and meetings between 1949 and 1988. With their help I have gained some understanding of the ways in which radically different needs in spectrum management may sometimes be reconciled.

The reproduction in this book of material taken from the publications of the International Telecommunication Union (ITU), Place des Nations, CH-1211 Geneva 20, Switzerland, has been authorised by the ITU as copyright holder.

Chapter 1

Introduction

Radio is the most versatile of the telecommunications transmission media. It is the only feasible medium for some kinds of communication system and for many others it is often the most economical medium. The operation of many navigation systems depends on the predictability of radio wave propagation.

New uses and many new users are being found for radio every year, adding to the pressure on the radio spectrum. Trunk telephone links, broadcasting facilities in variety, navigation systems and military radars must, for example, compete with cordless telephones and children's radio-controlled toys for a share of the same radio spectrum. We have become familiar with the radiotelephone antenna at the policeman's shoulder and on the motor car roof and we are becoming familiar with the cellular radiotelephone in the city executive's pocket. Military vehicles bristle with antennas. And these are just a few examples where the application of radio is obvious; many devices used, for example, for road vehicle fleet management, for theft and burglar alarms, in scientific apparatus, in industrial processes and for medical diagnosis and treatment do not reveal their use of the radio spectrum so openly.

Compared with guided transmission media like coaxial cables and optical fibre cables, radio transmission has important advantages but it also has four significant potential disadvantages. The operation of radio systems may be degraded or interrupted by natural events in the propagation path. Radio is not intrinsically private; compared with a message sent by cable, a radio message can more easily be intercepted by someone for whom it was not intended, leaving no trace. There is a risk of interruption or degradation of the functioning of a radio system due to interference from another radio system. Finally, while the radio spectrum is very wide and it can support enormous numbers of systems, its capacity is not limitless; indeed the radio propagation characteristics which are desirable for some kinds of system are confined to quite limited frequency ranges and careful optimisation of spectrum use, especially of those parts of the spectrum, is essential if the use of radio is to continue to grow.

Good system design can do much to minimise interruptions due to natural causes and to protect the privacy of radio circuits where this is required; it is the business of the system engineer to design accordingly. The control of interference and the management of the use of the spectrum to ensure that there is enough capacity at an acceptable cost are the objectives of the radio spectrum manager. These latter functions are the subject of this book.

Radio spectrum management is a responsibility of governments. The importance of adequate spectrum management is generally recognised and all governments set up agencies to regulate the use of radio within their jurisdiction, to ensure that the radio systems which they consider that their countries require can operate efficiently, in sufficient quantity and at least cost. This process takes the form of licensing the use of radio stations, together with all the back-up functions which licensing requires.

Governments are also concerned to ensure that public utilities such as telecommunications, broadcasting, radionavigation and other facilities using radio which are available to their citizens are of an appropriate standard of quality, availability and price. Accordingly, governments control the provision of such facilities; this is usually called 'regulation'. A licensing process is commonly used for this purpose also.

These two licensing functions of governments are obviously interrelated. For example, if the regulation of broadcasting or telecommunications facilities is liberal, there is likely to be a heavy demand for radio station licences and the liberal policy will be frustrated if licences to operate radio stations are not forthcoming. Conversely, the need for radio station licences is likely to be small if the government's policies on the licensing of facilities are restrictive. Despite this interrelation, the objectives of the two kinds of licensing process are quite different and this book does not consider the regulation of the provision of facilities.

Management of the use of the radio spectrum is usually performed or supervised by a government ministry, often a ministry which is responsible for communications, transport, industry or commerce. These national radio spectrum management authorities call themselves *administrations* and this usage will be followed in this book.

It is sometimes found to be convenient to divide spectrum management between two co-operating organisations, one responsible for oversight of the use of the radio spectrum by the organs of the government itself and the other regulating its use for non-government, typically commercial, purposes. In countries where telecommunications facilities used by the public are operated as a monopoly by a government-owned organisation, the regulation of the use of radio for all non-government purposes is sometimes made a function of the same operating organisation. In some countries a quasi-autonomous body responsible to the government or the legislature may be set up to perform some of these regulatory functions.

Radio station licensing

For regulatory purposes, any radio transmitter or receiver is called a *radio station* whether it is large or small, occupying premises on land, aboard a ship or a satellite or carried in a pocket of the user's jacket. In general it is illegal to operate a transmitting station of any kind without a licence issued by the appropriate administration, and a licence is often required for a receiving station. A government may however dispense with formal licensing for its own radio stations and a single licence is sometimes issued for a group of stations owned by a single non-government user.

Exceptionally the administration may determine that no licence is required for radio transmitters and receivers that are used for specified purposes in specified frequency bands. This is done typically if the power radiated by the transmitters is so low that interference is unlikely to be caused to any other radio system outside the premises of the user and if the receivers are so insensitive to interference that the system needs no protection from other radio systems. Some intruder alarms, limited-area paging systems, cordless microphones, cordless telephones and simple short-distance telecommand devices might, for example, be treated in this way.

Typically a licence will specify:

- the owner of the station as the licensee,
- the location of the station if it is fixed or the identity of the vehicle in which it is installed if it is mobile,
- the purpose for which the use of the station is licensed, and
- the period of validity of the licence.

In addition the licence will usually identify, directly or indirectly, the radio frequency or frequencies on which transmission or reception is permitted. Other technical characteristics and parameters may also be specified. The licence may impose statutory responsibilities on the licensee, for example to maintain the confidentiality of messages received at the station and to follow specified procedures in an emergency involving danger to human life. Finally, the issue of a licence to operate a radio station may be made conditional on the licensee having shown competence to operate it.

An administration may charge a fee for a licence. The charge may be a small one, to cover some appropriate part of the cost of radio spectrum management. Other charges may, however, be included in the licence fee. For example, in countries where broadcasting is financed by a levy on people who receive programmes, it may be convenient to add the levy to the fee for a broadcasting receiving station licence and to collect both together. A charge may be included, graduated so as to stimulate efficient use of the spectrum or to discourage the use of radio for purposes which are economic only if the opportunity cost of using the spectrum for the intended purpose is disregarded; this is sometimes called spectrum pricing. Finally there may be an element of taxation in the licence fee.

Frequency assignment

As is indicated above, a radio station licence will, in many cases, specify the radio frequency or frequencies on which the station is licensed to operate. Operating frequencies are said to be *assigned* to a station.

The assigned frequencies will have been chosen by the administration from a part of the spectrum that is technically suitable for the user's purpose and from a frequency band that had been designated for the use of stations in the category to which the station in question belongs. These categories are called *services* and a frequency band is said to be *allocated* for a service.

In assigning frequencies to a station, administrations aim to ensure that transmitted signals reach the intended receiving stations:

- with sufficient strength,
- without unacceptable degradation of reception by interference from other licensed transmitting stations,
- without causing interference at an unacceptable level to another receiving station that is licensed to receive from a different transmitting station at or near the same frequency, and
- without needlessly occupying spectrum in locations where it may be needed for another purpose later.

To achieve these aims the administration will usually specify in the licence the power that may be emitted at the assigned frequency. Other parameters affecting the extent of the interference that the emission could cause, such as the bandwidth occupied by the emission and key characteristics of the transmitting antenna, may also be specified in the licence or imposed, for example, through transmitter type approval requirements. In addition, in determining whether emissions from other transmitting stations would be likely to cause interference, the administration will assume that receiving equipment achieves specified standards of technical performance.

Whether signal and interference levels will or will not be acceptable can be determined with sufficient accuracy, in most cases, by computation, given the necessary radio propagation data and a knowledge of the topography of the transmission paths. Frequency assignments will, however, often be regarded as provisional pending confirmation that adequate performance is obtained and excessive interference does not arise in operation. If excessive interference arises subsequently, the administration will be involved in eliminating it.

The frequencies assigned to each transmitting station and some basic emission and equipment parameters are recorded in a national frequency assignment register. Some appropriate collective form of registration will be used for fleets of mobile stations. Assignments to some receiving stations are registered similarly. This register provides the basis for calculating interference potential to and from other radio stations that may be assigned the same or an adjacent frequency in the future.

This process of calculating prospective levels of interference is readily feasible only to the extent that all of the radio stations involved, transmitting and receiving, are within the jurisdiction of the administration making the frequency assignments. However, the possibility of interference to or from radio stations in other countries must often also be considered at the time of making a frequency assignment. There is a preliminary discussion of the international dimension of frequency assignment later in this Chapter.

The technical requirements which may be imposed upon licensees are intended to enhance the efficiency with which the radio spectrum is used, thereby increasing the number of radio systems that can operate satisfactorily in a limited radio spectrum. For the same reason an administration may go to great lengths to ensure that frequencies are chosen for assignment to stations in an effective manner. However, equipment designed to stringent specifications is usually more expensive than less advanced equipment and the careful planning which an administration might have to devote to effective frequency assignment is often the most costly and technically demanding aspect of the spectrum manager's task. Furthermore, a trade-off is possible; for a given level of demand, insistence on stringent equipment performance can reduce the need for effective assignment planning, and vice versa. This economic dimension of spectrum management is important.

The pressure of demand for radio facilities is growing everywhere, but it also varies from place to place, from service to service and from one part of the radio spectrum to another. Each administration must assess the pressure of domestic demand that its management methods and technical requirements must supply, whilst bearing in mind the effect that foreign use of the spectrum will have on its own country's freedom of action, and find the optimum management policy.

Frequency allocation

To assign frequencies to transmitters at random, disregarding the nature of the interference problems that they pose, their technical characteristics and the facilities that they provide, would lead to inefficient use of spectrum, inconvenience for both the users and the manufacturers of radio equipment and aggravation of spectrum management problems. To provide a basis for more orderly and efficient use of the spectrum, radio stations are categorised into services and an administration divides the frequency spectrum into discrete bands, each of which is allocated for one, or more rarely two, of these services.

The basis of the categorisation of radio stations by services is operational and administrative rather than technical. One important criterion involves the geographical or topographical aspects of protecting frequency assignments from interference. Thus for example:

- If an assignment is to be used to link a transmitter at one fixed location to a receiver at another fixed location, the transmitting station can be defined very precisely as the sole source of interference to other links, due to that assignment, and the receiving station can be defined very precisely as the one location where interference from other transmitters must be kept to a low level. These stations provide a basic example of the *fixed service.*

- If an assignment is for two-way simplex communication between base stations and mobile stations operating within a specified area, the transmitters of the system, considered as a source of interference, may be located anywhere within that specified area and the level of interference from other stations must be kept low over the whole of the specified area. These base and mobile stations illustrate the *mobile service.* For most purposes the mobile service is sub-divided into the *maritime mobile service,* the *aeronautical mobile service* and the *land mobile service.*

and so on. Other criteria are functional. Thus:

- If a transmission is intended for direct reception by the general public, the transmitting and receiving stations involved belong to the *broadcasting service* or, if satellite-borne transmitters are used, the *broadcasting-satellite service.*
- Stations using radio wave propagation for navigation or radiolocation purposes belong to the *radiodetermination service* or the *radiodetermination-satellite service.*
- The reception of radio waves of cosmic origin for the purposes of astronomy is the function of the *radio astronomy service.*

In all, at present, radio stations are categorised into 33 services.

This primary categorisation of stations into services is by no means exhaustive, and some services include many kinds of station which differ from one another in important ways. Stations of very different technical characteristics may be part of the same service, and it is sometimes desirable to operate these different kinds of station in different frequency bands. Thus an administration draws up a national frequency allocation table showing the allocations of defined frequency bands to the various services, but in addition:

- Some bands are likely to be designated for systems operated by the government, such as military systems, and others may be designated for commercial systems.
- Some bands allocated to broadcasting would be designated for sound broadcasting whilst others would be designated for television.
- Some bands allocated to the fixed service might be designated for the use of a specified major commercial user, such as a major public telecommunications carrier.
- Some bands or parts of bands allocated for the mobile service may be designated, for example, for cellular car radio systems giving access to the public telephone network, others for private mobile radio networks, cordless telephones, wide-area paging systems or Citizens' Band.

and so on. The bandwidth associated with each of these elements in the national frequency allocation table will have been determined having regard to the extent of the foreseen future needs of the country for radio systems of each kind.

Forecasts of future national needs for the various services and for the various kinds of facility that make up a service, expressed in terms of bandwidth, change from time to time as national demands for facilities change, as new facilities using radio are developed and old ones are superseded and as new transmission techniques emerge. Therefore it may be necessary, every few years, to revise the allocation of frequency bands to services or the designation of bands for different kinds of facility within a service. The prospect of changes of this kind may also be indicated in the national frequency allocation table. Thus, the national frequency allocation table can be an important planning document.

International co-operation in spectrum management

Administrations cannot manage the radio spectrum in isolation. For example:

- Information must be exchanged about intended frequency assignments that might cause trans-frontier interference problems.
- There must be agreed means for resolving trans-frontier interference problems when they arise.
- Many mobile stations, on ships, aircraft and road vehicles, move from country to country and need to use the same communication and navigation equipment wherever they may be.
- Alignment of national frequency allocation tables facilitates the provision of international communication links.

Thus, administrations must co-operate. Much of this co-operation is informal and bi-lateral, taking place between the administrations of neighbouring countries and dealing with relatively local frequency assignment problems.

Multi-lateral co-operation between the countries of a region is also found to be useful. In various parts of the world, regional organisations have been set up under which public posts and telecommunications undertakings and government ministries with responsibilities for communications meet to discuss internationally a wide range of topics and problems that may be solvable by consultation or joint effort. In Europe the European Conference of Posts and Telecommunications Administrations (CEPT) serves this purpose. In other parts of the world, for example, the Inter-American Telecommunications Conference (CITEL), the Asia-Pacific Telecommunity (APT) and the Arab Telecommunication Union (ATU) have this role, among many others.

In addition to this consultation and co-operation between administrations, some major international organisations set up by major users of radio find it useful to include radio spectrum policy among the many matters that they study jointly, in order to present co-ordinated proposals to the administrations. Perhaps the International Civil Aviation Organisation (ICAO), the International Maritime Organisation (IMO), the World Meteorological Organisation (WMO), the International Amateur Radio Union (IARU) and the various regional unions of broadcasting organisations are the most significant of these bodies.

However, the International Telecommunication Union (ITU), concerned with international consultation on all aspects of telecommunication, is the global forum for the administrations and it is by far the most important of the international organisations concerned with radio spectrum management. It may even seem that the ITU member-countries have surrendered control of the spectrum to the Union. This is not so; governments attach so much importance to radio that they will not allow this to happen. Nations have, however, undertaken to co-operate through the ITU in spectrum management in six main ways and these undertakings are enshrined in agreements having treaty status which virtually all governments have ratified. These six main ways are as follows:

- The same basic principles and directly equivalent terminology are used by all administrations in carrying out their spectrum management responsibilities.
- An international table of frequency allocations, within which there are various options, has been agreed, and administrations find it possible to construct national tables from amongst these options which meet most of their national needs.
- International frequency plans have been developed and formally agreed by administrations for radio stations operating in several frequency bands which are allocated to services which, it is thought, stand in particular need of the protection that such agreements provide. However, these plans relate to only a few services and a small part of the radio frequency spectrum.
- Procedures have been developed to assist administrations in the selection of interference-free frequencies for assignment to radio stations in bands and for services for which frequency assignment plans have not been agreed. A practical basis for resolving trans-frontier interference problems that arise has also been developed.
- Various technical constraints on the parameters of emissions and equipment have been agreed, for general implementation, in order to foster efficient spectrum utilisation.
- Technical data on matters relevant to spectrum management are developed co-operatively for the use of all nations.

The ITU is not involved in domestic spectrum management.

As these frequency allocation and frequency assignment agreements have evolved from year to year to take new needs into account, a great variety of specific provisions have been devised and agreed to accommodate the special requirements of particular services in particular frequency bands and in various geographical localities. Many compromises have been accepted to accommodate the various needs and aspirations of different countries. Consequently, the provisions of the agreements are, in general, well suited to the purpose, but they have also become extremely complex.

Chapter 2

The International Telecommunication Union

2.1 The Union

The International Telecommunication Union (ITU) was set up in 1932. The new organisation incorporated a number of predecessor bodies, most notably the International Telegraphic Union and various consultative bodies, which had been used to foster telecommunication and in particular international telecommunication, each dealing with particular aspects of the field. Accounts of these predecessor organisations are provided, for example, by Codding[1] and Leive[2]. The Union was reorganised at the Plenipotentiary Conference which took place at Atlantic City in 1947, at which time the Union also became a specialised agency of the United Nations.

The purposes of the Union, as stated in Article 1 of its new Constitution[3], (see p. 17) are:

- To maintain and extend international co-operation between all Members of the Union for the improvement and rational use of telecommunications of all kinds, as well as to promote and to offer technical assistance to developing countries in the field of telecommunications.
- To promote the development of technical facilities and their most efficient operation with a view to improving the efficiency of telecommunications services, increasing their usefulness and making them, as far as possible, generally available to the public.
- To promote the use of telecommunication services with the objective of facilitating peaceful relations.
- To harmonise the actions of Members in the attainment of those ends.

These purposes are pursued, in the words of the preamble to the Constitution, 'while fully recognising the sovereign right of each State to regulate its telecommunication . . .'

The activities of the Union fall into three main areas. One area is radio spectrum management, and with this is associated an involvement, discharged in collaboration with IMO and ICAO, in the regulation of the operating procedures used on telecommunications links with ships and aircraft. Secondly, the Union provides a forum for consultation between administrations and telecommunications operating organisations on the telecommunications facilities which are supplied to the public. An increasingly important aspect of this work lies in the definition of world-wide technical standards for telecommunications equipment. Thirdly, the Union is a channel for technical assistance to developing countries. This technical assistance is financed partly from the Union's budget, partly from the United Nations Development Programme and partly from voluntary contributions made by operating organisations and telecommunications equipment manufacturers.

Any state that is a member of the United Nations may become a member of the Union by acceding to the Constitution and the Convention[3]. Any other sovereign state may join the Union in the same way, given the approval of two-thirds of the members. All but a few very small countries have joined the Union and there are currently 166 members. No entity that is not a sovereign state may become a member of the Union, but there is provision for entities with an acknowledged interest in its work, such as other international organisations, telecommunications operating agencies, and scientific and industrial organisations (SIOs), to participate in some aspects of the work of the Union. Given the approval of the recognising government, a recognised private operating agency (RPOA) or an SIO can become a formal member of the international consultative committees which are part of the Union.

To carry out its radio spectrum management functions, Article 1 of the Constitution directs the Union to:

- Effect allocation of the radio frequency spectrum, the allotment of radio frequencies and registration of radio frequency assignments and any associated orbital positions in the geostationary-satellite orbit in order to avoid harmful interference between radio stations of different countries.
- Co-ordinate efforts to eliminate harmful interference between radio stations of different countries and to improve the use made of the radio frequency spectrum and the geostationary-satellite orbit . . .
- Co-ordinate efforts to harmonise the development of telecommunication facilities, notably those using space techniques, with a view to full advantage being taken of their possibilities.
- Promote the adoption of measures for ensuring the safety of life through the co-operation of telecommunication services.
- Undertake studies, make regulations, adopt resolutions, formulate recommendations and opinions, and collect and publish information concerning telecommunication matters.

For the purposes of spectrum management, the word 'telecommunication' is used broadly. As shown in Appendix A2, the formal definitions of 'telecommunication' and 'radiocommunication' which are used in the Union embrace such applications of radio as radionavigation and radio astronomy, which do not involve communication in the ordinary sense of the word. For some other applications of radio-frequency energy, typically for industrial processes, some scientific applications and medical treatment, the responsibility for international co-ordination lies elsewhere. Nevertheless, the Union has a general competence regarding all parts of the radio spectrum (up to a frequency limit set arbitrarily by paragraph 6 of the Radio Regulations at 3000 GHz) and, with the foregoing minor exceptions, all uses of the spectrum.

The present structure of the ITU consists of representative organs (the plenipotentiary conferences, the Administrative Council, administrative conferences and the consultative committees) and various headquarters organs (the International Frequency Registration Board, a General Secretariat, the Telecommunications Development Bureau and the specialised secretariats).

The representative organs

The *Plenipotentiary Conferences* are the supreme organ of the Union. Conferences are convened, typically, at intervals of five or six years to determine the main lines of Union policy and business. These conferences alone have authority to amend the Constitution and the Convention of the ITU, which are respectively the basic instrument and the procedural rule-book of the Union. They also elect the countries which will supply councillors to form the Administrative Council. The functions of plenipotentiary conferences are considered further on pp. 16–18.

The *Administrative Council* meets, usually once a year, to direct the work of the Union in the intervals between plenipotentiary conferences. The Council is responsible mainly for relations between the Union and major external bodies, for authorising the holding of administrative conferences and determining their agenda in consultation with the members, for oversight of the management of the secretariat and for control of the Union's budget. The Atlantic City (1947) Convention called for Councillors from 18 member-states, but this number has risen with successive revisions of the basic instrument and the 1989 Plenipotentiary Conference authorised 43 member-countries to provide councillors, eight from the Americas, seven from Western Europe, four from Eastern Europe with Northern Asia, 12 from Australasia with the remainder of Asia and 12 from Africa.

Administrative conferences are convened, as required, to consider major specific telecommunications matters and in particular to revise the Administrative Regulations, namely the International Telecommunication Regulations and the Radio Regulations. Some administrative conferences are global in authority and representation; they are called world administrative conferences. Others are Regional, that is, their agenda relates to only one or two of the three geographical Regions into which the Union divides the world for some purposes (see pp. 36–37). Administrative conferences dealing with radio matters are fairly frequent, there having been one in most years recently and two in some years. Administrative telegraph and telephone conferences have been less frequent, averaging about one per decade. Administrative radio conferences (ARCs) are considered further on pp. 19–21.

There are two *international consultative committees*, on radio (CCIR) and on telegraphy and telephony (CCITT), respectively. These bodies have their roots in the 1920s and they have retained a measure of organisational independence within the Union. Each consists of a number of working groups, functioning under the control of the CCIR and CCITT Plenary Assemblies and supported by specialised secretariats headed by the Directors of the CCIR and the CCITT.

The essential functional differences between the two consultative committees can be summed up as follows. The CCITT studies a wide range of questions related to telecommunications technology, operations and tariffs pertaining to telecommunications facilities offered for the use of the public but excluding matters which relate specifically to radio. The CCIR confines itself mainly to technical matters concerned with radio, athough not disregarding operational issues which have an impact on technical considerations, but its involvement with radio is comprehensive, covering for example broadcasting, private radio communication systems and the use of radio for navigation and for various research applications, in addition to the use of radio as a transmission medium for telecommunications facilities offered to the public.

There are areas in which the two consultative committees co-operate closely. The CCITT establishes channel performance objectives which are applied to links in the international telecommunications network regardless of the transmission medium which carries the links. One study group, dealing with the performance of international audio and video channels used for signals which are to be broadcast, is operated jointly by the two committees and reports to the plenary assemblies of both the CCIR and the CCITT. Other joint groups produce plans for the development of the world telecommunications network. There are special autonomous groups, reporting to both committees, that are responsible for the drafting of handbooks on the technology and economics of the various telecommunications options. There is provision for close co-operation on an *ad hoc* basis in other areas also, such as for example in the definition of standards for digital radio systems. On most areas of CCIR work however, the CCITT has little immediate impact.

The functions and working methods of the CCIR are considered further on pp. 21–27.

The headquarters organs

The *Secretary-General*, supported by the *Deputy Secretary-General* and the *General Secretariat*, is responsible to the plenipotentiary conferences and the Administrative Council for the management of the Union and for co-ordinating the activities of its various organs. He is the chief executive officer of the Union. He prepares the annual budget of the Union, he provides guidance on legal and procedural matters to all organs of the Union and he is responsible for a wide range of ITU publications.

The *International Frequency Registration Board* (IFRB) consists of five Board members, supported by a specialised secretariat. The principal functions of the Board are to supervise the working of the international process of registering frequency assignments, to advise members on the application of the radio regulations and to provide expert assistance to administrative radio conferences. The Board and its functions are considered in more detail on pp. 27–29.

The *Telecommunications Development Bureau* is responsible for the work that the Union does in the field of technical assistance.

The Secretary-General, the Deputy Secretary-General, the Chairman and Vice-Chairman of the Board, the Directors of the two consultative committees and the Director of the Telecommunications Development Bureau participate in the discussions of the administrative Council as of right but have no vote. These officials, together with the other Board members, are appointed by plenipotentiary conferences. By tradition, these ten principal officials are drawn from states which are widely spread, both geographically and in economic situation.

Decision-taking in the Union

If the resolution of differences of view in the Union can only be achieved by voting, the administration of every qualified member-state of the Union has a single vote. Under certain circumstances an RPOA which is a member of the CCIR or the CCITT may vote in the plenary assemblies of that committee. No other participant in the work of the Union may vote. Decisions are usually taken by simple majority of the votes cast.

However, states are not compelled to join the Union and they are free to leave it if they decide that the disadvantage of accepting unwelcome decisions exceeds the benefits of remaining a member. All members recognise that the usefulness of the Union to them is reduced if other countries find its decisions intolerable.

Consequently, decisions are reached, wherever possible, by consensus. Policies are sought that are broad enough and mild enough to enjoy universal toleration whilst serving well enough the essential purposes of the Union. This diminishes the risk that member governments will refuse to implement majority decisions which do not satisfy them. It leads to regulatory flexibility, which is essential in an organisation comprising nations in every state of development and with a wide range of telecommunications problems and needs. It does, however, also lead to regulatory complexity.

The funding of the Union.

A permanent staff of about 750 works at the headquarters of the Union in Geneva and a considerable number of additional staff are employed temporarily when major conferences are in progress. There is also a small ITU field force consisting of specialists deployed elsewhere in the world to assist in the execution of various telecommunications projects. The cost of running the Union, disregarding funds supplied through the United Nations Development Programme and used for technical assistance, is about 110 million Swiss Francs per annum. Some increase in these figures is to be expected over the next few years as technical assistance to developing countries, chargeable to the budget of the Union, increases in response to decisions taken at the Plenipotentiary Conference in 1989. Almost all of this cost must be met by contributions from the governments which are the members.

Most of the specialised agencies of the United Nations calculate the share which each member-state should contribute to the cost of running the agency in proportion to the Gross National Product of that country. However, the ITU still uses the method which it used before the United Nations was formed. Each member chooses freely the number of 'contributory units' it is willing to pay, in a scale ranging from 40 units down to one quarter of a unit; the least-developed states have the option of choosing to contribute one eighth or one sixteenth of a unit. RPOAs and SIOs active in the work of the Union also contribute to its expenses, and likewise international organisations if not specifically exempted from doing so. The total number of units, or the equivalent, offered by members and other participants is calculated and the total revenue requirement of the Union is divided by the total number of units to determine the amount of money represented by one contributory unit. This sum varies from year to year with the expenditure of the Union, but it is currently around SF 250 000 per year.

Review of the structure and functioning of the Union

There have been no radical changes in the structure of the Union since the decisions of the Plenipotentiary Conference in 1947 were implemented, but the Union's task has changed greatly in nature and scale during that period. Recent major conferences have identified organisational problems, suggesting that there may be a need for changes to be made. The Plenipotentiary Conference held at Nice in 1989 resolved that a committee composed of delegates of the highest personal reputation in international telecommunications and having broad experience of ITU affairs should be set up to carry out an in-depth review of the structure and functioning of the Union and to report its finding to the Administrative Council.

It was foreseen that it may be necessary to convene an additional plenipotentiary conference in the early 1990s to consider options for change defined by the committee and, if thought appropriate, to implement them.

2.2 Plenipotentiary conferences, the Constitution and the Convention

Plenipotentiary conferences are composed of delegations representing the members of the Union. Observers from the United Nations and its specialised agencies, from the International Atomic Energy Agency and from regional groupings of members may also attend. The most recent Plenipotentiary Conference (PC-89) was at Nice in 1989, the sixth since the present basic structure of the Union was established at Atlantic City in 1947. The previous conference was held in Nairobi in 1982. It is, however, intended that the period between plenipotentiary conferences should normally be five years and never more than six years; the next conference of the regular series is planned for 1994.

The main functions of these conferences as defined in the new Constitution can be summarised as follows:

- To review the state of the Union's work and the management of its secretariat, to plan the course of its future activities and to fix limits for its budget.
- To plan the programme of administrative and other main conferences foreseen to be required up to the time of the next plenipotentary conference.
- To elect the member-states that will supply councillors for the Administrative Council.
- To choose the ten elected officials to serve the Union.
- To revise, if necessary, the Constitution and the Convention of the ITU.
- To conclude or, if necessary, revise agreements between the Union and other international organisations.

Until recently the basic instrument of the ITU was the *International Telecommunications Convention*. The last version of this text[4] was agreed in 1982 in Nairobi. It set out the ways in which the Union was to function as an institution, defining the responsibilities of its organs, the duties of its officials, the procedures to be followed at its meetings, the basis for its finances and the status of the Administrative Regulations and of the Convention itself. In addition, some of the more important principles governing the conduct of telecommunication facilities were enshrined in it, for example:

- The public has a right to use international telecommunication services without discrimination or loss of confidentiality, subject to conditions which safeguard the security of states (Articles 18 to 22).
- Members are to ensure that international telecommunications facilities are up-to-date and efficient (Article 23).
- Absolute priority is to be given to telecommunications concerning the safety of life (Articles 25 and 36).

- Members are to use the radio spectrum and the geostationary satellite orbit with due regard for the rights of access of other members and the limitations of these natural resources (Article 33).
- Members are to ensure that radio stations under their jurisdiction do not cause harmful interference to other stations (Article 35).
  But
- Members retain freedom with regard to military radio stations, subject to certain constraints (Article 38).

However, revision of the Convention at every plenipotentiary conference, subject to decisions by a simple majority of votes, gave an inappropriate volatility to such an important text and occupied a considerable part of the time of each conference. All of the other specialised agencies of the United Nations have enshrined the essential elements of their constitutions in a more stable form. Resolution 62[5] of the 1982 Conference reflected a concern felt by many members about the process by which the Union maintained its basic instrument. The Resolution resolved:

- That the provisions of the present Convention should be separated into two instruments.

  (i) A Constitution containing the provisions which are of a fundamental character.

  (ii) A Convention comprising the other provisions which by definition might require revision at periodic intervals.

- That each of these instruments should contain its own amendment procedure, it being stipulated that amendment of the Constitution shall require a special majority.

The Administrative Council was instructed to set up a group of experts to draft a constitution and a convention, based on the 1982 Convention, for consideration at the following plenipotentiary conference. PC-89 reviewed the results of this work, amended the text in detail and approved the outcome as the Constitution of the ITU and the Convention of the ITU[3].

The Constitution of the ITU, subject to ratification by governments, sets out the objectives of the Union, defines the frame within which it is to operate and states major principles governing the conduct of telecommunications services, such as those quoted above from the International Telecommunications Convention of 1982. It is to become the basic instrument of the Union. Amendment will require the vote of two thirds of the delegations accredited to a plenipotentiary conference.

The Convention of the ITU, subject to ratification by governments, sets out the way in which the Union is to carry out its work. Amendment of its provisions will require the votes of a simple majority of the delegations accredited to a plenipotentiary conference.

The Constitution, the Convention and the Administrative Regulations are the 'instruments of the Union'. Once the government of a member state has ratified these texts, it is bound to abide by them (except where exemption from this requirement for its national defence stations is operative) in its own use of radio for international links and for emissions that could cause harmful interference to the radio facilities of other states. The same conditions must also be applied by an administration to authorised non-government stations.

2.3 Administrative radio conferences and the Radio Regulations

The Radio Regulations

The Radio Regulations (RR) are extensive and they touch on aspects of most radio systems. The most recent complete re-editing of the RR was carried out and approved at the World Administrative Radio Conference held in 1979 (WARC-79) and this new edition[6] was published in 1982. As revised then, the RR consist of 69 articles, 44 appendices, 87 resolutions and 90 recommendations, about 2000 pages in all for each language version, published in loose-leaf ring-file form in French, English and Spanish. The RR have been amended and greatly extended by the administrative radio conferences which have met since 1979. The text of these amendments and additions is to be found in the Final Acts of the various conferences. Every few years a set of additional and replacement leaves is published, and these can be used to up-date the edition of 1982. The Plenipotentiary Conference held in 1989 decided that the RR and the other important texts of the Union shoud be published in Russian, Chinese and Arabic, in addition to French, English and Spanish.

Every paragraph of the 69 articles of the RR is numbered in a single series and can be identified uniquely by that number. In this book the many references to paragraphs within the articles and to articles, appendices, resolutions and recommendations as a whole are in the form 'RR 1234', 'RR Article 8', 'RR Appendix 28' and so on. To avoid ambiguity, these references are related to the RR edition of 1982 and to final acts of administrative conferences, and not to the up-dated versions of the edition of 1982, but readers may find it more convenient, in most cases, to refer to the most recent up-dated version of the RR.

About half of the articles in the RR and most of the appendices set out the agreements and procedures regulating the allocation of frequency bands to services, the international registration of frequency assignments and the various technical measures that have been given mandatory force in order to increase the efficiency with which the radio spectrum can be used. The remainder of the articles and several of the appendices provide a procedural framework for the operation of the mobile radio services, and in particular for public telegraph and telephone services to and from ships. The remainder of the appendices contain current frequency assignment plans and frequency allotment plans that have been agreed for various radio services. The resolutions and recommendations cover a wide range of subjects. Some are temporary. In general they declare the decisions and intentions agreed at administrative radio conferences which could not well be expressed in permanent regulatory form or they call for action to be taken in preparation for future conferences.

The IFRB has published a handbook on the regulatory procedures[7], to assist administrations, which presents the regulations contained in the RR in full detail but in a more readily understandable form.

Administrative Radio Conferences

Administrative Conferences are convened to consider major and specific telecommunication matters. The end product of an Administrative Radio Conference (ARC) is typically expressed in the form of additions to or amendments of the RR. As indicated above, the text containing these changes to the regulations is published as the 'Final Acts' of the conference. No entity other than a WARC may amend or add to the RR.

Some conferences, typically frequency assignment planning conferences, must take such complex decisions that the conference has to be held in two sessions, separated by a year or longer, to allow planning approaches to be conceived, discussed and approved in principle, then developed in detail for applications and implementation after a period for reflection; in such cases the output of the first session is likely to include a report to the second session.

As with plenipotentiary conferences, the principal participants at an ARC are the delegations of the members, but observers from various international organisations may also attend, and RPOAs may send delegations if so authorised by the member which recognises them.

The decision to hold a World Administrative Radio Conference (WARC) or a Regional Administrative Radio Conference (RARC) is usually presented in the form of a resolution of a plenipotentiary conference or the Administrative Council. There is also provision for an ARC to be convened if at least one quarter of the members of the Union request it.

The resolution calling for a WARC or a RARC will usually:

- identify the circumstances which create the need for a conference,
- list the foreseen objectives,
- indicate approximately when the conference should be held, and
- invite appropriate bodies to do any necessary preliminary or preparatory work or to take any desirable interim action.

The agenda for an ARC and the arrangements made for holding it are determined by the Administrative Council in consultation with the member-states.

The preparatory work usually includes the preparation of briefing reports by the IFRB and the CCIR on the operational, regulatory and technical framework in which the conference will work.

The Administrative Radio Conference at Atlantic City in 1947 revised the pre-war Radio Regulations comprehensively. However, the use of the spectrum changed greatly in the next decade, with a vast expansion of point-to-point systems below 30 MHz and the opening-up of the VHF band to broadcasting and of the UHF and SHF bands to commercial radio relay systems and new radio navigation systems. These developments led to many regulatory changes. By 1959 it had become necessary to hold a general WARC to carry out another complete re-editing of the RR, amongst much else. The introduction of satellite radio systems in the 1960s and 1970s, bringing with them extensive new regulations and major changes to the international Table of Frequency Allocations, was the principal circumstance that made necessary the holding of another general WARC in 1979, which carried out another complete re-editing of the RR.

However, WARC-79 also identified many new problems needing urgent attention. As a result, the period between the plenipotentiary conferences in 1982 and 1989 was a time of intense activity in the further revision of the RR, including:

- WARCs on mobile services in 1983 and 1987,
- a group of three interlinked conferences (RARC-SAT-83, WARC-ORB-85 and WARC-ORB-88) concerned mainly with planning access to the GSO for the fixed-satellite and broadcasting-satellite services,
- a RARC in 1984 on sound broadcasting at VHF in Regions 1 and 3,
- a WARC held in two sessions (1984 and 1987) on HF broadcasting,
- a RARC held in two sessions (1986 and 1988) on MF broadcasting in Region 2, and
- the first session, held in 1987, of a two-session RARC on VHF and UHF broadcasting in Africa.

This succession of conferences dealt with most of the problems identified at WARC-79. PC-89 approved a somewhat shorter programme of ARCs for the period up to the plenipotentiary conference planned for 1994, as follows:

- The second session of the RARC on VHF/UHF broadcasting in Africa, to be held in 1989,
- a WARC to be held in 1992 to resolve various frequency allocation problems identified at WARC-HFBC-87, WARC-MOB-87 and WARC-ORB-88, plus various other specific spectrum management problems identified by PC-89,
- a WARC to be held in 1993 to resolve frequency assignment planning problems identified but not solved by the HF broadcasting conferences in 1984 and 1987, and
- a RARC to establish criteria for the shared use of VHF and UHF bands allocated in Region 3 for the mobile, broadcasting and fixed services.

PC-89 also decided that a group of experts should be established to study ways of improving the mechanisms of frequency allocation and, more generally, of simplifying the operation of the RR. The group would be asked to report to the Administrative Council meeting in 1992, and this might lead in turn to the addition of proposals for change to the agenda of appropriate WARCs.

The ITU has functions which are almost entirely regulatory or technical but it works in an economic and political environment and pressures from this environment are not absent from its deliberations. Accounts of WARC-79 which treat the resolution of international differences of view at a general level are provided, for example, by Robinson (1980)[8] and Codding and Rutkowski (1982)[9].

2.4 The International Radio Consultative Committee

All members of the Union are full members of the CCIR as of right. RPOAs and SIOs may also become members of the CCIR, given the approval of the administration concerned. Furthermore, in the absence of the administration at a Plenary Assembly but with its approval, an RPOA may vote. In addition, and subject to appropriate conditions, relevant international organisations and regional telecommunications organisations may participate in the work of the CCIR in an advisory capacity.

CCIR terms of reference

The terms of reference of the CCIR are given in Article 13 of the Constitution, as follows:

> The duties of the International Radio Consultative Committee (CCIR) shall be to study technical and operating questions relating specifically to radiocommunication without limit of frequency range, and to issue recommendations on them with a view to standardising telecommunications on a world-wide basis; these studies shall not generally address economic questions but where they involve comparing technical alternatives, economic factors may be taken into consideration.
>
> In the performance of its studies, each International Consultative Committee shall pay due attention to the study of questions and to the formulation of recommendations directly connected with the establishment, development and improvement of telecommunications in developing countries in both the regional and international fields. Each International Consultative Committee shall conduct its work with due consideration for the work of national and regional standardisation bodies keeping in mind the need for the ITU to maintain its pre-eminent position in the field of world-wide standardisation for telecommunications.

It may be noted that the upper frequency limit of 3000 GHz, applied to the work of the ITU in general and stipulated by RR 6, does not apply to the work of the CCIR.

The most fundamental of the duties of the CCIR lies in the preparation of recommendations which will be used by administrations, administrative radio conferences and the IFRB, for example through the further development of the RR and in spectrum management. However, the CCIR is also an important forum for the international standardisation of radio systems. Some of its work is conceived from the outset in the form of technical reports, handbooks and network development plans which are intended to assist administrations and operating organisations in the design of networks and the provision of facilities.

CCIR Organisation

The plenary assemblies of the CCIR supervise all of its work. Plenary assemblies are held at intervals of about four years to consider for approval the work done in the previous 4-year study period and to plan for future work. Major changes in CCIR organisation are currently in progress, but as of 1989 the work will have been done in:

- 11 CCIR study groups and their associated working groups, interim working parties and joint interim working parties.
- One study group which operates jointly with the CCITT under CCIR administration, namely the Joint Study Group for Television and Sound Transmission (CMTT).
- A number of Plan Committees, administered by the CCITT but which have CCIR participation, which produce plans for the development of the international public telecommunications network in the various regions of the world, using radio and any other suitable transmission media.
- Some Special Autonomous Groups, also administered by the CCITT and with CCIR participation, which are set up as required to produce handbooks on the technology and economics of various kinds of telecommunication system for the assistance of developing countries.
- Special meetings, called Conference Preparatory Meetings, which are convened as required to prepare briefing reports for administrative radio conferences.

An important activity of the CCIR and the CCITT is the definition and standardisation of the terminology used in the ITU and in telecommunication in general. This activity is reviewed briefly in Appendix A. Until recently this activity was co-ordinated by a second study group operated jointly with the CCITT, namely the Joint Study Group for Vocabulary (CMV). CCITT withdrew from CMV by a decision of its Plenary Assembly in 1988 and alternative methods of co-ordinating terminology usage are currently being devised.

The 11 CCIR study groups and the CMTT partitioned the study area which is the responsibility of the CCIR in two ways. Eight of the study groups specialised in the particular problems of one or more of the various services into which radio systems are categorised. The other three study groups and the joint study group were responsible for more broadly based work which affects several or all services. The formal terms of reference of these groups are to be found in CCIR Resolution 61[10]. In brief, the eight service study groups had the following areas of responsibility:

Study Group 2: The space research, earth exploration-satellite and radio astronomy services.
Study Group 3: The fixed service operating below about 30 MHz.
Study Group 4: The fixed-satellite service.
Study Group 7: The standard frequency and time signal service.
Study Group 8: The mobile, radiodetermination and amateur services.
Study Group 9: Radio relay systems of the fixed service operating above about 30 MHz.
Study Group 10: The broadcasting service used for sound broadcasting.
Study Group 11: The broadcasting service used for television.

It will be noted that the fixed service was the responsibility of Study Groups 3 and 9 and the fixed-satellite service fell to Study Group 4; in all other cases satellite services were studied by the study group which also studies the corresponding terrestrial service.

The four remaining study groups were concerned with the following areas of work:

Study Group 1: General studies of spectrum management techniques.
Study Group 5: Radio propagation in non-ionised media and associated radio noise.
Study Group 6: Radio propagation in ionised media and associated radio noise.
CMTT: Performance of sound and video channels used for long distance transmission of signals intended to be broadcast.

CCIR working methods

The work of the CCIR study groups is complex and multifarious. Study and discussion of a new problem usually extends over several years; studies extending over 10 or 15 years are not uncommon and a few topics are under virtually perpetual study. An ever-changing succession of experts in spectrum management, radio propagation, system design, system operation and equipment manufacture from many countries, viewing problems from many viewpoints, participate in the work. To ensure that such studies and discussions do not lose coherence and dynamism, the plenary assemblies appoint an experienced chairman and one or more vice-chairmen for each study group and lay down detailed rules for the conduct of CCIR business;

see CCIR Resolution 24[11]. The essence of this procedural discipline lies in the seven kinds of formal text through which the CCIR expresses the objectives and the outcome of its work, namely Questions, Study Programmes, Decisions, Opinions, Resolutions, Reports and Recommendations. The roles of these various kinds of text are as follows:

A *Resolution* is a text giving instructions on the organisation, methods or programmes of CCIR work; most resolutions are prepared and approved by plenary assemblies.
A *Decision* is a text giving instructions on the organisation of the work of a study group within the framework of its terms of reference; most formal decisions specify the terms of reference of an interim working party.
A *Question* is a statement of a technical or operational problem, to which an answer is required.
A *Study Programme* is used, if necessary, for definining with greater precision some part of the work called for by a question.
A *Report* may be a provisional answer to a question or a study programme. Alternatively it may be a statement, for information, on studies carried out by a study group on a given subject.
A *Recommendation* is an answer to a question or a study programme which the CCIR considers to be sufficiently complete to serve as a basis for international co-operation.
An *Opinion* is a proposal or a request addressed to another organisation, typically another organ of the ITU, such as the CCITT or the IFRB. A formal opinion does not necessarily relate to a technical subject.

The CCIR does its basic work by assigning questions and study programmes for the appropriate study group to answer. If the question asks for technical information or advice, the answer usually takes the form of a formal report. If the question asks for a decision on a technical issue, the ultimate answer is usually a formal recommendation, but several years may pass before the necessary study has been completed and a sufficient degree of convergence of opinion has been achieved, during which time the state of the work may be made available through the publication of successive revisions of a formal report.

When CCIR meets in plenary assembly, one of its most important tasks is to consider for approval the texts, and especially the recommendations, which have been prepared by the study groups during the previous four-year study period. These may be new texts or amended versions of texts which had been approved by a previous plenary assembly. The plenary assembly also considers the cancellation of texts which had been previously approved and published but which may no longer be useful. The approved texts are then published (see p. 25).

As with its organisation, the CCIR's working methods are currently evolving, primarily in order to respond more quickly to changes in technology, but up to 1989 work was done in the study groups broadly as follows. Eighteen months to two years after a plenary assembly the study groups held working meetings, called 'Interim Meetings', typically lasting from one to three weeks according to the size of the work load of the study group. National contributions responding to the formal questions and study programmes were discussed at these meetings. The outcome of the discussions, in the form of draft new texts and draft amendments to texts which had already been published, was distributed to participants after the meeting in what was called an 'interim booklet'.

After the lapse of a further one and a half to two years, each of the Study Groups had another working meeting, called the 'Final Meeting', also lasting from one to three weeks. Once again national contributions were discussed and draft new texts and draft amendments to previously-approved texts and to texts from the interim booklet were prepared. After the Final Meetings the results of the four years' work in the form of draft new and revised recommendations, reports etc. were referred for approval to the plenary assembly which met a few months later. The four-year cycle was then repeated.

All approved texts, consisting of texts newly drafted during the most recent four-year study period and texts maintained with or without amendment from earlier study periods, are published in French, English and Spanish after each plenary assembly. PC-89 decided that Russian, Chinese and Arabic versions should also be published in future. Each language set consists of 12 volumes of study group texts, one per study group, plus another volume containing minutes, resolutions etc. from the plenary assembly. For the time being there is a further volume for the texts of the CMV. Some of the volumes are bound in several parts, either because the extent of the study group's work requires it or in order to provide for convenient reference to texts which have been produced by two study groups in collaboration. In addition to these volumes, revised and republished every four years, a few very long reports and recommendations have been published as separate volumes; these are not republished until they need major revision.

All texts receive an identifying number when they are published after first being formally approved. Recommendations originating from any CCIR source are numbered in a single series which is common to the whole of the CCIR. Reports, opinions, resolutions and decisions are treated in the same way, each kind of text having its own single numbering series. When a text has been revised and re-approved a digit is associated with the number to identify the new version of the text. Thus Recommendation 465-3 is the third revision of Recommendation 465. Questions and study programmes are treated in a broadly similar way, but for these texts each study group has its own number series.

In this book CCIR texts are identified in lists of references at the end of each chapter by the full designation of the latest published version of the text at the time of going to press; elsewhere in the book the generic identification is quoted, without the symbol which indicates whether the text has been revised.

Some recommendations are precursors of new elements of the RR. Others typically define elements of the technical basis of the engineering of radio systems and telecommunications networks or for the control of interference between systems using the same radio frequency. CCIR recommendations which affect frequency sharing between space and terrestrial radio systems and between different space systems may effectively be given the same status as the RR between countries which agree to be bound by them, in advance of their incorporation in the RR by a WARC, through the procedure established by RR Resolution 703[12]. Each CCIR plenary assembly lists new or revised sharing recommendations for treatment in this way; see CCIR Resolution 86[13]. The list is forwarded to all administrations, and the administrations are asked to indicate which of the new provisions they accept for immediate application. The results of these enquiries are made known to all administrations, for guidance in subsequent co-ordination processes.

The basic working methods of the CCIR study groups, set out above, are broadly satisfactory but two problems remain:

- The study groups have no corporate existence except for a few weeks every second year, during their interim and final meetings, but there may be work to be done which cannot conveniently be treated at short biennial meetings, typically because it is urgent or particularly extensive.
- Problems do not always fit neatly into the terms of reference of the separate study groups. Consultation or collaboration between study groups is sometimes necessary. Nor is it always possible to come to a good conclusion in the CCIR without liaison with the CCITT.

The following means are used to solve these problems:

- An *Interim Working Party* (IWP) can function at any time. An IWP may be set up to deal with a specific urgent or extensive task. A chairman is appointed, the member-states wishing to participate nominate their delegates and terms of reference are agreed and published in a formal decision. The IWP chairman will arrange for the work to be done, if possible by correspondence, but if not by whatever means the IWP considers to be appropriate, reporting back to the study group to indicate progress and to deliver the final report.
- A *Joint Interim Working Party* (JIWP) comprising participants from more than one study group (including CCITT study groups if necessary) may be an effective way of establishing collaboration between different study groups in order to carry out a single, major task. Where the requirement is for collaboration between CCIR study groups on an on-going basis, it is more usual, however, to arrange joint sessions of the study groups or their working groups at their regular biennial meetings.

CCIR preparations for Administrative Radio Conferences

The formal resolution of a plenipotentiary conference or the Administrative Council which convenes an administrative radio conference usually also asks the CCIR to carry out any new studies that the prospective ARC will require, and to provide a report on all the technical and operational factors which should be taken into account in the work of the conference. This report takes the form of a summary of relevant CCIR texts and the results of new work, prepared in a way that the delegates to the ARC will find convenient for their use.

At the appropriate time the Director of the CCIR calls a meeting of the chairmen and vice-chairmen of the study groups, together with representatives of member-countries wishing to participate, to advise him on how the CCIR should prepare this report; this meeting is called a Conference Consultative Group. If the agenda of the ARC is such that only one CCIR study group needs to be involved in the preparation of the report, it may be possible to prepare the report at one of the regular meetings of that study group. Alternatively the study group may set up an IWP for the purpose. If the interests of several study groups are involved, it may be necessary to set up a JIWP or a special-purpose meeting, called a Conference Preparatory Meeting (CPM).

2.5 The IFRB and the Master International Frequency Register

The five members of the IFRB, all experts in radio spectrum management, will have been offered by their countries for election as Board members, but they are elected, in the words of Article 10 of the Constitution, to: 'serve, not as representing their respective Member states or a region, but as custodians of an international public trust.' The historical background, the traditions, the status and the functions of the Board are discussed in detail by Leive (1970)[2].

The essential duties of the Board are set out in broad principle in the Constitution, and a more detailed statement of the Board's functions and methods of working is to be found in RR 991 to RR 1016. In essence, the functions are as follows:

- To process, in accordance with the Radio Regulations and using appropriate recommendations of the CCIR or the Board's own technical standards, the notices of frequency assignments and satellite orbital location notified to it by administrations, and to record these notified assignments, if appropriate, in the Master International Frequency Register (MIFR).
- To maintain and review the contents of the MIFR.
- To participate, actively or passively as prescribed, in the various frequency management procedures defined in the RR.
- To assist, on request, in the resolution of cases of harmful interference between frequency assignments.
- To study means of making the use of the spectrum more effective and to make appropriate recommendations.
- To assist administrations, particularly those in special need of assistance,

to make better use of the spectrum.

- To provide technical assistance in the preparations for administrative radio conferences.
- To collect monitoring data and arrange for its publication.
- To provide assistance by training in spectrum management the staff of administrations, especially those of countries in special need.

In principle, the role of the Board is executive, not judicial; it implements the regulations that successive administrative radio conferences have agreed. However, while the regulations impose many duties upon administrations, and set out the Board's foreseen tasks in minute detail, they do not always determine what the Board should do if an administration fails to fulfil its obligations. Furthermore, the regulations, extensive, far-sighted and comprehensive as they are, do not always provide for new situations which a rapidly evolving medium throws up. Thus, the Board must sometimes exercise judgement in deciding issues that the RR do not resolve, pending the determination of a regulatory solution by a subsequent ARC; the Board's decisions in such cases are recorded in the 'IFRB Rules of Procedure' and made available to adminstrations for information.

The *MIFR* is a computer data base containing details of all frequency assignments notified to, and found admissible by, the Board; it also contains details of some assignments, registered with suitable annotations at the insistence of the notifying administration, which the Board considers to be in breach of the Convention, the RR or the Board's technical standards.

Administrations need information on the contents of the MIFR and so the *International Frequency List* (IFL)[14], containing the more important details of the entries in the MIFR, is published and revised at intervals. The number of entries in the MIFR is rising rapidly. In the mid-1980s the number of assignments in the Register passed one million and it was considered that publication of the IFL in printed form was no longer practicable. Since 1985 the complete IFL has been published six-monthly in microfiche form.

Administrations may need more up-to-date information on the assignments notified to the IFRB by other countries than can be provided by the six-monthly re-issues of the IFL. Accordingly, the IFRB publishes a *weekly circular* containing full information on all new frequency assignment notifications.

In addition, the IFRB keeps up-to-date the lists of various categories of radio stations which are published for the information of Members. Some of these lists are of particular value to stations operating in the mobile services. Details of all of these lists, called 'service documents', are to be found in RR Article 26 and RR Appendices 9 and 10.

2.6 References

1 CODDING, G.A.: 'The International Telecommunication Union – an experiment in international co-operation' (E.J. Brill, Leyden, Netherlands, 1952)
2 LEIVE, D.M.: 'International Telecommunications and International Law: the Regulation of the Radio Spectrum' (Oceana Publications Inc, Dobbs Ferry, NY, 1970)
3 The Final Acts of the Plenipotentiary Conference, Nice, 1989, containing the Constitution of the ITU, the Convention of the ITU (both subject to ratification) and the other texts agreed at the Conference; to be published in 1990
4 International Telecommunication Convention (ITU, Geneva, 1982)
5 Basic instrument of the Union; Resolution 62. ITU Plenipotentiary Conference, Nairobi 1982, bound with the ITC (Ref. 4)
6 Radio Regulations, edition of 1982 (ITU, Geneva, 1982)
7 IFRB Handbook on radio regulatory procedures. ITU, Geneva, 1982)
8 ROBINSON, G.O.: 'Regulating international airwaves: the 1979 WARC, *Virginia J. Int. Law*, Fall 1980, **21**, pp. 1–54
9 CODDING, G.A. and RUTKOWSKI, A.M.: 'The International Telecommunication Union in a changing world' (Artech House Inc, 1982)
10 Terms of reference and structure of CCIR Study Groups; Resolution 61–3; Recommendations and Reports of the CCIR, 1986, Volume XIV–1 (ITU, Geneva, 1982)
11 Organisation of CCIR work; Resolution 24–6. *ibid*, Volume XIV–1
12 'Relating to the calculation methods and interference criteria recommended by the CCIR for sharing frequency bands between space radiocommunication and terrestrial radiocommunication services or between space radiocommunication services'. Resolution 703; Radio Regulations, edition of 1982 (ITU, Geneva, 1982)
13 CCIR recommendations to be drawn to the attention of the Secretary-General in accordance with the provisions of Resolution No 703 of the WARC-79; Resolution 86–1; Recommendations and Reports of the CCIR, 1986, Volume XIV–1 (ITU, Geneva, 1986)
14 International Frequency List. ITU, Geneva, published twice per year.

Chapter 3

Frequency allocation

3.1 The need for frequency allocation

Administrations divide the radio frequency spectrum into hundreds of frequency bands, some wide and some narrow, and each band is allocated for the use of stations of specific radio services. To a considerable degree the pattern of frequency allocations, established through the ITU, is the same in all countries.

Such practices have disadvantages. In particular, within a limited geographical area, a band allocated for one service may be under-used, with no evident prospect of a substantial increase in use, whereas an adjacent band allocated for another service may be overloaded. Such seemingly avoidable congestion would arise sometimes if the allocations were determined by each country, without regard for allocations abroad, because the needs for spectrum for the various services had been wrongly forecast or because the relative demand for services varied from one part of the country to another. The mismatch between demand for spectrum and bandwidth allocated is likely to be aggravated by international uniformity in frequency allocations because different countries have different patterns of need for radio facilities and these patterns evolve at different rates in different countries.

It might seem that it would be better to assign frequencies to stations as needs arise without regard to service. There are, however, countervailing arguments which amply justify frequency allocation, co-ordinated internationally, the chief ones being as follows:

(a) Mobile stations: Mobile stations move from place to place and often from country to country, and their communication and radio navigation equipment would be unacceptably complicated if there were not a large measure of international uniformity in the principal frequency bands allocated for these purposes.

(b) International communication systems: A substantial part of the HF band is used for international fixed and broadcasting systems. A large part of

the microwave spectrum is used for radio relay and fixed-satellite links, many of which cross frontiers. These systems could not operate without a large measure of international uniformity in frequency allocations.

(c) Treatment of harmful interference: Every effort is made by administrations to prevent interference that would put human life in danger and to suppress such interference quickly if it does arise; the stations concerned provide the so-called safety services. Elaborate measures have been devised to provide for the elimination of harmful interference to most other stations, although with less priority. However, administrations may not concern themselves at all with interference arising between, for example, stations of the amateur services. This selective treatment of interference would be difficult to implement in the absence of frequency allocation.

(d) Efficient use of spectrum: More radio stations can operate in the same band of frequency if they are similar to one another in system parameters, operating methods and so on. Stations of the same radio service may not be completely uniform in these respects, but the stations of a single radio service are more homogeneous than stations in general.

(e) Specialisation of equipment design: For many applications of radio and perhaps most of all for the mobile and broadcasting services, the concentration of particular kinds of radio system into specific frequency bands, brought about by frequency allocation, has facilitated standardisation of the characteristics of systems and the optimisation of equipment design for the purpose for which it is to be used. Such equipment is more convenient and more effective in operation than equipment designed for general purposes, and international trade in equipment is facilitated by international standardisation of allocations, so reducing unit costs.

(f) Apportionment of spectrum: Countries are not all alike in the priorities that they give to the various radio facilities. In many parts of the radio spectrum, where radio propagation characteristics or system configurations permit cross-frontier interference, there is not enough bandwidth to meet all demands and in these circumstances there must be international co-operation in the elimination of interference. Frequency allocation provides a basis for international agreement on the apportionment of spectrum resources and the elimination of interference.

(g) Devolution of spectrum management: In some circumstances spectrum is managed more efficiently, and serves its users more effectively, if frequency assignments are planned with expert knowledge of the application for which it is used. Frequency allocation makes feasible, if required, the devolution of the planning function to the experts concerned with the application. For example, band management might be devolved to a big user at a national level, to an international meeting of specialist delegations at a WARC or a RARC or to a body like ICAO which is quite outside the ITU.

Despite these strong arguments for frequency allocation and a sufficient measure of international uniformity, the impossibility of finding a single and simple scheme of allocations which is sufficiently close to the needs of all countries is a real and important disadvantage. However, the disadvantage can be mitigated in various ways. The definitions of the major radio services are broad; for example, an administration can choose among many kinds of system in deciding how to assign frequencies in a band allocated to the mobile service. Some bands are allocated to more than one service in the international Table of Frequency Allocations, a practice called band sharing. Also, administration retain the option to use frequency bands in any way they think fit provided that this has no adverse effect on systems operated within the jurisdiction of other administrations which are in accordance with the ITU Radio Regulations. These options are considered further on pp. 43–48.

3.2 International frequency allocation

3.2.1 The ITU Table of Frequency Allocations

The international Table of Frequency Allocations is in RR Article 8. It covers the spectrum from 9 kHz to 275 GHz, and there are some preliminary planning indications in RR 927 for the frequency range from 275 to 400 GHz. The characteristics of radio propagation vary enormously over this frequency range. Frequency bands in different parts of the spectrum have been allocated for each of the major radio services, giving administrations the opportunity to assign frequencies with the propagation characteristics that suit each station best. Other services with more specialised applications have been allocated a few bands only or one. Consequently the Table is extensive; with its 483 footnotes it filled 174 pages in the 1982 edition of the RR.

It would not be appropriate to reproduce all of this detailed material in this book, although summaries of many of the more important elements of the Table are presented, service by service, in Chapters 5 to 14. However, the main features of the Table can be seen from a sample page. Fig. 3.1, which shows the section of the Table which covers 415–1606·5 kHz as it was in 1982, illustrates most of the main features of the Table, but it should be noted that a few minor amendments have been made to these allocations since 1982. The principal features to note are:

- The services for which frequency bands are allocated.
- Many frequency bands are allocated, not for one service but for two services or more; these bands are said to be 'shared' by the various services.
- Some allocations are world-wide, but for other bands there are Regional differences in the allocations.
- In bands which are shared, the various allocations may not enjoy equal status.
- Many allocations are qualified by footnotes.

Band sharing is considered in some detail on pp. 40–43. The other principal features of the Table are discussed below.

The services

The concept of 'service' which is used in frequency allocation is discussed in Appendix A3. The formal definitions of these services are to be found in Section III of RR Article 1 and many of the definitions are quoted and discussed in later chapters of this book. The following summary gives a more general view of the services.

The *fixed service* (FS) comprises stations at fixed locations which communicate with other fixed stations. If the links are made via satellites, both the earth stations and the space stations belong to the *fixed-satellite service* (FSS). A few narrow bands around 20 MHz have been allocated specifically for stations at fixed locations used to support aeronautical operations, forming the *aeronautical fixed service* (AeFS).

The *broadcasting service* (BS) covers transmitting stations which provide signals for direct reception by members of the public. The domestic receiving stations are also part of the broadcasting service. When the transmitters are on satellites, the service becomes the *broadcasting-satellite service* (BSS).

The *mobile service* (MS) comprises stations which operate whilst moving or when at rest at unspecified points, and the stations, fixed or mobile, with which they communicate. If these communication links are relayed at a satellite, then the mobile earth stations and the associated space stations are part of the *mobile-satellite service* (MSS). In addition to these general categories for mobile and mobile-satellite systems, three specialised categories have been created in the MS and MSS, for maritime, aeronautical and land mobile stations respectively. Thus for example a station on an aeroplane, operating without the use of a satellite, can be considered as part of either the MS or the *aeronautical mobile service* (AeM) and an earth station on a ship and the associated space station can be considered as part of either the MSS or the *maritime mobile-satellite service (MMSS)*.

The amateur (AmS) and *amateur-satellite* (AmSS) *services* are made up of the stations of radio amateurs.

Radio stations which are used for a variety of technical and scientific purposes form a group of services. In most cases the names of the services indicate clearly the nature of the application involved. They are as follows:

Meteorological aids service (MetA)
Standard frequency and time signal service (SFS)
Standard frequency and time signal-satellite service (SFSS)
Space operation service (SpO)
Earth exploration-satellite service (EES) and its specialised sub-category, the *meteorological-satellite service* (MetS)
Space research service (SRS)
Radio astronomy service (RA)

kHz
415 – 1 606.5

<table>
<tr><td colspan="3">Allocation to Services</td></tr>
<tr><td>Region 1</td><td>Region 2</td><td>Region 3</td></tr>
<tr><td>415 – 435
AERONAUTICAL RADIONAVIGATION
/ MARITIME MOBILE / 470
465</td><td colspan="2" rowspan="2">415 – 495
MARITIME MOBILE 470
469 471</td></tr>
<tr><td>435 – 495
MARITIME MOBILE 470
Aeronautical Radionavigation
465 471</td></tr>
<tr><td>495 – 505</td><td colspan="2">MOBILE (distress and calling)
472</td></tr>
<tr><td rowspan="3">505 – 526.5
MARITIME MOBILE 470
/ AERONAUTICAL RADIONAVIGATION / 473
465 471 474 475 476</td><td>505 – 510
MARITIME MOBILE 470
471</td><td rowspan="3">505 – 526.5
MARITIME MOBILE 470
/ AERONAUTICAL RADIONAVIGATION /
Aeronautical Mobile
Land Mobile
471</td></tr>
<tr><td>510 – 525
MOBILE
AERONAUTICAL RADIONAVIGATION</td></tr>
<tr><td rowspan="2">525 – 535
BROADCASTING 477
AERONAUTICAL RADIONAVIGATION</td></tr>
<tr><td rowspan="2">526.5 – 1 606.5
BROADCASTING
478</td><td>526.5 – 535
BROADCASTING
Mobile
479</td></tr>
<tr><td>535 – 1 605
BROADCASTING</td><td>535 – 1 606.5
BROADCASTING</td></tr>
</table>

Fig. 3.1 Specimen pages from the international Table of Frequency Allocations (Article 8 of the ITU Radio Regulations, edition of 1982)

465 Norwegian stations of the fixed service situated in northern areas (north of 60°N) subject to auroral disturbances are allowed to continue operation on four frequencies in the bands 283.5–490 kHz and 510–526.5 kHz.

469 *Additional allocation:* in Afghanistan, Australia, China, the Overseas French Territories of Region 3, India, Japan and Papua New Guinea, the band 415–495 kHz is also allocated to the aeronautical radionavigation service on a permitted basis.

470 The use of the bands 415–495 kHz and 505–526.5 kHz (505–510 kHz in Region 2) by the maritime mobile service is limited to radiotelegraphy.

471 The bands 490–495 kHz and 505–510 kHz shall be subject to the provisions of No. 3018 until the provisions of Recommendation 200 have been implemented.

472 The frequency 500 kHz is the international distress and calling frequency for radiotelegraphy. The conditions for its use are prescribed in Article 38.

473 In Region 1, in the band 505–526.5 kHz, the administrations which operate stations of the aeronautical radionavigation service shall take the technical steps necessary to avoid harmful interference to the maritime mobile service.

474 In the Federal Republic of Germany, Belgium, Spain, France, Iceland, Italy, Norway, the Netherlands, the United Kingdom, Sweden and Yugoslavia, the frequency 518 kHz is used on an experimental basis for the transmission by coast stations of meteorological and navigational warnings to ships, by means of narrow-band direct-printing telegraphy.

475 In the band 515.5–526.5 kHz, Austria may continue to operate only those broadcasting stations listed in Additional Protocol III to the Final Acts of the Regional Administrative LF/MF Broadcasting Conference (Regions 1 and 3), Geneva, 1975. This operation is allowed until the entry into force of a revision of the Geneva Plan, 1975, and subject to not causing harmful interference to the maritime mobile and aeronautical radionavigation services.

476 *Additional allocation:* in the United Kingdon, the band 519.5–526.5 kHz is also allocated to the broadcasting service on a secondary basis for the transmission of public utility information.

477 In Region 2, in the band 525–535 kHz the carrier power of broadcasting stations shall not exceed 1 kW during the day and 250 W at night.

478 *Additional allocation:* in Angola, Botswana, Lesotho, Malawi, Mozambique, Namibia, South Africa, Swaziland, Zambia and Zimbabwe, the band 526.5–535 kHz is also allocated to the mobile service on a secondary basis.

479 *Additional allocation:* in China, the band 526.5–535 kHz is also allocated to the aeronautical radionavigation service on a secondary basis.

Several frequency bands above 20 GHz has been allocated to the *inter-satellite service* (ISS), specifically to be used for links between satellites, whatever may be the service to which the satellites belong.

Finally, another group of services covers the use of radio, not for communication, but for radionavigation, radar and similar purposes. The *radiodetermination service* (RD) and the *radiodetermination-satellite service* (RDSS) embrace all of these applications, using terrestrial and satellite techniques respectively. Two sub-divisions of the RD are recognised. The *radionavigation service* (RN) comprises those applications of radiodetermination which involve the navigation of ships and aircraft, on which the safety of human life may depend. Other applications of radiodetermination, in which the safety of life aspect does not arise, comprise the *radiolocation service* (RL). In a similar way RDSS systems which involve the safety of life may be categorised as part of the *radionavigation-satellite service* (RNSS). Specialised categories of the RN and RNSS have also been created to identify systems used for maritime and aeronautical purposes.

For each service which has been named above, an abbreviation has been given, and there is a complete list of services, with abbreviations, in Table A.1 (Appendix A3). These abbreviations will normally be used for identifying a service in this book. However, whilst most of these abbreviations are widely used in the literature, it should be noted that there are no standardised abbreviations for service names.

There are 33 of these services in all. Those that involve the use of satellites comprise *space radiocommunication* (RR 9) and the others comprise *terrestrial radiocommunication* (RR 8), except for radio astronomy, which is not proper to either group.

The term *safety service*, defined in RR 56, is applied to a system of any of these services when it is used for the safeguarding of human life or property.

The Regions

It has sometimes been found convenient to allocate frequency bands to different services in different parts of the world. For this purpose the ITU divides the world into three Regions. The Regions are defined in RR 392–RR 399 and they are indicated approximately in Fig. 3.2.

Many frequency bands at LF, MF, VHF and UHF and some SHF bands have been allocated to different services in the three Regions. This creates a situation analogous to frequency band sharing, since RR 346 states as a basic principle that different services in different Regions have an equal right to operate if they have equal allocation status, each within its own Region.

Regional differences in frequency allocations are of historical origin. The practice gave a helpful measure of flexibility at a time when intercontinental interference was limited, in large part, to the HF and VLF parts of the spectrum. It raises many sharing problems now that space radio systems operate in many parts of the spectrum.

In this book the word 'regional' is printed with a capital initial letter where it refers to these three ITU Regions.

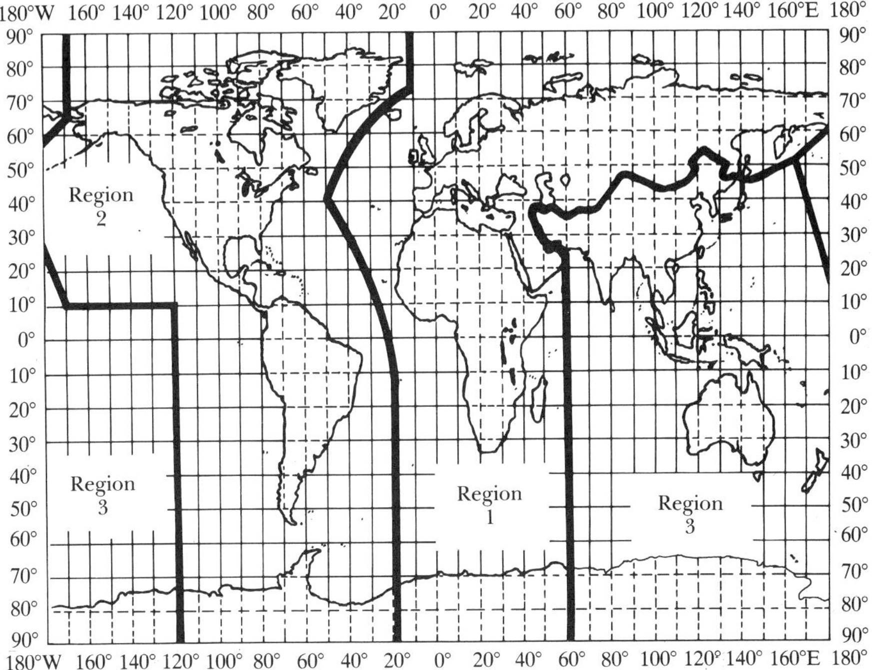

Fig. 3.2 The three ITU Regions

The status of allocations

The ITU recognises three categories of frequency allocation, 'primary', 'permitted' and 'secondary'. These categories are depicted in the international Table of Frequency Allocations as follows. For a primary allocation the name of the service is printed in capital letters. Allocations with permitted status are also printed in capital letters but with an oblique stroke added before and after the service name. Secondary status is indicated by printing the name of the service with a capital initial letter, the rest of the name being printed in lower case letters (see RR 413–RR 417). In Fig. 3.1 the allocations for Region 3 in the band 505–526·5 kHz include examples of all three grades.

Differentiation of status arises from the need to provide for the resolution of problems of interference between stations of different services when they occur. The significance of these differences of categories is explained in RR 419–RR 425 but the essence is as follows:

(*a*) No station belonging to a service with a secondary allocation may cause

harmful interference to a station of a service with a primary or permitted allocation.
(b) A station of a secondary service has no regulatory protection from interference from a station belonging to a service with primary or permitted status.
(c) If interference arises between two stations which belong to different services which have the same allocation status in the frequency band in question, then other factors (such as priority of use, notified to the IFRB) are taken into account when considering which station should give way.
(d) Finally, if an allocation has permitted status, then a frequency plan has been drawn up for a service having a primary allocation in the same band or the drawing-up of such a plan has been foreseen. The service to be planned has prior choice of frequencies for assignment at the time of planning; thereafter, permitted and primary allocations have equal rights.

The resolution of problems in interference is considered further on pp. 64–65.

The footnotes

There are hundreds of footnotes to the Table; the small segment of the spectrum covered by Fig. 3.1 is qualified by no fewer than twelve. They are all numbered in the same series as the paragraphs in the RR articles and can be identified in the same way. These footnotes have various functions, the following being typical:

(a) Additional allocations: Footnotes like RR 469, RR 476, RR 478 and RR 479 in Fig. 3.1 create sharing allocations within specified countries which are additional to the allocations in the framed part of the Table. Such areas consisting of two or more countries within the same Region are called 'sub-Regions'. In some cases these sub-Regional allocations are qualified by technical constraints or administrative procedures intended to limit the impact of these variant allocations on other countries.
(b) Alternative allocations: Many footnotes, although none shown in Fig. 3.1, have the purpose of up-grading, down-grading or cancelling within a country or sub-Region one or more of the allocations which are to be found in the framed part of the Table. Such footnotes sometimes also allocate the band, within the sub-Region, to a service which does not have an allocation in the framed part of the Table. As before, the footnote may include technical or administrative constraints to limit the effect of alternative allocations on the allocations of other countries.
(c) General footnotes: These may be technical, operational or administrative. RR 470, RR 471, RR 472, RR 473, RR 475 and RR 477 in Fig. 3.1 are typical examples of footnotes that define specific purposes for frequency bands or spot frequencies, place specific limitations on the operation of systems or specify procedures for dealing with problems which are thought likely to arise.

(d) Advisory footnotes: Information on the use to which some countries plan to put a band or a spot frequency within the band is sometimes given in a footnote. An example of this in Fig. 3.1 is RR 474 (this note was amended at WARC-MOB-83).

An advisory footnote gives no internationally recognised status to the particular spectrum use which it announces. However, footnotes giving 'additional allocations' and 'alternative allocations' do not merely record the preferences or the intentions of the countries named. Subject to any qualifications also recorded in the footnote, the intention of the specified countries, declared in this way, to depart from the scheme of allocations indicated in the framed part of the Table has been accepted by the administrations that have ratified the RR and these 'footnote allocations' have as much validity within the stated geographical limits as the allocations in the framed part of the Table have outside those limits (see RR 428 and RR 432).

3.2.2 Interference from industrial, scientific and medical equipment

Broad-spectrum noise-like electromagnetic radiation is generated by many kinds of electrical equipment and this 'man-made noise' may be a significant limitation on the performance of radio links operating between 10 MHz and 1 GHz unless the receiving station is remote from industrial plant and busy thoroughfares (see Appendix B5.5).

However, the electromagnetic energy radiated from much industrial, scientific and medical (ISM) plant does not closely resemble broad-band noise; instead it is mostly concentrated within relatively narrow frequency bands. To limit the impact of these radiations on radio services, the ITU is collaborating with the IEC and CISPR to ensure that ISM equipment is designed to concentrate interference into a few specified narrow frequency bands and to limit as much as possible the level of the unintended radiation, inside and outside those bands. CCIR Recommendation 433[1] lists numerous CISPR reports on this topic.

Footnotes to the international Table of Frequency Allocations designate the following frequency bands for ISM use:

6·765–6·795 MHz
13·553–13·567 MHz
26·957–27·283 MHz
40·66–40·70 MHz
433·05–434·79 MHz (Region 1 only; for variations see RR 661 and RR 662)
902–928 MHz (Region 2 only)
2400–2500 MHz
5725–5875 MHz
24·0–24·25 GHz
61·0–61·5 GHz

122–123 GHz
244–246 GHz

RR 1815 binds administrations to:

> . . . take all practical and necessary steps to ensure that radiation from equipment used for industrial, scientific and medical applications is minimal and that, outside the bands designated for use by this equipment, radiation from such equipment is at a level that does not cause harmful interference to a radiocommunication service . . .

The frequency bands designated for ISM are also allocated for radio services, the FS and RL in most cases, and most of the footnotes designating ISM bands indicate that radio services operating in these bands must accept whatever interference may arise from ISM equipment. Nevertheless, when several new bands were designated for ISM use at WARC-79, it was decided that more positive measures should be taken, in particular for the new bands, to protect radio services from ISM interference. RR Resolution 63[2] invites the CCIR to pursue further studies of the matter with CISPR and IEC, in particular in order that limits might be recommended for the level of radiation from ISM equipment in the frequency bands at 7 and 433 MHz and at 61, 122 and 245 GHz (see also CCIR Question 4/1[3], CCIR Study Programme 4D/1[4] and CCIR Decision 54[5]). Meanwhile, the footnotes which designate these five bands for ISM use require an administration authorising the use of ISM installations to consult first with other administrations that might be affected by interference to radio systems.

3.2.3 Growth of services, innovation and allocation sharing

In 1947, 1959 and 1971, world administrative radio conferences agreed frequency allocations for parts of the spectrum higher in frequency than any that had been allocated to services previously. These allocations were bold and uncomplicated. Subsequently, as substantial use of the spectrum advanced upwards, changing needs and new needs were accommodated by a process of *ad hoc* modification to the international Table of Frequency Allocations. The table has become complex. These complexities, necessary as they are, place burdens on spectrum management and constraints on spectrum users. It is of interest to consider how the process of modification has developed.

The scale and nature of the demand for radio facilities is always changing. The demand for most established facilities is growing and new facilities emerge, creating unforeseen spectrum needs. Usually it is feasible in most locations for new assignments to be made to established facilities in frequency bands which are already in use, perhaps after changes have been made in radio station technology or in spectrum management practice to increase the efficiency with which the bands are used. Other ways must, however, be found for new facilities and when growing services can no longer be economically contained within bands already in use.

One solution to a problem of this kind, and the easiest solution for the spectrum manager, is to take into use a frequency band which is already allocated for the growing service but is not yet being exploited. It almost always happens that an empty band that can be used in this way is higher in frequency than any band already in use for the service in question. Three kinds of problem arise for radio station owners when such bands are taken into use:

- Key components and equipment may have to be developed and put into manufacture to operate in the new band. This equipment tend to be more expensive, initially, than comparable equipment designed to operate in the bands which are already in general use, although prices usually fall eventually.
- As the frequency rises, radio propagation deteriorates. Absorption and scattering of signals by rain, cloud and foliage becomes more severe and ray path obstruction by terrain and buildings has a greater effect on systems which are required to provide area coverage. Systems can be designed to overcome this deterioration, but these measures further increase costs.
- The aperture of a receiving antenna of a given gain falls as the frequency rises. The power of the transmitter may have to be increased to compensate for this unless the nature of the system permits the gain of the receiving antenna to be increased. The need for higher transmitter power is always to be regretted and it has a very considerable economic impact on battery-driven hand-portable equipment.

Despite these disadvantages, bands above 30 GHz are already being taken into use for short-distance applications in, for example, the FS and RL, and the FSS will assuredly exploit bands above 15 GHz within a few years. Frequencies much in excess of 1 GHz seem unlikely to be satisfactory for most BS and some MS applications.

If migration up the spectrum to a vacant band is not feasible, other solutions must be sought. There may be a frequency band, at a point in the spectrum which is suitable for the growing service, which is allocated by international agreement to another service but is lightly loaded. Then:

- The allocation to the other service might be cancelled, frequencies in other bands being assigned to its stations and the band so cleared being re-allocated for the growing service. This is seldom done because it usually happens that more countries wish to maintain existing allocations than experience a sufficiently strong need for change.
- The band might be divided, part remaining allocated for the original service and part becoming a new allocation for the growing

service. Any such action must, of course, be planned so as to give stations of the original service enough time to transfer their operations, where necessary, to new frequency assignments.

- The whole band might be allocated for the growing service without cancelling the previous allocation, creating a 'shared' allocation. Sharing is not usually agreed unless it has been shown, by studies in the CCIR, that there is a substantial degree of technical or operational compatibility between the services involved, with or without geographical separation of the stations of the two services.

Shared frequency allocations are used in three different ways. In the most literal sense of the word administrations might assign frequencies from a shared band to stations of the sharing services without discrimination. Alternatively, one administration might use a band for assignments to stations of one of the services to which the band is allocated whilst another administration uses it for assignments to stations of another of the allocated services. Clearly, the administration of a large country may vary its use of a band in this way in the different parts of its own territory. For the third way, by careful constraint of system parameters, it may be possible to use the same spectrum of systems of the sharing services within the same geographical area, neither service significantly diminishing the usability of the band for the stations of the other service.

Often a new sharing allocation proposed by some administrations at a WARC is opposed by others. Sometimes no way can be found to reconcile these opposing views, there is no voting majority in favour of the change and the proposal falls. It may, however, be found possible to find spectrum management devices, acceptable to a majority, to minimise the effect that the establishment of stations of the new service will have on the use of the band for the service or services that are already established in it. Much may depend on the distance over which interference problems are expected to arise. If interference will be short-range and can therefore be controlled by the administration concerned, perhaps in consultation with the administrations of neighbouring countries, simple arrangements can usually be agreed without difficulty. If interference from or to the new service may arise at great distances, acceptable arrangements may be hard to negotiate and complex to implement, especially if the established services are safety services or if they are the subject of frequency assignment planning. If it is foreseen that the new service will involve only a small number of stations, it may be possible to negotiate individual agreements to cover the operation of each station. The following methods for protecting one service from another are typical:

- Limits are placed on parameters of systems, such as the power of emissions, if the interference from transmitters of one service to receivers of another service can be reduced to a negligible level thereby. These limits are called *sharing constraints*.

- The establishment of a station of the new service may be made conditional upon agreement with the administrations of neighbouring countries that their existing stations of the old service will not suffer interference from, nor cause interference to, the new station. This is called *frequency co-ordination.*
- A sharing allocation, world-wide or Regional, may be limited to secondary status.
- A sharing allocation may be made sub-Regional, being confined to a footnote to the international Table and limited to specified countries.

If no other solution can be found, a procedure set out in *RR Article 14* may be used for case by case consultation with prescribed baseline conditions (see pp. 58–60).

The use of sharing in the international Table has grown greatly in recent decades and especially since the introduction of satellite services. Nowadays few frequency bands, except in the HF part of the spectrum, are allocated exclusively to one service and many bands are shared by several services, some by as many as half a dozen. Sharing is a very valuable device in spectrum management, providing flexibility to meet differences in requirements from place to place, and allowing major frequency allocations to be made to new services, particularly space services, without the cancellation of previously-existing allocations in the same part of the spectrum. In some circumstances, sharing can be said to have allowed the spectrum to be used twice over.

However, sharing causes the loss, partial or complete, of some of the advantages of frequency allocation listed on pp. 30–31. These advantages have been regained to some degree by the implementation of various spectrum management devices, such as constraints on system parameters and various formal and informal procedures for the co-ordination of frequency assignments. Nonetheless, the means that have been used to manage sharing exact a price, possibly a high price, in system cost, system limitations and spectrum management effort.

No doubt many new sharing arrangements will be agreed in the future; international management of spectrum in a changing world depends to a considerable extent on them. Above all, sharing can provide a quick fix at a WARC to an allocation problem. However, the disadvantages of sharing should not be forgotten. Sharing arrangements involving unsatisfactory sharing constraints or onerous co-ordination may be replaceable by rearrangements of the international Table which make them unnecessary if sufficient time is allowed for frequency assignments to be changed; the quick fix and the slow solution should go hand in hand. However, the time needed to permit the implementation of a solution that involves the cancellation of an existing frequency allocation may be substantial. See for example pp. 80, 154 and 189 for references to the transfer of HF frequency bands from the fixed service to the broadcasting and maritime mobile services, involving delays of 10 and 15 years to give time for fixed stations to be found assignments in other frequency bands.

3.3 National frequency allocation

National obligations

When a ITU member government ratifies the instruments of the ITU, it does not undertake to implement the international Table of Frequency Allocations in its entirety. Its obligations in this respect are set out in the RR as follows:

> *RR 340:* Members undertake that in assigning frequencies to stations which are capable of causing harmful interference to the services rendered by the stations of another country, such assignments are to be made in accordance with the Table of Frequency Allocations and other provisions of these Regulations.
> *RR 342:* Administrations of the Members shall not assign to a station any frequency in derogation of either the Table of Frequency Allocations given in this Chapter or the other provisions of these Regulations, except on the express condition that harmful interference shall not be caused to services carried on by stations operating in accordance with the provisions of the Convention and of these Regulations.

Moreover, governments do not bind themselves to observe the RR at military radio stations. Article 37 of the 1989 Constitution reads as follows:

> 1. Members retain their entire freedom with regard to military radio installations.
> 2. Nevertheless, these installations must, so far as possible, observe statutory provisions relative to giving assistance in case of distress and to the measures to be taken to prevent harmful interference, and the provisions of the Administrative Regulations concerning the types of emission and the frequencies to be used, according to the nature of the service performed by such installations.
> 3. Moreover, when these installations take part in the service of public correspondence or other services governed by the Administrative Regulations, they must, in general, comply with the regulatory provisions for the conduct of such services.

Nevertheless, an ITU member country undertakes to respect the right of other member countries to implement the international Table when making assignments to non-military stations.

Major advantages flow from adopting national frequency allocations which lie within the range of options contained in the international Table, and an administration usually does so. However, an administration can depart to a considerable extent from the international Table if it considers that it is in its national interest to do so. Thus:

- The administration is quite free to refrain from making assignments in a band, or in a part of a band, to transmitting and receiving stations of any of the services which have allocations in that band in the international Table. Such an option will often be taken, for example, in bands at VHF and UHF which are shared in the international Table by the BS and the LM.
- The administration is free to make assignments to stations of services which are not included in the international Table for the band in question, provided that the use of such out-of-band assignments does not cause harmful interference to a foreign station which is operating in accordance with the international Table. However, no remedy is provided by the RR if an out-of-band assignment suffers interference from a foreign station with an assignment which conforms to the Table.

This selection of the stations which an administration will permit to use a frequency band within its jurisdiction constitutes a large part of the production of a national frequency allocation table.

Despite these various options for optimising national spectrum use, national usage which differs significantly from the usage of other countries, and especially trading partners and neighbouring countries, has many of the disadvantages associated with sharing (see p. 43). It may be unavoidable in the short term but it should be discontinued when the opportunity arises.

The national frequency allocation table

Many administrations draw up their own national frequency allocation table. National requirements and the international Table both change from time to time and national allocation tables should be kept under review, and revised where necessary, if they are to continue to serve well the needs of the country. Factors such as the following have to be taken into account when developing a national table from the international Table and when revising an existing national table:

- Forecasts of national needs for the facilities provided by systems of the various radio services, convertible into requirements for assignable bandwidth. These forecasts should look ahead at least ten years, and preferably twenty years.
- An awareness of the views of major national spectrum users and radio equipment suppliers as to the national allocations that would be desirable.
- Some appreciation of the existing pattern of spectrum utilisation in neighbouring countries and of the plans that administrations have for its development.
- An appreciation of the order of cost of applying spectrum utilisation efficiency measures to reduce the bandwidth required for the forecast facilities; in particular in cases where demand is growing to the point where the bands already nationally allocated for a service are approaching saturation.

The possible need to adopt spectrum-conserving methods, by up-grading equipment specifications or by improving the spectrum-management techniques used within the administration, is considered further in Chapter 4. In the present context it should merely be noted that an improvement in the efficiency with which one frequency band is used may keep open valuable options for the use of other bands.

Starting from these factors, a national table of frequency allocations can be drawn up and an existing one can be revised. Many international allocations will no doubt be incorporated in the national table without substantial change, such as:

- Those that will be needed for the country's international links.
- Those that will be needed by ships and aircraft, civil and military, for terrestrial and satellite communication and radiodetermination.
- Those under which ITU frequency plans, of interest to the country, have been or are going to be drawn up.
- Those which are allocated in the international Table to one service only.

Elsewhere in the spectrum an administration has more choice as to which of the services sharing a band in the international Table should become the national allocation. Clearly, existing practice will be a dominant consideration. It will frequently be found that valuable flexibility can be built into the national table by a shared allocation; for example it may enable a frequency band to be shared equitably between all of the stations having special need of the propagation characteristics which that band provides, even though those stations may belong to several different services.

On the other hand, it may often happen that national needs are best served by selecting only one of the allocations in the international Table for inclusion in the national table. Some exclusive allocations, for example, improve efficiency of spectrum utilisation due to homogeneity. Others may eliminate or greatly reduce the need for costly frequency management procedures such as the international co-ordination of space and terrestrial systems. Decisions of this kind should preferably be made in consultation with the administrations of neighbouring countries, since the full benefit of the decision may not be obtained if a country acts alone.

Some kinds of radio system radiate very little power. If it is clear that interference will not be caused to systems operating in other countries, an administration is free to allocate any band to any service regardless of the international allocation. However, such systems will have no protection under the RR from interference from abroad, and this factor must be taken into account in making national allocations of this kind.

With all of these options it is usually possible for an administration to draw up the national table within the provisions of the international Table, but sometimes it is perceived that a change in the international Table would be desirable. From time to time the ITU convenes WARCs with authority to amend all, or some specified parts, of the international Table.

A national delegation to such a conference may succeed in gathering enough support to ensure that the desired change to the Table will be made, particularly if other delegations want that same new option and if the impact of the new allocation on existing systems is foreseen to be small. If a new world-wide allocation is not agreed, it may be possible to secure a Regional allocation, a national footnote allocation or a footnote allocation subject to special constraints or 'Article 14' consultation procedures. Any of these outcomes may help in the optimisation of the national table.

Finally, in preparing or revising its national frequency allocation table, an administration is likely to find it desirable to make some of its allocations specific, not merely to services as the ITU defines them but to particular applications of services. For example, some bands allocated in the international Table to the land mobile service might be allocated nationally, in whole or in part, for use for land mobile stations operated by the police, by the armed services, by private mobile radio systems, for conventional car radiotelephone systems having access to the public telephone system, for cellular radio systems, for 'citizens band' and so on.

Changes of national allocations

As the demand for radio facilities has grown in recent decades, and particularly as the number of broadcasting stations, fixed links and land mobile systems has grown, there has been a marked tendency in many countries for frequency bands allocated for such purposes in the VHF and UHF parts of the spectrum to become saturated with assignments. Saturation can be delayed by improvements in the efficiency with which the spectrum is used, but ultimately it becomes necessary to assign more bandwidth to these growing services. Expansion has usually been into bands higher in the spectrum.

This migration up the spectrum has been relatively straightforward for the FS. Bands previously unused between 10 and 15 GHz have been widely taken into use. Bands still higher in the spectrum are also being taken into use, mainly for short links, with any necessary adaptations of system architecture to take account of less favourable propagation conditions. However, this option is not so freely available for the LM and BS. The use of frequencies as high as 1 GHz for, for example, land mobile systems presents significant problems owing to the characteristics of radio propagation in an urban environment and it is not yet clear that frequencies higher than 2 GHz are usable for cost effective facilities serving road vehicles. The spectrum from 200 MHz up to 1 GHz and beyond is heavily used already for other services, and it has been found necessary in some countries to phase out the use of this part of the spectrum for the FS to make room for expansion of the LM.

Services have also been displaced from the bands in which they have been long established by changes that have been made in the international Table. For example, WARC-79 made substantial readjustments to the proportions of the HF band that are allocated exclusively to the FS, MM and BS. Also, several administrative radio conferences since 1963 have provided narrow exclusive allocations for totally new services, typically space services, at the cost of displacing existing users of those bands, although some years' notice of such changes is usually provided for.

For various reasons like these, it may be necessary for an administration to stop the use of a band by stations of one service in order that it may be made available to the stations of another service. Administrations must exercise care and forethought if such changes are not to cause substantial expense or serious operational disruption to the parties which operate the service being displaced. It may be possible to foresee the need for such changes five or ten years in advance, in which case an administration should, as early as possible, cease making new assignments that will not be sustainable. Preparations may also be necessary for providing assignments in other bands for stations that will have to relinquish their existing assignments. For this reason it is important that radio station licences, including specified frequency assignments, should be for a fixed term, with an assured prospect of renewal, perhaps for a shorter period, on expiry, provided that no change of allocation is then imminent.

3.4 References

1 'Methods for the measurement of radio interference and the determination of tolerable levels of interference'. CCIR Recommendation 433–3; Recommendations and Reports of the CCIR, 1986, Volume I (ITU, Geneva, 1986)

2 'Relating to the protection of radiocommunication services against interference caused by radiation from industrial, scientific and medical equipment'. Resolution 63; Radio Regulations, edition of 1982 (ITU, Geneva, 1982)

3 'Limitation of unwanted radiation from electrical apparatus and installations'. CCIR Question 4–3/1; Recommendations and Reports of the CCIR, 1986, Volume I (ITU, Geneva, 1986).

4 'Limitation of radiation from industrial, scientific and medical (ISM) equipment'. CCIR Study Programme 4D–1/1; *ibid.* Volume I.

5 'Radiation from ISM equipment'. CCIR Decision 54–1; *ibid.* Volume I

Chapter 4

Frequency assignment

4.1 The elements of frequency assignment

For most radio services, the licence for a transmitting station states the frequency or frequencies assigned to the station and various parameters of the emissions that may be radiated, such as the transmitter power, the bandwidth and basic antenna characteristics. Various technical, operational and regulatory stipulations may also be included. There are corresponding provisions in the licences of many receiving stations. The selection of frequencies for assignment, the determination of the technical and operational stipulations and the various managerial processes which support these functions of an administration involve the following elements:

(*a*) An assigned frequency must be usable with the licensee's equipment. It is therefore usually necessary for the licensee, before buying equipment, to be aware of the frequency band from which the frequency will be assigned.

(*b*) An assigned frequency will, with rare exceptions, be in a frequency band which has been allocated nationally for the service for which the assignment is to be used.

(*c*) If a frequency is to be used for an international link or if the emission will be capable of causing interference to, or is susceptible to interference from, a station within the jurisdiction of another administration, the assignment will be made from a frequency band which is allocated in the international Table of Frequency Allocations (RR Article 8) for the service for which the assignment is going to be used.

(*d*) Taking into account the predicted propagation properties over the signal path of the frequency being assigned, the characteristics of the licensee's equipment (in particular the antenna characteristics) and the parameters of the emission being authorised should be such that the system can be expected to attain the required standard of performance in operation, an appropriate margin being left for low-level interference.

(*e*) For most radio services, the selection of a frequency for assignment must take foreseeable high-level interference adequately into account. Use of a frequency assigned for transmission should not cause an unacceptable increase in interference at other stations that are already authorised to receive from other transmitting stations at or near that frequency. Similarly, use of a frequency assigned for reception should not involve the reception of an unacceptable level of interference from unwanted stations that are already authorised to transmit at or near that frequency. If interference problems arise after operation has begun, they must be resolved, typically by the assignment of a different frequency for the newcomer.

(*f*) Having assigned a frequency to a station, an administration includes details of the assignment in a national register.

(*g*) If appropriate, administrations take measures to establish internationally the right of the station to use the frequencies that have been assigned.

(*h*) A requirement is placed upon transmitting stations to identify their emissions, to the extent that this is feasible, chiefly to facilitate the clearance of interference if it should arise.

(*i*) Regulatory and operational stipulations are included by the administration in the licence, typically in order to discharge international undertakings which the administration has given (for example, on the confidentiality of communications or the protection of frequencies used for emergency or distress communications) or in order that governmental objectives connected with the quality of the telecommunications facilities offered to the public may be achieved.

(*j*) The selection of frequencies for assignment often has amongst its objectives the achievement, not only of effective operation with a level of interference which is not unacceptably high but also of an appropriate level of efficiency of spectrum utilisation. Equipment performance standards and limits on the parameters of emissions may be imposed for this purpose by the administration as terms of the licence.

(*k*) An administration ensures that transmitting stations do not operate within its jurisdiction without its permission. Conversely, action may also be taken to verify that a licensee makes use of the permission he has been given to use spectrum, so that other parties with a real need are not unnecessarily denied the use of the frequency that has been assigned to the licensee.

Most frequency assignments are for emissions from terrestrial stations with low transmitter power, operating in parts of the spectrum where line-of-sight propagation conditions prevail, and for receivers which do not have an exceptionally high sensitivity to interference. In all but the smallest countries, most assignments of this kind can be managed by the responsible authority without consideration for the assignments made by other administrations. For such assignments, the elements listed above, excluding (*c*) and (*g*), constitute the national frequency assignment function which administrations perform. This function is considered further below.

When assignments are made for international links or for emissions which can cause or suffer trans-frontier interference, all of the above elements (*a*) to (*k*), must be taken into account, making the administration's task much more complex. Some of these elements become subject to international regulation and the oversight of the IFRB. The extension of the national frequency assignment function to include consideration of the risk of interference to and from foreign stations (element (*e*)), the acquisition of international acceptance of the right of a station to use a frequency assigned by the administration (element (*g*)) and the resolution of international interference problems if they arise after operation starts (part of element (*h*)) are considered on pp. 53–65.

Attention needs to be given to the efficient use of spectrum (item (*j*)) in connection with several aspects of both national and international management of frequency assignments. Ways of achieving efficient spectrum use in the context of specific services are considered in later chapters of this book. However, it is of interest to examine this topic in a more general context and it is reviewed briefly on pp. 65–76.

4.2 Frequency assignment in a national context

The service to which a radio station belongs, the provisions of the national frequency allocation table and the extent to which suitable frequency bands are already occupied by assignments to existing systems will usually determine the frequency band from which the frequencies required for a new facility should be assigned.

If it may be assumed that interference to or from foreign countries does not need to be taken into account, the specific frequencies which are to be assigned from the chosen band are selected by one or other of two basic methods, as follows:

(*i*) *National frequency planning:* In many frequency bands a national radio channel plan is drawn up. For example, in a VHF band allocated for LM which is to be used for private mobile radio facilities, the channel plan might consist of channels 25 kHz or 12·5 kHz wide occupying the whole available band, associated in go-and-return pairs. Similarly, in a microwave band used for short-distance links between fixed stations, each link being capable of carrying 30 multiplexed PCM telephone channels using 4-phase PSK modulation, the basis of the plan would be similar but the channels might be 2 MHz wide. A non-mandatory radio channel plan agreed in the CCIR might be adopted in such a case.

When assignments are required for an LM base station and the associated mobile stations, a channel-pair would be chosen and assigned which is not already used, or is little used, in the locality in which the new licensee will operate.

When an assignment is required for a link between specified FS stations, a channel pair would be identified and assigned, subject to three criteria:

- the signal level of the wanted emission at each receiving station would be adequate,
- the total interference at the new receiving stations from assignments that had been made previously to other stations would be acceptably low, and
- the interference that would be caused by a new transmitting station at any previously-licensed receiving station would be acceptably low.

Predictions of signal strength and interference levels would normally be used; the CCIR has made an extensive study of radio propagation data and these studies are reviewed in Appendix B. All new assignments would be entered in a national register, to enable these predictions of interference level to remain effective.

For other kinds of facility and mixtures of facilities, planning and selection might be more complicated in practice but they would be similar in principle. Procedures might need to be devised to ensure that the process produced a suitable geographical distribution of assignments, leading to efficient spectrum utilisation.

(*ii*) *The inspection method:* Efficient implementation of a national plan might not be feasible in a frequency band which was to be used for a heterogenous population of emissions. In such a case, an inspection of the national register of frequency assignments may allow the identification of frequencies which might be suitable when a new assignment is required. The calculation of signal levels and interference levels would then identify which, if any, of the possible assignments would be satisfactory if implemented.

These methods, and especially the inspection method, put a substantial management burden on the administration. A variation applicable to either method, which transfers much of the work load from the administration to the users of radio, involves the announcement of details of tentative new assignments, proposed by the potential user after inspection of the national register, to all other national users of the band. These other users are required to object to confirmation of the assignment if they consider that their own established assignments would interfere with its reception or would suffer interference from the emission.

National planning is used where it is feasible. The inspection process is laborious in a country making extensive use of radio, but it has the advantage of versatility. In a location where the demand for radio facilities is intense, neither process can be used in a way that makes efficient use of spectrum without the aid of computers and suitable software. Such techniques are discussed, for example, in CCIR Report 841[1]. There is also an ITU handbook on this subject.[2].

The assignment having been made, a call sign or some other form of identification signal may be assigned (see pp. 74–75), and the frequency is brought into use at the station or stations to which it is assigned. In some cases it may be thought prudent to notify the assignment to the IFRB for inclusion in the MIFR (see below).

If it is found, when a new frequency assignment has been taken into use, that the levels of interference caused or suffered by the new assignment are unacceptably high, it will rest with the administration to resolve the matter, most usually by changing the most recent frequency assignment.

4.3 International management of frequency assignments

4.3.1 International registration of assignments

RR 1214–RR 1216 (applicable to terrestrial services) and RR 1488–RR 1490 (applicable to space services) require an administration to notify the IFRB of frequency assignments which it makes to most kinds of radio station (although not those to mobile stations):

> . . . if the use of the frequency is capable of causing harmful interference to any service of another administration or if the frequency is to be used for international communication.

The notification is to include the location of stations involved, various data on the parameters of the emission and the date when the assignment is taken into service. This is required even if the system for which the frequency has been assigned lies entirely within the jurisdiction of the administration making the assignment.

Notification is an essential part of various procedures which are used to determine whether a new assignment is not merely capable of causing interference to unspecified foreign assignments but is likely to cause interference to an actual, registered foreign assignment. If these procedures show that interference is unlikely, the new assignment can be registered in the MIFR and dated.

Registration of an assignment in the MIFR may be of great importance, particularly in congested frequency bands. RR 341 reads as follows:

> Any new assignment or any change of frequency or other basic characteristic of an existing assignment shall be made in such a way as to avoid causing harmful interference to services rendered by stations using frequencies assigned in accordance with the Table of Frequency Allocations ... and the other provisions of these Regulations, the characteristics of which assignments are recorded in the Master International Frequency Register.

Thus, the recording of an assignment in the MIFR confers status on the assignment, protecting it to some extent from harmful interference from subsequent new assignments (see pp. 64–65).

An administration may also notify to the IFRB for registration the details of assignments which are not capable of causing harmful interference to foreign stations:

> 'if it is desired to obtain international recognition of the use of the frequency'

see RR 1217 and RR 1491. This option might be exercised, for example, to protect from interference an assignment to a receiving station if the associated transmitting station is beyond interference range of the frontier. Notification and registration of receiving assignments to land stations in the mobile services can be used to provide a degree of protection for the transmitting assignments which are made to mobile stations but which cannot be registered. Nevertheless, large numbers of assignments which are not thought likely to cause or suffer international interference are not notified to the IFRB.

The term 'harmful interference' is defined in RR 163, as follows:

> Interference which endangers the functioning of a radio-navigation service or of other safety services or seriously degrades, obstructs, or repeatedly interrupts a radiocommunication service operating in accordance with these Regulations.

There are several points to note about this definition. The protection provided for safety services is of a higher order than the protection provided for other services. Secondly, the degree of interference which is judged to be harmful is subjective but the extent of the degradation of a wanted signal by interference which may exist without the condition being judged to be 'harmful' is worse than that caused by the perceptible background of interference (sometimes called 'permissible interference') that is accepted, where it arises, as a price to be paid for efficient spectrum utilisation. Finally, this definition totally disregards interference that occurs to an emission that is not in accordance with the RR.

The IFRB informs all administrations promptly of all the notifications of frequency assignments that it receives by publishing the details in the IFRB weekly circular. This information alerts the administrations of countries neighbouring the notifying administration to the possibility of future interference to systems within their own jurisdiction and it provides helpful guidance to administrations that are themselves seeking frequencies to assign to stations.

The Board then verifies that the notified assignments are in accordance with the Radio Regulations. The Board may also be required to examine the notice, and the assignments already registered in the MIFR, to ascertain whether the new assignment would interfere with an established assignment which has been registered on behalf of another administration. Much depends on the regulations which apply to the frequency band in question, but to generalise:

- At frequencies below 28 MHz the Board makes a technical examination to determine the risk of causing interference.
- For terrestrial services above 28 MHz, it is assumed that the administration submitting the notice will have examined the risk of interference in collaboration with any concerned neighbouring administrations and will have come to a satisfactory conclusion before making the notification.
- For space services, and in bands in which interference may occur between stations of a space service and a terrestrial service, the RR lay down detailed procedures by which the administrations concerned are required to collaborate to determine the extent of any interference risk, and to report the result to the Board.
- Special conditions apply to assignments made in accordance with agreed frequency assignment and frequency allotment plans.

If it is confirmed that all of the requirements of the RR have been met, including satisfactory interference prospects, the details of the new assignment, including the date of bringing it into use, are registered in the MIFR.

There is provision for registering a notification, with an appropriate endorsement, in some cases where not all of the requirements of the RR are met. The most significant of these cases below 28 MHz is where an administration is not satisfied with the IFRB's conclusion that interference will be caused to an existing assignment (see p. 86). Another significant case arises where the station to which a frequency has been assigned belongs to a service which has no allocation in the frequency band in question; such an emission is said to be 'out-of-band'. An out-of-band assignment may be registered, provided that an undertaking is given to cease the out-of-band emission immediately if harmful interference occurs to any emission that is operating in accordance with the RR; see RR 342 and, for example, RR 1267 and RR 1521.

The MIFR is intended to be an accurate register of all assignments of international significance that are in current use. It is important that assignments which are no longer in use should be deleted from the Register and administrations are required to notify the withdrawal of assignments from service; see RR 1430 and RR 1573. Inevitably deletion is pursued by administrations with less zeal than registration. There are various provisions in the RR which direct the IFRB to ensure, as completely as possible, that the MIFR remains an accurate and up-to-date record.

4.3.2 Selection of frequencies for assignment

In some circumstances an international conference may be convened to draw up an international plan of frequency assignments or allotments, the plan being subsequently ratified by governments as part of an international agreement. This formal planning process is considered on pp. 60–64. For some services, and in particular the MM, it has been found to be desirable to agree radio channel plans for the various kinds of facility that are required but to refrain from the further step of formal international frequency assignment planning. More commonly, however, the principle of the inspection method described on pp. 52–53 for national frequency selection is applied in the international context also.

When using the inspection method to choose a frequency for assignment in a situation where international interference may arise, an administration usually needs more information on the use already made of the spectrum within the frequency band of interest than its national list of frequency assignments can provide. The IFL, up-dated six-monthly, shows the assignments that other administrations have had registered. The IFRB weekly circular shows current notifications to the Board. Neither source contains as complete a set of information on the use made of a frequency band in other countries than the national list of those countries would provide, but most administrations do not give access to their national list to foreign administrations. Much more information than any of these sources contain might be needed to determine whether interference between stations at frequencies subject to tropospheric wave propagation would exceed acceptable limits. Monitoring may be helpful in making an initial choice (see pp. 75–76) but its value is likely to be limited.

Consultation with the other administrations involved is often essential before a final decision is made in choosing a frequency for assignment. Without it, much time may be lost in notifying assignments to the IFRB which are subsequently refused registration because of interference risks. Other assignments, although registered, are found to be untenable because of unforeseen interference situations. And on the other hand, without consultation, many opportunities to operate two stations on the same frequency would not be identified because foreseen interference levels are pessimistic.

In general, such consultation is informal. In some frequency bands, however, for some services and in some sharing situations, formal procedures for consultation have been laid down in the RR and the IFRB is required to ensure that these procedures are implemented. Thus:

- All assignments to space services, except those contained in frequency assignment or frequency allotment plans, must be co-ordinated with other space service assignments in accordance with procedures set out in RR Article 11 (see pp. 281–288).
- For most space services operating in frequency bands which are shared with terrestrial services, assignments to earth stations must be co-ordinated with assignments to terrestrial stations within interference range, this procedure also being set out in RR Article 11. Once the initial co-ordination of such an earth station has been done, the procedure becomes reciprocal, assignments to terrestrial stations being

co-ordinated with the earth station assignments also (see pp. 118–131).

- In a substantial number of cases, administrative radio conferences have agreed to a new sharing allocation provided that every assignment made under it is subject to agreement with all administrations likely to be affected. The procedure laid down for this purpose is set out in RR Article 14 (see pp. 58–60).

This somewhat unstructured frequency selection process, involving variously the inspection of lists of existing assignments, monitoring, national planning and *ad hoc* or formal discussions between administrations is used for finding frequencies for new assignments for most services and in most frequency bands. Inevitably it is to be compared with the international planning process discussed on pp. 60–64. The chief conclusions to be drawn from such a comparison are as follows:

- The inspection method, with consultation, is effective and well proved in practice, but it is sometimes laborious and difficult to carry out, although the assistance of the IFRB can be requested if the process is beyond the means of an administration; for example see RR 1218.
- The inspection method is empirical and, to some degree, random. Where very large numbers of assignments have to be crowded into a frequency band, it may be very efficient. Planning tends to be less efficient.
- For some facilities it is desirable for radio carriers or communication channels to be spaced at uniform intervals in the frequency domain. Where the stations of different countries are involved, this uniformity is easily achieved by international planning but it is unlikely to be achieved by a random process of selection, such as the inspection method.
- The inspection method is experimental and the resulting assignments may lack permanence. When a new frequency assignment is brought into service, it may be found to cause unacceptable interference to another system, the assignments of which had already been registered in the MIFR. This interference may start immediately, but if it is caused, for example, by slow changes in propagation conditions, the interference may not appear until long after the new assignment has been brought into use. The possible need to change the operating frequency after a period of time may be of relatively little importance in, for example, some systems of the FS or the MS but it would be extremely inconvenient in some others, such as broadcasting.
- In planning, provision is made for requirements which are firm and for other requirements which are not firm, and which indeed may never come to maturity. In order to make it possible for these latter, unrealistic, requirements to be included in the plan, an unnecessarily high standard of spectrum utilisation efficiency may have to be set for all stations that go into service, requiring high technical standards of all systems and needlessly raising costs. Furthermore, some firm requirements may not find a place in the plan.

- The inspection method, coupled with the use of priority of the date of entry into service as the principal criterion used in the resolution of interference problems, gives an advantage to countries which establish their radio systems early. Countries which have started to set up radio systems only recently, typically developing countries, have had great difficulty in establishing registrable assignments in some frequency bands and for some services.

4.3.3 Consultation in accordance with RR Article 14

There has sometimes been strong opposition at a WARC to a proposal to allocate a frequency band for a service, sharing with the service or services for which the band was already allocated. In some such cases the outcome has been a requirement, associated with the new allocation, that an administration obtain the agreement of every other administration that may be affected, whenever it wishes to assign a frequency to a station of the new service in that band. Such allocations are not usually made by an entry in the framed part of the international Table of Frequency Allocations (RR Article 8) but by a footnote to the Table. The procedure to be followed in carrying out this consultation process is set out in RR Article 14.

There are now about 65 footnotes to the Table which make allocations subject to consultation on assignments in accordance with RR Article 14. They mostly involve services having stations which are few in number but which are capable of causing interference at great distances, or they share bands with services with receiving stations which are very sensitive to interference. Most of the footnotes apply to bands above 100 MHz and the majority are for space radio services. In some cases the footnotes also apply sharing constraints to stations of the new allocation or place limitations on the options for discussion in the consultations.

A few examples of these allocations will illustrate how this option has been applied:

(*a*) In Region 2 the band 942–960 MHz is allocated for the FS (primary) and the MS (secondary) but footnote RR 708 states as follows:

> Different category of service; in the United States, the allocation of the bands 942–947 MHz and 952–960 MHz to the mobile service is on a primary basis ... and subject to agreement obtained under the procedure set forth in Article 14.

(*b*) In Regions 2 and 3 the band 1710–2290 MHz is allocated for the FS and the MS (both primary) and in Region 1 the same band is allocated for the same services, but the MS allocation is secondary. However, footnote RR 747 states as follows:

> Subject to agreement obtained under the procedure set forth in Article 14, the band 2025–2110 MHz may also be used for

> Earth-to-space and space-to-space transmissions in the space research, space operation and earth exploration-satellite services. The services using space-to-space transmissions shall operate in accordance with the provisions of Nos, 2557 to 2560 and shall not cause harmful interference to the other space services.

(RR 2557–RR 2560 contain sharing constraints, in the form of a maximum value of spectral power flux density at the Earth's surface, applicable to the Earth-to-space links mentioned in RR 747, designed to reduce to an acceptable level the interference from those links to receiving stations of the FS and MS.)

(*c*) The band 74·8–75·2 MHz is allocated for AeRN and used worldwide for the marker beacons of ILS installations; it is, however, foreseen that ILS will eventually be replaced by a new aircraft landing system, MLS, operating in a different frequency band. Accordingly the following footnote RR 572A was added to the Table at WARC-ORB-87:

> Additional allocation; in Afghanistan, the Federal Republic of Germany ... Syria and Turkey, the band 74·8–75·2 MHz is also allocated to the mobile service on a secondary basis subject to agreement obtained under the procedure set forth in Article 14. In order to ensure that harmful interference is not caused to stations of the aeronautical radionavigation service, stations of the mobile service shall not be introduced in the band until it is no longer required for the aeronautical radionavigation service by any administration which may be identified in the application of Article 14.

In outline, the procedure set forth in RR Article 14 is as follows:

(i) An administration proposing to make an assignment which would be subject to consultation under the Article sends the details to the IFRB in advance of formal notification. The Board endeavours to identify the other administrations whose facilities might be affected, then it publishes the details in the IFRB Circular and alerts the other administrations which it has identified.
(ii) Any administration which considers that its facilities might indeed be affected, whether identified by the IFRB or not, must so inform the administration making the proposal within four months.
(iii) These other administrations are required to give details of their systems which might be affected to the administration making the proposal and to enter into a co-operative consultation, to determine whether interference would indeed arise and if so, to find ways of avoiding it. The IFRB's assistance is available at this stage if it should be required.
(iv) When solutions to all problems have been found, the IFRB is advised and the assignment is notified for registration in the MIFR.
This very flexible and basically simple procedure was agreed at WARC-79.

The IFRB has reported that relatively few assignments are proposed with respect to most of the footnotes, perhaps one or two assignments per year per footnote. This may indicate that the procedure is working effectively in most cases, meeting needs that arise seldom. However, substantial numbers of assignments are proposed in connection with a few of these footnotes, and in particular:

RR 747 (SRS, SpO and EES at 2025–2110 MHz)
RR 750 (SRS, SpO and EES at 2200–2290 MHz)
RR 811 (SRS at 7145–7235 MHz)
RR 839 (FSS and BSS at 11·7–12·7 GHz in Region 2)

and various difficulties in implementing the provisions of the Article are reported. In particular:

(*a*) There may be no agreed criteria for use in determining which administrations may be affected by a proposed assignment, nor for determining whether the predicted level of interference is to be considered acceptable.
(*b*) Having declared an interest, an administration is under no regulatory pressure to provide information on his systems quickly, or to discuss problems within a reasonable period of time.
(*c*) The IFRB has no authority to resolve deadlocked discussions, even if the position taken up by one or other of the parties appears to be unreasonable.
(*d*) Space radio systems may have to be co-ordinated under RR Article 11 in addition to the RR Article 14 procedure. This may seem to involve the duplication of substantial effort.

No WARC with powers to make radical amendments to the Article has met since 1979, although WARC-ORB-88 agreed provisions in the form of RR 1060·1[3] and RR 1619A[3] to RR1619D[3] to ease difficulties (*b*) and (*d*) above. There is a clear need for the Article to be revised more generally, to make it more effective. However, it may also be argued that a procedure of this kind is not appropriate for allocations that are to be used by substantial numbers of systems; the best solution may lie in devising better allocations in the international Table for the services involved.

4.3.4 International frequency planning

The term 'frequency planning' is used in international spectrum management in the sense of drawing up an international agreement to enable administrations to make interference-free frequency assignments to specified stations or to stations within defined geographical areas to meet stated requirements. The agreement will consist mainly of a schedule of planned assignments ('the plan') and regulations for its management. Frequency planning of this kind is used instead of the methods discussed on pp. 56–57 for several services in certain frequency bands, in particular in some broadcasting bands and in some mobile service bands and radionavigation bands and most especially in Region 1.

A typical frequency plan

The process of generating a plan and the various provisions that go with it varies from case to case but the following is typical:

(i) The idea that a frequency plan would be desirable for a service in a band allocated for it with primary status might receive general support at a WARC which had authority to amend the international Table of Frequency Allocations. If so, the fact might be put on record by adding a footnote to the Table; see for example RR 480. If the band was shared with another service having equal allocation status, which would impede planning, the Table would be amended, perhaps by reducing the sharing allocation to permitted or secondary status, or by the addition of a specific footnote to the Table; see for example RR 838. A resolution of the WARC might initiate work in, for example, the CCIR or the IFRB to prepare for the planning conference.
(ii) A decision to implement the idea would most likely be taken at the next plenipotentiary conference. Thus, Resolution 1 of the 1982 Plenipotentiary Conference lists 12 administrative radio conferences which were to be held before 1989 and nine of them had frequency planning as a main element of their work. At the appropriate time the Administrative Council would determine the agenda for each conference and authorise the Secretary General to arrange it. In many cases frequency planning conferences are held in two sessions separated by one or two years, the first session to determine the principles to be followed and the second to collect statements of requirements, produce the plan and approve it.
(iii) At or before the conference the minimum period for which the plan is to be valid is determined; 10 to 20 years would be typical. Administrations state the assignments which they foresee making in the course of that period. Performance characteristics to be required or assumed and standard emission parameters are adopted. Note will be taken of any assignments already registered in the MIFR which may not be disregarded in preparing the plan.
(iv) At the conference a schedule of frequency assignments would be drawn up, in keeping with these criteria, within the available frequency band, if this is possible. If it is not possible, agreement must be sought to a reduction of the requirements or the adoption of more stringent criteria until a plan can be produced or the conference breaks down.
(v) Procedures are devised for use when the plan is implemented if the general provisions of the RR are not fully sufficient. Typically these provisions would consist of articles of the agreement which:

- provide for modifications to the plan, with the agreement of all administrations that would be affected,
- state a procedure to be used by an administration in notifying the IFRB when it assigns to a station an assignment contained within the plan,
- direct the IFRB in the handling of such notifications,
- provide, if necessary, a technical and regulatory basis for the use of the band by stations of the services for which the band is also allocated.

Other articles and annexes to the agreement record such matters as the technical basis of the plan, the date of entry into force, the minimum period of validity, a list of delegations signing the agreement and various protocols of a diplomatic nature.

(vi) The agreement is lodged with the IFRB and governments ratify the agreement.
(vii) When the agreement has entered into force, and an administration notifies to the IFRB an assignment which is completely in accordance with the plan, the assignment is registered in the MIFR.
(viii) If the plan has been drawn up for a shared band, the assignments made to stations of sharing services will be constrained so that interference to stations of the planned service will be very low.

Frequency plans tend to fall into two groups. *Frequency assignment plans* consist of a scheme of assignments, narrowly defined in all the parameters proper to an assignment, including the location of the radio stations. In *frequency allotment plans* some substantial element of flexibility has been built into the plan, the location of transmitting stations most typically being left somewhat indeterminate. Reference may be made to the definitions in Appendix A.4.

Special agreements

Administrative radio conferences, with worldwide or Regional responsibilities, are often convened for drawing up frequency assignment and frequency allotment plans. However, Article 27 of the 1989 Constitution of the ITU and RR Article 7 also provide for frequency planning to be done at meetings outside the ITU. The outcome of such meetings, called Special Agreements, are recognised by the ITU, provided that the agreement is in keeping with the provisions of the RR and provided that all administrations with interests that could be affected had been invited to participate in the meeting. This option is sometimes used, for example, when the planning process relates to a geographical area which is less than one of the ITU Regions.

Frequency planning and equitable access to resources

Frequency assignment plans have been used for many years in the LF and MF BS bands in Regions 1 and 2 and in the VHF and UHF bands in Region 1. Frequency allotment plans have been used in certain bands for other services and in particular for the MM, AeM, MRN and AeRN. More recently, plans have been drawn up in certain bands for the BSS and the FSS. These plans are discussed in later chapters as appropriate.

Frequency planning can avoid some problems raised by frequency selection by inspection which are listed on pp. 57–58. It is evident that systematic spectrum use and a large measure of permanence of assignments of emissions which are capable of causing interference at great distances can be achieved by frequency planning, and perhaps in no other way. Planning may also achieve efficient and economical spectrum use, although it sometimes fails to do so; much depends on whether the forecast requirements are realistic. However, it is chiefly the desire to achieve equitable distribution between the nations of access to the spectrum and to the geostationary satellite orbit (GSO) that has created a strong interest in frequency planning in recent decades.

In 1947, as part of the process of reconstruction after the Second World War, a Resolution[4] of the International Radio Conference set up the Provisional Frequency Board and charged it with responsibility for drawing up global frequency assignment plans for the FS and BS in the bands allocated to these services between 14 and 150 kHz and between 2·85 and 27·5 MHz. Regional conferences were convened to produce frequency assignment plans for all of the other services then operating below 27·5 MHz except the AmS and SFS.

Satisfactory plans were produced and implemented for a few bands and a few services. For most bands, however, and most notably for the FS and BS bands between 4 and 27·5 MHz, the attempt failed. The reason for this failure was that the requirements stated by administrations for assignments for facilities, current and foreseen, exceeded by far the total capacity of the HF part of the spectrum and no acceptable way was found for scaling down the stated requirements to what the bandwidth available could accommodate. Many countries were dissatisfied with this outcome.

The issue was reopened in the 1960s. Previously a majority of ITU members had considered management of access to spectrum by priority of notification and registration of frequency assignments to be the best available option, although not fully satisfactory. However, the accession to the ITU of many newly independent countries considerably increased the number of developing countries among ITU members. By 1965 a majority of members favoured a change of regulatory methods for the bands below about 30 MHz, seeing planning as a way for new countries to get equitable access to spectrum. By the mid-1970s, when satellite systems began to be perceived as an important potential medium for domestic communication, the same concerns arose regarding, in particular, the FSS and BSS. A considerable number of attempts to draw up frequency assignment plans have been made since 1965. Some have been relatively successful. Other attempts have failed, mainly for the same reasons as had led to the failure of the Provisional Frequency Board and the associated Regional planning conferences. There are references to these plans and planning attempts, service by service, in later chapters.

4.3.5 The elimination of harmful interference

The methods for selecting frequencies for assignment, discussed on pp. 53–64, are designed to ensure that harmful interference does not occur between stations for which different administrations are responsible. Nevertheless, interference may arise, sometimes after a long period of satisfactory operation.

If interference occurs between two stations, both of which are using assignments planned for them in an agreed frequency assignment plan or frequency allotment plan, and if the interference can be attributed to a departure from the system characteristics assumed in planning for the interfering station, that station will stop using the assignment until its characteristics can be brought into accord with the planning assumptions. If the interference is attributable to under-performance on the part of the station suffering interference, the interference will usually have to be endured. If interference occurs although the characteristics of the stations are in accord with those assumed in planning, it will usually be necessary to get the plan amended to eliminate the interference.

If a station using an assignment in accordance with a frequency plan suffers harmful interference from a station using a frequency which is not in the plan, and both of the administrations involved are parties to the plan agreement, then the station with the unplanned assignment will stop using it.

If a station belonging to a service which has secondary allocation status in the frequency band in question causes harmful interference to a station with primary allocation status, the former station must stop using the frequency or modify its emission so as to eliminate the harmful interference. If, on the other hand, the station with secondary allocation status suffers interference from the station with primary allocation status, it has no remedy.

RR Article 22 contains such rules as have been agreed for dealing with cases of international harmful interference in other situations. These rules are expressed in rather broad terms but their essential features and the customary practices that have developed in implementing them are briefly as follows:

(*a*) The responsibility for action rests with the parties which are suffering and causing the interference. The ITU usually plays no direct part in the settlement of the problem, although it will assist an administration with advice if requested to do so.
(*b*) Contact with the transmitting station causing the interference is usually initiated by the manager of the station which transmits the signal which is

suffering interference, following a complaint from a receiving station. If a solution is not found quickly, discussion may continue between the head offices of the operating organisations involved and ultimately, if necessary, between the administrations concerned.

(*c*) If the signal suffering interference is used for a safety service, the effective resolution of the problem is treated as a matter of urgency.

(*d*) If no safety service is involved, a search will be made for methods of eliminating the interference or reducing it to a tolerable level. The methods that are considered will depend on the order of frequency and the nature of the facility, but they may include, for example, rearrangement of frequency plans or frequency complements, small changes to carrier frequencies, adjustments to the parameters of emissions, changes of antenna characteristics, an acceptable constraint on daily hours of use, and so on. In the maritime and aeronautical mobile services it may be feasible to devise operational arrangements which reduce or eliminate the interference.

(*e*) However, if non-radical changes such as these do not solve the problem, it may be necessary for one or other of the transmitting stations to stop using the frequency. But which station should yield? It is customary for the station with the assignment with the later date of entry into service as registered in the MIFR to give way to the station with the assignment having priority. This is a simple rule, it accords with common sense and it is usually honoured. However, this custom, whilst it is often implied by the terms of the RR, has not been given regulatory force. Furthermore, a claim of priority might by challenged if the use being made of the frequency were not precisely the same as the registered assignment. Thus, there may be room for negotiation, and factors other than the regulatory situation may be involved in the resolution of the problem.

In general, the parties to an interference problem show good will in solving it; this is to be expected, because the good operation of radio systems depends on co-operation between administrations and between operating organisations, However, it may be no small matter to change an established working frequency. Where the interference is suffered by a safety service, action to stop the interference is always very prompt. Of other cases it may be said that international interference problems are not always solved quickly or to the satisfaction of all parties.

4.4 Efficient spectrum utilisation

4.4.1 The need for efficient spectrum utilisation

If there were plenty of bandwidth for all requirements, frequencies would be assigned to stations so that interference did not occur and the equipment used for systems would be designed for the least cost consistent with achieving the required transmission capacity and the channel performance objectives.

If long distance interference could be disregarded, this state of affairs would be obtained in many countries over much of the spectrum, although there might be congestion, for example, in the frequency ranges that are optimum for broadcasting, land mobile and radio relay systems. However, long distance interference cannot be ignored below 30 MHz nor in any frequency range which is used by major space services. Prospective growth of the latter adds another dimension to the problem, namely the possibility that the capacity of the GSO might become exhausted. Thus, there is a need for the GSO and much of the spectrum to be used efficiently, and this need has an international dimension.

This need for efficiency and its international dimension are addressed in ITU texts at many levels. Thus Article 33 of the 1989 Constitution of the ITU reads as follows:

> Members shall endeavour to limit the number of frequencies and the spectrum used to the minimum essential to provide in a satisfactory manner the necessary services. To that end they shall endeavour to apply the latest technical advances as soon as possible.
>
> In using frequency bands for radio services, Members shall bear in mind that radio frequencies and the geostationary-satellite orbit are limited natural resources and that they must be used rationally, efficiently and economically, in conformity with the provisions of the Radio Regulations, so that countries or groups of countries may have equitable access to both, taking into account the special needs of the developing countries and the geographical situation of particular countries.

These themes are repeated and expanded in RR Articles 5, 6 and 9. Note in particular that:

- RR 307 urges that the bandwidth of emissions should be kept, in general, as small as the state of the technique and the nature of the service permits,
- RR 309–RR 311 stress the importance of high performance in radio receivers, and
- RR 954 urges that the HF band should be reserved, as much as possible, for long-distance communication.

Much of the work of the CCIR is directed towards the design of systems which occupy less spectrum and towards efficient spectrum management. Efficient spectrum utilisation demands efficient frequency assignment practices and it may involve operational or technical constraints on licensees. Occasionally, requirements designed to encourage the use of efficient system architecture have to be imposed by administrations on system designers. The cost of such requirements is not necessarily high. There are usually alternative ways of achieving an improvement in spectrum utilisation efficiency and an administration will consider the options carefully, no doubt in consultation with operating organisations and equipment manufacturers, before choosing the way or combination of ways that provides enough improvement at the least cost in the relevant circumstances. Often the essential step is to recognise that efficient spectrum utilisation is indeed a factor to be taken into account in the specification, design and procurement of equipment; given that, an improvement in spectrum management or in the characteristics or the architecture of equipment which will improve spectrum utilisation efficiency may cost little.

It should also be noted that efficient spectrum utilisation can yield advantages even when the achievement of higher efficiency is not made essential by the local or global pressure of demand. For the bandwidth required for a given application of radio to be allowed to grow until it overflows unsystematically into a new band may not constitute good spectrum management. Also efficient spectrum utilisation may give the spectrum manager options for postponing band usage decisions until those decisions can be based on more complete data. Such benefits derived from higher efficiency may be expressible in money terms and set off against the cost of obtaining them. Thus:

- A new band is usually higher in frequency than bands in established use and its propagation characteristics are often less satisfactory. Such inferiority can usually be overcome by appropriate equipment design or system configuration, but the cost may be higher.
- Taking into use a new frequency band, owing to the exhaustion of the bands already in use, may raise design and inventory problems for radio facility providers and equipment suppliers.
- Efficient use of frequency bands gives an administration more freedom to rationalise the use of frequency bands and specified parts of bands for particular purposes. This may facilitate the devolution of the management of frequency assignment in specified bands to major users, relieving the administration of the duty of performing that task and possibly further increasing spectrum use efficiency. It may also make it easier to provide bandwidth for new radio products when they arise.
- The international Table of Frequency Allocations (RR Article 8) includes a number of sharing situations which, though technically feasible, involve much careful spectrum management effort and significant constraints on system deployment. More efficient use of bands which are allocated nationally to one service only may make national sharing of such allocations unnecessary.

Various regulatory ways of achieving efficient use of the spectrum are reviewed briefly on pp. 68–72. The best way to apply such methods depends on the service, and these methods are discussed in the context of the relevant services in later chapters of this book.

Raising the efficiency of use of the spectrum by good spectrum management or regulation has disadvantages. It may involve more management effort than an administration wishes to give to it and it may lead to inflexibility, system complexity or higher equipment costs. It has been argued that equivalent benefits can be obtained in a more satisfactory manner by allowing market forces to determine how the spectrum should be used, and by whom. There is a brief review of spectrum pricing, as this approach is called, on pp. 72–74.

4.4.2 Spectrum utilisation efficiency through regulation

Efficient spectrum utilisation involves two broad concepts. First there is the design of radio systems so that they are able to perform their intended function whilst denying to other radio systems the use of the least amount of spectrum within the smallest geographical area. This concept might be called optimised system isolation. Secondly, there is the combination of good spectrum management practices and good system architecture that enables a large number of stations in a limited geographical area to transmit a large amount of traffic through a limited bandwidth. This concept might be called maximised spectrum occupancy.

Optimisation of system isolation turns mainly on the passive elements of radio equipment design. Its significance can be modelled well by a microwave radio relay link. Spectrum occupancy is complex and dynamic, greatly influenced by the ways in which the demand for radio facilities varies from time to time and from place to place. This concept is probably modelled most clearly in VHF and UHF land mobile systems. Nevertheless, both concepts probably contribute to some degree to the total spectrum utilisation efficiency of most radio systems. The efficiency of spectrum utilisation, analysed through both of these concepts, can be enhanced to some degree by regulatory means.

An elementary examination of the two concepts follows.

Optimised system isolation

At an elementary level it is evident that system isolation may be increased, for example, through the use of antennas with better directive properties and more selective intermediate frequency filters at transmitting and receiving stations. Many licences lay down standards for such characteristics.

As a first step in considering how the isolation of radio systems can be raised, it is helpful to consider how the isolation of a single link in a radio relay network might be quantified.

Studies in progress are searching for a means for assessing objectively the efficiency with which spectrum is used by a radio system; see CCIR Report 662[5] and Berry (1977)[6]. The problem is complex and the final outcome may not be a unique concept that can be applied to all kinds of radio system. Nevertheless, helpful concepts of more limited applicability are already emerging from this work, for example in the form of a measure of the efficiency with which a microwave radio relay link uses the spectrum. Thus:

if C = measure of the communication capacity of the link
L = length of the link
B = required radio frequency bandwidth
S = measure of the space which the emission occupies

then CL is a measure of the telecommunications facilities provided by the link and BS is a measure of the spectrum and space which the link occupies. Then:

$$\text{Efficiency} = \frac{CL}{BS} \text{ in arbitrary units} \tag{4.1}$$

S_T might be interpreted as the volume of the space within which the power flux density of the emission exceeds a certain threshold value, or the area of the land which underlies that volume. The threshold value might be chosen, for example, as the maximum level of interference that the receiver of a hypothetical second link could tolerate if its antenna were omnidirectional. Then S_T would be the space in which the transmitting station of the subject link denies access to the spectrum which it occupies to the receiving station of the other link, the receiving antenna of that other link being assumed to be omnidirectional. In a complementary way it would be possible to define another volume S_R within which the receiving station of the subject link could not tolerate the presence of transmitting station of a third link, assumed to have an omnidirectional antenna, without the level of interference at the subject receiving station exceeding the threshold of acceptability. CCIR Report 662 proceeds to extend this approach to the evaluation of the spectrum utilisation efficiency of a radio relay chain.

Much work remains to be done to make this treatment of spectral efficiency more generally applicable. Nevertheless, it already provides a useful starting point for the identification of general principles of efficient spectrum use, and these may be illustrated by the following examples:

(*a*) If the link performance targets are being exceeded, then the power of the transmitter could be reduced and this would cause S_T to shrink. If the performance of the system is limited by receiver thermal noise, then a further reduction in the transmitter power, accompanied by a further shrinkage of S_T, would become feasible if the noise factor of the receiver were reduced, although this would make the receiver more susceptible to interference, causing S_R to expand.
(*b*) An improvement in the characteristics of the antennas may change the shape of S_T and S_R and reduce their size. If the gains of the antennas in the wanted directions are increased, it becomes feasible to reduce the power of the transmitter, with the same benefit as was indicated in (*a*).
(*c*) In particular locations, the presence of natural barriers such as hills which do not obstruct the wanted ray path may cut off parts of S_T and S_R which are not useful, so reducing the space denied to other potential users for the frequency band.

(*d*) The spectral efficiency of the system may be increased by optimising the ratio between C and B. This may involve, for example, changes in the characteristics of the communication channels, the use of baseband signal processing, changes in the nature or the parameters of the modulation process or the introduction of dual polarisation.
(*e*) The optimisation of C/B may affect the propensity of the transmitted signal to cause interference or the susceptibility of the receiver to suffer interference, thus causing S_T and S_R to expand or contract. For example, if a 4-phase digital modulation system were replaced by a 16-phase system with the same information capacity, then C and L would be unchanged, B would be reduced, but S_T or S_R or both of them would increase, depending upon the extent to which the carrier power of the transmitter was increased to compensate for the higher susceptibility of the system to interference. Conversely, if forward error correction (FEC) were incorporated into the system, B would be increased but S_T or S_R or both might be reduced.

Through studies of this kind it becomes possible to identify technical characteristics of systems which favour efficient spectrum utilisation, and this may provide a basis for desirable technical constraints.

Maximisation of spectrum occupancy

If a frequency band is considered to be divided into a number of radio channels, each wide enough for a single emission, and if these channels are assigned to radio stations which operate continuously, then the band could be said to be 100% occupied at a given location when every channel is assigned and implemented. If, however, the radio channels are assigned to radio stations which operate intermittently, the band could approach 100% occupation at a given location, corresponding to every channel being occupied all of the time, only if there were more stations than channels and every channel could be used by more than one station.

These two aspects of spectrum occupancy relate to a single location. To bring in the geographical dimension it would be necessary to consider to what extent the area defined by factor S for a single system in expression (4.1) is itself minimal and how completely the sum of all the areas associated with all of the systems assigned a given channel fills the space in question.

It must, of course, be recognised that real radio systems have to serve the needs of real user communities and that variations of user demand with time and place are likely to limit the spectrum occupancy that is, in practice, usefully achievable. Nevertheless, if the problems that this non-uniformity of demand creates are clearly recognised, means may be found for minimising the impact of such limitations.

The frequency band occupancy criterion is particularly relevant to radio facilities which occupy channels intermittently; this is typical of the mobile services. If each channel was assigned to only one transmitting station, occupancy might be quite low. If several or many transmitting stations used each channel, the occupancy might become quite high at peak traffic times but interference or the non-availability of the channel when it was required would be objectionable to users. Where circuits are operator-controlled, protocols for sharing a frequency assignment, perhaps on a demand basis, may be helpful. If a large number of stations had access to whichever of a large group of channels happened to be free when access was required, very high occupancy could be achieved at peak traffic times with good availability of a channel when required, demonstrating the advantages of good system architecture. A good understanding of the occupancy of a frequency band used in such ways requires a statistical approach; see for example Spaulding and Hagn (1977)[7].

It is characteristic of the mobile services that traffic tends to be geographically non-uniform. In the land mobile service traffic is concentrated in the metropolitan areas. In the aeronautical mobile and maritime mobile services concentrations of traffic are less marked but they tend to occur in the approaches to major airports and in heavily used shipping lanes near to seaports respectively. Frequency assignment planning for mobile radio services must be optimised for these points of concentration; elsewhere, efficient use of spectrum for these facilities may not be important.

Other services do not necessarily show this strong tendency towards geographical concentration. For example, broadcasting stations and radio relay stations of the fixed service tend to be more evenly distributed. The dominant problem of efficient frequency assignment in such cases is to achieve a satisfactory geographical occupancy. One general rule is to assign frequencies so that interference from all sources at one of the less favourably located receiving stations does not exceed, but sometimes approaches, the level corresponding to the minimum acceptable carrier-to-interference ratio. In order to maximise the number of times that a frequency can be 're-used' in this way, it is important that interference should contribute a substantial fraction of the noise-plus-interference level which is one of the channel performance objectives of systems.

The achievement by an administration of effective frequency assignment in the land mobile service may be particularly onerous, and it may be found desirable to devolve some part of the function, perhaps to a trade association; see, for example Gianessi (1981)[8]. In modern systems providing flexible channel access, such as trunked and cellular systems, the system architecture provides a technical solution to this problem. High geographical occupancy of broadcasting and fixed service channels may also place a heavy burden on an administration. This burden can sometimes be lightened by allotting a complete allocated frequency band or a sub-band to a major radio user and devolving the management of assignments to the user. Major computer aids are also available to assist in this work; see CCIR Report 841[1].

A variety of operational requirements and technical parameters arise to complicate the process of assigning frequencies to stations. However, the concept of spectrum occupancy, taking geographical considerations into account, may provide a valuable check on the efficiency with which the frequency assignment process has been carried out. CCIR Reports 842[9] and 976[10] may be of assistance in making such a comparison. Hale (1980)[11], Zoellner (1973)[12] and Struzak (1982)[13] provide approaches to the optimum assignment of frequencies to an array of stations.

Finally, an important factor which is complementary to efficient radio spectrum utilisation is a benign electromagnetic environment. By specifying radio equipment and designing radio stations so that spurious emissions are minimised, the radio-using community can do much to ease its own radio noise problems. By the exertion of pressure through many channels of influence, administrative and industrial, the radio frequency noise generated by electrical plant of all kinds and radiated in the vicinity of radio stations can be reduced. In the absence of these precautions, the effectiveness of other measures designed to optimise the use of the spectrum and the performance of radio systems will be considerably reduced.

4.4.3 Spectrum pricing

Economists, in particular in the USA, have considered the application of market forces to the allocation and assignment of spectrum for radio services. Levin (1971)[14], for example, points out that the radio spectrum is exceptional among factors of substantial economic importance in that use of it is made available virtually without charge, and he argues that in consequence there is no satisfactory mechanism for ensuring that spectrum is distributed between users in an economically optimum way. He recognises that the establishment of a market in radio spectrum presents special problems which would probably make a completely free market impracticable, but he concludes that some substantial application of market forces in frequency management would be beneficial. Furthermore, the cost of administering the spectrum might be reduced by such a change; see also Coase (1959)[15], Webbink (1977)[16], Merriman (1983)[17], Rudd (1986)[18] and Levin (1988)[19].

A recent study performed by CSP International for the United Kingdom government[20] has argued that the introduction of the profit motive into the process of assigning frequencies to radio station licensees would ensure an increase in spectrum occupancy. Another recent study, by NERA for the New Zealand government[21], advocates the grant or auction of country-wide leases of bands of frequency, valid for 20 years, usable in whatever way the lessee wishes, including sub-leasing of bands as a whole or after sub-division. Frequency bands for which cross-frontier interference would raise problems are excluded from the NERA proposal.

From a starting assumption that there is not enough radio spectrum with suitable propagation characteristics to meet the needs of all would-be users, the arguments that have been offered in favour of spectrum pricing may be summarised as follows:

1. One of the factors which influence the choice of transmission medium for any communication facility is the cost to the user of the medium itself. If the price of the medium is set below its economic value, the opportunity cost, then the choice made is not a truly economic one. Some facilities currently provided by radio could operate instead through cable networks of various kinds. Most operators of systems of the fixed and broadcasting services, for example, have this choice and together they now occupy a large part of the radio spectrum. There are other facilities for which radio is the only feasible transmission medium. The use of radio for facilities which could use other media is made too attractive by virtually free access to the medium. Consequently, access to the spectrum may become more difficult for facilities for which radio would be the true economic choice or the only feasible medium. Some facilities for which radio would be the only feasible medium may be suppressed altogether. The remedy proposed is to make spectrum available to the highest bidder in a market which is as free as it may be.
2. Spectrum pricing would provide an economic basis for optimising the apportionment of spectrum between the various users. In its most extreme form spectrum pricing would make a frequency assignment a kind of property, like land, that could be bought and sold and used by its owner or lessee in whatever way he chose within the limits of his title.
3. Whilst frequency assignments are free or virtually free, there is no economic pressure on a user to use spectrum efficiently. When an existing assignment is no longer needed, the fact may not be declared, whether from negligence, to provide a reserve against an unsuccessful application for another assignment later when needs may have risen or in order to deny access to spectrum to a competitor. There is thus no effective pressure upon a user to adopt the many techniques open to him to reduce his occupation of spectrum, once he has access to it. It is argued that a spectrum user would ask for no more spectrum than he needed if he had to pay a substantial price for it, related to the bandwidth to which he had access, and that he would surrender the surplus, making it available for use by others, when his needs declined.
4. Finally, if the selection of frequencies for assignment is a purely bureaucratic process, it is likely to be done inefficiently and spectrum occupancy will be low. The addition of the profit motive to spectrum management would increase spectrum occupancy.

It would seem to be unprofitable at present to discuss spectrum pricing in its most extreme form, in which the right to use some combination of spectrum and geographical space becomes a saleable commodity. With the exception of relatively minor parts of the frequency spectrum, most countries are not free to use the spectrum without regard for the interests of other countries. The problems of international frequency management, for which governments now have authority to act on behalf of their country, do not seem to be solvable in any other way.

Even if the international dimension is disregarded, some basic problems that this concept presents have not been solved. The following problems are typical:

- How are usable packages of spectrum/space, appropriate to the generality of radio systems, to be defined and traded and how would unforeseen interference problems be resolved, without the use of such large margins of safety that the efficiency of spectrum utilisation is seriously reduced?
- How is government policy in the regulation of the provision of services, such as broadcasting and telecommunications, to be harmonised with a free market in radio spectrum?
- Could any meaningful application of market forces be used to discipline government use of spectrum, military or civil.

It may be concluded that the concept of using unconstrained economic forces to determine the pattern of frequency allocation, if indeed any coherent pattern of frequency allocation survived, has not been developed yet to the point where it could be evaluated.

The use of spectrum pricing in the leasing or sale of bands of frequency within a framework of conventional frequency allocation is a less revolutionary concept although some of the unsolved problems of the totally free market remain with it. Further dilution of the concept leads to payment in some competitive form for the right to use a band of frequency in a specified way, but this seems to be indistinguishable from the sale of licences to provide facilities, without any necessary change in spectrum management.

It may be concluded that the visible way ahead does not lie with the transfer to users of absolute rights to use and trade spectrum. Assignments should be for defined purposes and time-limited, in many cases specific to emissions with defined parameters. Given that foundation, there are various ways in which economic pressures could be added to sound judgment, administrative skill, perceptive planning, technical constraints and an adequate monitoring system to ensure that the spectrum is managed well and at an acceptably low management cost to provide the facilities that society needs. Substantial licence fees, bandwidth-related licence fees and devolved frequency assignment management within constraints which leave ultimate control with government are just three options which seem likely to be used in appropriate situations. The optimum combination of measures will depend on many factors, including the geographical situation of the country.

4.5 Identification of stations

If a receiving station suffers interference it needs to be able to identify the source, so that the interference problem can be resolved. For this reason there is an internationally accepted obligation for radio emissions to carry signals which identify the transmitting station, although it is recognised that suitable means have not yet been devised for adding usable identification signals to some kinds of emission; see RR 2055. Identification signals are also sometimes necessary to help a receiving station to establish a wanted communication link.

No single way of identifying a transmitting station or an emission is satisfactory for all kinds of system and various means are used, as may be convenient. Typically these means involve the transmission of an identifying message at frequent intervals, usually using the same form of signalling and modulation as is used for the normal operation of the emission. The identifying message, in some form which is easy to read, would typically consist of one of the following:

- the name or location of the transmitting station,
- a call sign in accordance with an internationally recognised format,
- a programme name or a video test card for a broadcasting station,
- a selective call number for a mobile station equipped for access to an automatic communication network, and
- the flight number of an aircraft.

RR Article 25 and RR Appendices 42, 43 and 44 contain internationally agreed formats for call signs and selective call numbering systems. Station identification in the maritime mobile service is discussed further on p. 194.

The process of identifying the station transmitting an emission which is causing interference but which offers no ready means of identification is sometimes very laborious. Studies continue in the CCIR, seeking ways of identifying stations with types of signal for which no satisfactory way has yet been found, such as radars and multi-channel radio relay systems, and seeking more convenient ways of identifying signals for which no fundamental difficulty arises; see for example CCIR Report 1061[22].

The convenient recognition of familiar emissions may also be assisted by precise measurement of modulation characteristics, regardless of the presence or absence of identification signals; see CCIR Report 978[23].

4.6 Monitoring of emissions

If it is to do its work effectively, an administration needs to have means for monitoring the use that is made of the spectrum within its jurisdiction. In this way it can ensure that the frequency assignments that it makes are put into use, that they are used in the way intended and, to the extent feasible, that the technical constraints that are included in licences are observed.

In addition to these basic functions of a monitoring facility, it is desirable for an administration to have available means for receiving and identifying signals which are interfering with authorised reception within its jurisdiction. Monitoring may also guide the selection of frequencies for assignment.

Some administrations have set up stations at fixed locations with comprehensive facilities for monitoring signals, over a wide frequency range, originating from both terrestrial and satellite transmitters. These fixed stations may be complemented by mobile stations that can observe emissions from stations operating at the higher frequencies, close to the transmitting stations.

The CCIR has made extensive studies of the techniques of radio signal monitoring, and reference should be made to the following texts.

Carrier frequency measurement: CCIR Report 272[24]
Bandwidth of emission measurement: CCIR Recommendation 443[25] and CCIR Report 275[26]
Spectrum occupancy: CCIR Recommendation 182[27] and CCIR Reports 279[28], 668[29] and 835[30].
Monitoring of spacecraft emissions: CCI Report 276[31]
Monitoring from mobile stations: CCIR Report 277[32]

There is also an ITU Handbook for Monitoring Stations[33].

There is a global dimension to radio emission monitoring, in that it could be advantageous to use reports from a wide range of monitoring stations as a means of testing the validity of the record presented by the MIFR. To this end, RR Recommendation 30[34] seeks to encourage the construction of more monitoring stations to provide comprehensive geographical coverage; see also RR Article 20.

4.7 References

1 'Spectrum management and computer-aided techniques'. CCIR Report 841–1; Recommendations and Reports of the CCIR, 1986, Volume I (ITU, Geneva, 1986)
2 'Spectrum management and computer-aided techniques'. (ITU, Geneva, 1983)
3 Final Acts of the second session of the World Administrative Radio Conference on the use of the geostationary satellite orbit and the planning of space services utilising it (WARC-ORB-88); (ITU, Geneva, 1988)
4 Resolution relating to the preparation of the new international frequency list. Final Acts of the International Radio Conference, Atlantic City, 1947
5 'Definition of spectrum use and efficiency'. CCIR Report 662–2; Recommendations and Reports of the CCIR, 1986, Volume I (ITU, Geneva 1986)
6 BERRY, L.A.: 'Spectrum metrics and spectrum efficiency – proposed definitions', *IEEE Trans.*, 1977, **EMC–19**, pp. 254–259
7 SPAULDING, A.D. and HAGN, G.H.: 'On the definition and estimation of spectrum occupancy', *IEEE Trans.*, 1977, **EMC–19**, pp. 269–280
8 GIANESSI, M.: 'NABER frequency co-ordination, the hows and whys of this process', *Commun.*, 1981, **18**, pp. 52–54 and 57
9 'Spectrum-conserving terrestrial frequency assignments for given frequency-distance separations'. CCIR Report 842–1; Recommendations and Reports of the CCIR, 1986, Volume I (ITU, Geneva, 1986)
10 'A systematic planning method for the use of unassigned frequency bands'. CCIR Report 976; *ibid.*, Volume I
11 HALE, W.K.: 'Frequency assignment, theory and applications', *Proc. IEEE,* 1980, **68**, pp. 1497–1514

12 ZOELLNER, J.A.: 'Frequency assignment games and strategies', *IEEE Trans.*, 1973 **EMC–15**, pp. 191–196
13 STRUZAK, R.G.: 'Optimum frequency planning, a new concept', *Telecommunication J.*, 1982, **49**, pp. 29–36
14 LEVIN, H.J.: 'The invisible resource, use and regulation of the radio spectrum' (The Johns Hopkins Press, 1971)
15 COASE, R.H.: 'The Federal Communications Commission', *Journal of Law and Economics,* October 1959, **2**, pp. 1–40
16 WEBBINK, D.W.: 'The value of the frequency spectrum allocated to specific uses', *IEEE Trans.*, 1977, **EMC–19**, pp. 343–351
17 MERRIMAN, J.H.H.: 'Report of the Independent Review of the Radio Spectrum (30–960 MHz)', Cmnd. 9000 (HMSO, London, 1983), pp. 50–54 and 112–135
18 RUDD, D.: 'A renting system for radio spectrum?', *Proc. IEE,* 1986, **133**, Pt. A, pp. 58–64
19 LEVIN, H.J.: 'Emergent markets for orbit spectrum assignments; an idea whose time has come', *Telecommunications Policy,* 1988, **12**, pp. 57–76
20 'Deregulation of the radio spectrum in the UK'. CSP International (HMSO, London, 1987)
21 'Management of the radio frequency spectrum in New Zealand'. National Economic Research Associates; Ministry of Commerce, New Zealand, 1988
22 'Transmission of supplementary information in amplitude modulation sound broadcasting'. CCIR Report 1061; Recommendations and Reports of the CCIR, 1986, Volume X–1 (ITU, Geneva, 1986)
23 'Automatic identification of radio emissions'. CCIR Report 978; *ibid.*, Volume I
24 'Frequency measurements at monitoring stations'. CCIR Report 272–5; *ibid.*, Volume I
25 'Bandwidth measurements at monitoring stations'. CCIR Recommendation 443–1; *ibid.*, Volume I
26 'Bandwidth measurement at monitoring stations'. CCIR Report 275–4; *ibid.*, Volume I
27 'Automatic monitoring of occupancy of the radio-frequency spectrum'. CCIR Recommendation 182–3; *ibid.*, Volume I
28 'Visual monitoring of the radio-frequency spectrum'. CCIR Report 279–3; *ibid.*, Volume I
29 'Automatic monitoring and measuring in the radio-frequency spectrum'. CCIR Report 668–2; *ibid.*, Volume I
30 'The automatic acquisition of radio-frequency spectrum occupancy data and its use for spectrum management purposes'. CCIR Report 835–1; *ibid.*, Volume I
31 'Monitoring of radio emissions from spacecraft at fixed monitoring stations'. CCIR Report 276–5; *ibid.*, Volume I
32 'Measurements at mobile monitoring stations'. CCIR Report 277–3; *ibid.*, Volume I
33 'Handbook for monitoring stations' (ITU, Geneva, 1963)
34 'Relating to international monitoring'. Recommendation 30; Radio Regulations, edition of 1982 (ITU, Geneva, 1982)

Chapter 5

The fixed services

5.1 Introduction to the fixed service

RR 21 defines the fixed service (FS) as 'a radiocommunication service between specified fixed points.' The service covers a very wide range of applications and technology and perhaps the only feature that its radio links have in common is that an operating frequency can be assigned with knowledge of the location of the transmitting station as a potential source of interference to other users of the same frequency and likewise that the receiving station is a known place to be protected from interference.

The FS has been allocated a large number of frequency bands and these can be divided into three groups. One group lies between 14 and 190 kHz, in the VLF and LF range. The second group starts at 1606·5 kHz and extends to 28 MHz; it can conveniently, although imprecisely, be called the HF range. The third group extends upwards from 28 MHz. The technology, the operational practices and the regulatory situation of the stations using these three groups of frequency bands differ radically, and the spectrum management of each group is considered separately on pp. 78–80, 80–90 and 90–134.

The aeronautical fixed service (AeFS) is defined in RR 23 as 'a radiocommunication service between specified fixed points provided primarily for the safety of air navigation and for the regular, efficient and economical operation of air transport'. Two frequency bands near 22 MHz have been allocated to the AeFS. Also RR 459 allocates the band 160–190 kHz to the AeFS in polar areas of Region 2. Except for their limitation to this one particular application, these bands can be considered to be allocated to the FS.

5.2 The fixed service at VLF and LF

In Regions 2 and 3 almost 80% of the spectrum between 14 and 190 kHz is allocated internationally to the FS. In Region 1 over 70% of the bandwidth between 14 and 148·5 kHz is so allocated, the spectrum immediately above 148·5 kHz being allocated in Region 1 to the broadcasting service (BS). However, most of these FS allocations in all three Regions are shared with the maritime mobile service (MM) or the radionavigation service (RN) or both, these latter services having equal or superior allocation status and being the main users of these bands.

In the early years of communication by radio, this part of the spectrum was used for long distance fixed links and a great demand developed for its very limited information capacity. Around 1930 ways were found for using the HF part of the spectrum for long distance links and the demand for VLF and LF assignments declined. Nevertheless, some use of these bands for fixed links continues. The characteristics of radio propagation in this part of the spectrum (see Appendices B.6.2 and B.6.3) will probably ensure that this use will continue, particularly in polar regions where ionospheric propagation at HF is erratic and access to geostationary satellites is difficult or impossible.

The easing of demand, due to the introduction of the use of HF bands, was so substantial that in 1949, in accordance with a decision of the ITU International Radio Conference in 1947, it was feasible to draw up a frequency assignment plan for 14–150 kHz which met the requirements of all countries for the FS and the sharing services. The planned frequency assignments, having been assigned to stations by the responsible administrations, brought into use and notified to the IFRB, were recorded in what is now called the Master International Frequency Register (MIFR). Since then, any desired additions to, deletions from and modifications of these entries in the MIFR are notified to the IFRB for registration, following the procedure given in RR Article 12. The same procedure is used for the FS at HF and it is reviewed on pp. 84–86.

Several technical requirements and constraints, intended to make more efficient the use of these allocations, have been agreed. In particular:

- RR 303 and RR Appendix 7 require the carrier frequencies of all FS transmitters to be maintained, after 1990, within 100 parts per million (p.p.m.) of the nominal value if the frequency is below 50 kHz and within 50 p.p.m. if the frequency is between 50 and 535 kHz,
- for limits on spurious emissions, see Appendix A5,
- RR 454 limits the use by FS stations of frequencies between 90 and 160 kHz to certain specified classes of emission, namely A1A, F1B, A2C, A3C, F1C and F3C. (See Appendix A6)

RN is a 'safety service' (see p. 414) and there is particular concern to avoid harmful interference to RN stations, which might endanger human life. WARC-MOB-87, which had no authority to amend regulations affecting the FS, included two relevant resolutions in its Finals Acts[1] for the consideration of a subsequent WARC that would have power to act on them. RR Resolution 705[2] seeks concerted action on the control of interference to station of the RN between 70 and 130 kHz from MM and

FS stations. RR Resolution 706[3] presses for the cancellation of the FS allocation between 90 and 110 kHz. These two issues may be included in the agenda of a WARC which PC-89 agreed should be held in 1992.

5.3 The fixed service at HF

5.3.1 Survey of the FS regulation at HF

The frequency allocations

The FS has been allocated a number of frequency bands between the upper end of the MF broadcasting band (1606·5 kHz in Regions 1 and 3; 1705 kHz in Region 2) and 4000 kHz but the allocations are all shared with mobile services. These bands are used, to a limited extent, for short distance FS links but the mobile services are the dominant users. The allocated band limits, the sharing situation and the terms under which sharing takes place differ in a complicated way from Region to Region and reference should be made for the details to RR Article 8 (see also pp. 211–213).

The spectrum between about 4 MHz and about 28 MHz is the only frequency range which is usable for regular long distance communication by ionospheric wave propagation. About 55% of this bandwidth is now allocated to the FS, in about 40 bands distributed between 4 and 28 MHz, a typical band being a few hundreds of kilohertz wide. All of these allocations are primary, many of them are exclusive to the FS or shared with a service with a secondary allocation and most allocations are world-wide. Some bands are shared with other services which also have primary allocation status, typically the MM, but this raises no fundamental regulatory problem since the same regulatory procedures are used for managing frequency assignments for all services in these bands and the technical parameters and operating conditions of the systems are broadly similar.

The first repeatered trans-oceanic coaxial submarine cable was laid in 1956 and the INTELSAT global satellite communication system began operation in 1965. By the early years of the 1970s these new transmission media were carrying much of the world's long distance telecommunications traffic but the demand for access to the HF radio spectrum for the FS before that stage was reached became very great. The development of these other media caused the need for long distance FS links at HF to start to decline and some spectrum became available for reallocation to other services with needs that were still growing.

Thus, seven frequency bands between 9 and 22 MHz, previously allocated exclusively to the FS, were transferred to the BS at WARC-79. Six more exclusive FS bands between 12 and 23 MHz were transferred to the MM at the same time. The precise band limits are quoted in RR 531 and RR 532 respectively. Three more bands previously shared by FS and MM were made exclusive to the MM at WARC-79; see RR 544. In all of these cases the FS retained the right to use the bands on an interim basis until July 1989 under the terms of RR Resolution 8[4], and until July 1994 for the one band that is below 10 MHz, to give time for replacement assignments to be found for the stations of the FS. The period of grace for the bands above 10 MHz listed in RR 531 for transfer to the BS has since been extended by RR Resolution 512[5] of WARC HFBC-87 pending decisions to be taken at a future WARC. However, WARC-MOB-87 drew up plans for the MM to use the bands listed in RR 532 and RR 544 starting on 1 July 1991 (see p. 198).

Typical systems in operation

Many HF FS emissions carry a single sideband analogue telephony channel, typically of class of emission R3E (see Appendix A.6). Most of the remainder carry telegraphy signals of between 20 and 200 bauds, typically class of emission F1A or F1B. Some emissions use frequency division multiplex for up to four telephone channels and some multi-tone telegraphy systems carry digital signals for several low-speed telegraph channels or one medium-speed data channel. The bandwidth occupied by emissions ranges typically from 12 kHz down to a few hundreds of hertz. Arrangements of channels in some multichannel emissions are defined in CCIR Recommendations 348[6] and 436[7].

Frequency assignments for HF links

Each medium-distance HF radio link (for example, 1000 km or longer) operating a long daily schedule will probably require the use of two or three frequency assignments distributed appropriately between 4 and about 16 MHz. This set of assignments is known collectively as a complement of frequencies. Each frequency is used when ionospheric propagation conditions make it the most suitable one of the complement (see Appendices B.3.4 and B.6.4). An intercontinental link may need four or five assignments in its complement, extending perhaps as high as 28 MHz, in order that, from hour to hour, its operating frequency may be matched sufficiently exactly to the prevailing propagation conditions.

The times at which the various frequencies of a complement will be used on a link are not random; they show a distinct pattern, although the pattern may be peculiar to a particular geographical path. This pattern of use is basically diurnal, but the diurnal pattern changes seasonally and in accordance with the solar activity cycle. It also varies sporadically due to naturally occurring disturbances in propagation conditions and in reaction to interference to one or more of the frequencies of the complement.

A single frequency below 4 MHz may be sufficient for a short link, perhaps a few hundred kilometres long, operating by ionospheric wave propagation or under favourable conditions, by ground wave.

HF assignments are sometimes used for quite short fixed links, perhaps only a few tens of kilometres long. If such links can be operated by ground wave propagation, it is desirable to assign and use frequencies so that ionospheric reflection does not occur. This may, for example, involve using a frequency above 20 MHz by night and a frequency below about 7 MHz by day; in this way interference will not be caused to distant receiving stations and the link will not be liable to interference from distant transmitters. If the ground wave field strength is too low to sustain the link, it will be necessary to use one or more frequencies which provide adequate ionospheric wave continuity (see Appendix B.6.4). However, there is agreement amongst administrations that the HF spectrum is to be reserved, as much as possible, for use on long distance links (see p. 87).

International spectrum management

Until the early years of the 1970s, HF radio was the transmission medium used for most intercontinental communication links. HF radio was also used for many shorter links, domestic and international. The total bandwidth of around 14 MHz which was available then for FS systems (although with some allocations shared with other services) was utterly inadequate for the purpose, and this led to an acute lack of long distance telecommunication facilities and to widespread interference.

Radio interference is limited, to the extent possible, by international registration of national frequency assignments, the right to use a frequency without interference being customarily conceded to a station with priority of registration (see p. 65). Deference to the station with priority of registration provides a practicable means for limiting interference, but it has serious disadvantages. In particular, it tends to give preference to countries which have been using HF radio extensively for a long time. Some of these countries have large numbers of assignments listed in the MIFR with early dates of registration. Consequently, other countries which are trying to build up their communications infrastructure may have great difficulty in getting new assignments established. The significance which is given to priority of registration also leads to a reluctance to cancel entries in the MIFR when the assignments which they represent are no longer in use. As a result, the Register tends to become out of date and its usefulness as a record of the real situation is reduced.

Recognition of a station's right to use a frequency is made particularly complex in the FS at HF by the vagaries of ionspheric propagation and the irregular nature of the use made of the frequency assignments which form the complement for a link. A station does not need the exclusive use of the frequencies in its complement; it may be possible to assign a frequency to many different stations for different links in different parts of the world. Under these conditions interference will not arise if high propagation loss protects a receiving station from unwanted emissions in use for other links at all times when the same frequency is required for use on the wanted link. However, a frequency assignment which has been satisfactory in use for months or years may suddenly begin to suffer severe interference from a station with a prior assignment, because of changes in propagation conditions or changes in the times when the frequency is used on either the wanted or the unwanted link. It is therefore necessary to rely, to a much greater extent than in other parts of the spectrum, on long trials to discover whether a new assignment is compatible with the use of the same frequency at other stations.

Attempts to find ways of managing FS HF assignments by frequency assignment planning were made after the Administrative Radio Conference in 1947 but without success (see p. 63).

The situation has been relieved by the availability of other telecommunications transmission media; many telecommunications routes which once depended on HF radio now use long radio relay systems, submarine cables or satellite systems. Nevertheless, there are still circumstances in which HF radio provides an economically expedient or operationally attractive transmission medium for medium- and long-distance links, and this part of the spectrum is still in heavy demand for the FS.

In order to enable international frequency assignment management by priority of registration to meet the pressing demand for spectrum as well as possible, the procedures used by the IFRB are intended to take the practical nature of the operation of HF systems realistically into account. These procedures recognise that several or even many stations may be able to use the same frequency, but they seek to exclude from the MIFR new assignments which are notified by administrations but which are likely to cause harmful interference to assignments already notified by other administrations. Allowance is made for the possibility that predictions of propagation conditions and the information provided by administrations as to the use that is made of assignments may not be precise enough for a reliable determination to be made of the risk of interference. It is recognised that there is a need to protect the rights of an assignment which is not, for the time being, in use because propagation conditions make more appropriate the use of other assignments in a complement. Furthermore, account is taken of the fact that some HF links are operated regularly but others are used only occasionally. The possibility that an assignment may have gone permanently out of use without the fact having been notified by the assigning administration is not overlooked. These procedures are reviewed briefly on pp. 85–86.

Strenuous efforts have been made to eliminate out-of-date entries from the MIFR. The most recent programme of work to this end was agreed at WARC-79 in RR Resolution 9[8]. In implementing this resolution, the IFRB called on administrations to cancel all entries in the MIFR between 3 and 27·5 MHz which are no longer required and to classify the remaining entries in accordance with various operational criteria. This should have facilitated the examination of notices of new assignments to ascertain whether interference is likely, and clarified the status of newly-registered assignments; see also CCIR Report 860[9].

Finally, the RR incorporate spectrum management principles which administrations undertake to observe and also technical requirements which they undertake to impose upon their radio stations, to enable the HF frequency bands allocated to the FS to meet more of the demands which the world makes on them. These technical measures are reinforced by recommendations and advice from the CCIR (see pp. 86–90).

National spectrum management

At HF an administration has little opportunity for independent action in managing the spectrum because ionospheric propagation of signals, with very low loss at times, makes international interference considerations dominant at these frequencies. The administration will carry out its responsibilities towards the HF stations within its jurisdiction, and one of the more difficult of its functions may be to co-operate with radio station operators in finding frequencies which may be suitable for newly-required assignments. The IFRB offers its assistance to administrations, and in particular those of developing countries, which are experiencing difficulty in identifying carrier frequencies that would be suitable for assigning to their stations; see in particular RR 1218 and RR Resolution 103[10]. An administration is also likely to become involved in the resolution of 'harmful interference' problems which the transmitting stations within its jurisdiction may have caused and those from which its receiving stations may suffer

An administration is also expected to ensure that the requirements and recommendations listed on pp. 86–90 are observed nationally and that the entries in the MIFR for the stations within its jurisdiction accurately reflect the use that is made of its HF assignments.

5.3.2 International registration of frequency assignments below 28 MHz

Priority of registration in the MIFR is the criterion customarily employed in the FS for determining the right to use a frequency assignment without harmful interference. For reasons which are reviewed on pp. 82–83, the procedure by means of which the IFRB makes that criterion available at HF must maximise opportunities for access to the spectrum for new assignments and this makes the procedure somewhat complex. Similar factors arise in the FS at frequencies below the HF range, although perhaps with less force, and the same procedures are used, down to the VLF band. The same procedure is also used for other services at HF and below, except where special procedures for specified services have been adopted as part of a frequency plan agreement.

At higher frequencies, radio propagation conditions are less favourable to trans-frontier interference and simpler registration procedures are considered sufficient to deal with it (see pp. 103–104). The transition point from one procedure to the other is set rather arbitrarily at 28 MHz.

The procedures are set out in full detail in RR Article 12. In short, they have the purpose of protecting the proper status of assignments which are already registered in the MIFR but also of ensuring that a newly notified assignment can obtain the favourable consideration which the terms of the RR intend, notwithstanding the possible neglect of other administrations to respond to enquiries or bring up to date or cancel entries in the MIFR which no longer represent current or planned usage.

The following six stages give only a simplified account of the procedures:

Stage 1. Notification: An administration notifies the IFRB of a frequency assignment, transmitting or receiving, to a station within its jurisdiction if the use of the assignment is considered to be capable of causing harmful interference to the radio links of another country or if the frequency is to be used for an international link. Other assignments may be notified to the IFRB if the administration desires to obtain international recognition for them, for example to obtain protection from interference for a receiving assignment see RR 1214–RR 1217. The information which is to be included in the notification is listed in Section A of RR Appendix 1. In addition, RR 1222 requires the 'class of operation' of the link concerned to be stated, that is, whether it is to be in regular operation, or in either standby or occasional use; RR Resolution 9[8] also refers. RR 1228 requires the notification to reach the IFRB not more than three months before, nor more than one month after, the assignment has been brought into use.

Stage 2. Publication: On receipt of a notification containing all of the required information, the IFRB publishes the details in the circular which is mailed every week to all administrations.

Stage 3. Examination: The IFRB examines the notification to verify that the new assignment is in accordance with the regulatory aspects of the RR (that is, the aspects which do not relate to interference). This is called the 'regulatory examination' (RR 1240). The IFRB also examines the notification to ascertain whether it is probable that the use of the new assignment would cause harmful interference to the facilities provided by assignments which are already recorded in the MIFR and which are in full accordance with the RR. This is called the 'technical examination' (RR 1241 and RR 1242). In making this examination the IFRB uses technical criteria at its discretion, but it may be noted that CCIR Recommendations 240[11] and 339[12] and Reports 989[13], 990[14] and 991[15] give CCIR data on protection ratios for systems of the FS. The relevant CCIR propagation data are reviewed in Appendices B.2, B.3 and B.6.2–B.6.4. The IFRB's action on the outcome of these examinations is as follows:

(*a*) If the results of the regulatory examination and the technical examination are both favourable, the assignment is dated and registered in the MIFR; see RR 1250.

(*b*) If the result of the regulatory examination is unfavourable but the technical examination shows no prospect of interference to a registered assignment, the assignment is dated and recorded in the MIFR, appropriately annotated. If however this irregular assignment should cause harmful interference to any assignment, already registered or subsequently registered, which is in full compliance with the regulatory terms of the RR, the use of the irregular assignment must cease immediately; see RR 342, RR 1271 and RR 1419.

(*c*) If the result of the regulatory examination is favourable but the technical examination indicates a significant possibility of interference to a registered assignment which is in full conformity with the RR etc., the notification is returned to the notifying administration, giving the reason and, if possible, suggesting ways of changing the assignment so that the interference will not be thought likely to occur; see RR 1253.

Stage 4. Resubmission with modification: If a notification is rejected by the IFRB, the notifying administration may amend the assignment and the notification and refer the latter back to the IFRB. If the notification, as modified, is satisfactory on both regulatory and technical examination, the assignment is recorded in the MIFR; see RR 1254.

Stage 5. Reconsideration on insistence: The administration may believe that a notification has been rejected without sufficient reason; for example it may be thought that the assignments recorded in the MIFR which the IFRB expect to suffer interference are no longer in use. If so, the administration may resubmit the notification after the assignment has been brought into use, and insist that it be reconsidered. The IFRB will then re-publish the details in the weekly circular, drawing the notification to the attention of all of the administrations which would seem likely to be adversely affected; see RR 1255–RR 1260. If a complaint of interference from the new assignment is received within two months of re-publication in the weekly circular, the notification will be returned once more to the administration, and the usual processes for clearing interference to a station with priority will follow at the initiative of the station which has suffered the interference. See RR 1262 and pp. 64–66. However, if no complaint of interference is received within two months of re-publication, the assignment will be recorded in the MIFR, appropriately dated and annotated. The MIFR record will be further annotated if there is a subsequent complaint of interference to a station with priority; see RR 1261 and RR 1263.

Stage 6. Modification or cancellation: The process of notifying and registering a modification to the entry for an assignment in the MIFR is similar to the process of securing the initial registration. It is a simple transaction if the change does not increase the likelihood and extent of interference but it involves a repetition of Stages 1 to 5 if interference prospects are expected to be made worse; see RR 1306–1308. The permanent cessation of an assignment is to be notified within three months; see RR 1430.

5.3.3 Efficient use of the HF spectrum

Most of the measures discussed on pp. 82–83 and 85–86 are intended to enable the MIFR, through the IFL, to be an effective aid for administrations that want to select frequencies for new assignments and for eliminating harmful interference. In addition, the following regulatory pressures and constraints are applied by the RR in broad terms to ensure that the best use is made of the total bandwidth available:

- In recognition of the special importance for long distance links of this quite limited frequency range, RR 954 records the agreement of the members of the ITU to make every possible effort to reserve the bands between 5 and 30 MHz for long distance communications. Whenever such frequencies are used for short- or medium-distance communications, the minimum power necessary is to be employed. RR 955 urges the use of other means of communication whenever practicable.
- Specific mention is made of the need for care to avoid wasteful use of frequencies in the operation of FS networks used for the international exchange of police and synoptic meteorological information; see RR 2702–RR 2705.
- RR 2700 urges administrations to discontinue the use in the FS of double sideband (DSB) amplitude modulated (AM) radiotelephony (that is class of emission A3E) and RR 2701 prohibits the use below 30 MHz of frequency modulation (FM) or phase modulation for radiotelephony (classes of emission F3E and G3E respectively).
- RR 436 has the effect of forbidding the use of ionospheric scatter propagation in the FS.

More specific requirements are applied to the technical characteristics of systems, also for the purpose of improving the efficiency with which spectrum is used. Thus:

- RR 303 and RR Appendix 7 impose mandatory standards of carrier frequency tolerance on all FS transmitters. Relatively mild tolerances applied up to 1 January 1990. From that date onwards, however, the carrier frequency error is not to exceed the values shown in Table 5.1. The stringent new requirements applying to the more important classes of emission will virtually eliminate wastage of spectrum due to carrier frequency instability. The requirements for those classes of emission are also strict enough to permit the use of receivers without automatic frequency control, provided that the stability of the receiver local oscillators is also good; see CCIR Recommendation 349[16].
- For constraints on spurious emissions, see Appendix A.5.
- The RR call, in general terms, for the use of the least transmitter power that is compatible with acceptable performance and for the use of efficient directive antennas where possible, in order to minimise interference. For example:

  All stations shall radiate only as much power as is necessary to ensure a satisfactory service. (RR 1804) and
  ... radiation in and reception from unnecessary directions shall

be minimised by taking the maximum practical advantage of the properties of directional antennae whenever the nature of the service permits. (RR 1807)

It is feasible to use antennas of considerable directivity for most HF radio links of the FS. CCIR Report 356[17] considers how much gain can be obtained economically on typical links, stressing the importance of matching the characteristics of the antenna in the vertical plane to the propagation modes that will give reliable link operation. CCIR Recommendation 162[18] puts the intention of RR 1807 into concrete terms for the FS at HF.

Table 5.1 Frequency tolerance for FS HF emissions (RR Appendix 7)

Class of emission	Power of emission (peak envelope power) (W)	Tolerance, in Hz or parts per million (p.p.m.)	
		below 4 MHz	above 4 MHz
SSB and ISB telephony	< 200	50 Hz	
(H3E, R3E, J3E and B8E)	> 200	20 Hz	
	< 500		50 Hz
	> 500		20 Hz
FSK telegraphy (F1B)		10 Hz	10 Hz
All other classes of	< 200	100 p.p.m.	
emission	> 200	50 p.p.m.	
	< 500		20 p.p.m.
	> 500		10 p.p.m.

Another aspect of optimising the efficiency of use of the HF spectrum for the FS involves improving the performance of circuits, since better circuit quality permits more traffic to be carried in the bandwidth occupied. The texts of the CCIR contain much that is relevant to HF circuit performance improvement.

The impairments of ionospherically propagated signals can be divided into long-term impairments and short-term impairments. Long-term impairments arise when the ionosphere temporarily loses the ability to propagate a signal from the transmitting station to the receiving station at the frequency in use, causing interruptions of the link lasting many minutes or hours. Short-term impairments, due primarily of the multipath nature of ionospheric propagation, include signal fading and intersymbol distortion.

The severity of the long-term impairments depends, to a considerable degree, on the length and geographical disposition of the great circle path between the ends of the link. In the absence of interference, short paths tend to be satisfactory for most the time, except those which are at high latitudes. For a path of inter-continental length, much depends on the design of the radio equipment, the performance of antennas at the angles of elevation at which significant propagation modes will be active (see CCIR Report 356[17]), the adequacy of the complement of assigned frequencies and the match, from hour to hour or from minute to minute, between the operating frequency and prevailing propagation conditions.

Predictions of optimum working frequencies and the skill of operating staff in selecting frequencies for operation can be augmented or perhaps replaced by various Real Time Channel Evaluation (RTCE) methods; see CCIR Report 357[19] and Annex II of Report 551[20]. However, most RTCE methods require test signals to be transmitted in addition to the signals which carry traffic and therefore they add to the general burden of interference. Accordingly, CCIR Recommendation 613[21] urges users of RTCE to have due regard for the need to minimise the interference that such systems cause (see also Appendix B.3.4).

CCIR Recommendation 455[22] recommends the use of linked compressor-and-expander devices on HF radiotelephone circuits. These devices reduce the perceived level of noise and interference in the signal received. On two-way circuits they also stabilise the loop gain and provide echo suppression. There is a performance specification of 'Lincompex', one such device, in Recommendation 455. CCIR Report 354[23] describes other designs, including 'Syncompex' and gives an account of the benefits which equipment of this kind has been shown to provide.

Fading is the main cause of short-term impairments of HF circuits. It is discussed briefly in Appendix B.3.4 and references will be found there to CCIR texts which examine the nature and causes of fading. For radiotelephony a major cause of performance degradation of double-sideband (DSB) emission arises from the use of envelope detection in the presence of selective fading. This effect can be greatly reduced at the receiver by demodulating in the single-sideband mode with a high-level reconditioned carrier or a locally generated carrier. Further improvement can be obtained by using a single-sideband (SSB) emission, with a reduced-level or suppressed carrier instead of DSB.

Diversity reception can do much to eliminate bit errors due to fading from radiotelegraph channels if the mean signal-to-noise ratio is satisfactory and the fading is not very deep; see CCIR Report 327[24]. Even so, a large number of incorrect characters will be recorded on a simple automatic telegraph system during periods when the signal is weak, making such systems unsuitable for use by any but skilled operators. Automatic error control systems, typically using automatic error correction by repetition on request (ARQ), have been developed for various operating configurations to reduce or virtually to eliminate these incorrect characters from the record. CCIR Recommendation 342[25] includes a performance specification for one such system and other systems are described in other texts of CCIR Study Group 3. Relatively little protection from fading is provided for radiotelephony by diversity reception; see CCIR Report 327[24].

Finally, the pressing need for economy in the use of bandwidth at HF has led to many studies of possible means of reducing the necessary bandwidth of emissions carrying telephone and telegraphy channels. Some results are to be found in CCIR Reports 176[26], 177[27] and 862[28].

5.4 The fixed service above 28 MHz

5.4.1 Survey of FS regulation above 28 MHz

Frequency allocations and sharing, 28 MHz–1 GHz

The FS has been allocated a large number of frequency bands between 28 and 1000 MHz in the international Table of Frequency Allocations (RR Article 8). The band limits are complex and there are many differences in the limits which apply in the three Regions. Footnotes to the international Table make additional sub-Regional allocations for the FS. Reference should be made to RR Article 8 for the full details of these allocations. A large proportion of the total bandwidth of this frequency range is allocated for the FS in Region 3; the proportion is less but still substantial in the other two Regions. Most of these allocations are primary.

However, virtually all of these FS allocations are shared with other services which also have primary allocations and these latter allocations are, in general, for services like BS and MS, for which this part of the spectrum has major technical advantages. In Regions 1 and 2 the FS allocations below 470 MHz are mostly shared with one or more of the mobile services, between 470 and 790 MHz the FS allocations are limited to a few sub-Regional allocations by footnote and above 790 MHz sharing is with the BS as well as the MS. In Region 3 all of the FS allocations are shared with the MS or the BS or both.

There are some regulatory constraints on the choices which administrations may make from amongst these internationally agreed FS allocations in drawing up their national frequency allocation tables. In particular:

- In Region 1 broadcasting assignments are formally planned and the broadcasting assignments are protected from interference by the terms of the plan agreements.
- In some bands which are allocated Region-wide for BS, some footnotes to the international Table of Frequency Allocations give primary status to the FS within sub-Regional areas, but the footnotes make the assignment of frequencies to stations of the FS subject to agreements reached using the procedures of RR Article 14; see for example RR 674 and RR 675.
- A radio channel plan has been agreed for the maritime mobile service between 156 and 162·05 MHz. Almost the whole of this band is shared with the FS and other mobile services, but administrations have given a qualified undertaking to respect the MM in those bands; see RR 613.

Consideration should be given to the possibility of sporadic sky-wave interference at long distances below about 100 MHz (see Appendices B.3.3 and B.6.5). Tropospheric scatter systems may also cause interference at long distances in the upper part of this frequency range (see Appendix B.6.9). In general, however, interference tends to have quite limited range above 28 MHz and the use that is made of this part of the spectrum is largely at the discretion of each administration. Co-ordination between the administrations of neighbouring countries is informal, except for a few footnote allocations, without the use of such devices as the internationally agreed sharing constraints and frequency co-ordination procedures which have been adopted at higher frequencies in bands shared between space and terrestrial services.

However, the factor which will often determine whether the FS is to get access to shared bands in this frequency range is competition for spectrum from the other sharing services within the same country, and in particular competition from the BS and the LM. In some parts of the world, and especially in Region 3, where the number of broadcasting stations within interference range may be relatively small, plenty of bandwidth between 28 MHz and 1 GHz may be available for assignments to the FS without limiting the development of the MS. Elsewhere, for example in Europe, where the use of VHF and UHF spectrum for broadcasting is heavy and use for the LM is large and growing rapidly, the quite limited share which the FS is likely to obtain of this part of the spectrum should probably be reserved for systems with propagation paths which are marginally non-optical, and which would not be well served by microwave assignments.

WARC-79 considered that a RARC should be convened to decide what administrative or technical measures would be desirable to ensure that Region 3 made good use of the bands in this frequency range that are shared between FS, MS and BS; see RR Resolution 702[29]. PC-82 and PC-89 took the same view. This formal conference has not yet been convened but progress is being made in dealing with the topic at less formal meetings.

Frequency allocations and sharing above 1 GHz

The following bands have been allocated worldwide to the FS with primary status:

1·7 –2·69 GHz	10·7 –12·5 GHz	21·2 –23·6 GHz
3·4 –4·2 GHz	12·75–13·25 GHz	25·25–29·5 GHz
4·4 –5·0 GHz	14·4 –15·35 GHz	36·0 –40·5 GHz
5·85–8·5 GHz	17·7 –19·7 GHz	47·2 –51·4 GHz

Together these twelve bands cover half of all the spectrum between 1 and 51 GHz. There are various other, narrower FS allocations, mostly sub-Regional, mostly below 15 GHz, and many footnotes to the international Table make yet more FS allocations with sub-Regional coverage.

Substantial allocations have also been made to the FS above 51 GHz, but these are little used at present.

Most of these FS allocations are shared with the MS and there are sharing allocations for other terrestrial services also, especially in the lower part of the frequency range. The MS allocations are little used at present except for transportable stations providing temporary links. These higher frequencies are not well suited to the low antenna gain and the obstructed propagation path of a typical mobile system. Consequently, most of the allocated bandwidth is available for assignment to the FS. However, it is likely that pressure will grow within the next ten years, particularly in metropolitan areas, for more bandwidth allocated to the MS below about 2 GHz to be used for links with road vehicles and hand-portable equipment. The need to transfer FS systems from bands below and around 2 GHz to bands higher in the spectrum, to make room for the LM, will clearly develop in some countries.

As is found between 28 MHz and 1 GHz, interference between stations of the FS and those of other terrestrial services (excluding the AeM) seldom occurs at great distances above 1 GHz and informal co-ordination between neighbouring administrations is sufficient to identify and resolve these sharing problems. However, many FS microwave bands are also allocated to space radio services and in particular to the FSS. Sharing between space services and terrestrial services raises complex problems.

WARC-59 made a few narrow secondary allocations for space services but WARC-63 was the first ARC that needed to allocate frequency bands with primary status for space services with major bandwidth requirements. The conference considered various alternatives:

- It would have been desirable to make exclusive allocations to space services below 10 GHz. However, extensive investment had already been made in terrestrial radio systems between 1 and 10 GHz and it was decided that the cost of cancelling existing allocations to terrestrial services, to make exclusive allocations available for space radio systems, would be unacceptable.
- Above 10 GHz there had been relatively little investment in terrestrial systems by 1963 and exclusive space allocations might perhaps have been made there, after the cancellation of allocations to terrestrial services. However, too little was known then about radio propagation between space and Earth at such high frequencies and it was feared that high propagation losses, caused by rain, might jeopardise the success of systems of the space services.
- Studies had shown that the geometry of wanted signal and interfering signal ray paths and the emission parameters then foreseen for FS, MS and FSS systems, would make it technically feasible to use the same frequency bands within a broad geographical area for both services, given some suitable administrative means of regulating interference.

Accordingly WARC-63 allocated a number of bands to space services below 10 GHz, sharing with the FS. WARC-71 substantially extended the practice to bands above 10 GHz and WARC-79 made many more such allocations. Much of the bandwidth allocated to the FS between 1 and 51 GHz is now shared with the FSS and there are also sharing allocations for the BSS and MSS.

When the FS shares a band with a space service, the short interference distance which are characteristic of terrestrial systems are far exceeded. The emissions of a typical FSS earth station have high spectral power density and they may cause interference at an FS station receiver many hundreds of kilometres distant. Similarly, the very sensitive receiver of a typical FSS earth station may suffer interference from a distant FS transmitter. A transmitter at a space station (that is, a radio station located on a satellite or spacecraft) can interfere with a receiving station of a terrestrial service which is located thousands of kilometres away from the earth station at which the space station emission is intended to be received. A receiver on a space station can likewise suffer interference from terrestrial transmitters located anywhere in an area continental in its extent (see Fig. 5.1). The simple arrangements which suffice to identify potential interference problems between terrestrial stations are not satisfactory where space services and terrestrial services share a frequency band. More complex methods, strongly international and formal in character, have been found necessary to control such interference. The various methods can be summarised as follows:

(*a*) A package of measures has been designed to deal with sharing between the FS and the FSS and the same package has also been applied to other sharing situations. The package consists of three elements:

- A sharing constraint is applied to the carrier power and the EIRP of transmitters of terrestrial systems, designed to limit to an acceptable level the total interference power entering a space station receiver on any geostationary satellite from all of these terrestrial station transmitters. Below 10 GHz the maximum EIRP is subject to a particularly severe constraint if the antenna beam is directed in azimuth towards one of the points where the GSO crosses the horizon.
- Similarly a sharing constraint is applied to the power flux density (PFD) which a space station transmitter may set up at the Earth's surface, designed to limit to an acceptable level the total interference entering the receiver of any line-of-sight radio-relay station from all of these space station transmitters.
- A co-ordination procedure is required to be performed to make sure that unacceptable levels of interference do not arise from an earth station in one country to a terrestrial station in another country, or vice versa.

This method is discussed in some detail on pp. 105–131. There may be a need for these sharing constraints and co-ordination procedures to be developed further in the future to deal with new situation arising from new modes of operation of space services (see pp. 131–133).

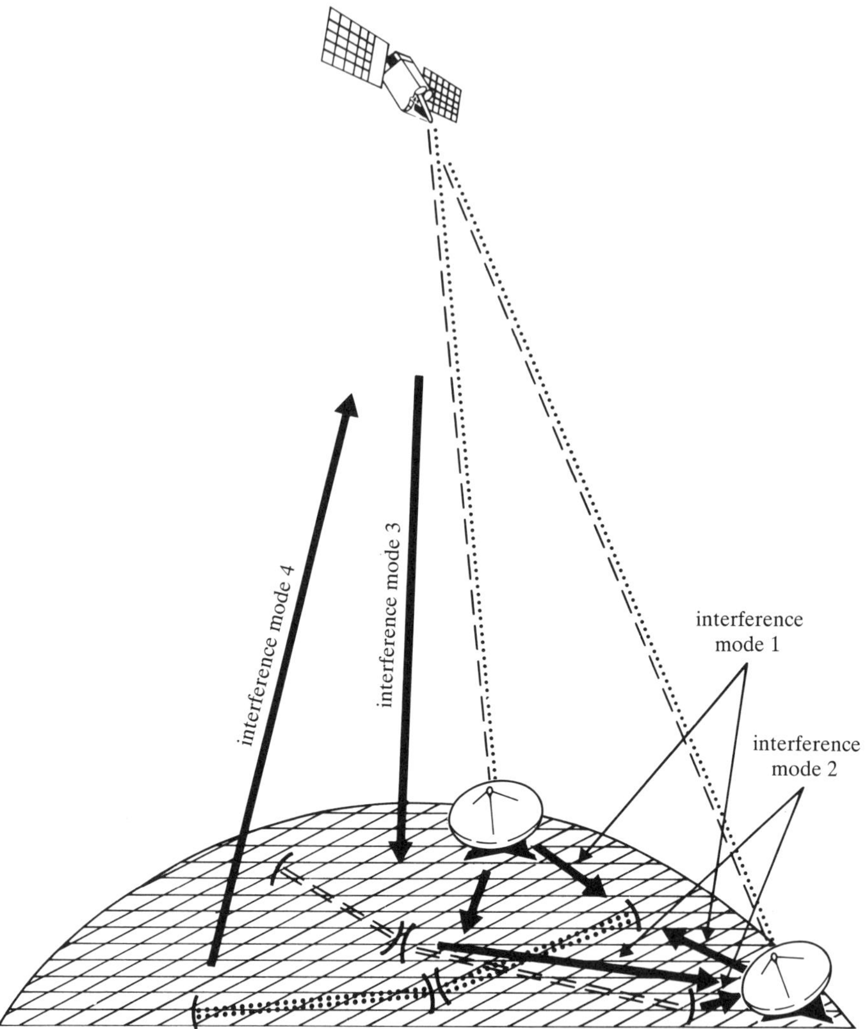

Fig. 5.1 Interference modes between satellite links and terrestrial links using the same frequencies. Wanted signal paths are in broken line
_ _ _ _ satellite down-link frequency
.... satellite up-link frequency

(*b*) The co-ordination procedure referred to in (*a*) and used to limit interference between stations of the FS and earth stations of the FSS is not readily applicable to the mobile earth stations of the MSS and no fully acceptable alternative method has been approved yet for use in its place, although see CCIR Report 773[30] and paragraph 7 of RR Appendix 28. Accordingly, assignments to the MSS in most bands shared with terrestrial services are made subject to agreements arrived at under the procedure of RR Article 14; see for example RR 641, RR 701 and RR 812, but see also RR 730, where RR Article 14 consultation has not been required.

(*c*) In certain 12 GHz BSS allocations for which frequency assignment plans have been agreed and in the associated bands at 15 and 18 GHz which are used for feeder links to the broadcasting satellites, the procedures discussed on pp. 131–133 are used. RR 913 prepares the way for a similar arrangement to protect BSS operations in the 85 GHz band from interference from FS stations when a BSS frequency assignment plan has been drawn up for that band.

(*d*) Finally, in a considerable number of FS bands shared with space services other than FSS, MSS or BSS, assignments to space services are subject to agreements reached under the procedure of RR Article 14.

It may be noted that sharing constraints can have significant impact on the capabilities of systems of both space and terrestrial services, and that both co-ordination and RR Article 14 consultation may impose a substantial managerial load on administrations.

Typical systems in operation

The most important kind of FS system used above 28 MHz is the line-of-sight link, spanning distances up to about 60 km, carrying telephone channels in a multiplexed baseband. Transmitter power is small, typically ranging from a fraction of one watt up to several watts but sometimes as high as 20 W. Typical antenna gains are in the range 25 to 40 dB, transmitting and receiving, for SHF systems but rather less for UHF systems. Below 1 GHz most RF carriers are frequency modulated by a baseband channelled by frequency division multiplex (FDM) with a capacity of between 24 and 300 telephone channels. FM and FDM are also widely used between 1 and 15 GHz for systems with baseband channel capacities up to 2700 channels. The basebands of FM system often carry video signals in addition to or instead of multiplexed telephone channels. There are also some multichannel single sideband FDM systems carrying very large numbers of channels.

Digital switching and digital transmission are coming into general use in telecommunications networks, creating a need for multi-channel line-of-sight links carry channels digitised by pulse code modulation (PCM) in time division multiplex (TDM). Most FS links operating above 15 GHz and many at lower frequencies are digital now. Most digital radio systems use 4-phase shift keying (PSK) but higher-order PSK modulation and hybrid modulation systems are also being used, in order to increase the information flow through a given occupied bandwidth.

Several or many line-of-sight links like these may be connected in tandem to form radio relay systems which may be hundreds or even thousands of kilometres long. Such radio relay systems provide an important alternative to underground cables, coaxial or optical fibre, for international or domestic trunk routes in the public telephone network. Some multi-channel links, usually digital, are used in city centres to provide high capacity connections between the public switched network and the premises of major telecommunications users. Also many single links and some radio relay systems form part of private communications networks.

Other FS links are narrow band, low power systems, typically carrying telemetry or telecommand circuits or a single telephone channel to remote locations. Some telemetry links are connected in a star formation by which a single central station polls in turn each of a large number of out-stations to gather locally measured data. Some links carrying a few channels provide direct access to the telephone network from substantial numbers of isolated subscribers in rural locations.

However, not all FS links operating above 28 MHz have line-of-sight propagation paths. Thus;

- A circuit capable of carrying a relatively low average data rate can be transmitted, using a carrier frequency about 50 MHz, over distances of up to about 2000 kilometres by reflection from meteor trails (see Appendices B.3.3 and B.6.5).
- Trans-horizon UHF and SHF radio links using tropospheric scatter propagation often have wide basebands, multiplexed if used for telephony (see Appendixes B.4.5 and B.6.9). Relatively high transmitter power, low-noise receiver input amplifiers and high antenna gain are needed to overcome the high transmission loss of long, wide-band tropospheric scatter systems, raising substantial interference problems. Typical hop lengths for tropospheric scatter links would be several hundreds of kilometres but links may be connected in tandem to provide long distance connections.

International spectrum management

Until WARC-63 allocated frequency bands to space services, sharing with terrestrial services, international management of frequency assignments above 28 MHz through the good offices of the ITU was minimal because little was necessary.

In these circumstances the principal protection against interference was provided by an administration when it selected a frequency for assignment in the knowledge of the other assignments it had made to other stations within its jurisdiction. If international interference arose it would almost certainly be to or from a station in a neighbouring country, and precautions were taken against this through direct informal co-ordination between neighbouring administrations. After any necessary international co-ordination had been done, the IFRB would be notified of the assignment for registration in the MIFR, if it was thought to be desirable. Some administrations notified assignments for 'typical stations' to be registered, representing perhaps many stations operating on the same frequency, in accordance with RR 1223. Many assignments were not registered in any way. The action to be taken was largely a matter for the judgement of each administration in the light of its own circumstances.

These principles are still applied above 28 MHz in the management of the FS and other terrestrial services but they are not adequate, in themselves, in bands shared with space services, in particular because of the great distances over which earth stations can cause interference to stations of the FS and vice versa. In these bands the established procedures for the international notification and registration of frequency assignments have been overlaid with additional procedures to prevent unacceptable trans-frontier interference from earth stations to terrestrial stations and vice versa, once it has been established that there is an earth station within interference range of a terrestrial station.

No need has been seen for severe standards of technical performance to be imposed through the ITU upon FS stations in order to ensure that the spectrum is used efficiently. In many parts of the world there is plenty of bandwidth above 28 MHz allocated to the FS. If stringent technical standards must be applied to stations because the demand for facilities is locally high, it is proper that national administrations should determine what those standards should be, and enforce them. Indeed the mandatory requirements in the RR above 28 MHz are minimal, namely;

Frequency tolerance: RR 303 and RR Appendix 7 impose mandatory standards of frequency tolerance on all FS transmitters up to 40 GHz. The current tolerances are given in Table 5.2; they are not severe. Still milder tolerances applied before 1990.
Spurious emissions: RR 304 and RR Appendix 8 place limits on the power of any spurious emission from any FS transmitter operating below 17·7 GHz (see Appendix A.5

Some precautions have, however, been taken to limit the impact which scatter-mode systems might have on the usability of frequency bands, in particular for line-of-sight systems. Thus:

- RR 436 denies classification in the fixed service to systems using ionospheric scatter propagation 'except if otherwise specified in a footnote to the Table of Frequency Allocations'. There are no such footnotes at present, the last, relating to various bands between 32 and 40 MHz, having been cancelled by WARC-79.
- Various footnotes to the international Table of Frequency Allocations seek to encourage the making of assignments to tropospheric scatter

stations in bands where their use will be least disadvantageous to other kinds of system; see, for example RR762–RR 764.

- CCIR Recommendation 302[31] makes various recommendations on the minimisation of interference from trans-horizon radio relay systems.
- RR Recommendation 100[32] calls for a further search for the best way to accommodate tropospheric scatter systems in the spectrum.

Table 5.2 Frequency tolerance for FS emissions above 28 MHz (RR Appendix 7)

Frequency range	Power of emission (peak envelope power for SSB transmitters, mean power for others) (W)	Frequency tolerance (parts per million)
28–29·7 MHz		as Table 5.1
29·7–100 MHz	< 50	30
	> 50	20
100–470 MHz	< 50	20
	> 50	10
470–2450 MHz	< 100	100
	> 100	50
2·45–10·5 GHz	< 100	200
	> 100	50
10·5–40 GHz		300

National spectrum management

The ITU role in the management of spectrum allocated to the FS above 28 MHz being quite limited, it is necessary for each administration to consider at what level it should set its own efforts to achieve efficient management of the spectrum in these bands. To some degree this decision will be affected by the various regulatory limits and technical constraints on sharing which have been incorporated into the Radio Regulations and the extent of use of the bands allocated to the FS in neighbouring countries but the principal factor will be the national demand for facilities in these bands. If there is heavy use of the bands within the jurisdiction of the authority, then intensive spectrum management will be justified. If, however, use of the bands is light, a relatively simple management regime will provide adequate protection of users at less cost.

Whether the spectrum management regime is intensive or simple, it should have four main elements:

- a strategy for the use to be made of the various frequency bands which are allocated internationally for the fixed service,

- a plan of the carrier frequencies and assigned bandwidths ('radio channels') which will be assigned in each band to be used for the FS,
- technical standards which radio stations will be required to achieve, in order that spectrum will be used sufficiently efficiently, and
- techniques for assigning a specific frequency to one, several or many stations, in such a way that link performance objectives are achievable but without a needlessly large margin of protection against interference. Such techniques will be simple in areas where the demand for assignments is small.

The development of a strategy for allocating frequency bands is a part of the preparation of a national frequency allocation table. The most important step is likely to be to decide upon the national allocation for bands which are shared internationally between the FS and one or more other services. The most significant of these choices is likely to be that between the BS, LM and FS in bands below 1 GHz. This is a long term decision, to be taken with great care, because of the particular difficulty of changing an allocation to the BS after it has been implemented. Consideration will also have to be given to agreements that may have been reached with the administrations of neighbouring countries regarding the use of such bands. For example, neighbouring countries may have a common specification for radiotelephone equipment for use on road vehicles or ships which circulate internationally, and this may exclude the FS from some bands. The impact of BS frequency assignment plans, and the policy determined for their implementation domestically and in neighbouring countries, on the use of spectrum for the FS may also be considerable.

There is another important choice which should be considered with great care in countries which foresee using the FSS to provide a domestic network linking substantial numbers of earth stations. It is feasible to share spectrum nationally between the FS and a domestic FSS network but many problems arise when this is done. It may be preferable to withhold a national FS allocation from the bands, or the parts of bands, which are to be used for a domestic FSS system, so avoiding an onerous domestic frequency co-ordination problem. Conversely it may be desirable to confine FSS earth stations using bands shared nationally with the FS to locations where co-ordination with stations of the FS will not be difficult.

Consideration should then be given to the kinds of FS system that are to be licensed to operate in each frequency band, or part of a band, which is allocated nationally to the FS. The following points, for example, need to be taken into account;

(*a*) Some spectrum at UHF should be reserved, if possible, for links which will obtain particular technical or economic benefit from operating in this part of the spectrum. For example:

- links with particularly long or somewhat obstructed propagation paths which benefit from the relatively greater diffraction around obstacles which occurs at the lower frequencies,
- links which can operate with greater spectrum economy at UHF (for example, narrow band links which can be required to maintain tight frequency tolerances at UHF), and
- systems which have relatively little information flow per station (such as some telemetry data collection networks and rural telephone subscriber access networks) for which the cost of station equipment must be kept very low.

(*b*) The frequency bands for tropospheric scatter links must be chosen having regard to the high sensitivity of the receiving stations to interference and the long distance at which the transmitting stations can cause interference. Sharing constraints (see p. 117) will exclude these links from bands shared with Earth-to-space links of space services and interference from space station transmitters is likely to be significant and may be unavoidable in bands shared with space-to-Earth links. The technical requirements of systems may demand a UHF assignment for long tropospheric scatter links. It will usually be desirable for neighbouring countries to decide together which bands should be used for tropospheric scatter systems.

(*c*) It will probably be found to be desirable to make a national allocation of bandwidth in several microwave bands for assignments to transportable stations used to provide temporary wideband links, including temporary television links for use in outside broadcasting, and mobile links, and to exclude the mobile service from other microwave bands. It will be desirable for these choices also to be harmonised with the choices of neighbouring countries.

(*d*) Relatively narrow bands within fixed service allocations are shared, internationally or nationally, with various other services, such as the SRS, MetS, SpO and RA. It may be necessary to exclude the FS from the immediate vicinity of the stations of these other services in these bands, and it would be desirable to consider whether such action would have any impact on the kind of FS facility that should operate in these bands.

(*e*) It will usually be found that a small number of organisations within a country make extensive use of the FS above 28 MHz, and a number, small or large, of other organisations use it on a relatively small scale. If so, it may be found expedient to allot wide blocks of spectrum, perhaps complete allocated bands, to each of the big users, and to hold these organisations responsible for efficient spectrum management within their own allotments. This delegation of responsibilities will enable the big users to plan their networks on a long term basis and to standardise their equipment. It may also give the administration the opportunity to concentrate possibly limited spectrum management resources on the more difficult task of achieving efficient utilisation of the bands made available to the other users.

When these various points have been resolved, the decisions are incorporated into the national frequency allocation table. Occasional reviews of these decisions will be necessary to identify the changes in national allocations which changing circumstances will have made desirable.

A plan for assigning carrier frequencies in each band, designed to accommodate various standard emission bandwidths, facilitates efficient spectrum utilisation. A band with defined limits, wide enough for one emission of specified parameters, is commonly called a 'radio channel'. The CCIR has developed standard plans of carrier frequency and polarisation mode assignment for many frequency bands and for various baseband capacities and types of modulation. These CCIR plans have been developed primarily to facilitate the interconnection of national telephone networks through high capacity radio relay system but they are widely used in domestic radio relay systems also. They provide a pattern of carrier frequency assignment that keeps to an acceptably low level the performance degradations that can arise from interference between the different links of the same radio relay system and they avoid the more difficult problems of waveguide multiplexer and duplexer design at stations which use common antennas for the transmission and reception of several carriers.

A simple example of such a plan, taken from CCIR Recommendation 382[33], is reproduced as Fig. 5.2*a*. Fig. 5.2*b* shows how carrier frequencies might be assigned, in accordance with this plan, at relay stations for a two-way through signal. However, the CCIR plans do not cover all frequency bands or very narrow emission bandwidths, and administrations may need to draw up their own plans for such applications and provide for a mix of assignable radio channels which is appropriate to their national needs.

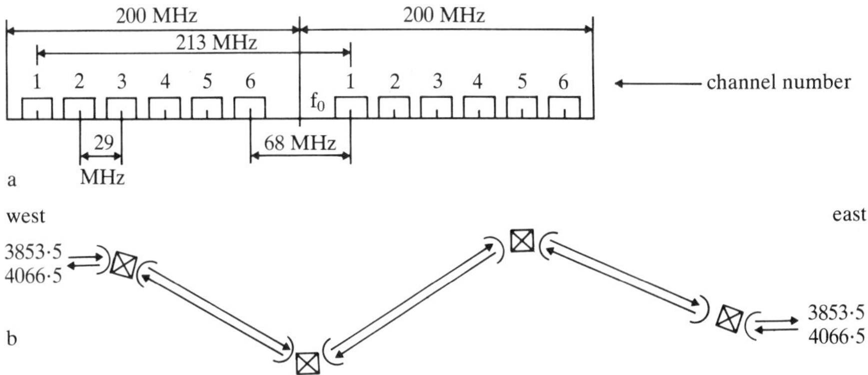

Fig. 5.2 Radio relay system radio channels for FDM emissions carrying 600 to 1800 telephone channels in the 4 GHz band
a A CCIR plan providing for up to 6 two-way FDM systems (CCIR Recommendation 382[33])
b An example of frequencies (in MHz) used for a single FDM system occupying Channel 2

In countries where large numbers of fixed systems are to operate above 28 MHz, the administration will need to impose national performance standards on licensees, by type approval of the equipment that is used or by inspection of stations. The parameters to be specified will probably include the following;

Transmitter frequency tolerance: For narrow-band links, a tolerance considerably more stringent than the mandatory one imposed by the Radio Regulations would help to promote efficient spectrum use.
Spurious emissions and out-of-band emissions: The mandatory limits on spurious emissions given in Appendix 8 of the Radio Regulations are not stringent enough to ensure efficient spectrum use in areas where many stations are operating, and the application of more stringent national standards should be considered. In addition, it may be desirable to apply stringent limits to the spread of spectral energy at the edges of the occupied bandwidth of the modulated emission, in order to limit interference between adjacent channels.
Receiver selectivity: It may be desirable to apply limits to the susceptibility of a receiver to interference from an emission in the adjacent radio channel, especially where a narrow-band emission is to be received.
Transmitter carrier power and transmitting antenna forward gain: The licensee should not be permitted to radiate much more EIRP than the adequate performance of his link requires.
Antenna off-beam gain envelope and cross-polar response: Finally, efficient spectrum management in the fixed service above 28 MHz depends, to a marked extent, on the number of times that the same frequency is assigned and used within a limited geographical area. Given good propagation data, adequate topographical information and knowledge of all significant existing frequency assignments, and given pre-determined link interference power limits and adequate information on the significant characteristics of all of the links involved, it becomes feasible to calculate whether any available radio channel would be satisfactory for assignment to a new link. It is, however, a time-consuming task if done by hand. Computer software is available for carrying out this channel selection process; its use is justified where the demand for new assignments is large enough.

5.4.2 International registration of frequency assignments above 28 MHz

The procedure that is followed when an administration wishes to have a frequency assignment below 28 MHz, made to a station of the FS, registered in the MIFR is reviewed on pp. 85–86. In the absence of certain sharing situations, a similar procedure is used for frequencies above 28 MHz. However, at these higher frequencies it is usually much easier for an administration to identify potential interference situations, mainly by informal co-ordination with neighbouring countries, and to avoid them.

Accordingly the procedure can be made rather simpler. This simpler procedure, set out in full in RR Article 12, is also used for assignments to stations of some other terrestrial services above 28 MHz. An outline of this basic procedure follows. It is followed in turn by a review of the ways in which the basic procedure is extended when sharing makes this necessary.

The basic procedure

Stage 1. Notification: After any necessary informal co-ordination of a new frequency assignment with neighbouring administrations, an administration notifies the assignment to the IFRB if the use of the frequency is capable of causing harmful interference to radio links of another country or if the frequency is to be used for an international link. Other assignments may also be notified to the IFRB for registration if the administration desires to obtain international recognition for them; see RR 1214–RR 1217. The information which is to be included in the notification is listed in Section A of RR Appendix 1. RR 1228 requires the notification to reach the IFRB not more than three months before, or one month after, the assignment has been brought into use.

When an administration assigns the same frequency above 28 MHz for use with the same emission parameters to a number of different stations of the FS (or of various other services), it is acceptable in most cases for a single notification to be supplied, in the form of an assignment to a 'typical station' in a stated area, to represent the assignments to all of the stations in the area; see RR 1223–RR 1227. The information required in the notice is listed in Section C of RR Appendix 1.

In considering whether the use of a frequency is capable of causing harmful interference to a foreign receiving station, allowance is made for the expected sensitivity to interference of typical terrestrial systems. A similar policy is usually followed in deciding whether to seek to protect the reception of an assigned frequency at an FS station by notifying the assignment under RR 1217. Consideration is not given to the higher sensitivity to interference of a foreign earth station that might be built in the future in a frequency band shared with a space service. A different procedure, indicated below, is used to take into account the risk of interference to known existing foreign earth stations.

Stage 2. Publication: On receipt of a notification containing all of the required information, the IFRB publishes the details in the circular which is mailed every week to all administrations.

Stage 3. Examination: The IFRB examines the notification to verify that the new assignment is in accordance with the regulatory aspects of the RR; see RR 1240. The technical examination which is applied to notifications below 28 MHz to verify whether interference is likely to arise is not applied at higher frequencies unless an administration asks for the examination to be made, informal co-ordination between administrations having not been possible; see RR 1244. The IFRB's action on the outcome of these examinations is as follows:

(*a*) If, in the absence of a technical examination, the result of the regulatory examination is favourable, the assignment is registered in the MIFR; see RR 1247–RR 1248.

(*b*) If the assignment is found, through the regulatory examination, to be not fully in keeping with the RR but the notifying administration gives a specific undertaking in accordance with RR 342 to cease operation on the frequency if harmful interference arises, then the assignment is registered in the MIFR with an appropriate endorsement; see RR 1266-RR 1267. In the absence of an undertaking to observe RR 342, the notification is rejected by the Board; see RR 1266 and RR 1268.

(*c*) If the results of the regulatory examination are favourable and a technical examination has been necessary, the procedures that can follow are similar to those for stages 3 to 5 for frequencies below 28 MHz, shown on pp. 85–86.

Stage 4. Modification or cancellation: The process of obtaining a modification to the entry for an assignment in the MIFR is similar to the process of obtaining the initial registration; see RR 1306–RR 1308. The permanent withdrawal of a registered assignment is to be notified within three months; see RR 1430.

Additional procedures required by sharing

In many frequency bands which the FS shares with other terrestrial services with equal primary status, frequency assignments are managed internationally for both services on the same basis, as set out above. Thus, no special spectrum management arrangements are made necessary by the existence of such sharing. In some bands, however, there are circumstances, regulatory or technical, which make it necessary to depart from, or add to, these management arrangements.

In a few frequency bands, allocated to the FS in sub-Regional areas by means of a footnote to the international Table of Frequency Allocations, there is a requirement that agreement to frequency assignments be reached with other concerned administrations in accordance with the provisions of RR Article 14. RR 1610 requires this agreement to be reached before an FS frequency assignment is notified to the IFRB for registration. The Board verifies that this has been done as part of its regulatory examination of a notification; see RR 1240.

Frequency assignment plans for the BSS have been drawn up in certain frequency bands at 12 GHz which are shared with the FS. Frequency assignment plans for the feeder links associated with these planned BSS facilities have been drawn up in bands allocated to the FSS (Earth-to-space) at 15 and 18 GHz and these bands are shared, in part, with the FS also. Special conditions apply to the operation of FS systems and the registration of FS assignments in all of these bands (see pp. 133-134).

Finally, many frequency bands above 1 GHz allocated to the FS, in addition to those to which the previous two paragraphs apply, are shared with space services. In these bands the procedures for the notification and registration of frequency assignments to the FS are not affected by the presence of the space service allocation if the location of the FS station is not within the co-ordination area already registered for a foreign earth station. However, assignments to FS stations that are within the registered co-ordination area of a foreign earth station must be co-ordinated with the administration responsible for the earth station before notification for registration.

5.4.3 ITU performance and interference objectives for telephone channels

Interference between terrestrial stations of the FS and earth stations of the FSS in the frequency bands which these services share (interference modes 1 and 2 in Fig. 5.1) is kept within acceptable limits by means of frequency co-ordination. Interference between terrestrial stations of the FS and space stations of the FSS (interference modes 3 and 4) is limited by means of sharing constraints. To enable either of these methods to be implemented, it is necessary first to determine what level of interference is to be accepted.

The transmission of trunk telephone channels, which may form part of the international public telecommunications network, is a major application for both terrestrial line-of-sight radio relay systems of the FS and satellite networks of the FSS. For many countries it is the dominant use for both services. Objectives for the power level of noise-plus-interference in such channels when used for international telephone connections have been established by CCITT. There is a considerable measure of uniformity in the radio emission parameters of these radio relay systems of the FS. The emission parameters of FSS satellite networks when used in this way are also rather uniform, although quite different from the parameters of FS systems.

Thus, for telephone channels carried by radio relay systems, it is feasible to allot a fraction of the noise-plus-interference power level to interference from FSS systems. This fraction can be sub-divided between interference mode 1 and interference mode 3. The mode 3 element can be expressed in the form of an approximate equivalent aggregate of interfering radio frequency spectral energy density reaching the Earth's surface from all satellite transmitters operating at the same frequency as the radio relay system, to form the basis of a sharing constraint on satellite transmitters. Also the mode 1 element can be used as a basis for frequency co-ordination between earth station transmitters and radio relay receivers. In analogous ways, a sharing constraint on radio relay transmitters and a basis for co-ordinating radio relay transmitters with FSS earth station receivers (interference modes 4 and 2 respectively) can de derived from the fraction of the noise-plus-interference power level in telephone channels carried by FSS networks which is allotted to interference from radio relay systems. The performance objectives and the interference allotments for international telephone connections are discussed below and on pp. 259–265 and the associated sharing constraints and co-ordination procedure are discussed on pp. 110–131.

These factors having been determined for a major application of the FS and the FSS, they have been applied with minimal changes to virtually all situations where these two services share frequency bands and indeed to most situations where a terrestrial service shares with a space service.

Noise and interference objectives in analogue systems

Many of the impairments which degrade telephone channels in analogue transmission systems, typically FDM/FM, can be expressed as equivalent noise in the channel and they can be aggregated sufficiently accurately as the sum of the noise powers of the various components. The CCITT recommends performance objectives for telephone channels which may form part of international connections. The recommendation takes the form that the total transmission impairments, expressed as noise, in the worst-affected channel of a multiplexed baseband, should not exceed w picowatts, psophometrically weighted, when measured at a point in the telephone transmission network of zero relative level (pW0p).

The transmission impairments expressed as noise consist of:

- the noise or noise-equivalent impairments arising within the system or entering the system from external noise sources,
- interference entering the channel from transmitters of other systems of the same service, and
- interference entering the system from transmitters of systems of other services.

The CCIR recommends that the noise in a telephone channel of an FS radio relay system due to interference from all of the transmitting stations of the FSS sharing the frequency allocation used for the wanted system should not exceed x_T% of w pW0p. For a telephone channel in an FSS system the recommended maximum noise level due to interference from all FS radio relay transmitters sharing the frequency allocations (up-link and down-link) used for the wanted system should not exceed x_s% of w pW0p.

For FSS systems the CCIR further recommends that the total noise power equivalent of the interference entering a channel from all other FSS networks should not exceed y% of w pW0p. Thus, the recommended limit on the noise power due to impairments other than interference from other systems is $(100-x_s-y)$% of w pW0p. For FS systems the interference from other FS links is much more completely under the control of the FS system designer and the responsible administration. Accordingly, this form of interference is treated as if it were part of the transmission impairments of the FS system and the ITU does not recommend on the apportionment of $(100-x_T)\%$ of w pW0p between in-system noise and interference from other FS links.

These performance objectives having been determined, multi-channel systems are designed so that the level of noise in the worst-affected channel, including the noise-equivalents of various transmission distortions arising within the system (and, for FS systems, including interference from other FS links) does not exceed $(100-x_T)\%$ of w picowatts for the FS, nor $(100-x_S-y)\%$ of w picowatts for the FSS. Spectrum management methods are applied and systems are configured to ensure that the percentages x and y are not exceeded by interference from the relevant sources. Thus, the channel performance objective should be attained. However, two factors complicate the application of these simple principles.

Firstly, variations in radio propagation conditions cause variations of the channel noise levels and the interference levels. These variations can be taken into account without loss of precision in the definition of performance objectives by expressing the objectives and the interference criteria in a statistical form. Typically for an analogue system;

- The performance objective calls for the total noise power to not exceed w_1 pW0p for more than 20% of any month, nor w_2 pW0p (a considerably larger value) for more than a small percentage (perhaps 0.1%) of any month.
- The interference objective calls for the total interference from transmitters of the other service to not exceed $x\%$ of w_1 pW0p for more than 20% of any month, nor w_2 pW0p for more than a very small percentage (perhaps $x\%$ of 0·1%) of any month.

The second complicating factor is as follows. A line-of-sight FS system may consist of a single link or many links in tandem, forming a radio relay chain. Telephone channels carried by such systems, single hop or multi-hop, may form part of a telephone connection thousands of kilometres long, every element of which will be adding its contribution to the noise of the connection. International telephone connections, the prime concern of the CCITT in this area, are sometimes very long. The ultimate criterion for the performance of a transmission channel, whether it is a line-of-sight FS link, a satellite link, a cable link or whatever, it whether its channels can form an acceptable part of a long connection. Channel performance objectives must take this into account.

To deal with this second factor the CCITT has defined Hypothetical Reference Circuits (HRC) for analogue telephony, typically 2500 km long, applicable to any transmission medium; see for example CCITT Recommendations G.212[34], G.222[35] and G.322[36]; see also CCIR Recommendation 390[37]. Corresponding with these definitions, the CCIR has defined HRCs for various analogue radio relay configurations,

2500 km long; see for example Fig. 5.3, taken from CCIR Recommendation 392[38]. Likewise there are HRCs for the FSS (see p. 260). The performance objectives, W_1 and W_2 are therefore set, not for a single link but for the complete 2500 km HRC. It will be noted from Fig. 5.3 that the HRC specifies the quantity of multiplex equipment that can be expected to be present in a terrestrial system 2500 km long but not the number of radio hops, of which there would probably be about 50. The contribution of channel noise acceptable from any one link of a radio relay system will, in the absence of reasons to the contrary, be related to W as the length of the link is related to the length of the HRC.

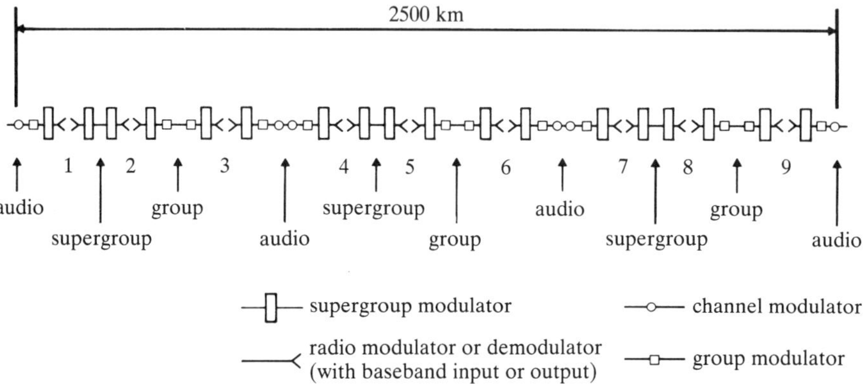

Fig. 5.3 HRC for radio relay systems using FDM with a capacity of more than 60 telephone channels per RF carrier (CCIR Recommendation 392[38])

The application of these principles for regulating sharing can be illustrated as follows:

- CCIR Recommendation 393[39] recommends, for the 2500 km HRC for analogue line-of-sight radio relay systems carrying telephony, that the mean noise power in any channel averaged over one minute should not exceed 7500 pW0p for more than 20% of any month, nor should it exceed 47 500 pW0p for more than 0.1% of any month. These recommended noise objectives exclude noise due to the FDM multiplexing equipment, for which the CCITT allows a further 2500 pW0p in a 2500 km HRC; see also CCIR Recommendation 395[40] and CCIR Report 288[41]
- CCIR Recommendation 353[42] makes corresponding recommendations for the FSS analogue telephone HRC but with 10 000 pW0p, 50 000 pW0p and 0.3% replacing the 7500 pW0p, 47 500 pW0p and 0.1% figures respectively of CCIR Recommendation 393. In the FSS HRC. no special provision is made for noise generated in the multiplexing equipment.
- CCIR Recommendation 357[43] recommends that the interference from FSS transmitters into the analogue radio relay HRC in shared

frequency bands, averaged over one minute, should not exceed 1000 pW0p for more than 20% of any month, nor 50 000 pW0p for more than 0·01% of any month.

- Similarly, CCIR Recommendation 356[44] recommends that the interference from FS transmitters into the analogue FSS HRC in shared frequency bands, averaged over one minute, should not exceed 1000 pW0p for more than 20% of any month, nor 50 000 pW0p for more than 0·03% of any month.

The noise and interference criteria for FSS networks are considered further on pp. 262–263.

Various other CCIR recommendations define HRCs and performance objectives for other kinds of FS system using analogue transmission. Thus:

- Whereas CCIR Recommendation 392 relates to a line-of-sight FS system equipped to carry more than 60 telephone channels, CCIR Recommendation 391[45] gives a corresponding HRC for systems with 60 channels or fewer. The performance and interference objectives referenced above in relation to Recommendation 392 apply to the Recommendation 391 HRC also.
- CCIR Recommendations 567[46] and 555[47] provide an HRC and various performance objectives for terrestrial systems carrying international video channels intended for broadcasting; see also CCIR Report 375[48]. No recommendation has been made yet for interference from FSS transmitters into the video channels to which Recommendations 555 and 567 apply.

An HRC and performance objectives have also been recommended by the CCIR for telephone channels carried by chains of trans-horizon links. However, the emission parameters of such links would not be compatible with the sharing constraints discussed on p. 117 below and so no sharing criteria have been considered for them.

Objectives for digital transmission systems

The development and international agreement of a system of performance objectives for international digital circuits is now in an advanced stage but it has not yet been completed. CCITT Recommendation G.821[49] provides the basic performance objectives for international digital connections forming part of an integrated services digital network (ISDN). For systems using digital transmission the Hypothetical Reference Digital Path (HRDP) takes the place of the HRC of analogue systems; see for example CCIR Recommendation 556[50]. Maximum bit error ratio (BER) objectives are recommended for the FS HRDP, depending on the purpose to which the system is to be put; see CCIR Recommendations 594[51] and 634[52] and also CCIR Reports 1052[53], and 1053[54]. There have been corresponding developments in setting performance objectives for an HRDP carried by an FSS network (see p. 262).

Possible ways of apportioning the bit-errors of the objectives between the system degradations of an FS system and interference from the FSS are discussed in CCIR Report 793[55]. Thus, for example, CCIR Recommendation 594[51] recommends that the BER due to all causes at the output of an FS HRDP which may form part of an ISDN should not exceed one bit error per million, with an integration time of 1 minute, during more than 0·4% of any month. CCIR Recommendation 615[56] adds that interference from transmitters of the FSS should not cause the BER to exceed one bit error per million for more than 0·04% of the time.

At many stations along a digital radio relay chain the received signal will be retransmitted without demodulation and regeneration of the modulating waveform. However, there will be regeneration at intervals, and the distance between regenerators affects the amount of interference that should be accepted, for example, from any one earth station into any one FS receiver. The performance objectives for the HRDPs have also to be scaled down to a BER appropriate to a single radio relay hop to facilitate system design. These issues present problems, not all of which have been resolved satisfactorily to date; see CCIR Report 930[57].

5.4.4 Sharing constraints on FSS space station and FS emissions

Much of the bandwidth allocated to the FS above 1 GHz is shared with the FSS, both services having primary status. When a band is used for both services, interference may occur between a space station and hundreds or even thousands of terrestrial stations, and some of the terrestrial stations involved might be thousands of kilometres away from the service area of the space station. The case-by-case frequency co-ordination procedure used to limit interference between earth stations and terrestrial stations is not feasible in this other context, where extremely large numbers of independent interference entries may arise.

In any frequency band there is a limit to the number of satellites that could cover with their transmitting and receiving antenna beams a given geographical area without unacceptable interference arising between the satellite networks. Given that limit, it has been shown empirically that the aggregate interference from all space station transmitters to any terrestrial station receiver will not be unacceptably high if the spectral power density of the emissions from each space station does not exceed a certain value. Likewise the number of FS stations expected to operate in a given large geographical area in any one frequency band may be large but it is not boundless and few of these stations will direct their transmitting antenna towards any specific satellite; in these circumstances the aggregate interference from all terrestrial transmitters to any space station receiver should not be unacceptably high if the carrier EIRP of the terrestrial stations does not exceed a certain value.

Fortunately these maximum power levels are high enough for most applications of the FS and the FSS. Accordingly, power limits have been applied as sharing constraints to stations of both services operating in frequency bands which they share with equal allocation status. The constraints are considered in some detail below; see also CCIR Report 209[58] and CCIR Recommendation 355[59].

Constraints on the power of space station emissions

CCIR Recommendation 358[60] recommends maximum permissible values for the power of emissions from space stations of the FSS, for use in frequency bands shared with equal allocation status with line-of-sight radio relay stations. These limits have been incorporated in RR Article 28 as a constraint on the emissions of FSS space stations in frequency bands allocated for FSS space-to-Earth links in the frequency bands and in the geographical areas where there is also an allocation for the FS with equal allocation status. The constraints so far defined cover the frequency range between 2·5 and 19·7 GHz, with provisional extension to 40·5 GHz pending further study in the CCIR (see RR 2582.1). (Exceptionally RR 873 withholds the constraint in the band 19·7–21·2 GHz in the sub-Regional areas where it allocates the band to the FS.) The limits also apply to various specified sharing situations involving other services; see RR 2552–RR 2585.

The constraints take the following form. The PFD at a point on the Earth's surface, Φ, produced by emissions from a space station, measured within a specified sampling bandwidth b, shall not exceed the values indicated by the following expressions when the angle of the satellite above the horizontal plane, δ (degrees) at the point in question, falls within the stated arcs

Φ_1 dB(W/m^2) when $0° < \delta < 5°$
$\Phi_1 + k(\delta - 5)$ dB(W/m^2) when $5° < \delta < 25°$
Φ_2 dB(W/m^2) when $25° < \delta < 90°$

where

Φ_1 has various values between -154 and -148 dB(W/m^2) in frequency bands below 12·75 GHz and is -115 dB(W/m^2) above 17·7 GHz,
$\Phi_2 = \Phi_1 + 15$ dB and k = 0·75 in the frequency band 2·5–2·69 GHz,
$\Phi_2 = \Phi_1 + 10$ dB and k = 0·5 in all other frequency bands,
b = 4 kHz below 12·75 GHz and 1 MHz above 17·7 GHz

The limiting values of Φ are shown in Fig. 5.4. RR 2585 allows these power flux densities to be exceeded on the territory of any country, given the consent of its administration.

CCIR Report 387[61] gives the rationale for the limits recommended in Recommendation 358. The elements of the rationale are as follows:

(i) CCIR Recommendation 357[43] recommends that the total interference power from all systems of the FSS entering any telephone channel of a 2500 km HRC carried by FDM analogue angle-modulated radio relay systems should not exceed 1000 pW0p for more than 20% of any month.

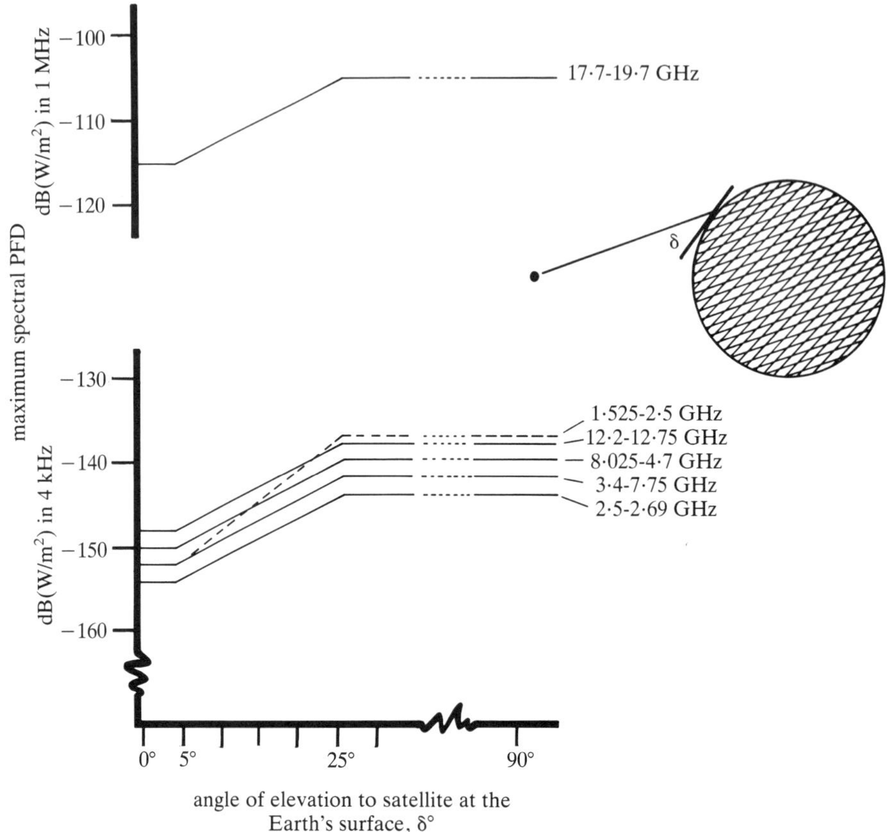

Fig. 5.4 Constraints on the spectral PFD at the Earth's surface from space station transmitters, applied to limit interference to terrestrial stations operating at the same frequency

(ii) The FSS will use a frequency band for either up-links or down-links but it will not use a band for both directions of transmission. Therefore there will be no interference from earth station transmitters in frequency bands in which space stations transmit and the whole of the interference budget identified in (i) above is available for interference from space station transmitters.

(iii) Various model line-of-sight radio relay systems were defined, compatible with the HRC, and operating at 2·5 and 4 GHz.

(iv) Models were defined for the FSS, which was assumed to use geostationary satellites. These models postulated that satellites would be located at geocentric intervals of 3° or 6° all round the GSO and that all of the satellites in sight of the area in which a model radio relay system was located would illuminate that area simultaneously with an emission at the same frequency as a radio channel used by the radio relay system (see Fig. 5.5).

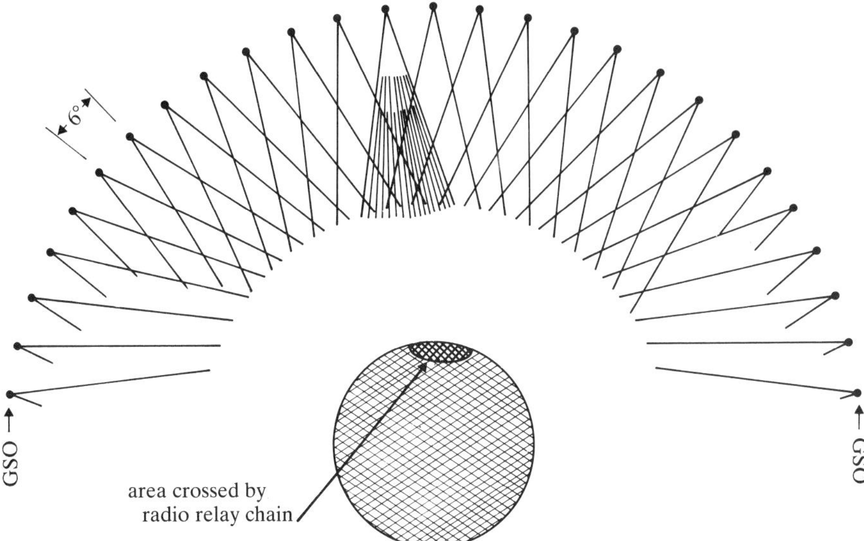

Fig. 5.5 A model of the geometry of interference to a radio relay system from an array of geostationary satellites at orbital intervals of 6° (geocentric), all interfering with the terrestrial system

This was one of the models used in the determination of sharing constraints to protect the FS

(v) Interference from space stations to terrestrial stations is more severe, for a given power flux density, if the angle of elevation of the interference on arrival at the terrestrial station is low, because the gain of the antenna of a typical terrestrial station is higher at low angles. Also the mission of a satellite is often such that the gain of its transmitting antenna may be made smaller in directions that intersect the Earth's surface at low angles of elevation. Thus, a relationship can be determined, of benefit to most FSS networks, whereby a satellite PFD of Φ_1 at an angle of elevation δ_1 is treated as the equivalent of a larger PFD Φ_2 at a higher angle of elevation δ_2. The relationship adopted is as shown in Fig. 5.6, where Φ_2 and $(\Phi_2-\Phi_1)$ depend on the frequency band, $\delta_1=5°$ and $\delta_2=25°$.

(vi) The interference level in the worst affected telephone channel in an FDM angle-modulated link with a low modulation index, typical of line-of-sight radio relay systems, is approximately proportional to the total spectral PFD of interference from all sources in the worst affected part of the first sideband of the emission, the spectral PFD being measured in a sampling bandwidth equal to the bandwidth of a telephone channel, nominally 4 kHz.

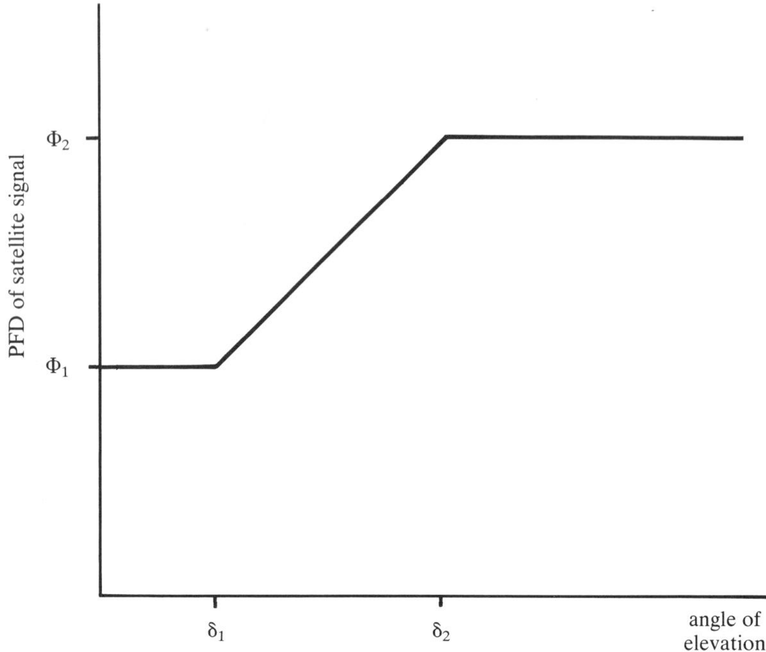

Fig. 5.6 The relationship used to take advantage of the equivalence between interference to a radio relay receiver produced by a weak space station signal (PFD = Φ_1) reaching the Earth at a low angle of elevation δ_1 and a stronger interfering signal (PFD = Φ_2) at a higher angle δ_2

(vii) Taking all these factors into account and having regard to the nature of typical satellite emission spectra, it was found that the interference from satellites into the worst affected radio relay channel would somewhat exceed the allowance in (i) above with some combinations of terrestrial system and space system models, and would be somewhat less than the allowance with other combinations of models, if the following constraints were applied to the down-path PFD of each satellite emission:

- At 2·5 GHz, if the highest interference power flux density from any one satellite, measured in a sampling bandwidth of 4 kHz, reaching the Earth's surface at an angle of elevation of zero degrees equals −152 dB(W/m^2) and $(\Phi_2 - \Phi_1)$ = +15 dB.

- At 4 GHz the corresponding values are −152 dB(W/m^2) and +10 dB.

The 1974 version of CCIR Report 387 refers[62]. Accordingly these limits were adopted for those frequency bands. Note the information in CCIR Report 792[63] on calculating the maximum power density, measured in a sampling bandwidth of 4 kHz, of the spectrum of an angle-modulated carrier.

(viii) The results in (vii) were extrapolated to frequency bands up to about 20 GHz, taking account of foreseen frequency-related differences in typical system parameters, propagation effects etc. It was assumed that most radio relay systems operating above 15 GHz would carry wide-band digital signals and allowance was made for the characteristics foreseen for such systems, including an increase in the interference sampling bandwidth to 1 MHz.

It cannot yet be said that the parameters of typical radio relay systems for operation above 20 GHz are well established, and meanwhile it would be premature for CCIR to make a technical recommendation as to suitable sharing constraints in these higher frequency bands. Nevertheless a start has been made in studying the matter; see CCIR Report 876[64].

Element (ii) above is an assumption that some frequency bands allocated for space services will be used in the space-to-Earth direction and other bands will be used in the Earth-to-space direction but none will be used in both directions. Situations in which this assumption is not valid are emerging and are discussed on pp. 131–133 and 259. At the present time such bi-directional use of bands is uncommon but it might be necessary to adjust the sharing constraints if the practice became more general in the future.

The sharing constraints in RR Article 28 have not been amended since they were adopted at a WARC in 1971. At that time there were few digital radio relay systems in service and CCITT performance objectives for digital connections had not been determined. The performance objectives which have since been determined are stringent, especially for systems which may form part of an ISDN, and the sharing constraints of RR Article 28 may not protect all such systems from unacceptable interference from satellites. Radio relay systems which include large numbers of radio relay hops with an east–west orientation are particularly susceptible to interference from geostationary satellites. There is a discussion in Annexe IV of CCIR Report 387[61] on the extent to which it may be desirable to obtain additional rejection of interference by ensuring that most hops are directed away from the azimuth bearings at which the GSO intersects the horizon at the terrestrial receiving station. CCIR Report 393[65] gives information on the intersection of radio relay antenna beams and various satellite orbits, including the GSO.

Constraints on terrestrial station emissions

CCIR Recommendation 406[66] recommends limits for the power of the emissions of radio relay stations operating in frequency bands between 1 and 30 GHz which are shared with the Earth-to-space links of networks of the FSS (see p. 117). The recommendation also urges that the sites for radio relay stations should be chosen so that the antenna beams are not avoidably directed towards the GSO. Fig. 5.7 shows schematically, for

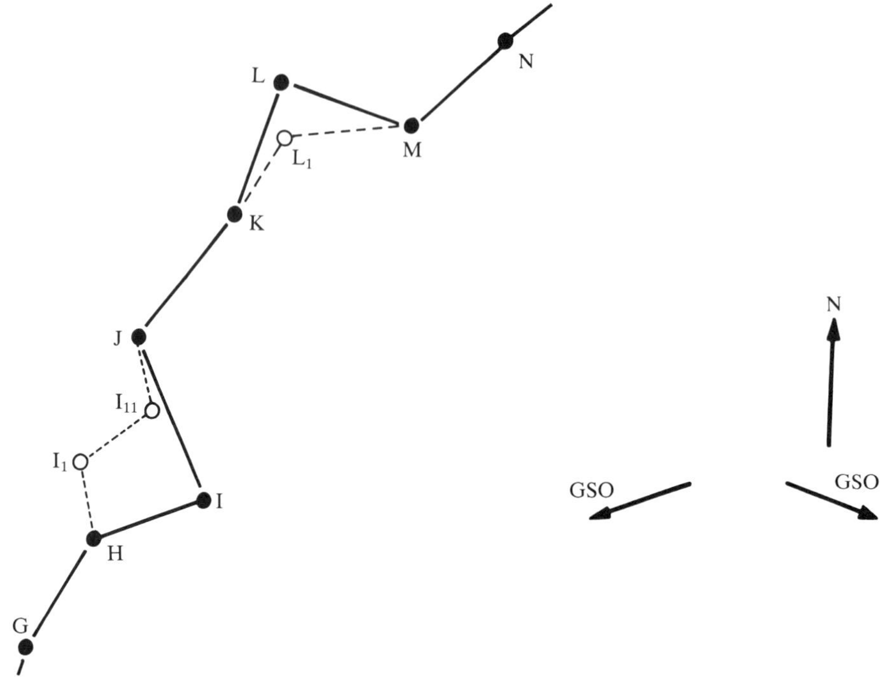

Fig. 5.7 The possible replacement of station I and L of a radio relay chain by stations at other sites in order to eliminate beams (such as I to H and L to M) which point towards the GSO

example, how the exposure of satellites in the GSO to interference from radio relay emissions (and vice versa) can sometimes be reduced by careful choice of station sites. The cost of so doing is not necessarily great when a radio relay system is initially planned. The cost penalty may be more severe when an existing route is to be extended and spare accommodation in station buildings and on masts, which may not have been located so as to avoid illuminating the GSO, is already available for the installation of additional equipment.

CCIR Recommendation 406 was formulated with the intention of giving adequate protection to channels in FSS networks which may be used for international connections. Adequate protection requires that the total interference entering the worst-affected channel in the FSS system via the space station receiver from all stations of the FS does not exceed 500 pW0p, another 500 pW0p being assumed to enter via the earth station receiver from FS stations nearby.

CCIR Recommendation 406 was the basis for the sharing constraints given in RR Article 27 for FS stations operating in frequency bands between 1 and 30 GHz which are also allocated with equal status for the Earth-to-space links of FSS networks. The same constraints are also applied in a number of other cases where terrestrial services share bands with the space-to-Earth links of space services having equal allocation status, see RR 2502–RR 2511.

The sharing constraints, applicable to all the relevant frequency bands between 1 and 30 GHz, take the following form:

(*a*) The EIRP of transmitting stations of the FS shall not exceed +55 dBW, (but see (*c*) below).
(*b*) The power input to the transmitting antenna of a station of the FS shall not exceed +13 dBW in bands below 10 GHz, nor +10 dBW in bands above 10 GHz.
(*c*) Between 1 and 10 GHz it is urged that sites for terrestrial stations should be chosen so that the transmitted beam is directed at least 2° off the direction of the GSO, unless the EIRP is not more than +35 dBW. If this is not feasible, the EIRP shall not exceed +47 dBW in any direction within 0·5° of the orbit, the limit rising linearly in decibels at a rate of 8 dB per degree to +55 dBW at an angle 1·5° away from the orbit.
(*d*) Between 10 and 15 GHz, it is urged, but not required, that sites for terrestrial stations with an EIRP above +45 dBW should be chosen so that the transmitted beam is directed at least 1.5° away from the geostationary satellite orbit.
(*e*) No guidance is offered on beam pointing direction above 15 GHz for terrestrial stations.

These constraints were arrived at as follows:

- The version of CCIR Report 209[67] which was available to WARC-63 assumed that satellites of the FSS would have medium altitude orbits. It was estimated that a radio relay emission with an EIRP of +55 dBW from a high-gain antenna would produce interference at a level of 2500 pW0p in channels of a satellite network when the satellite was passing through the middle of the beam of the terrestrial station. It was concluded that this peak interference level should be tolerable because satellites would move quickly through the beams of radio relay systems and the interference level would be much less for most of the time.
- The initial version of CCIR Recommendation 406, also available to WARC-63, took into account the prospect of large numbers of low-level interference entries at satellite receivers arising from the side-lobes of high-gain terrestrial station antennas and from terrestrial stations with low gain antennas. It recommended that the power input to the antenna of a terrestrial station should not exceed 13 dBW.

WARC-63 accepted and implemented both of these recommended constraints.

Geostationary satellites soon established their dominance in the FSS, negating the amelioration of the impact of high levels of interference due to satellite movement on which WARC-63 had relied. However, the basic transmission loss between terrestrial antennas and the GSO is significantly greater than the loss to lower orbits. Also, many radio relay stations have less EIRP than +55 dBW. Accordingly, the maximum power levels set for radio relay stations in 1963 were not changed. To improve regulatory protection to space services using geostationary satellites, specific measures were recommended in the 1967 version of CCIR Recommendation 406 to discourage the construction of new radio relay chains with transmitting antennas directed towards the GSO.

The use in the FSS of frequencies above 10 GHz was also in sight in 1970 and CCIR Report 450[68] included an estimate of the interference that a parametric model population of radio relay stations operating in a band between 10 and 15 GHz would cause to geostationary satellites carrying telephone circuits in FDM/FM systems. The conclusion reached was that interference might exceed 500 pW0p if the power input to terrestrial station antennas exceeded +8 dBW and if the EIRP of FS transmitting stations directed within 0·5° of the orbit exceeded +51 dBW, although an EIRP as high as +59 dBW would be acceptable if the radio relay antenna was directed at least 1·5° away for the GSO.

WARC-71 considered both Recommendation 406 and Report 450 and added to the Radio Regulations the elements which are summarised in sub-paragraphs (*c*) and (*d*) above. Thus, existing FS stations were permitted to be equipped for new radio channels and new frequency bands even if antenna beams would be directed at the GSO, although subject to a constraint on EIRP below 10 GHz, but it was urged that new stations should be sited so as to avoid this orientation.

Two further studies in this area have been carried out more recently. CCIR Report 790[69], using with greater sophistication the methodology of CCIR Report 450, came to two conclusions. It justified in general the absence of constraints on the pointing direction of terrestrial transmitters operating at 30 GHz (sub-paragraph (*e*) above). However, it showed that the existing constraints, while sufficient to protect digital networks of the FSS in most circumstances, may allow excessive interference to occur if a high gain satellite beam is directed towards the edge of the Earth's disc and a transmitting terrestrial station within the footprint of the satellite beam is directed towards the satellite. The other study found that the existing constraints gave sufficient protection to BSS feeder links around 18 GHz; see CCIR Report 1006[70].

5.4.5 Frequency co-ordination between FSS earth stations and FS stations

The basic free space transmission loss between an earth station and a geostationary satellite is typically about 60 dB greater than the corresponding loss for a terrestrial line-of-sight link. Some of this difference has to be made good by the use of earth station transmitters of much higher power than FS stations use for emissions occupying comparable bandwidth, and low noise receivers are used at earth stations. Consequently, earth station emissions may interfere with FS station receivers several hundreds of kilometres away operating in the same frequency band, much further than the interference range between line-of-sight links under normal propagation conditions. Earth station receivers are susceptible to interference from FS station transmitters at even greater distances.

The consequences of these long interference distances are mitigated by sharing constraints on earth stations and by the care usually taken in the selection of earth station sites. The interference levels remaining must be determined through frequency co-ordination and, if necessary, reduced.

Constraints on earth station emissions

To facilitate frequency band sharing between earth station transmitters and terrestrial station receivers, sharing constraints are used to limit the spectral power density which earth stations radiate towards the local horizon in shared bands. Another sharing constraint puts a lower limit on the angle of elevation of the centre of a transmitting earth station's beam; see RR 2540–RR 2551 and CCIR Report 386[71]. These constraints are discussed further on pp. 265–268.

Choice of earth station sites

RR 2501 and RR 2539 draw attention to the need to locate stations of both the FS and the FSS having regard to the needs of the stations of the other service. CCIR Report 385[72] provides general advice on the subject and CCIR Report 390[73] contains information on the protection from interference that can be obtained from obstacles, such as hills and buildings, in the path of an interfering signal. Locating earth stations in large pits in the ground and in disused quarries may considerably reduce the interference level. CCIR Report 709[74] considers the reduction of interference that can be obtained by the careful positioning of FS and earth station antennas which are unavoidably in close proximity to one another.

For international satellite networks, involving one or few earth stations per country, the avoidance of interference with stations of the FS can be made a prime consideration in choosing sites for earth stations and the avoidance of major interference problems may not be technically difficult. It is often possible to find suitable sites far removed from terrestrial stations, existing and planned, or well shielded from them by hills, except perhaps for the radio relay stations providing terrestrial links for the earth station itself. However, when large numbers of earth stations are required and these are to be located in or near towns, interference is likely to be serious if the spectra of the emissions of the two services overlap.

The basis of frequency co-ordination

Sharing constraints and care in choosing station sites having facilitated sharing, it is necessary to estimate how much interference the frequency assignments to a new earth station would nevertheless cause to FS stations and vice versa before a new station is built or a new frequency assignment for an existing system is put into service. This basically technical study is followed by discussions, domestic or international, to resolve interference problems if they arise. Finally, the internationally recognised right to use a frequency assignment without excessive interference is established by priority of registration in the MIFR. A somewhat simpler process is required if new FS station frequency assignments are planned within potential interference range of an established earth station.

The technical practices and the procedures which have been agreed through the ITU to implement internationally the processes set out in the previous paragraph when a new earth station is planned are complex. This complexity arises because:

- The specific frequency assignments to foreign terrestrial stations that may cause or suffer interference may not have been registered in the MIFR, use having been made, for example, of the option under RR 1223 whereby assignments are notified to the IFRB for 'typical stations within a stated area'. Registration under RR 1223 is nevertheless respected as a basis for the establishment of priority of use.
- Administrations may wish to minimise their disclosure of details of the location and characteristics of their terrestrial stations.
- The practices and procedures are burdensome for administrations and network operators; to limit the burden, they should not be applied more generally than is necessary.
- Some administrations may be unwilling or unable to play their part in carrying out the process.

In presenting these complex processes here, it is convenient to divide them into five stages. It should be noted that it may in practice be desirable for some of these stages to be carried out simultaneously. The stages are as follows:

Stage 1. The identification and resolution of potential interference problems within the jurisdiction of the administration which is responsible for the earth station.
Stage 2: The publication of information on potential cross-frontier interference to or from the earth station.
Stage 3: The declaration by foreign administrations of terrestrial stations which might cause or suffer interference. The administrations exchange detailed information on station characteristics and frequency assignments. Prospective interference levels are calculated. Interference problems are resolved. This stage is called frequency co-ordination.

Stage 4: The IFRB is notified of the co-ordinated frequency assignments to the earth station and the terrestrial stations and registers them in the MIFR.

Stage 5: Changes to assignments to any of these stations, of the FS or the FSS, which increase the risk of interference are re-co-ordinated. Assignments to new terrestrial stations capable of causing interference to the earth station must be notified to the IFRB and are subject to a technical examination for compatibility with registered assignments to the earth station.

Stage 1: National interference surveys

A site and the principal equipment parameters, emission parameters and frequency assignments are provisionally selected for the earth station. Then all of the terrestrial stations which might cause or suffer significant interference within the jurisdiction of Administration A, the administration responsible for the earth station, are identified.

A calculation is made of the interference power due to emissions from the terrestrial station transmitters that would enter the earth station receiver when it is tuned to each of the frequencies which are assigned to it. Similar calculations are made for earth station emissions entering terrestrial station receivers. The impact of these calculated interference levels on the performance of the worst-affected channel of each wanted emission is calculated and compared with the interference objectives. If the interference entry would not be compatible with the interference objectives, taking into account other foreseeable interference entries, ways must be found for reducing the interference power received so that the objectives can be met, unless justification is found for accepting interference in excess of the objective.

All of the necessary radio emission parameters, geographical data and topographical data should be available to the administration, since all of the stations involved would be operating within its territory and should be operating in accordance with its frequency assignments. The off-beam gains assumed for the earth station and terrestrial station antennas should ideally be based on measurements; in the absence of more specific data, the reference radiation patterns in CCIR Recommendation 465[75] and Report 614[76] may be used. Propagation data in CCIR Report 569[77] cover all of the important propagation modes (see also Appendix B.6.11 below) but the administration is, of course, free to use other propagation data, based perhaps on local measurements, if it is thought that these would give more reliable results. For example, local propagation studies may provide the most reliable forecast available of rainfall-rate statistics and the probability of tropospheric ducting. The interference power at the input to receivers having been calculated, CCIR Report 388[78] shows how the interference noise in the worst-affected channel can be deduced.

It is then necessary to consider what level of interference can be accepted in channels of terrestrial systems from the earth station and in channels of the satellite system from terrestrial stations.

The most stringent conditions will probably arise when channels with CCITT performance objectives are involved. Thus CCIR Recommendations 357[43] and 615[56] recommend limits for the total impact of interference on the performance of HRCs and HRDPs of FS systems due to interference from all systems of the FSS entering at any terrestrial receiver in a 2500 km chain, real or hypothetical (see pp. 108–110). In almost all circumstances may be assumed that there will be no interference from space stations of the FSS in frequency bands in which earth stations cause interference to terrestrial stations. Therefore the whole of the recommended allowance of interference (for example 1000 pW0p for up to 20% of any month in the worst-affected channel of an analogue system) is available for distribution between all of the earth stations, domestic or foreign, which may cause interference at any of the terrestrial stations, real or hypothetical, which may be considered to form part of an HRC or HRDP. Just what share of the total amount should be allotted to a specific emission from the particular earth station under co-ordination at one particular FS station will be a matter for informed judgement; the CCIR does not advise on this point.

Similarly CCIR Recommendations 356[44] and 558[79] recommend limits for the total impact of interference from FS stations on HRCs or HRDPs of the FSS. See CCIR Report 917[80] for provisional interference criteria for telephone channels of the maritime mobile-satellite service, the feeder links of which may be part of the FSS. Half of the recommended total (for example, half of 1000 pW0p, not to be exceeded for more than 20% of any month in an analogue FSS system) is allotted to interference entering the satellite system at space station receivers. The rest may be distributed between all of the interfering signals from FS transmitters, domestic and foreign, entering the worst-affected channel at the earth station under co-ordination, since, as is shown on p. 260, there is only one receiving earth station in an FSS HRC or HRDP. Once again, careful judgement will be required in deciding how much interference can be accepted from any one domestic FS station.

There are no ITU performance objectives for telephone channels which will not form part of the international public telephone network. Such channels, particularly analogue telephone channels, may not have to attain the very high objectives which the CCITT recommends and it may be considered that higher levels of interference can be accepted by such systems.

If the forecast level of interference, measured in the worst-affected channel, is too high when judged by these criteria, the administration may consider whether the criteria could be relaxed somewhat, without failure to attain circuit performance objectives, because, for example, the level of noise generated within the wanted system is less, or can be made less, than the standard impairment budget allows. In general, however, if interference is worse than the maximum that the CCIR recommendation indicates, it will be necessary to find ways of reducing it. Rearrangement of frequency assignments, adjustment of emission parameters or reduction of transmitter power levels where existing system performance margins are unnecessarily large are three possible ways of reducing interference at relatively low cost. In some situations adaptive interference cancellers can be effective; see CCIR Report 875[81]. More radical expedients, such as the reduction of the sidelobe gain of antennas, the selection of a different site for the new earth station or the resiting of a radio relay station, may however be necessary, particularly when a prospect of severe interference arises involving several stations.

The co-ordination process which may have to be conducted with other administrations (Stage 3) may identify similar needs to modify or resite stations and it is desirable for Stages 1 and 3 to proceed in parallel, in order that solutions to interference problems may be reached which serve both needs.

Use of co-ordination contours in Stage 1

The comprehensive survey of interference levels indicated above is readily feasible when only a few FS stations in the same broad geographical area as the earth station are assigned frequencies in bands which the earth station also uses. Software is available to make less laborious the substantial numbers of calculations that are involved. However, in many countries a comprehensive survey for an earth station which is to operate in frequency bands which are widely used for the FS, typically the 4 and 6 GHz bands, will involve making very large numbers of calculations, a large proportion of which will indicate negligible prospective interference. To reduce the extent of the survey work without running any risk of overlooking a significant interference prospect, a 'sieving' technique is required to eliminate most of the interference sources which will prove to be negligible. That sieve is provided by application of the concept known as the 'co-ordination distance'. The co-ordination distance is the calculable distance from a radio station beyond which interference to, or from, other radio stations sharing the same frequency band can safely be neglected for planning purposes.

Consider an FSS earth station, transmitting at 6 GHz and receiving at 4 GHz, the centre of its antenna beam being directed at azimuth 110° east of north and at an angle of elevation 10° above the horizontal plane. Let it be assumed that large numbers of radio relay stations are operating in both the 6 and the 4 GHz bands in the same broad geographical area. Co-ordination distances, transmitting and receiving, for that earth station can be calculated as follows:

(*a*) If the performance of the earth station antenna, in-beam and out-of-beam, has been measured comprehensively, a polar diagram can be drawn showing the gain relative to isotropic in the direction of the horizon for all azimuths at typical transmitting and receiving frequencies. If the characteristics of the antenna are not known in sufficient detail, the reference radiation pattern in CCIR Recommendation 465[75] can be used instead, as has been done for Fig. 5.8*a*.

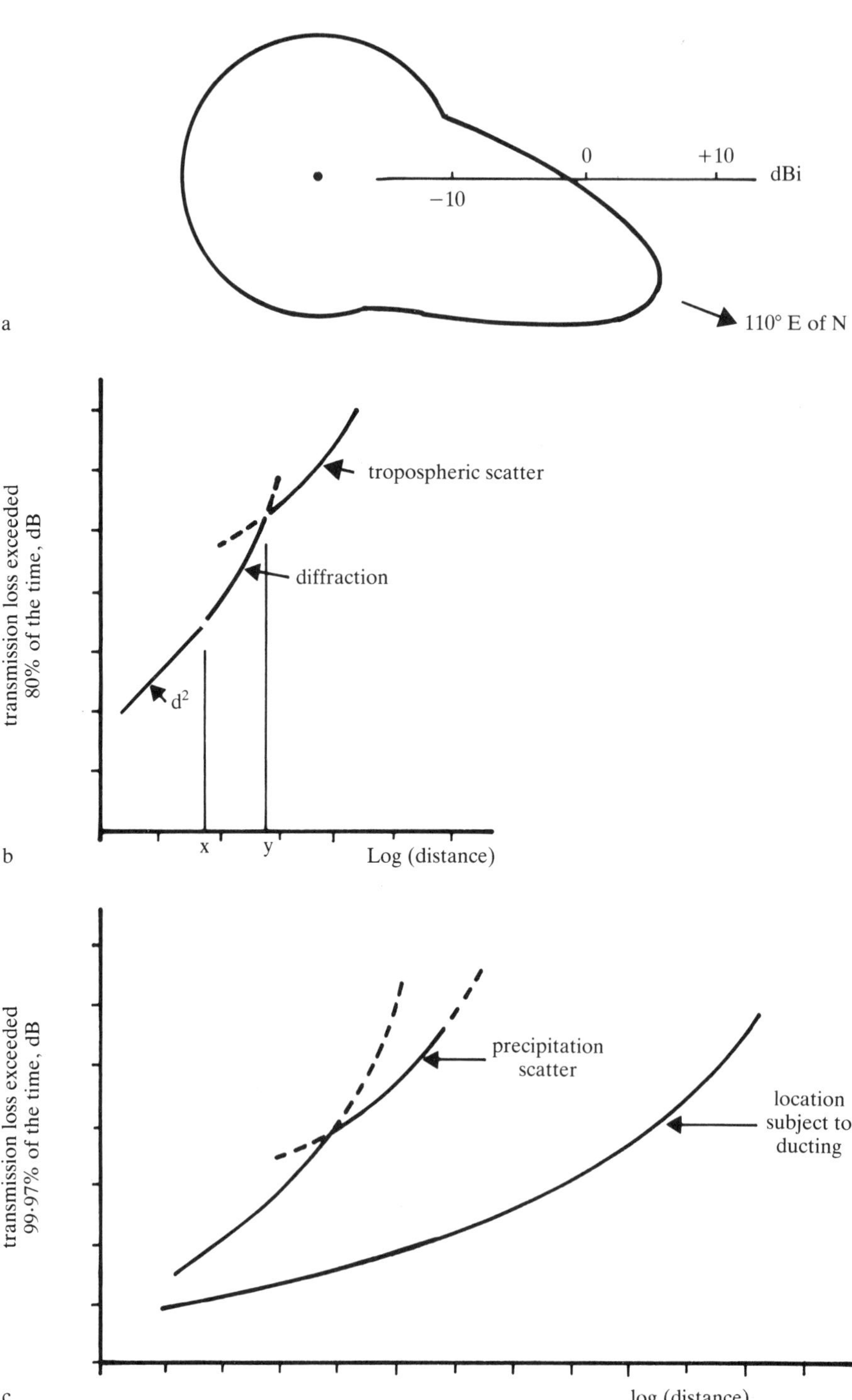
0
+10
dBi
−10
110° E of N
a
tropospheric scatter
diffraction
d^2
transmission loss exceeded 80% of the time, dB
x
y
Log (distance)
b
precipitation scatter
location subject to ducting
transmission loss exceeded 99·97% of the time, dB
log (distance)
c

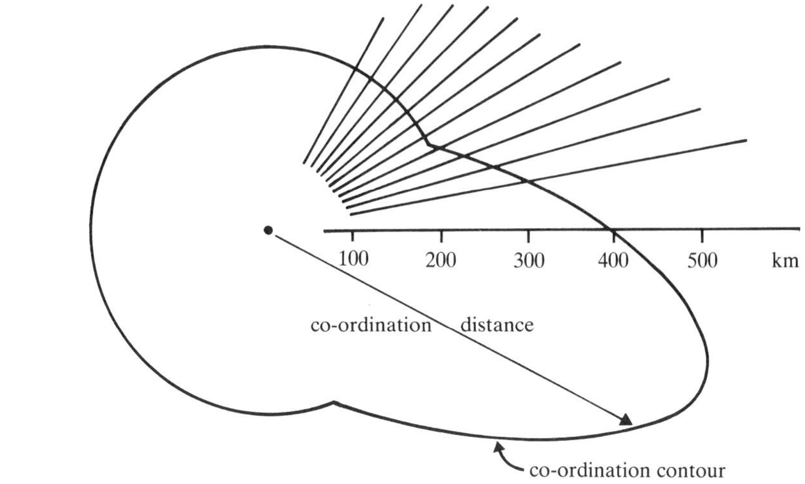

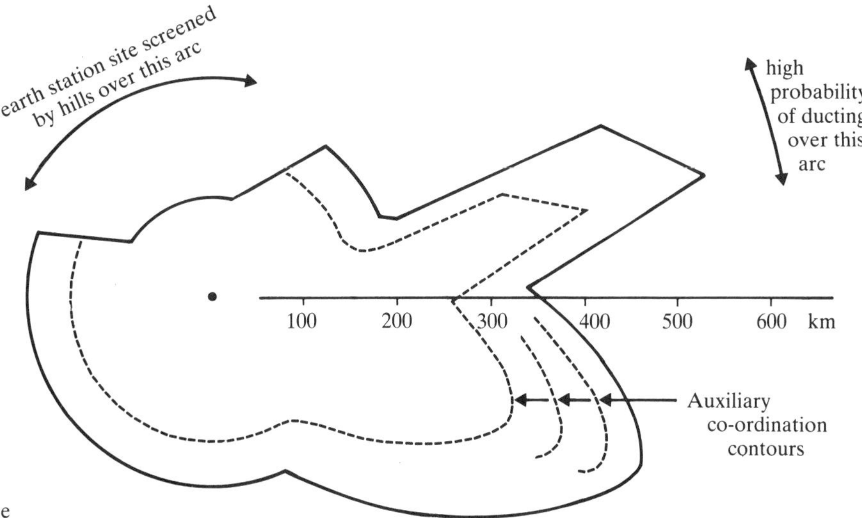

Fig. 5.8 Sketches illustrating steps in the construction of a map of the co-ordination area round an earth station

a A polar diagram of the gain of the earth station antenna in the direction of the horizon

b and *c* The increase with distance of the propagation loss occurring for at least 80% and 99·97% of any month respectively

d The co-ordination area, assuming smooth, uniform surroundings

e The co-ordination area, making allowance for site screening and the enhanced likelihood of ducting over water. Auxiliary contours are also shown (*b*) Consider a hypothetical radio relay station transmitting at 4 Ghz with its antenna 1 km distant from the earth station antenna. It is illuminating the latter with an EIRP or +55 dBW (that is, it is using the maximum power permitted by RR 2505 and RR 2509), its antenna is directed straight at the earth station antenna and there are no obstructions. The interference power in dBW, delivered at the output of the earth station antenna will be W_0, where

$$W_0 = P_{RR} + G_{RR} - L_{BF0} + G_{ES}\,\text{dBW} \qquad (5.1)$$

P_{RR} and G_{RR} = power input to the radio relay antenna (dBW) and the gain relative to isotropic of the radio relay antenna (dBi) respectively ($P_{RR} + G_{RR}$ = 55 dBW)
L_{BF0} = free space basic transmission loss (dB) between the two stations (= 104·5 dB for a path of 1 km at 4 GHz; see eq. (B.9) of Appendix B.1)
G_{ES} = gain relative to isotropic of the earth station antenna in the direction of the radio relay station (dBi)

(*c*) Let is be assumed that interference power not exceeding W_{20} and $W_{0\cdot03}$ dBW arising for up to 20% and 0·03% of any month respectively, having its source in any single FS transmitter, can be completely disregarded, but higher levels of interference need to be investigated. Appropriate relationships between W_{20} and $W_{0\cdot03}$ on the one hand and the interference levels in telephone channels recommended in CCIR Recommendations 356[44] and 558[79] on the other hand may be deduced with the help of CCIR Report 382[82].

(*d*) Let it be assumed that an actual radio relay station, also directing an EIRP of 55 dBW towards the earth station, is distant *d* kilometres from it. Fig. 5.8*b* shows schematically how the loss between the two antennas, assumed to be located above a smooth Earth surface, increases as the distance between them increases. Between 1 km and x km the loss increases from L_{BF0} at a rate of 20 dB per tenfold increase in distance; this is the conventional free-space inverse-distance-squared relationship. When *d* is between *x* and *y* the loss increases more rapidly, the transmission path being no longer optical and the interfering signal being sustained by diffraction over the spherical surface assumed for the earth. Starting at *y* and continuing for a considerable further distance, there will probably be some reduction in the rate of increase of loss with distance, owing to propagation of interference by tropospheric scatter. The loss due to these various propagation modes is approximately stable with time and will be present for at least 80% of the time, unless the interference path is particularly liable to tropospheric ducting. Data on these propagation factors are to be found in CCIR Report 569[77] (see also Appendix B.6.11 below).

(*e*) Using the curve in Fig. 5.8*b* it is possible to define the co-ordination distance, operative for 80% of any month, measured along a straight line from the earth station antenna, associated with the gain of the earth station antenna in that direction, namely G_{ES}. It is the distance at which the increase of the loss between the antennas is $(W_0 - W_{20})$ dB greater than L_{BF0}. If there were a radio relay station located less than the co-ordination distance from the earth station in the same direction, it might set up interference exceeding W_{20} for more than 20% of any month; to determine whether the permissible interference level would indeed be exceeded it would be necessary to consider all of the other relevant parameters of the emissions of the two stations. Any radio relay station more remote from the earth station could be disregarded. The locus of points, at the co-ordination distance from the earth station, measured along a series of radials as shown in Fig. 5.8*d* is a closed curve called the co-ordination contour and it encloses the co-ordination area.

(*f*) In a way similar to that of stage (*d*) above, a curve of the increase of transmission loss with increasing distance can be drawn, showing the loss in addition to L_{BF0} that is exceeded for at least 99·97% of the time (see Fig. 5.8*c*). This curve also assumes a smooth Earth surface between the two stations, but it recognises that the inverse-distance-squared and diffraction modes of attentuation will be bypassed for part of the time by tropospheric ducting or precipitation scatter, leading to a lower rate of increase of loss with distance.

(*g*) A second co-ordination contour can be plotted using the distance/loss curve of Fig. 5.8*c*. This will enclose the radio relay stations, interference from which may be significant for 0·03% of some months.

(*h*) It will be noted that propagation conditions may differ on the various azimuthal lines from the earth station. Sea areas and large lakes may substantially affect the prevalence of tropospheric ducting on some azimuths. Hills and obstructions may increase losses, despite diffraction, on other azimuths; this is called 'site shielding', Furthermore, as indicated above, the co-ordination contour defines an extreme case; radio relay stations located beyond that contour will not cause significant interference regardless of their technical characteristics, provided only that they conform to the EIRP limits of RR Article 27. By choosing co-ordination distances related to values of additional loss (Fig. 5.8*b*) which are 10 dB, 20 dB, 30 dB etc. less than the value that reduces W_0 to W_{20}, auxiliary contours may be plotted within the main contour, beyond which stations can be disregarded if their technical parameters (antenna gain in the direction of the earth station, EIRP and so on) would ensure that their interference would be 10 dB, 20 dB, 30 dB etc. less than the extreme case. A more efficient sieving process can be obtained by taking these various factors into account (see Fig. 5.8*e*).

(*i*) A process analogous to steps (*a*) to (*h*) is then carried out to identify co-ordination contours beyond which interference from the earth station emissions to radio relay stations operating at 6 GHz will not exceed the level which the latter could accept under worst-case conditions.

A method for defining co-ordination contours, using standardised system parameters and standardised propagation data and including detailed instructions for its use, is to be found in RR Appendix 28.

The various co-ordination contours having been defined, radio relay stations located outside the contours can be disregarded as possible sources or victims of interference involving the earth station. The calculation of actual interference levels can be concentrated on stations within the contours.

Stage 2: Publication of information

Administration A, the administration having jurisdiction over the new earth station, must also set in motion an examination of the prospect of significant interference between the new earth station and FS stations which are under the jurisdiction of other administrations. It is convenient to identify these other administrations, responsible for neighbouring countries, as Administration B1, Administration B2 etc.

The examination of prospective interference levels involving FS stations in foreign countries involves problems which do not arise in Stage 1. In Stage 2, Administration A has all the necessary information on the earth station and the extent of the site shielding that obstructions, such as hills and buildings in the vicinity of the earth station, will provide. However Administration A cannot know with certainty the location of foreign terrestrial stations, the frequencies assigned to them, the gain of their antennas in the direction of the earth station, any site shielding in the vicinity of the radio relay station, the susceptibility to interference of the emissions which they receive or the level of interference which a B administration will accept from the earth station. The co-ordination contour technique, suggested for use in Stage 1 if large numbers of radio relay stations must be considered and set out in RR Appendix 28, was developed to solve the problem that this lack of knowledge of foreign stations creates.

However, Stages 2 and 3 are not primarily technical. These stages have important regulatory purposes. They provide Administration A with the opportunity to establish the right of an earth station to use the FSS allocations with equal status with stations of the FS, once co-ordination has been successfully completed. They also provide the B administrations with the opportunity to assert the priority of the frequency assignments they have made to stations of the FS, even when these have not been previously registered specifically for the station in question, if the new FSS assignments are not compatible with the old FS assignments. In view of these important regulatory aspects, it is necessary for Stage 2 to be carried out precisely in accordance with rules and criteria which have been agreed in the ITU and are set out in the RR.

The regulatory aspects of Stages 2 and 3 are set out in Section III of RR Article 11, the key paragraphs being RR 1107 and RR 1113. When Administration A is ready to initiate the process, it prepares two sets of documents:

- A set of maps of the co-ordination areas of the earth station. These maps must be drawn up strictly in accordance with RR Appendix 28.
- A set of notifications of the frequency assignments which it proposes to make to the earth station, containing the information called for in RR Appendix 3. RR Appendix 3 was re-edited by WARC-ORB-88[83].

RR Appendix 28 contains detailed instructions and technical data for drawing maps of the co-ordination area around an earth station, showing the area beyond which it may be assumed that interference from a terrestrial station to the earth station would be negligible and the area beyond which it may be assumed that interference from the earth station to a terrestrial station would be negligible. The technical data which relate to the emission parameters of the space service and the terrestrial service are drawn, in particular, from the FS and the FSS, although the same procedure is also used when stations of other pairs of services are co-ordinated.

If no part of any of the co-ordination areas lies within the territory of another administration, no further action under stages 2 and 3 is required and Administration A is free to proceed with the notification of assignments to the IFRB, which is part of Stage 4. If the co-ordination area encloses any foreign territory, the responsibility of administration B1, B2 or B3 etc., then Administration A must send a copy of the maps and the details of the proposed assignments to each of the B authorities concerned, requesting co-ordination. This publication of information to the interested administrations prepares the way for Stage 3. The same material is also sent to the IFRB.

Complex technical problems lie behind the methods and the data contained in RR Appendix 28 and some of these problems are not yet completely solved. It has been necessary to use pessimistic approximations in the propagation data which are used for relating transmission loss to distance in order to make sure that no potential interference risk is overlooked. This, coupled with the pessimistic assumptions about the characteristics of the terrestrial stations, ensures that many of the terrestrial stations which are identified in Stage 3 as possible sources of or sufferers from interference are found after further study to have negligible interference potential; indeed some B administrations may be drawn into the co-ordination process although none of their terrestrial stations prove, after detailed study, to raise interference problems.

The CCIR continues to study how the methods and data given in RR Appendix 28 could be made more precise, discriminating with greater certainty between interference situations that will and will not prove to be significant, responding to the specific request in RR Resolution 60[84]. The results of this work are being accumulated in CCIR Report 382[82], and this report can be seen as an interim draft for the revision of Appendix 28 by some future WARC. CCIR Recommendation 359[85] withholds formal endorsement of the material in CCIR Report 382 as an alternative to RR Appendix 28 at the present stage, pending further progress in the studies. However, it is recognised that a time will come when a Plenary Assembly of the CCIR will decide that the then-current version of Report 382 is technically satisfactory and is such a significant improvement on the existing text in RR Appendix 28 that it should recommend that a competent WARC should revise the Appendix, taking this material into account. When that recommendation is made, RR Resolution 703[86] provides a mechanism whereby administrations that wish to bring the new material into use immediately where their own interests are concerned without waiting for a competent WARC to be convened will be able to do so.

Stage 3: International co-ordination

Administration B1, B2, B3 etc. having received co-ordination area maps and details of proposed frequency assignments from Administration A, will make a preliminary examination of the probability that their terrestrial stations within the earth station transmitting co-ordination area will suffer significant interference. Likewise they will consider the probability that their terrestrial stations within the earth station receiving co-ordination area will cause significant interference to the earth station. CCIR Report 448[87] gives advice on this process. Many of the terrestrial stations will probably be found to have negligible interference potential because, for example, their antenna beams are directed away from the earth station, there is good site shielding in the relevant direction or the EIRP of the radio relay station is less than 55 dBW. Some stations may however seem to have some potential for causing or suffering interference, in which case detailed study will be needed. The B administration concerned will give Administration A all relevant information on these selected terrestrial stations (RR 1116–RR 1128 refer).

If necessary Administration A will then meet with each of the B administrations to calculate accurately what the interference levels would be. At this stage administrations are free to use whatever propagation data and whatever methods for calculating channel interference levels from interfering carrier levels that they agree, between themselves, to use. If significant interference is foreseen, the same processes, the same criteria and the same kinds of solution as were used in Stage 1 to deal with national interference will be applied to international interference (see pp. 121–123). The RR provide a regulatory time-frame for the execution of the co-ordination procedure and the IFRB has authority to assist in the achievement of that time-frame; see RR 1121–RR 1144. However, the ITU offers no technical guidance for the resolution of the problems arising in co-ordination.

When all of the potential interference problems have been solved, co-ordination is complete and Stage 4 can begin.

Stage 4: Notification and registration of assignments

Administration A notifies the IFRB of the earth station frequency assignments, giving the information listed in RR Appendix 3[83], if no foreign territory falls within the co-ordination contours; a copy of the co-ordination maps is also supplied. If co-ordination with other administrations has been necessary, Administration A reports the successful outcome to the IFRB and notifies its earth station frequency assignments, modified in whatever ways have been agreed in the course of the co-ordination process. The IFRB's examination of these notifications is integrated with the corresponding process arising from co-ordination between satellite networks (see p. 293).

The B authorities also notify the IFRB of the frequency assignments to their terrestrial stations which have been co-ordinated with Authority A; see RR 1127. These assignments are registered in the MIFR if they are otherwise in order, and are not registered already, after scrutiny by the IFRB.

Stage 5: New and amended frequency assignments in the co-ordination area

The process of co-ordination has to be repeated if Administration A wishes to change the assignments to the earth station in a way that could aggravate interference to foreign terrestrial stations; see RR 1107 and RR 1110.

Conversely, co-ordination is required with Administration A if one of the B administrations wishes to change assignments to one of the terrestrial stations that had been co-ordinated in a way that aggravates interference to the earth station. Additionally, if one of the B administrations wishes to make an assignment to a terrestrial station, within the published co-ordination area of an earth station, which has not been co-ordinated with the earth station, the assignment must be notified to the IFRB; the assignment will be subjected to a technical assessment by the IFRB to ascertain whether co-ordination is necessary, before it is registered in the MIFR; see RR 1147–RR 1167 and RR 1351–RR 1385.

5.4.6 Sharing frequency bands with space radio services operating in both directions of transmission

At present, with the few exceptions indicated below, the frequency bands allocated for space radio services are allocated and used for those services for up-links or down-links but not for both directions of transmission. Thus, for example, most FSS networks use the 6, 8 or 14 GHz bands for Earth-to-space links and the 4, 7 or 11/12 GHz bands respectively for space-to-Earth links.

There is no fundamental reason why properly co-ordinated space radio networks should not use frequency bands in either direction, some networks using band X for up-links and band Y for down-links while other networks use band Y for up-links and band X for down-links. Nevertheless, when this is done in bands shared with terrestrial services, one of the basic assumptions on which the sharing criteria have been constructed is no longer valid.

The international Table of Frequency Allocations (RR Article 8) makes some provision for space service allocations to be used in both directions already. The following allocations are all primary:

2655–2690 MHz: in Regions 2 and 3 the FSS has an Earth-to-space allocation and the band is also allocated to the BSS, which operates only in the space-to-Earth direction. In Region 2 only there is also an FSS (space-to-Earth) allocation in this band.
8025–8400 MHz (Region 2 only) is allocated to the FSS (Earth-to-space) and the EES (space-to-Earth). The band 8175–8215 MHz is also allocated to the MetS (Earth-to-space).
10·7–11·7 GHz (Region 1 only): The FSS has allocations for transmission in either direction, the use of the upward direction being limited by RR 835 to feeder links for the BSS.
12·5–12·75 GHz (Region 1 only): The FSS has allocations for transmission in either direction.
17·7–18·1 GHz (all Regions): The FSS has allocations for transmission in either direction, the use of the upward direction being limited by RR 869 to feeder links for the BSS.

In each of these bands there are also primary allocations for the FS world-wide, Region-wide or through a footnote for a substantial sub-Regional area.

In some of these frequency bands, where interference with satellite broadcasting systems operating in accordance with 12 GHz frequency assignment plans may arise, special regulatory provisions apply (see pp. 133–134). In all of these bands, terrestrial station receivers are exposed to interference from additional sources; they may suffer interference from space station transmitters with down-links in the band (to which the standard down-path PFD limit, discussed on p. 111, applies) and also from earth station transmitters in their vicinity.

At the present time the use by space services of frequency bands for both up-links and down-links is virtually limited to operations of the FSS and EES at 8 GHz. Most EES satellites, operating in low orbits, would pass rapidly through the beam of an FS receiving antenna and the interference would have little effect on the long term performance of the channels in the terrestrial system.

Thus, the incidence of high levels of interference due to this cause has not arisen yet in a significant way, However, the bi-directional use of bands by space services is likely to develop; in particular, frequency assignment plans in the band 17·7–18·1 GHz have been drawn up for feeder links to broadcasting satellites and the use of the same band for down-links of FSS networks may develop soon. Consideration has also been given to the use of pairs of frequency bands in alternate directions of transmission by different networks of the FSS; see CCIR Report 557[88] and p. 259. Studies are in progress to find ways of dealing with the interference problems that these new modes of operation would create; see CCIR Report 1005[89].

5.4.7 Sharing with the BSS and its feeder links

At 12 GHz there are primary BSS allocations as follows:

Region 1; 11·7–12.5 GHz
Region 2; 12·2–12.7 GHz
Region 3; 11·7–12.2 GHz

Frequency assignment plans have been drawn up for the BSS in these bands, by WARC-77 for Regions 1 and 3 and by RARC-SAT-83 for Region 2. Frequency assignment plans for the associated feeder links have also been drawn up, in the band 17·3–17·8 GHz by RARC-SAT-83 for Region 2 and in the bands 17·3–18·1 GHz and 14·5–14·8 GHz by WARC-ORB-88 for Regions 1 and 3. For information on these plans, see Chapter 9.

The FS has primary allocations also in all of these bands except 17·3–17·7 GHz but various measures have been agreed to ensure that FS transmitters do not cause harmful interference to either a planned feeder link or a planned BSS down-link, even if the FS assignment is taken into use before the satellite broadcasting assignments; see below.

Other frequency bands have been allocated for the BSS and its feeder links, some bands being shared with the FS. Preparatory provision has been made, in some of these bands, to protect the BSS against interference from the FS at some future time. However, in no case are these measures in effect at present.

The FS in the BSS planned bands

RR 838 and RR 844[90] forbid stations of the FS to cause harmful interference to BSS systems operating in accordance with agreed frequency assignment plans. This advantage to the BSS is regardless of other considerations, such as priority of bringing the assignment into use; it is implemented by amendments to RR Articles 11, 12, 13 and 15 and RR Appendix 30 that were agreed at WARC-ORB-85 (see the Final Acts of that Conference[90]).

If an administration wishes to have registered in the MIFR a frequency assignment to an FS station in one of the 12 GHz bands which have been planned for the BSS, then RR A.12.1[90] (a footnote to the title of RR Article 12) and RR 1656[90] require the provisions of RR Appendix 30[90] to be applied. Article 6 of RR Appendix 30[90] requires a co-ordination procedure to be carried out, before any other action is taken to secure the registration of the FS assignment, if the necessary bandwidth of the FS emission overlaps the necessary bandwidth of a planned BSS assignment and the PFD of the former exceeds, within the BSS service area, a threshold value, based on worst case assumptions, which indicates the possibility of harmful inteference. The threshold value is determined as in Annexe 3 of RR Appendix 30[90]. The purpose of this co-ordination is to determine whether harmful interference would indeed occur, if and when the BSS assignment is taken into use; the notification will not be progressed to registration if the appropriate protection ratio is not provided, unless the proposing administration undertakes to discontinue operation on the frequency if interference occurs, in accordance with RR 342; see also CCIR Report 631[91].

In general, if a BSS emission interferes with an FS emission, there is no regulatory remedy provided that the BSS satellite emission is operating within the planned parameters. However, the terms of RR Appendix 30[90] are not entirely one-sided. It has been accepted that the protection of the FS in Regions 1 and 3 from interference from BSS emissions from satellites serving a Region 2 country may not have been given sufficient consideration in the preparation of the Region 2 BSS plan. To provide the FS with some additional protection, Article 9 and Annexe 1 of RR Appendix 30[90] set a sharing constraint on off-beam radiation from Region 2 BSS satellites.

The FS in planned BSS feeder link bands

The agreement of frequency assignment plans for satellite broadcasting at 12 GHz at conferences in 1977 and 1983 has given high status to BSS assignment which are in accordance with the plan. Much of that high status is inevitably shared by assignments to the feeder links of those broadcasting satellites, which are also the subject of agreed plans. Feeder link assignments are protected from interference by procedures similar to those protecting planned BSS down-links through RR A 12.4[83], RR 1668[83] and Article 6 of RR Appendix 30A[83]. As before, the provisions of the plan reflect the intention that assignments to the FS in the feeder link plan bands shall not frustrate the implementation of the satellite broadcasting plan, even if that implementation does not occur for many years.

Nevertheless it has been recognised that long delay in deciding where a feeder link earth station should be built might deny a neighbouring country the opportunity to make use of a considerable amount of spectrum over a large geographical area. Article 6 of RR Appendix 30A[83] established a simple form of co-ordination procedure to limit this potential impediment.

5.5 References

1 Final Acts of the World Administrative Radio Conference for the Mobile Services (MOB-87), Geneva 1987 (ITU, Geneva, 1988).
2 'Mutual protection of radio services operating in the band 70–130 kHz'. Resolution 705; Final Acts of WARC-MOB-87; see [1] above
3 'Operation of the fixed and maritime mobile services in the band 90–110 kHz'. Resolution 706; Final Acts of WARC-MOB-87; see [1] above
4 'Relating to implementation of the changes in allocations in the bands between 4000 kHz and 27 500 kHz'. Resolution 8; Radio Regulations, edition of 1982 (ITU, Geneva, 1982)
5 'Operation of HFBC transmitters in the extended bands above 10 MHz'. Resolution 512; Final Acts of WARC HFBC-87 (ITU, Geneva, 1987)
6 'Arrangement of channels in multi-channel single-sideband and independent-sideband transmitters for long-range circuits operating at frequencies below about 30 MHz'. CCIR Recommendation 348–2; Recommendations and Reports of the CCIR, 1986, Volume III (ITU, Geneva, 1986)
7 'Arrangement of voice-frequency telegraph channels working at a modulation rate of about 100 bauds over HF radio circuits'. CCIR Recommendation 436–2; *ibid.*, Volume 111
8 'Relating to the revision of entries in the Master International Frequency Register in the bands allocated to the fixed service between 3000 kHz and 27 500 kHz'. Resolution 9; Radio Regulations, edition of 1982 (ITU, Geneva, 1982)
9 'Criteria to be used in differentiating between classes of operation'. CCIR Report 860; Recommendations and Reports of the CCIR, 1986, Volume III (ITU, Geneva, 1986)
10 Relating to improvements in assistance to developing countries in securing access to the HF bands for their fixed services and in ensuring protection of their assignments from harmful interference'. Resolution 103; Radio Regulations, edition of 1982 (ITU, Geneva, 1982)
11 'Signal-to-interference protection ratios'. CCIR Recommendation 240–4; Recommendations and Reports of the CCIR, 1986, Volume III (ITU, Geneva, 1986)
12 'Bandwidths, signal-to-noise ratios and fading allowances in complete systems'. CCIR Recommendation 339–6; *ibid.*, Volume III
13 'Measurement of protection ratios for J3E emissions'. CCIR Report 989; *ibid.*, Volume III
14 'Conversion factors for deriving the protection ratios for J3E, R3E, H3E, A3E and B8E emissions'. CCIR Report 990; *ibid.*, Volume III
15 'Measurement of protection ratios and minimum necessary frequency separation for class of emission J7B'. CCIR Report 991; *ibid.*, Volume III
16 'Frequency stability required for systems operating in the HF fixed service to make the use of automatic frequency control superfluous'. CCIR Recommendation 349–9; *ibid.*, Volume III
17 'Use of directional antennas in the bands 4 to 28 MHz'. CCIR Report 356–2; *ibid.*, Volume III
18 "Use of directional antennas in the bands 4 to 28 MHz'. CCIR Recommendation 162–2; *ibid.*, Volume III
19 'Operational ionospheric-sounding systems at oblique incidence'. CCIR Report 357–1; *ibid.*, Volume III
20 'Automatically controlled HF radio systems'. CCIR Report 551–1; *ibid.*, Volume III
21 'The use of ionospheric channel sounding systems operating in the fixed service at frequencies below about 30 MHz'; CCIR Recommendation 613; *ibid.*, Volume III
22 'Improved transmission system for HF radiotelephone circuits'. CCIR Recommendation 455–1; *ibid.*, Volume III

23 "Improved transmission systems for use over HF radiotelephone circuits'. CCIR Report 345–5; *ibid.*, Volume III
24 'Diversity reception'. CCIR Report 327–3; *ibid.*, Volume III
25 'Automatic error-correcting system for telegraph signals transmitted over radio circuits'. CCIR Recommendation 342–2; *ibid.*, Volume III
26 'Compression of the radiotelephone signal spectrum in the HF bands'. CCIR Report 176–5; *ibid.*, Volume III
27 'Compression of the radiotelegraph signal spectrum in the HF bands'. CCIR Report 177–1; *ibid.*, Volume III
28 'Rejection of interference by using frequency-band compression technique'. CCIR Report 862; *ibid.*, Volume III
29 'Relating to the convening of a Regional Administrative Radio Conference to establish criteria for the shared use of the VHF and UHF bands allocated to fixed, broadcasting and mobile services in Region 3'. Resolution 702; Radio Regulations, edition of 1982 (ITU, Geneva, 1982)
30 'A concept of co-ordination and protection contours for use in the co-ordination of mobile earth stations'. CCIR Report 773; Recommendations and Reports of the CCIR, 1986, Volume VIII–3
31 'Limitation of interference from trans-horizon radio-relay systems'. CCIR Recommendation 302–1; *ibid*, Vol. IX–1 (ITU, Geneva, 1986)
32 'Relating to preferred frequency bands for systems using tropospheric scatter'. RR Recommendation 100; Radio Regulations, edition of 1982 (ITU, Geneva, 1982)
33 'Radio-frequency channel arrangement for radio-relay systems for television and telephony for 600 to 1800 telephone channels or the equivalent, operating in the 2 and 4 GHz bands or for medium-capacity digital radio-relay systems operating in the 4 GHz band'. CCIR Recommendation 382–4; Recommendations and Reports of the CCIR, 1986, Volume IX–1 (ITU, Geneva, 1986)
34 'Hypothetical reference circuits for analogue systems'. CCITT Recommendation G. 212; CCITT Blue Book, Fascicle III.2 (ITU, Geneva, 1989)
35 'Noise objectives for design of carrier-transmission systems of 2500 km'. CCITT Recommendation G. 222; *ibid.*
36 'General characteristics recommended for systems on symmetric pair cables'. CCITT Recommendation G. 322; *ibid.*
37 'Definitions of terms and references concerning Hypothetical Reference Circuits and Hypothetical Reference Digital Paths for radio relay systems'. CCIR Recommendation 390–4; Recommendations and Reports of the CCIR, 1986, Volume IX–1 (ITU, Geneva, 1986)
38 'Hypothetical reference circuit for radio-relay systems for telephony using frequency-division multiplex with a capacity of more than 60 telephone channels'. CCIR Recommendation 392; *ibid.*, Volume IX–1
39 'Allowable noise power in the hypothetical reference circuit for radio-relay systems for telephony using frequency-division multiplex'. CCIR Recommendation 393–4; *ibid.*, Volume IX–1
40 'Noise in the radio portion of circuits to be established over real radio-relay links for FDM telephony'. CCIR Recommendation 395–2; *ibid.*, Volume IX–1
41 'Noise in circuits forming part of very long telephone connections set up over radio-relay systems for telephony using frequency-division multiplex'. CCIR Report 288–2; *ibid.*, Volume IX–1
42 'Allowable noise power in the hypothetical reference circuit for frequency-division multiplex telephony in the fixed-satellite service'. CCIR Recommendation 353–5; *ibid.*, Volume IV–1
43 'Maximum allowable values of interference in a telephone channel of an analogue angle-modulated radio-relay system sharing the same frequency bands as systems in the fixed-satellite service'. CCIR Recommendation 357–3; *ibid.*, Volume IV/IX–2
44 'Maximum allowable values of interference from line-of-sight radio-relay systems in a telephone channel of a system in the fixed-satellite service

employing frequency modulation when the same frequency bands are shared by both systems'. CCIR Recommendation 356–4; *ibid.*, Volume IV/IX–2

45 'Hypothetical reference circuit for radio-relay systems for telephony using frequency-division multiplex with a capacity of 12 to 60 telephone channels'. CCIR Recommendation 391; *ibid.*, Volume IX–1

46 'Transmission performance of television circuits designed for use in international connections'. CCIR Recommendation 567–2; *ibid.*, Volume XII

47 'Permissible noise in the hypothetical reference circuit of radio-relay systems for television'. CCIR Recommendation 555; *ibid.*, Volume IX–1

48 'Noise objectives for sound-programme circuits 2500 km long provided by means of analogue radio-relay systems'. CCIR Report 375–3; *ibid.*, Volume IX–1

49 'Error performance of an international digital connection forming part of an integrated services digital network'. CCITT Recommendation G.821; CCITT Red Book, Fascicle III.3 (ITU, Geneva, 1985)

50 'Hypothetical reference digital path for radio-relay systems which may form part of an integrated services digital network with a capacity above the second hierarchical level'. CCIR Recommendation 556–1; Recommendations and Reports of the CCIR, 1986, Volume IX–1 (ITU, Geneva, 1986)

51 'Allowable bit error ratios at the output of the hypothetical reference digital path for radio-relay systems which may form part of an integrated services digital network'. CCIR Recommendation 594–1; *ibid.*, Volume IX–1

52 'Error performance objectives for real digital radio-relay links forming part of a high grade circuit within an integrated services digital network'. CCIR Recommendation 634; *ibid.*, Volume IX–1

53 'Error performance and availability objectives for digital radio-relay systems used in the 'medium grade' portion of an ISDN connection'. CCIR Report 1052; *ibid.*, Volume IX–1

54 'Error performance and availability objectives for digital radio-relay systems used in the local grade portion of an ISDN connection'. CCIR Report 1053; *ibid.*, Volume IX–1

55 'Derivation of interference criteria for digital systems in the fixed-satellite service sharing bands with terrestrial systems'. CCIR Report 793–1; *ibid.*, Volume IV/IX–2

56 'Maximum allowable values of interference from the fixed-satellite service into terrestrial radio-relay systems which may form part of an ISDN and share the same frequency band below 15 GHz'. CCIR Recommendation 615; *ibid.*, Volume IV/IX–2

57 'Performance objectives for digital radio-relay systems'. CCIR Report 930–1; *ibid.*, Volume IX–1

58 'Frequency sharing between systems in the fixed-satellite service and terrestrial services'. CCIR Report 209–5; Volume IV/IX–2

59 'Frequency sharing between systems in the fixed-satellite service and terrestrial radio services in the same frequency bands'. CCIR Recommendation 355–3; *ibid.*, Volume IV/IX–2

60 'Maximum permissible values of power flux-density at the surface of the Earth produced by satellites of the fixed-satellite service using the same frequency bands above 1 GHz as line-of-sight radio-relay systems'. CCIR Recommendation 358–3; *ibid.*, Volume IV/IX–2

61 'Protection of terrestrial line-of-sight radio-relay systems against interference due to emissions from space stations in the fixed-satellite service in shared frequency bands between 1 and 23 GHz'. CCIR Report 387–5, *ibid.*, Volume IV/IX–2

62 'Protection of terrestrial line-of-sight radio-relay systems against interference due to emissions from space stations in the fixed-satellite service in shared frequency bands between 1 and 23 GHz'. CCIR Report 387–2; Texts of the CCIR XIIIth Plenary Assembly, Geneva 1974 (ITU, Geneva, 1975)

63 'Calculation of the maximum power density averaged over 4 kHz of an angle-modulated carrier'. CCIR Report 792–2; Recommendations and Reports of the

CCIR, 1986, Volume IV/IX–2 (ITU, Geneva, 1986)
64 'Frequency sharing between systems in the fixed-satellite service and the fixed service in frequency bands above 40 GHz'. CCIR Report 876; *ibid.*, Volume IV/IX–2
65 'Intersections of radio-relay antenna beams with orbits used by space stations in the fixed-satellite service'. CCIR Report 393–3; *ibid.*, Volume IV/IX–2
66 'Maximum equivalent isotropically radiated power of line-of-sight radio-relay system transmitters operating in the frequency bands shared with the fixed-satellite service'. CCIR Recommendation 406–5; *ibid.*, Volume IV/IX–2
67 'Communication-satellite systems – Frequency sharing between communication-satellite systems and terrestrial services'. CCIR Report 209; Documents of the Xth Plenary Assembly, Geneva 1963, Volume IV (CCIR, ITU, Geneva, 1963)
68 'Frequency sharing between line-of-sight radio-relay systems and communication-satellite systems at frequencies above 10 GHz'. CCIR Report 450; CCIR XIIth Plenary Assembly, New Delhi, 1970, Volume IV Part 1 (ITU, Geneva, 1970)
69 'EIRP and power limits for terrestrial radio-relay transmitters sharing with digital satellite systems in bands between 11 and 14 GHz and around 30 GHz'. CCIR Report 790–1; Recommendations and Reports of the CCIR, 1986, Volume IV/IX–2 (ITU, Geneva, 1986)
70 'Fixed service EIRP limits for the protection of the broadcasting satellite feeder links around 18 GHz'. CCIR Report 1006; *ibid.*, Volume IV/IX–2
71 'Determination of the power in any 4 kHz band radiated towards the horizon by earth stations of the fixed-satellite service sharing frequency bands below 15 GHz with the terrestrial services'. CCIR Report 386–3; *ibid.*, Volume IV/IX–2
72 'Feasibility of frequency sharing between systems in the fixed-satellite service and terrestrial radio services – Criteria for the selection of sites for earth stations in the fixed-satellite service'. CCIR Report 385–1; *ibid.*, Volume IV–1
73 'Earth-station antennas for the fixed-satellite service'. CCIR Report 390–5; *ibid.*, Volume IV–1
74 'Consideration of the coupling between an earth-station antenna and a terrestrial link antenna'. CCIR Report 709–1; *ibid.*, Volume IV/IX–2
75 'Reference earth-station radiation pattern for use in co-ordination and interference assessment in the frequency range from 2 to about 30 GHz'. CCIR Recommendation 465–2; *ibid.*, Volume IV–1
76 'Reference radiation patterns for radio-relay system antennas'. CCIR Report 614–2; *ibid.*, Volume IX–1
77 'The evaluation of propagation factors in interference problems between stations on the surface of the Earth at frequencies above about 0·5 GHz'. CCIR Report 569–3; *ibid.*, Volume V
78 'Methods for determining interference in terrestrial radio relay systems and systems in the fixed-satellite service'. CCIR Report 388–5; *ibid.*, Volume IV/IX–2
79 'Maximum allowable values of interference from terrestrial radio links to systems in the fixed-satellite service employing 8-bit PCM encoded telephony and sharing the same frequency bands'. CCIR Recommendation 558–2; *ibid.*, Volume IV/IX–2
80 'Permissible levels of interference into telephone channels in the maritime mobile-satellite service'. CCIR Report 917–1; *ibid.*, Volume VIII–3
81 'A survey of interference cancellers for application in the fixed-satellite service'. CCIR Report 875; *ibid.*, Volume IV–1
82 'Determination of co-ordination area'. CCIR Report 382–5; *ibid.*, Volume IV/IX–2
83 Final Acts of the World Administrative Radio Conference on the use of the geostationary satellite orbit and the planning of space services utilizing it, Second Session, Geneva 1988 (WARC-ORB-88) (ITU, Geneva, 1988)
84 'Relating to information on the propagation of radio waves used in the determination of the co-ordination area'. Resolution 60, Radio Regulations,

edition of 1982 (ITU, Geneva, 1982)

85 'Determination of the co-ordination area of earth stations in the fixed-satellite service using the same frequency bands as the systems in the fixed terrestrial service'. CCIR Recommendation 359–5; Recommendations and Reports of the CCIR, 1986, Volume IV/IX–2 (ITU, Geneva, 1986)

86 'Relating to the calculation methods and interference criteria recommended by the CCIR for sharing frequency bands between space radiocommunication and terrestrial radiocommunication services or between space radiocommunication services'. Resolution 703, Radio Regulations, edition of 1982 (ITU, Geneva, 1982)

87 'Determination of the interference potential between earth stations and terrestrial stations'. CCIR Report 448–4; Recommendations and Reports of the CCIR, 1986, Volume IV/IX–2 (ITU, Geneva, 1986)

88 'The use of frequency bands allocated to the fixed-satellite service for both the up link and down link of geostationary-satellite systems'. CCIR Report 557–2; Recommendations and Reports of the CCIR, 1986, Volume IV–1 (ITU, Geneva, 1986)

89 'Frequency sharing between systems of the fixed service and systems of the fixed-satellite service comprising forward band working (FBW) networks and reverse band working (RBW) networks'. CCIR Report 1005; *ibid.,* Volume IV/IX–2

90 Final Acts of the World Administrative Radio Conference on the use of the geostationary satellite orbit and the planning of space services utilizing it, First Session, Geneva 1985 (WARC-ORB-85) (ITU, Geneva, 1985)

91 'Frequency sharing between the broadcasting-satellite service (sound and television) and terrestrial services'. CCIR Report 631–3; Recommendations and Reports of the CCIR, 1986, Volume X/XI–2 (ITU, Geneva, 1986)

Chapter 6

The broadcasting service

6.1 Introduction to the broadcasting service

The BS is defined in RR 36 as:

> 'a radiocommunication service in which the transmissions are intended for direct reception by the general public . . .'

The expectation that broadcasting signals will be received by the general public is unique among major radio services and it leads to special practices in designing systems and managing frequency assignments. Systems tend towards the use of high transmitter power, and accept high complexity at the programme source or the transmitting station, if necessary, in order that receivers may be simple to operate and low in unit cost. Frequency assignments are planned with care, at a national or international level, and plans are not avoidably changed, also for the benefit of the receiving public.

Broadcasting mainly involves sound radio and television, although broadcasting of other kinds of signal is not excluded. Already the transmission of teletext material in systems intended primarily for television of moving images is well established and the transmission of signals on narrow-band broadcasting channels which are to be displayed for reading, in combination with or instead of sound programme material, began recently. Systems designed to transmit multiple sound channels are being developed. For the present, however, the management of frequency bands allocated for broadcasting is based on the assumption that narrow-band frequency assignments will be used for the transmission of a single audio frequency programme and that wide-band frequency assignments will be used for the transmission of a single television signal including one or more associated audio channels.

Broadcasting frequency bands are not allocated specifically for sound radio or television. However, it is technically impossible to use the bands below 30 MHz for television with moving images; these bands are used for sound radio broadcasting. Frequency assignment plans which presuppose application for either sound radio or television have been drawn up as each band above 30 MHz is taken into substantial use and in general each VHF or UHF broadcasting band has been planned and is used for one purpose or the other but not for both. There are, however, exceptions; the predominant use of the various BS bands between 47 and 72 MHz is for television but some countries use this part of the spectrum for sound broadcasting and the contrary applies between 76 and 108 MHz.

From a spectrum management viewpoint, the main frequency allocations for broadcasting fall conveniently into four groups:

- the LF and MF bands between 148·5 kHz and 1705 kHz considered on pp. 142–153,
- the 'tropical broadcasting' bands between 2300 and 5060 kHz, which, with other allocations between 3900 and 4000 kHz which are not limited to use in the tropical zone, are considered on pp. 153–154,
- the HF bands between 5950 and 26 100 kHz, considered on pp. 154–163, and
- the four VHF/UHF frequency bands between 41 and 960 MHz, considered on pp. 164–178.

There are additional allocations for terrestrial broadcasting which coincide with the three main broadcasting-satellite allocations at 12, 42 and 85 GHz. Experimental terrestrial television transmissions have been made in the 12 GHz band in several countries, leading to regular operation in one country; see CCIR Report 961[1]. However, it would be premature to consider here the spectrum management situations arising in these bands.

Some administrations have assigned frequencies in the microwave part of the spectrum for use in a wideband mode for the transmission of several or many channels of television direct to the public in limited areas. Such facilities are often called Multi-point Video Distribution Systems (MVDS) and they are to be seen as a radio alternative to cable broadcasting. These emissions are properly categorised as part of the BS, although assignments for them are sometimes made in frequency bands allocated to the FS, for example in the bands around 2·5 GHz.

The European VHF/UHF Broadcasting Conference at Stockholm in 1961 designated the frequency bands which it was planning as follows:

Band I 41–68 MHz
Band II 87·5–100 MHz
Band III 162–230 MHz
Band IV 470–582 MHz
Band V 582–960 MHz

These frequency limits differed from the BS allocations in Regions 2 and 3 in 1961, and they no longer represent accurately the allocations for any Region. Nevertheless these designations are still found to be a convenient way of identifying VHF and UHF BS allocations in broad terms and they are widely used. Recently the BS allocations at 12 GHz have come to be known as Band VI.

It is accepted that much HF broadcasting is intended for reception in foreign countries. In general, however, the RR assume that broadcasting is a national, domestic facility, responsible to governments. Thus:

> *RR 2665:* The establishment and use of broadcasting stations (sound broadcasting and television broadcasting stations) on board ships, aircraft or any other floating or airborne objects outside national territories is prohibited.
> *RR 2666:* In principle, except in the frequency band 3900–4000 kHz, broadcasting stations using frequencies below 5060 kHz or above 41 MHz shall not employ power exceeding that necessary to maintain economically an effective national service of good quality within the frontiers of the country concerned.

6.2 Broadcasting at LF and MF

6.2.1 The frequency allocations

The LF allocation

In Region 1 the band 148·5–283·5 kHz is allocated with primary status for broadcasting. There is no corresponding allocation in the other Regions. This, the so called 'long wave' broadcasting band, is used extensively in Europe, the Mediterranean Basin and Northern Asia for sound radio broadcasting. Elsewhere in Region 1 the atmosphere noise level tends to be too high for economic use for broadcasting.

WARC-79 changed the lower limit of the LF broadcasting band from 150 kHz to 148·5 kHz and a change of the upper limit from 285 kHz to 283·5 kHz became effective on 1 February 1990; see RR 458. These changes were made to permit the frequencies assigned to all transmitting stations to be changed slightly, to become integral multiples of 9 kHz, which is the frequency separation between adjacent carriers in the current frequency assignment plan; see RR Resolution 500[2]. This was done to reduce interference to reception, due to beats between carriers.

Parts of the band are shared in Region 1 with the AeRN or the FS. Where these sharing allocations are Region-wide, being shown in the framed part of the international Table of Frequency Allocations (RR Article 8), they have permitted status. However, the sub-Regional allocations for these services, made by RR 460–RR 463 for countries in parts of the Region where the band is not used for BS, are primary. Sharing between these services does not present major problems, in part no doubt because a frequency assignment plan has been agreed for the BS (see pp. 149–153).

The MF allocations

In Regions 1 and 3 the band 526·5–1606·5 kHz is allocated with primary status for broadcasting. In Region 2 the band limits were 525 kHz and 1605 kHz until recently, but a staged extension to 1705 kHz was agreed at WARC-79, the last stage entering into force on 1 July 1990; see RR Recommendation 504[3]. This, the so called 'medium wave' broadcasting band, is used by many thousands of transmitting stations all over the world for broadcasting sound programmes.

There are some sharing allocations for other services, mostly with permitted or secondary status, covering relatively narrow bands at the upper and lower ends of the broadcasting band. Exceptionally the band 525–535 kHz is shared in Region 2 between AeRN and BS with equal primary status and RR 477 imposes a sharing constraint on the BS, limiting transmitter power to 1 kW by day and 250 W by night. MF frequency assignment plans have been agreed for the BS in all Regions (see pp. 149–153) and these provide a framework within which the sharing services can get controlled access to the spectrum without interference problems arising. For most purposes however these MF bands can be considered to be exclusive to the BS.

6.2.2 Spectrum management factors in system design

In some radio services which have a large demand for spectrum, international spectrum management has involved the adoption of various globally-applicable technical standards and constraints on systems, to create a foundation of efficient spectrum utilisation upon which other management methods, such as frequency co-ordination, can if necessary be built. This pattern has not been followed in broadcasting at LF and MF. Here only two general constraints have been applied. RR 303 and RR Appendix 7 require broadcasting stations, in general, to maintain their carrier frequencies within 10 Hz of their assigned frequency, although tighter control of frequency is often required for operational reasons. RR 304 and RR Appendix 8 require the power of spurious emissions to be limited to 50 mW and urge that a better performance be achieved.

However, in most parts of the world there is a heavy and growing demand for MF frequency assignments for broadcasting and co-channel interference can occur at night at great distances. The demand is being met in the frequency assignment plans, partly by the adoption of additional constraints specific to each plan, but mainly by a reduction in the average separation distance between stations which are assigned the same frequency, raising the general level of interference. The way in which the field strength of a LF or MF transmitter is distributed within its coverage area tends to limit the effect of this increase in interference to fringe locations where reception is also degraded by natural propagation phenomena.

The constraints which the frequency assignment plans apply to systems are indicated on pp. 149–153. The present section considers the system characteristics which are of concern to spectrum managers but are not necessarily constrained by the plans.

Channel bandwidth requirements

DSB AM with full level carrier (A3E) is used without exception in the LF and MF bands. There is no visible trend at present towards spectrum-saving modulation systems, in particular SSB AM, although see Resolution 8 of the Final Acts of the Regional Administrative LF/MF Broadcasting Conference, 1975[4].

When frequency assignments are planned, an assumption has to be made about the separation that there should be between the carrier frequencies of adjacent channels. In practice the carrier separation is standardised for all assignments of a given plan and it is customary to speak of the carrier separation as numerically equal to the radio channel bandwidth. To provide the required number of assignments, the carrier frequency separation chosen must not be avoidably large. On the other hand, if typical domestic receivers are not selective enough to reject the carrier wave of emissions in the adjacent channels, the separation distance required between stations transmitting in adjacent channels may have to be made very large to suppress strong heterodyne whistles at night; this also limits the number of assignments that a plan can accommodate. Interference from energy in the outer parts of the sidebands of the emissions to which the adjacent channels are assigned may also be troublesome if the occupied bandwidth of emissions exceeds the channel bandwidth.

The occupied bandwidth of an A3E emission is equal to the necessary bandwidth (that is, the width of the radio frequency channel which is just wide enough to ensure the transmission of the information carried by the emission with the quality required (RR 146)), plus a margin of bandwidth, desirably small, which encloses the out-of-band emission:

- The necessary bandwidth is equal to twice the highest modulating frequency that is transmitted at a significant level. However, a severe limitation on the spectrum of the modulating signal at the transmitter and high receiver selectivity may degrade the intelligibility of received speech and the aesthetic value of music, the main elements of quality implied by RR 146 in defining 'necessary bandwidth'. Thus, the choice of the necessary bandwidth involves compromises between the channel quality objective, the prevailing interference level and the pressure of demand for broadcasting facilities.
- The extent of the 'out-of-band emission' (that is, the radiated power lying outside the necessary bandwidth) depends not only on the presence in the modulating signal of low-level high-frequency components but also on the effectiveness of the measures taken to prevent overmodulation.

These various factors are considered in CCIR Recommendations 639[5] and 328[6].

In Regions 1 and 3 a considerable number of LF and MF transmitters limit their necessary bandwidth to 9 kHz and have effective means for preventing overmodulation. The corresponding figure for Region 2 is 10 kHz. The channel bandwidths used in the frequency assignment plans are also 9 kHz and 10 kHz respectively. However, large numbers of transmitters in use have necessary bandwidths up to 20 kHz and the devices that they have for preventing overmodulation are of only limited effectiveness.

Coverage area and interference at LF and MF

For any broadcasting station, depending on its operating frequency, there is a threshold field strength at which the quality of the signal it provides to a receiving installation of acceptable characteristics is just satisfactory. Where the signal has a lower field strength, the performance objectives would not be attained. This 'usable field strength' is determined mainly by the level of natural and man-made noise, the level of interference which has to be accepted, the characteristics assumed for the receiving equipment and the quality of reception that is sought. In broadcasting, the area around a transmitting station which is enclosed within the 'usable field strength' contour is called the coverage area by some people and the service area by others. Distinctions between these terms are discussed on p. 151.

Examples of the 'usable field strength' levels applied in drawing up recent frequency assignment plans cover the range from 77 to 87 dB relative to 1 microvolt per metre at LF and from 40 to 88 dB relative to 1 microvolt per metre at MF. Formal definitions of the various terms involved and a discussion of the criteria are to be found in CCIR Recommendations 638[7] and 560[8] respectively; see also CCIR Recommendation 559[9].

The field strength at a distance from a transmitting station depends greatly on the propagation conditions arising. The propagation phenomena operative in these frequency bands are reviewed in Appendix B.6.3. Chapters 2 and 3 of Annex 2 of the Final Acts of both of two recent frequency assignment planning conferences[4,10] provide a convenient source of data for predicting field strengths. In brief, the signal is propagated from the transmitting antenna by two main modes, the ground wave (at all times); and the sky wave (night-time only) (see Fig. 6.1). By day in the LF band the sky wave is greatly attenuated by absorption in the D layer of the ionosphere, and in the MF band the daytime sky wave is almost totally absorbed. The principal phenomena shown by the radiation from a transmitter are as follows:

- The strength of the ground wave at a distance from the transmitter depends mainly on the frequency, the conductivity of the ground beneath the transmission path and the power effectively radiated by the antenna with vertical polarisation. The rate of decline of field strength with distance is less for lower frequencies and higher ground conductivities. For a high power transmitter operating in the LF band over high conductivity ground, the ground wave may exceed the 'usable field strength' over a radius of several hundreds of kilometres. The area within which the field strength exceeds this criterion is sometimes called the 'primary service area'. A low power transmitter operating at the higher end of the MF band may have a primary service area of less than 10 km. (These figures for the radius of the service or coverage area, and those which follow below, are given for illustrative purposes only.)

- The sky wave returns to the ground at night after reflection at the E layer or the F layer. Wave interference between the various components of the sky wave causes fading. The sky wave from a powerful MF transmitter may exceed the 'usable field strength' at a range between 500 and 1000 km, although it will be degraded by fading to

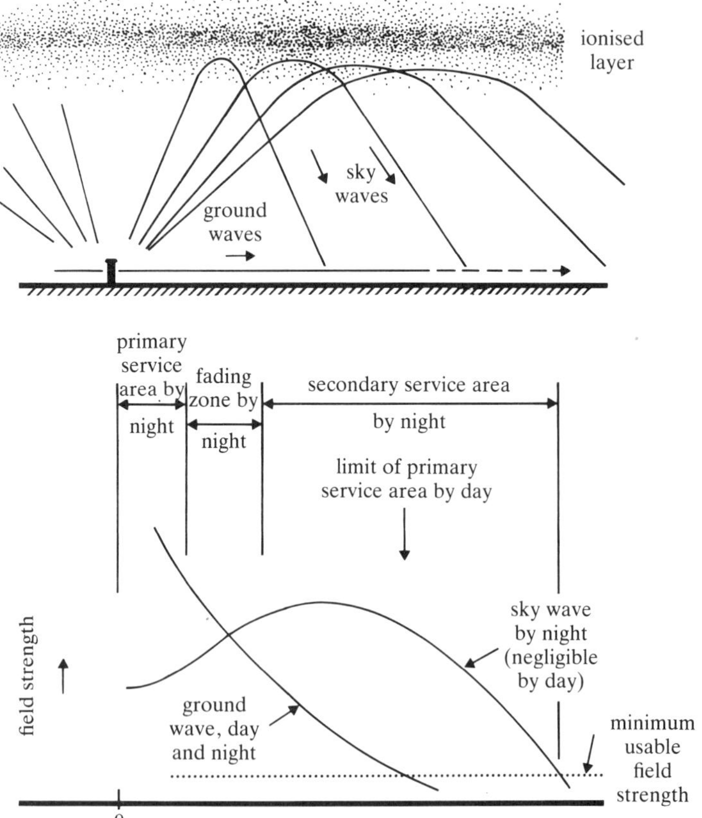

Fig. 6.1 MF propagation modes and the primary and secondary service areas of a powerful broadcasting transmitter, by day and by night

some degree. Thus, at night the sky wave from a transmitting station may provide coverage of a substantial area which receives no signal by day. However, along a contour somewhere near to the limit of daytime ground wave coverage, the sky wave and the ground wave are received with approximately equal field strength at night. This causes selective fading, severe in places, in an annular area surrounding the primary service area, seriously degrading the quality of the signal received at night in places which may be served well by day. In the outer parts of the area served by the sky wave, there may be interference at night from distant transmitters assigned the same frequency or an adjacent one. The area in which the sky wave field strength exceeds the usable field strength and interference is not excessive is sometimes called the 'secondary service area'.

- A perceptible field strength, too weak for satisfactory reception as a wanted signal but capable of causing severe interference, may be propagated at night for some thousands of kilometres by multiple reflections between the ionosphere and the Earth's surface.

These various factors bearing on the efficiency with which a broadcasting station can perform its function are examined in CCIR Report 616[11] and CCIR Recommendation 598[12].

Transmitter radiation pattern optimisation

Four techniques for optimising the coverage of broadcasting stations, of significance to spectrum managers, emerge from consideration of these basic propagation phenomena.

(i) A typical transmitting antenna operating in the LF band consists of a vertical monopole, fed at the base, the height of which is small compared with the wavelength. In the MF band, vertical quarter-wave antennas, also fed at the base, are typical. Such antennas radiate a substantial amount of power at high angles of elevation. Various antenna designs have been developed, typically with greater physical or electrical height relative to the wavelength, which have higher gain at low angles of elevation and less gain at high angles. This configuration extends the ground wave coverage area by day and the area which is substantially free from fading due to the sky wave at night and it also increases the sky wave field strength at night in the region beyond ground wave coverage; see Fig. 6.2*a* and CCIR Reports 401[13] and 616[11]. Reduction in this way of the power which is radiated at a high angle of elevation has the incidental further benefit of reducing the intensity of the power flux which illuminates the ionosphere above a powerful transmitting station, so reducing the occurrence of ionospheric cross-modulation; see CCIR Recommendation 498[14].

(ii) If horizontal elements form the antenna, the signal radiated may consist of a large horizontally polarised component but only a small vertically polarised component. The strength of the ground wave declines rapidly with distance, giving a small ground wave coverage area. Consequently, there is fading at night in a small annular area close to the transmitting station due to wave interference between the ground wave and the sky wave, the latter having its origin mainly in the horizontally polarised component of the radiated signal, but this area of fading is surrounded by a large area covered by the sky wave. If this technique is used to provide night-time coverage to a large area, it may be necessary to use a different antenna by day, with a substantial vertically polarised component in its radiation, to provide adequate ground wave coverage near the transmitting station; see Fig. 6.2*b* and CCIR Report 616[11].

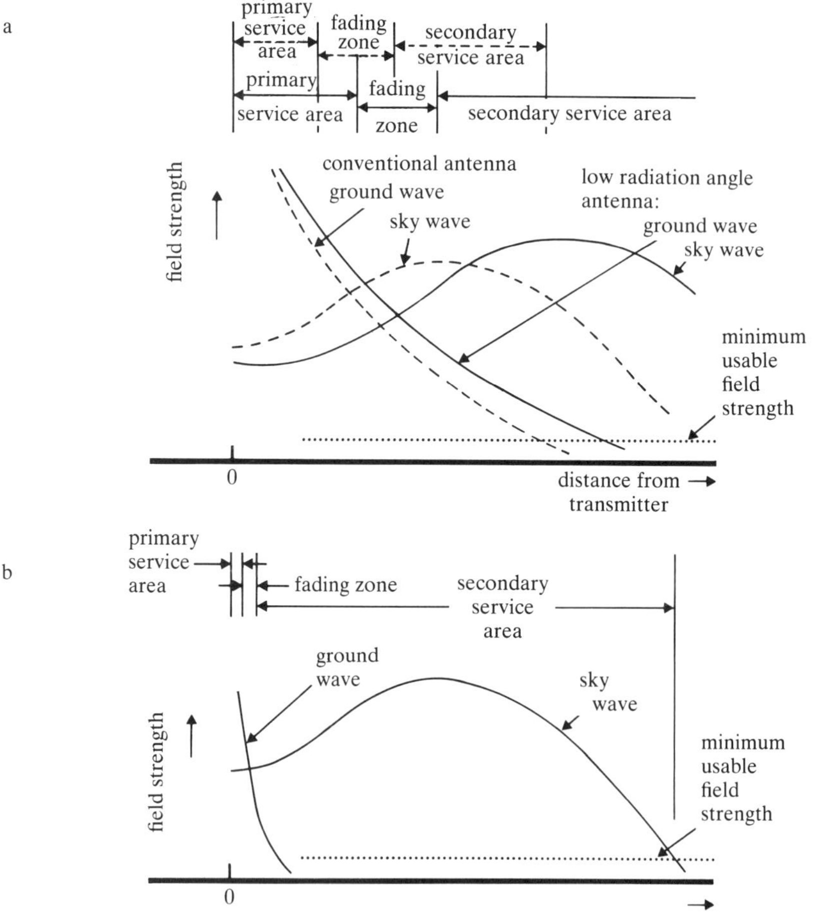

Fig. 6.2 Effect of MF transmitting antenna design on the night-time service area
a Compares a conventional antenna with a 'low-angle' antenna
b Illustrates the coverage provided by an antenna with a large horizontally polarised component

(iii) Depending on the geographical pattern of distribution of the population which receives the broadcasting signals, there are two techniques for using several or many transmitting stations of low or medium power, all assigned the same frequency. These techniques may cost less and use spectrum more efficiently than a single very powerful station:

(*a*) With carrier frequencies made very precisely equal and preferably phase-locked, with common programmes and with attention given to the synchronisation of the common modulating signals, it becomes possible to overlap the ground wave coverage areas of several stations, producing a single extensive coverage area with little fading due to wave interference, even at night; see CCIR Recommendation 560[8]. This technique is suitable for serving a large contiguous area with a rather evenly distributed population using medium-power transmitters, typically 10 to 50 kW.

(*b*) Alternatively, each of a number of towns may have its own low-power transmitter (typically 1 kW) to serve it. The ground wave field strength will be low in between towns, and there will be fading at night in such locations due to wave interference between the various ground waves and sky waves, but the ground wave will be strong enough within the towns to prevent fading even at night. By day it may be feasible to modulate each transmitter with different programme material without any significant loss of coverage, but at night the service area of each transmitter may be significantly reduced if a common programme is not used.

(iv) Finally there are various means by which the horizontal radiation pattern of a transmitting antenna may be shaped so as to serve efficiently an area which is not symmetrical about the station. The horizontal radiation pattern is determined typically by control of the phase of the feeds to two or more horizontally-spaced radiating elements. Similar techniques may be employed to reduce the interference which the station would otherwise cause in the coverage area of another, distant, station assigned the same frequency.

6.2.3 Frequency assignment planning

CCIR Report 616[11] and CCIR Recommendation 598[12] consider the various factors referred to on pp. 143–149 and others, and draw general conclusions on desirable principles for efficient planning of the use of spectrum for LF and MF broadcasting.

The current frequency assignment plan for Regions 1 and 3 is to be found in the Final Acts of the Regional Administrative LF/MF Broadcasting Conference, 1975[4]. That conference found that it had insufficient time to prepare a fully-developed plan for the large number of assignments in the MF low-power channels, so it was arranged that the provisional plan for them contained in Appendix 1 of the Final Acts should be replaced by a new version to be prepared by the IFRB; this was published in 1978[15].

The current Region 2 plan covering the band 535–1605 kHz was drawn up at the Regional Administrative MF Broadcasting Conference, 1981[10]. The band 1605–1705 kHz, having been newly allocated in Region 2 to the BS at WARC-1979, was not included in the 1981 plan. A frequency assignment plan to cover this extension band was prepared at a two-session RARC meeting in 1986 and 1988, in accordance with RR Recommendation 504[3].

These various plans are similar in structure although they differ in detail. Their regulatory provisions are implemented in conjunction with RR Article 12. They are fairly complicated and reference should be made to the plan agreements for the details, Nevertheless, a summary of a few of the main aspects of these arrangements may be helpful here.

Administrative aspects

The 1975 Agreement for Regions 1 and 3 entered into force on 23 November 1978. The corresponding date for the main Region 2 Agreement was 1 July 1983. The Agreements were intended to meet the needs of the administrations for 11 and 10 years respectively from the dates of entry into force and they remain in force until replaced by other agreements produced by competent conferences.

Technical aspects

In the 1975 Plan, the LF band (Region 1 only) is divided into 15 channels, each 9 kHz wide. Each channel is assigned, on average, to about 12 stations. The same plan divides the MF band (Regions 1 and 3) into 120 channels, also 9 kHz wide. Three channels, centred on 1485, 1584 and 1602 kHz, are reserved for stations of low transmitter power with small service areas; about 2500 stations have assignments in these channels. The remaining 117 channels are used mostly by stations with medium or high power; each channel is assigned, on average, to about 60 stations.

The Plan for Region 2 divides the MF band below 1605 kHz into 107 channels, each 10 kHz wide, each assigned, on average, to about 90 stations.

The stations to which each channel is to be assigned in a plan have been chosen after calculations have shown that the interference at the edge of the service area from all unwanted stations will not exceed the level determined by the nominal usable field strength (a generalised value of the usable field strength) and the agreed co-channel protection ratio. Likewise it will have been verified that the level of interference at the edge of the service area due to unwanted stations in the adjacent channels will not exceed the agreed adjacent channel protection ratio. The propagation data that have been used in making these calculations are set out in detail in the agreements.

There are differences of detail in the nominal usable field strengths and the protection ratios that have been used in the two plans for broadcasting stations fulfilling various kinds of operational function. The objective has been to ensure that a sufficiently large number of stations can make use of every channel, thus permitting assignments to be made to meet all of the requirements declared by administrations. This may have created a rather severe night-time interference environment, but it has been found possible to give each station the minimum protection which it must have to carry out its basic function. The results of applying these different criteria have been quite similar in the two plans; stations with an extensive rural night-time sky wave service area have been assigned channels in which the level of interference will be low, while stations with a small urban ground wave service area will often have to operate in a high-interference environment.

However, there are a few significant differences in the ways in which some of these concepts have been interpreted in drawing up the two Plans. In particular:

- The Plan for Regions 1 and 3 assumes that the transmitter power and antenna characteristics which the responsible administration declares to be used at a broadcasting station will indeed be used. The Region 2 Plan places limits on the power that may be radiated by stations which are not intended to provide sky wave service to a rural area at night, and forbids the introduction of any new transmitters with an output power in excess of 100 kW by day or 50 kW by night.
- The Region 2 Plan excludes from a service area any territory which is outside the jurisdiction of the administration for which the assignment is planned, whereas the Plan for Regions 1 and 3 does not make this distinction.

It may be noted that this second difference reflects an issue in ITU terminology which is not yet completely resolved as well as a difference in practice which may be a consequence of developments in planning concepts between 1975 and 1981. There is agreement that the term 'coverage area' is purely technical in significance. In broadcasting, a coverage area may be delineated by a contour of specified field strength; see CCIR Recommendation 638[7]. The view is gaining ground, however, that the definition of the term 'service area' should have a regulatory element in addition to the technical one. In this view any part of a broadcasting coverage area (as defined above) which lies outside the jurisdiction of the administration responsible for the station should be excluded from the area over which the administration may claim protection from interference; the coverage area, curtailed in this way, would be called the service area according to this view; see CCIR Recommendation 573[16].

Regulatory aspects

The Plans identify channels for assignment to a broadcasting station of stated characteristics at a stated location. However, only the administration responsible for that station has the authority to make the assignment. If and when the administration makes an assignment to a station corresponding to an entry in the appropriate plan and the station prepares to bring the frequency into use, the details of the assignment which the administration has made are notified to the IFRB in accordance with the procedure in RR Article 12, RR 1239–RR 1314. After verifying that the assignment is in conformity with the plan, and subject to certain safeguards, the IFRB registers the assignment in the MIFR. If the assignment is not in conformity with the plan, the IFRB requires the administration to secure an amendment of the plan, to bring the plan into line with the assignment which the administration has made, before further consideration is given to registration.

The safeguards referred to above have two purposes. Firstly, some countries did not sign their Regional Agreement; indeed some may not have participated in the Conference. Some of the countries which did not sign the agreement for their Region have, no doubt, acceded to it since, but others may have not done so. If those latter countries are members of the ITU, they retain the right to have assignments registered in the MIFR, subject to standard conditions, and the IFRB respects such registrations when a planned assignment is notified for registration. Secondly, there are narrow frequency bands at both ends of the exclusive MF broadcasting band which are allocated to other services with equal primary status, in the same Region or in another Region. The IFRB will investigate the risk of interference between a newly notified planned broadcasting assignment and a registered assignment for one of these other services.

The procedure by which an administration which has signed one of the Regional MF broadcasting plans can seek to have that plan amended, to change a planned assignment, or to add a new assignment is set out in Article 4 of both agreements. In essence, the administration wishing to make a change which would be significantly detrimental, according to agreed criteria, to the operation of other assignments must obtain the agreement of the administrations responsible for those other assignments. The IFRB maintains oversight of the progress of these discussions and publishes progress reports in its weekly circular. Given a successful outcome of these discussions, the master plan maintained by the IFRB is amended and the amendment is incorporated in the revised version of the plan which is published from time to time; see, for example, Section 3 of Article 4 of the 1975 Agreement[4].

An administration wishing to cancel a planned assignment merely informs the IFRB, which publishes the fact and deletes the assignment when the next revision of the plan is published, see, for example, Section 4 of Article 4 of the 1975 Agreement[4].

Finally, an administration which is not a party to the appropriate Regional plan will notify new assignments to the IFRB for registration. The

IFRB will deal with such notifications in accordance with Section IIA of RR Article 12, that is in much the same way as a notified assignment for the FS below 28 MHz (see pp. 85–86).

6.3 Tropical broadcasting at 2 to 5 MHz

The level of atmospheric noise in tropical countries around 200 kHz is very high, so the Region 1 LF BS allocation is not used in the tropical zone. The atmospheric noise level is also high in the MF broadcasting band in the tropical zone; this band is widely used to serve towns and the vicinity of towns in tropical countries in all three Regions but very high transmitter power is needed for serving a large area, making costly the extension of MF coverage to rural areas, The use with sky-wave propagation of frequencies above the MF broadcasting band, starting at around 2 MHz and extending into the lower part of the HF range, is an alternative for wide-area night-time domestic coverage.

The following frequency bands are allocated for broadcasting in a defined Tropical Zone:

2300–2495 kHz (extended to 2498 kHz in Region 1)
3200–3400 kHz (all Regions)
4750–4995 kHz (all Regions)
5005–5060 kHz (all Regions)

The Tropical Zone is defined in RR 406–RR 410. In brief it consists of all of the area between the Tropics of Cancer and Capricorn, extended in Regions 1 and 3 to 35° South and 30° North latitude, with certain other extensions to the North between longitudes 10° and 80° East. RR 411 provides an option for the Tropical Zone in Region 2 to be extended northwards to 33° North, subject to special agreements between the countries concerned. The use of these bands for broadcasting outside the defined zone is forbidden (see RR 503 and RR 2669). Use for broadcasting within the tropical zone is expressly limited to internal, national broadcasting (see RR 2668) and the maximum transmitter power is 50 kW (see also CCIR Recommendation 215[17]).

One or other of these bands will be suitable for providing sky-wave service by night with zero skip distance in the Tropical Zone at any time of year. Absorption in the D or E layer will be considerable by day, making it necessary to complement assignments below 5 MHz with others at higher frequencies if daytime operation is also required. The management of these higher frequency bands is discussed on pp. 157–158. The various propagation phenomena which arise are discussed in CCIR Report 304[18].

There is another group of allocations for broadcasting in this part of the spectrum, as follows:

3900–3950 kHz: Region 3 and, in accordance with RR 513, a sub-Regional area in southern Africa

3950–4000 kHz: Regions 1 and 3, and, in accordance with RR 514 and RR 515, Canada and Greenland

The use of these two bands for broadcasting is not limited to the Tropical Zone but in other respects their spectrum management is similar and it is convenient to group them here with the tropical broadcasting bands.

Sound radio programmes are transmitted in these various bands, using A3E emissions. The frequency tolerance and spurious emission constraints which apply in the MF band (see p. 143) apply here also, with minor variations. Frequencies suitable for assignment are found by inspection and are notified to the IFRB for registration in accordance with Section IIA of RR Article 12, as for the FS below 28 MHz (see pp. 84–86).

However, all of these six bands are shared with other services, typically the FS and various mobile services, with equal primary status. Exceptionally the sub-Regional allocation made at 3900 kHz by RR 513 is exclusive although assignments are subject to agreements with other administrations concerned in accordance with the procedures of RR Article 14.

CCIR Recommendation 48[19] urges that administrations should avoid using the tropical broadcasting bands for assignments to mobile stations; where the bands are not needed for broadcasting, it would be preferable to use them for stations of the FS, because the interference which such stations generate would be less harmful to the reception of broadcasting. Reference should be made to CCIR Recommendation 216[20] and Reports 302[21] and 303[22] for information on the protection ratios required to protect broadcasting reception against interference and atmospheric noise.

These six frequency bands below 5 MHz are in heavy use, even in the Tropical Zone, for the FS and the mobile services to which they are also allocated. Use for broadcasting is on a relatively small scale. RR 2671 states that the BS has priority over other services within the Tropical Zone in the tropical broadcasting bands but it seems that it has not been found necessary for this priority to be implemented.

6.4 Broadcasting at HF

6.4.1 Survey of HF BS regulation

The frequency allocations and sharing

Eight frequency bands at HF were already allocated for the BS prior to WARC-79, seven of them being world-wide and the eighth being limited to Regions 1 and 3. It was agreed at WARC-79 that the width of one of these bands should be reduced somewhat at a date to be determined, this being part of a package of allocation changes leading to the provision of an allocation for RA at 25 MHz; see RR 545. However, it was also agreed that seven bands, previously allocated for the FS and mostly adjacent to previously-existing BS bands, should be re-allocated world-wide for the BS.

Table 6.1 Exclusive frequency allocations for HF broadcasting

The allocations at the start of WARC-79 (kHz)	The 'extension bands' which were newly allocated at WARC-79 (note 1) (kHz)
5950–6200	
7100–7300 (note 2)	
9500–9775	9775–9900
11 700–11 975	11 650–11 700 11 975–12 050
	13 600–13 800
15 100–15 450	15 450–15 600
17 700–17 900	17 750–17 700
21 450–21 750	21 750–21 850
25 600–26 100 (note 3)	

Notes

1 The date of entry into force of these new allocations is discussed in the text.

2 Regions 1 and 3 only

3 WARC-79 agreed that the allocation of the band 25 600–25 670 kHz should be transferred from BS to the radio astronomy service at some date to be fixed; see text.

These additional bands have become known as the 'extension bands'. The frequency limits of all of these bands are shown in Table 6.1.

As a result of the changes agreed at WARC-79, the total bandwidth allocated for the BS between 6 and 26 MHz in Regions 1 and 3 became 15·6% of all of the spectrum in this frequency range. In Region 2 the figure became 14·6%.

Three of the extension bands, namely those at 9·8, 11·65 and 12·0 MHz, may also be used for domestic FS links, the radiated power being limited to 24 dBW, provided that harmful interference is not caused to the BS; see RR 530. The 7·2 MHz band, allocated for BS in Regions 1 and 3 only, is allocated for AmS in Region 2 but RR 528 requires that this use by amateurs does not impose constraints on the BS in Regions 1 and 3. All of the other bands listed in Table 6.1 are exclusively for the BS.

The arrangements whereby the seven extension bands were to be transferred from the FS to the BS are set out in outline in RR 531 and in detail in RR Resolution 8[23]. These arrangements allowed administrations the period up to 1 July 1989 to find new frequencies to assign to FS stations that were then operating in the extension bands above 10 MHz, after which date the use of the bands by the FS was to have ceased, in order that the BS might use them, in the words of RR 531, 'subject to provisions to be established by the world administrative radio conference for the planning of HF bands . . . (see Resolution 508)'. For the one extension band that is below 10 MHz the handover date was 1 July 1994. The WARC called for by RR Resolution 508[24] has since taken place, in two sessions in 1984 and 1987, and the outcome is discussed on pp. 162–163 below. However, it may be noted here that WARC-HFBC-87 did not bring to a conclusion its consideration of how the extension bands should be taken into use; RR

Resolution 512[25], to be found in the Final Acts of that conference[26], has the effect of postponing the handover date for extension bands above 10 MHz from 1989 to a date to be fixed by a later conference.

WARC-HFBC-87 was also to make arrangements for releasing the band 25 600–25 670 kHz to the RA, as agreed in principle at WARC-79 (see RR 545) but did not do so.

Systems, constraints and ITU recommendations

There are two quite distinct applications for broadcasting at HF. The bands below about 12 MHz are used for domestic broadcasting in many countries, in particular big tropical countries. In addition, all of the bands, from 6 to 26 MHz, are used for international broadcasting.

The demand for the opportunity to broadcast at HF is heavy. The pressure on such bandwidth as is available is aggravated by the use of higher transmitter power than is technically necessary and a spectrally inefficient type of modulation and also by the simultaneous transmission of programmes in several different frequency bands in order to overcome the vagaries of HF radio propagation and interference. Furthermore, only limited use is made of the substantial bandwidth allocated to the BS above 20 MHz, because these bands are above the tuning range of many broadcasting receivers. Whilst these factors persist and having regard to the needs of other services for the HF range of spectrum, it is probably not feasible to satisfy the spectrum demand for broadcasting at HF.

Virtually all currently operational systems in HF broadcasting use DSB AM with high-level carrier (class of emission A3E). DSB permits the use of simple low-cost receivers with envelope demodulation. Low-cost receivers are a basic requirement in broadcasting but in most other respects the arrangement is a bad one. Thus:

- Severe distortion arises in the output signal of a receiver when the amplitude of the received carrier falls, relative to the sideband amplitude, because of selective fading, which occurs commonly at HF.
- The radiation of both sidebands halves the number of programmes that could be transmitted in a spectrum of finite width.
- The radiation of a high-level carrier adds considerably to potential interference levels at distant receivers, further reducing opportunities for satisfactory frequency re-use.

Various means have been devised for improving the quality of reception and for increasing the efficiency with which spectrum is used for broadcasting at HF and some of these means have become the subject of ITU requirements or recommendations; they are reviewed on pp. 158–160.

Two mandatory constraints for BS transmitters at HF are as follows:

Carrier frequency tolerance: RR 303 and RR Appendix 7[26] require the carrier frequency of emissions to be maintained within 10 Hz of the nominal value.

Appendix 7 also recommends that departures from the nominal frequency of a few hertz should be avoided, since this produces effects due to beats between carriers which resemble fading. Frequency differences approaching 10 Hz are less objectionable but the implementation of a tolerance of about 0·1 Hz would be even better; see also the reference to synchronised transmitters on pp. 160 below.

Transmitter spurious emissions: RR 304 and RR Appendix 8 place limits on spurious emissions from transmitters (see Appendix A.5 below).

International spectrum management

The Radio Regulations agreed by the International Radio Conference in 1947 provided a simple interim procedure[27] for registering HF BS frequency assignments, pending the devising of a more satisfactory procedure. Assignments were to be notified to the then-newly-formed IFRB, which would publish the notifications in its weekly circular. If an assignment was in accordance with the RR and if no other administration claiming prior use of the frequency objected, it would be registered in the MIFR. If either of these two conditions was not fulfilled but the notifying administration insisted on registration, the assignment would be registered but with an appropriate annotation.

The need of the public for stability in the frequencies and schedules used for HF broadcasting made desirable a more stable arrangement for choosing frequencies for assignment. Frequency assignment planning was favoured by many countries, partly for the convenience of listeners but mainly because many administrations had then, and still have, a concern that they would not be able to secure an equitable share of spectrum, without interference, in any other way.

The first attempt to produce a frequency assignment plan for the BS at HF began at an administrative conference in Mexico City in 1948 and continued at other meetings extending over several years. The attempt was unsuccessful. No way could be found for planning assignments with acceptably low foreseen interference levels to meet all of the requirements that were stated, using the bandwidth then allocated to the BS, nor could agreement be reached on scaling down the stated requirements to a level at which planning was feasible.

The matter was considered further at WARC-59. No progress was made towards a usable planning technique. However, the procedure to be used by administrations and the IFRB for notifying and registering assignments was changed, with two purposes in view. The procedure was made more responsive to the realities of the annual cycle of radio propagation conditions. Secondly, the process of registration in the MIFR was changed to diminish the degree of international recognition which registration could be seen to give to an assignment that had been used. The procedure as revised in 1959 is still in use, after relatively minor further amendments in 1979 and 1987; see RR Article 17[26] and pp. 160–161 below.

WARC-79, having increased the bandwidth allocated to the BS at HF, determined that another attempt should be made to introduce frequency assignment planning; see RR Resolution 508[24]. This resolution called for a WARC to be convened in two sessions, the first session to establish planning principles and the second session to implement them. The two sessions were held in 1984 and 1987. WARC-HFBC-87 came to no firm decision on the introduction of frequency assignment planning but a tentative scheme for doing so was drawn up for further study. PC-89 decided that a third WARC session should be convened in 1993.

6.4.2 Spectrum utilisation efficiency in HF broadcasting

HF broadcasting using DSB and a high level carrier occupies spectrum inefficiently and the quality of the signal reaching listeners is often poor. There are technical and operational means of enhancing spectrum utilisation efficiency, so reducing and conceivably eliminating the present disparity between the number of channels that are available and the number that are required, and some of these means also improve the quality of service to listeners, The means which have been studied within the ITU are reviewed briefly in this section; see also RR Recommendation 503[28].

System parameters

Standard system specifications for both DSB and SSB emissions were agreed at WARC-HFBC-87 and are set out in RR Appendix 45[26]. RR 2673A[26] and RR 2673B[26], taken with the Preamble to the Final Acts of the Conference[26], require that the DSB specification be implemented by 1 September 1988; see also CCIR Report 458[29].

A tentative date has been set for the general implementation of SSB and the cessation of the use of DSB, namely 31 December 2015; see RR Resolution 517[30], also CCIR Recommendation 640[31] and CCIR Report 1059[32]. However, the period of transition to SSB has already begun, and SSB may already be used instead of DSB provided that the level of interference to DSB emissions is not made greater by the use of SSB than that caused by the DSB emission that it replaces. Meanwhile, in estimating the probability of interference between emissions during the transition period, it is to be assumed that DSB is being used for all emissions.

The essential features of the SSB specification are as follows:

- the necessary bandwidth shall not exceed 4·5 kHz,
- the upper sideband shall be emitted, the lower sideband being suppressed by at least 35 dB relative to the upper sideband,
- during the transition phase the radio channel width is 10 kHz, although provision is made for the use of a 10 kHz channel for two emissions, one with an 'interleaved' carrier frequency, with appropriate safeguards; when transition to SSB has been completed, the radio

channel width will be 5 kHz (see also CCIR Recommendation 597[33]), and

- nominal carrier frequencies shall be integral multiples of 5 kHz, and the carrier level should be minus 12 dB relative to peak envelope power (PEP), although minus 6 dB relative to PEP is to be used during the period of transition from DSB to SSB (class of emission R3E).

A reference receiver is defined in RR Appendix 45[26] as a basis for determining performance and interference criteria. This receiver is assumed to be equipped with a synchronous demodulator, using for carrier acquisition a method whereby the carrier is regenerated by means of a control loop which pulls the tuning of the receiver to the frequency of the incoming carrier. Relatively stringent selectivity characteristics are stipulated.

Propagation forecasting

In most other services which use ionospheric propagation, the operators receive reports in real time, or soon afterwards, of the condition of the signal being received in the intended reception area; operating frequencies may be changed quickly if this is thought likely to be beneficial. Broadcasting organisations however must predict, some months in advance, which of the frequency assignments available to them will be most successfully propagated over the desired signal path and they may wish to forecast whether another signal of known frequency and source is likely to cause interference.

The basic CCIR texts on the prediction of HF propagation conditions are referenced in Appendix B.6.4. Recent further studies directed in particular at the use of predictions of sky-wave field strength in frequency assignment planning are reported in RR Recommendation 512[34]. CCIR Recommendations 411[35] and 559[9] and CCIR Report 1058[36] provide basic information on protection ratios required against interference and noise; see also RR Recommendation 517[37] which relates in particular to the protection ratios required in the reception of SSB signals.

The prediction processes are complex and the phenomena are not yet completely characterised. Furthermore, the variation of propagation conditions from hour to hour and from day to day, particularly those involving F2 layer reflection, shows a substantial statistical scatter and forecasts are made less trustworthy by sporadic ionospheric disturbances (see Appendices B.3.3, B.3.4 and B.6.4.). Lacking reliable predictions, or given predictions that indicate variable propagation conditions, HF broadcasting operators may think it to be desirable to transmit programmes simultaneously on two frequencies or more. However, this aggravates the general shortage of spectrum for HF broadcasting. The avoidable use of multiple frequencies is deprecated; see RR 1743 [26] and also CCIR Recommendation 410[38].

Directional transmitting antennas and transmitter power

Transmitting antennas used for HF broadcasting should be designed to serve the intended coverage area efficiently. In particular, where the transmitting station is not located within the coverage area, and especially where the coverage is not domestic, antennas of appropriate horizontal directivity should be used. This permits transmitter power to be reduced and reduces overspill outside the coverage area whilst maintaining an adequate field strength within the coverage area.

The CCIR has made extensive studies of the technology and the characteristics of directive antennas for HF broadcasting; see CCIR Recommendations 80[39] and 414[40], CCIR Reports 32[41] and 1062[42] and the CCIR Book of Antenna Diagrams[43]. So far, however, neither the use of appropriate transmitting antennas nor the limitation of transmitter power in HF BS systems has been given the force of a recommendation by the ITU.

Synchronisation of carriers

There are circumstances in which a broadcasting operator wishes to transmit the same programme material to different reception areas at about the same time, or to a single reception area which is too large to be served satisfactorily by a single beam. In such circumstances, propagation conditions may permit the use of the same carrier frequency for two or more simultaneous transmissions. This practice is recommended in CCIR Recommendation 205[44] as a means of economising in the use of spectrum. It leads to no significant degradation of reception provided that the several carriers are derived from a common source or are quasi-synchronised with a frequency error of not more than 0·1 Hz. The modulating signals must, of course, be identical and where possible they should be correctly phased, one with another; see also RR Recommendation 516[45].

6.4.3 The RR Article 17 procedure

WARC-59 devised a procedure which provided some of the advantages sought from frequency assignment planning. In particular the procedure took into account the seasonal variation of optimum operating frequencies and opened up lines of communication between administrations, via the IFRB, for consultation on the HF assignments to be made to broadcasting stations. It also ensured the publication of more accurate information than had previously been available on the assignments which are actually in use at any given time, permitting knowledge to be accumulated that would enable more reliable forecasts of optimum frequencies and interference prospects to be made on subsequent occasions. Finally, the use of this procedure could be expected to lead to a diminution of the significance of priority of registration as a determinant of the right to use a frequency assignment. The procedure is set out in RR Article 17, most recently revised in the Final Acts of WARC-HFBC-87[26].

In essence the procedure is as follows. The year is divided into four seasons, namely March/April, May-August, September/October and November-February. Then:

- Up to one year before the start of each season, an administration responsible for HF broadcasting transmissions informs the IFRB of the frequencies and schedules which it intends to assign for use on each transmission path during that season.
- The IFRB publishes this information two months before the start of the season as a Tentative High Frequency Broadcasting Schedule[46], for the information of administrations. The IFRB studies the general pattern of frequency utilisation which the schedule presents and advises administrations if it considers that changes of frequency or schedule would be desirable to reduce the likelihood of interference. Any changes of intention flowing from this correspondence are published in the weekly circular.
- After the end of each season the IFRB circulates a revised version of the tentative schedule, called the High Frequency Broadcasting Schedule[47], showing the frequencies actually used and indicating which assignments were reported to be unsatisfactory.

For a number of years the information in the revised schedule was consolidated by the IFRB into a High Frequency Broadcasting Frequency List[48], annually up-dated, and published as a supplement to the IFL. Assignments present in this list were also included automatically in the MIFR. WARC-79 decided that this latter practice should cease and there is now no provision in the RR for the addition of HF BS assignments to the MIFR; see RR 1350. WARC-HFBC-87 decided that the preparation of a High Frequency Broadcasting Frequency List should be discontinued.

6.4.4 Frequency assignment planning and WARC-HFBC-84/87

The report of the first session (WARC-HFBC-84) of the conference which was set up in accordance with RR Resolution 508[24], addressed to the second session (WARC-HFBC-87), contained planning principles and technical parameters which were proposed as a basis for frequency assignment planning in the frequency bands allocated exclusively to the BS at HF. WARC-HFBC-84 also asked the IFRB to develop software that could be used for planning, in accordance with principles and parameters which the conference had chosen, and to prepare trial plans for the consideration of WARC-HFBC-87.

WARC-HFBC-87 did not agree that the trial plans produced by the IFRB in accordance with the instructions of WARC-HFBC-84 met the requirements for domestic and international HF broadcasting in an adequate way. Instead, three objectives were agreed for the way ahead:

- The present spectrum management method (RR Article 17) should continue to be used in the short term with relatively minor changes.
- For the medium term, there should be a transitional stage when both RR Article 17 and an improved planning method would be used, side-by-side, for the management of different allocated sub-bands.
- In the long term, the management of all exclusive HF BS allocations should be by a planning method of demonstrated effectiveness.

The details of the package of measures which was agreed to achieve those objectives are to be found in the Final Acts of WARC-HFBC-87[26]. The principal elements of the package can be summarised as follows:

(i) Various relatively minor changes were made to the RR Article 17 procedure and, as amended, this procedure will continue to be used for the time being for the management of the HF bands allocated to the BS prior to WARC-79 with minor exceptions (see p. 161). Other amendments made to RR Article 17 define the intentions of the intended new procedure, stressing in particular the principle that all nations, and both domestic and international broadcasting, were to be treated equally; see RR 1737[26] to RR 1746[26].

(ii) Various modifications were made to the planning principles and parameters put forward by WARC-HFBC-84. As modified these principles and parameters are to be found in RR Resolution 515[49]. The IFRB is to develop methods for planning, using these revised criteria, and will prepare trial plans in the following sub-bands:

9775–9900 kHz
11 650–11 700 kHz
11 975–12 050 kHz
13 600–13 800 kHz
15 400–15 600 kHz
17 550–17 750 kHz
21 650–21 850 kHz
25 900–26 100 kHz

See RR Resolution 511[50]. These sub-bands are almost identical with the extension bands allocated to the BS by WARC-79. A group of experts drawn from administrations in all parts of the world is to be set up to participate with the IFRB in developing these trial plans; see RR Recommendation 509[51].

(iii) It was recommended that another WARC should be convened not later than 1992 to progress the matter; see RR Resolution 511[50]. This conference would consider the trial plans produced by the IFRB and the effectiveness of the management being provided by the revised RR Article

17 procedure, with a view to the introduction of a dual system of management in the full bandwidth available at HF to the BS, including the extension bands. It was foreseen that a frequency assignment planning procedure of proved effectiveness would eventually be taken into general use for managing all HF spectrum allocated to the BS. Meanwhile, the extension bands were not to be used for broadcasting; see Resolution 512[25].

(iv) In addition to the improvements to the planning criteria referred to in the previous paragraph, various other actions were initiated by WARC-HFBC-87 to increase the probability that future attempts at assignment planning would be successful. In particular:

- The use of SSB is to be stimulated; see RR Appendix 45[26], RR Resolution 517[30] and RR Recommendation 515[52].
- Monitoring programmes were initiated to identify sources of interference in the HF BS bands and to get the interference eliminated; see RR Resolution 513[53].
- The study and use of optimised transmitting antennas is to be stimulated; see RR Resolution 516[54].
- The need for further study of propagation prediction methods was emphasised; see RR Recommendation 514[55].
- Administrations were asked to draw the attention of HF broadcasting receiver manufacturers to the need to market low cost SSB receivers and also receivers covering all HF frequency bands allocated to the BS, including the higher bands; see RR Recommendations 515[52] and 518[56].
- WARC-HFBC-87 recommended that the Administrative Council ask the Plenipotentiary Conference, meeting in 1989, to consider action that might lead to the allocation of additional spectrum for HF broadcasting; see RR Recommendation 511[57].

(v) WARC-HFBC-87 acknowledged that it had concentrated its attention mainly on international HF broadcasting and noted the need for the future conference referred to in item (iii) above to take the needs of national broadcasting at HF fully into account; see RR Recommendation 513[58].

The Plenipotentiary Conference meeting in 1989 took account of these measures. It agreed that a WARC should take place in 1992, at which various frequency allocation problems identified at recent WARCs should be considered; no doubt the need for more bandwidth for HF broadcasting will be included in its agenda. It also agreed that a WARC should be convened in 1993 to progress further the work on HF broadcasting started at WARC-HFBC-84/87. Thirdly, it endorsed the decision, taken at WARC-HFBC-87, that the extension bands allocated to BS at WARC-79, should not be taken into use for broadcasting before a frequency assignment planning system had been implemented.

6.5 Sound and television broadcasting at VHF and UHF

6.5.1 Survey of broadcasting regulation at VHF and UHF

The frequency allocations and sharing

The international frequency allocations of primary status for broadcasting are shown in Table 6.2. For a few countries the limits of Bands I, II and III are extended somewhat, upwards or downwards, by footnotes to the international Table of Frequency Allocations (RR Article 8).

There are also allocations of almost all of this spectrum to other services by entries within the framed part of the international Table or by footnotes to it. The situation is too complex to be treated here in detail, but the following outline summarises the primary or permitted sharing allocations. There are also many secondary allocations, Regional or sub-Regional, mostly for the FS and MS:

Table 6.2 Broadcasting service allocations at VHF and UHF

	Region 1 (MHz)	Region 2 (MHz)	Region 3 (MHz)
Band I	47–68	54–72	47–50 and 54–68
Band II	87·5–108	76–108	87–108
Band III	174–230	174–216	174–230
Bands IV and V	470–960 (note)	470–608 and 614–890	470–960

Note: In accordance with RR 703, the broadcasting allocation in Region 1 in the band 862–960 MHz is limited to the African Broadcasting Area, excluding Algeria, Egypt, Libya and Morocco.

(i) The following primary allocations in the framed part of the international Table share with the BS on equal terms:

Region 1: The FS at 790–862 MHz (although the BS retains its exclusive allocation in a number of countries; see RR 695, RR 696 and RR 702). In the African Broadcasting Area, where the band 862–960 MHz is allocated to the BS (see RR 703) this band is shared with the FS and the MS (except AeM).
Region 2: The FS and MS at 806–890 MHz.
Region 3: The FS and MS share all of the BS allocations on equal terms, except 100–108 MHz. Also AeRN at 223–230 MHz and RN at 585–610 MHz.

(ii) A number of footnotes to the international Table allocate bands in one country or a few countries to services other than BS with equal primary status. Thus:

> *AmS:* 50–54 MHz, in several countries in southern Africa; see RR 559.
> *AeRN:* 54–68 MHz, in several countries in southern Africa and in various bands between 200 and 230 MHz in various countries in Region 3; see RR 561, RR 624–RR 626 and RR 636.
> *FS and some or all mobile services:* 54–68 MHz in several countries in southern Africa and Central America, 76–88 MHz in several countries in Region 2 and 174–216 MHz in Mexico; see RR 561, RR 562, RR 576 and RR 620.
> *RL:* 890–942 MHz in Australia; see RR 706.

(iii) There are many other footnote allocations to a wide range of services, some globally applicable and others geographically restricted. These include, for example:

> the BSS at 620–790 MHz,
> various mobile-satellite allocations,
> various aeronautical radionavigation allocations,
> radio astronomy at 608–614 MHz,
> numerous FS and MS allocations in all parts of these bands:
> the mobile service allocations in Region 1 at 87·5–88·0 MHz (RR 581) and 104–108 MHz (RR 587–RR 590) are particularly significant.

However, in all of this third group of sharing allocations, provision is made to protect the status of the BS or the frequency assignment planning of the BS, through the use of permitted allocation status for the sharing service, sharing constraints, consultation in accordance with RR Article 14, limitations on the dates of validity of the allocations or undertakings that harmful interference will not be caused to the BS nor protection claimed from it.

The CCIR has not determined the signal-to-interference ratios required to protect sound broadcasting signals at VHF from the signals typical of other services with sharing allocations, the most important of which are the FS and the MS. However, Chapters 4 and 5 of Annex 4 of the Final Acts of the European VHF/UHF Broadcasting Conference, Stockholm 1961[59] provide co-ordination criteria considered appropriate to interfering FS and MS stations and Chapters 2 and 3 of Annex 5 of the same document offer protection ratios which might be used.

Protection ratios have been agreed for use in implementing sharing between television broadcasting and radionavigation systems around 600 MHz; see CCIR Recommendation 565[60]. A general examination of concepts of sharing between television broadcasting and the FS and MS is to be found in CCIR Report 1087[61] and CCIR Recommendation 655[62] provides information on which protection ratios might be based. However, much remains to be done in this field.

It will be seen from the foregoing that the sharing of BS allocations with other services, and especially the FS and MS, is much more general in Region 3 than in the other Regions. WARC-79 considered that the unco-ordinated development of these various services in these bands might be disadvantageous; the Conference also noted that there were no well-established criteria for sharing spectrum between these services. Accordingly, it was decided that an RARC should be convened to establish technical criteria for sharing between the BS, FS and MS in this part of the spectrum; see RR Resolution 702[63]. This resolution was endorsed by PC-82 and PC-89. This RARC has not yet been convened but progress is being made in developing criteria through less formal meetings.

Typical sound broadcasting systems in service and in prospect

Some countries use part of Band I for sound broadcasting, and some use part of Band II for television. However, the preponderant use of Band II is for sound broadcasting, whilst the preponderant broadcasting use of Band I and the only broadcasting use of Bands III, IV and V, is for television.

With few exceptions, perhaps none, sound broadcasting systems at VHF use frequency modulation (FM) (class of emission F3E) and most systems are designed to provide facilities of high audio-frequency quality. Some stations transmit stereophonic signals. Various new methods of achieving stereophony are under study. Systems capable of transmitting several channels, each carrying sound programme signals or other broadcasting signals, are under study in many countries; see CCIR Recommendation 450[64] and CCIR Reports 300[65] and 463[66].

Typical television broadcasting systems in service and in prospect

All terrestrial television broadcasting at present uses one of two sets of scanning parameters, either 625-lines/25-frames per second or 525-lines/30-frames per second; in both cases each frame is presented as two fields with lines interlaced. The 525-lines/60-fields per second standard is called System M. There are, however, a number of different system specifications for the scanning systems and synchronisation waveforms of systems operating at 625-iines/50-fields per second; at present Systems B, D, G, H, I, K K1, L and N are recognised and in use.

All systems are transmitted using vestigial-sideband amplitude modulation of the vision or luminance carrier (class of emission C3F). The upper sideband is radiated at full level and the lower sideband is limited in bandwidth and partially suppressed. Another carrier, modulated by the sound signal, is located between 4·5 MHz and 6·5 MHz above the vision carrier, depending on the system specification used. Exceptionally in system L the spectrum is the inverse of the normal arrangement, the upper sideband being curtailed and the sound carrier being below the vision/luminance carrier. The sound carrier is usually frequency modulated, but amplitude modulation is used with System L. A number of countries use, or contemplate using, one or more additional sound carriers, typically for signals with digital modulation or alternative languages; these additional carriers are located somewhat below the prime sound carrier. Systems M and N require a nominal radio frequency bandwidth of 6 MHz, System B requires 7 MHz and the remaining systems require 8 MHz.

All colour television systems now in use derive two colour-difference signals for transmission from the three primary colour signals, red, green and blue, which are obtained when the picture is initially scanned and are used when it is finally displayed. All systems except system L use a chrominance sub-carrier located between 3·58 and 4·4 MHz above the vision carrier, the frequency difference between the carrier and the sub-carrier being precisely related to the line frequency. However, three different techniques are used for modulating the sub-carrier with the two colour-difference signals; these are identified as the NTSC, SECAM, and PAL systems. Picture standards and colour information transmission techniques are coupled in various ways, as a result of which 16 sets of television signal standards are in use at present, as follows:

NTSC	*PAL*	*SECAM*
System M	Systems B, D, G I, H, M, N and a variant of N/PAL	Systems B, D, G, H, K, K1 and L

The specifications of these standards are outlined in CCIR Report 624[67], together with a table showing which standard is used, or is planned to be used, in each of the countries in which television is currently broadcast or where intentions to begin broadcasting have been declared.

Having so many different standards for television signals in service at once has disadvantages. One situation in which such disadvantages emerge is when programmes originating in one country are to be shown in another country where different technical standards are used. CCIR Recommendation 472[68] offers a set of video signal parameters for 625-line/50-fields per second scanning which might be used with least inconvenience when exchanging programmes between countries where different 625-line systems are broadcast. CCIR Recommendation 470[69] seeks to prevent further proliferation of system types by recommending that countries starting television broadcasting facilities should choose from systems defined in CCIR Report 624 or CCIR Recommendation 472, or compatible improvements of those systems.

Teletext facilities, supplying information typically in digital form, are sometimes transmitted in time division multiplex with the vision signal of a television broadcast. Alternatively teletext may be transmitted in the same format as a television broadcast signal and be displayed on a suitably equipped television receiver, completely replacing the video signal. A further alternative is to transmit information signals of this kind in a special purpose, narrow-band format; see CCIR Recommendation 653[70] and CCIR Reports 802[71] and 956[72].

In the design of present-day television broadcasting systems, and especially colour television systems, a number of compromises have been made, chiefly to enable colour emissions to be received on monochrome receivers, to limit the radio-frequency bandwidth required and to keep receiver prices within acceptable limits using state-of-the-art manufacturing technology. Development work in progress, having the objective of escaping from the disadvantages of these compromises and improving picture and audio quality, is summarised in CCIR Reports 312[73], 801[74], 1072[75] and 1077[76]; see also system developments addressed in particular to satellite broadcasting and referenced in Chapter 9.

International spectrum management: Region 1

In Region 1 it has become customary for the management of broadcasting allocations to involve frequency assignment planning wherever it has been found to be feasible to do so. The terrestrial use of VHF and UHF bands has been found well suited to this treatment and the Region 1 BS allocations are not encumbered with sharing allocations to an extent that would make planning for the BS difficult to implement.

A conference held in Stockholm in 1961 produced an Agreement[59] containing a frequency assignment plan for sound and television broadcasting in Bands I, II, III, IV and V in the European Broadcasting Area. (This area is defined formally in RR 404 but it can be defined approximately as that part of Region 1 which lies north of 30° North latitude and west of 40° East longitude.) This plan replaced an earlier plan produced at a conference in 1952, also in Stockholm. A conference held in Geneva in 1963 produced an Agreement[77] containing a corresponding plan for the African Broadcasting Area. (This latter area is defined formally in RR 400–RR 403 but it can be defined approximately as Africa south of 30° North, together with the islands in the Atlantic Ocean and Indian Ocean which are in Region 1 and lie between the parallels of 30° North and 40° South.)

Much of the 1961 and 1963 plans, especially the planned television assignments, is still in effect. However, a conference convened in accordance with RR Resolution 510[78] met in 1984 and prepared a frequency assignment plan for FM sound broadcasting in the band 87·5–108 MHz which covered the whole of Region 1 and certain Region 3 countries with land boundaries adjoining Region 1[79]. The 1984 Agreement formally replaced the corresponding elements in the 1961 and 1963 Agreements with effect from 1 July 1987 through the decisions of subsequent conferences in 1985[80, 81]. The technical bases for the frequency assignment plans for television and sound broadcasting, and the existing plans themselves, are reviewed on pp. 170–174 and 174–178 respectively.

A RARC to revise the frequency assignment plan for television in the African Broadcasting Area has been convened late in 1989; see also RR Resolution 509[82].

International spectrum management: Regions 2 and 3

At VHF and UHF it has not been thought necessary to use the mechanism of formal international frequency assignment planning for broadcasting in Region 2. With the exception of the participation of the administrations of a few countries neighbouring Region 1 in the 1984 planning conference, the same is true of Region 3. Apart from the constraints on frequency stability and spurious emissions, mentioned below, the only international element of spectrum management is, therefore, the process whereby the IFRB registers frequency assignments for broadcasting stations; see Sections I, II and IIA of RR Article 12. For Regions 2 and 3, this process is similar to that used for FS assignments above 28 MHz; see pp. 102–104. In short, a notified assignment is registered and dated in accordance with RR 1248 if it does not contravene the RR, no consideration begin given by the IFRB to possible interference prospects. Thus, responsibility for efficient spectrum management rests entirely with the administrations, individually and in collaboration with neighbours.

Technical constraints

RR 303 and RR Appendix 7 require the carrier frequencies of all sound broadcasting transmitters with mean power in excess of 50 W operating in Bands I and II to be maintained within 2 kHz of the nominal frequency. For television transmitters operating in Bands I to V, RR Appendix 7 requires the carrier frequency to be maintained within 500 Hz of the nominal frequency, the tolerance being relaxed to 1 kHz for stations using system M picture standards (NTSC). More severe tolerances are imposed by plan agreements on some Region 1 television emissions.

RR 304 and RR Appendix 8 require the worst spurious emission of any VHF transmitter (Bands I–III) having a mean output power greater than 25 W to be at least 60 dB below the output power level, up to a maximum of 1 mW. In Bands IV and V, the maximum level for spurious emissions is raised to 20 mW, and this constraint is not mandatory until 1994 for transmitters installed before 1985. Somewhat less stringent requirements apply to transmitters of lower power (see also Appendix A.5).

However, it should be noted that a potentially serious spurious emission problem has been identified which may require more stringent constraints to be applied in Band II in the future. The growth of broadcasting in the sub-band 100–108 MHz leads to concern that interference may be caused to radionavigation receivers in aircraft in the adjacent band, 108–117·975 MHz. The possibility of interference to aircraft receivers used for communication (AeM) and operating in the band 117·975–136 MHz has also been noted. The measures that were adopted to deal with this problem at the Geneva conference in 1984 are reviewed on pp. 177–178 below.

National spectrum management

In Region 1 the existence of frequency assignment plans will relieve administrations of all but routine tasks in managing the frequency assignments for broadcasting stations that were included in the appropriate plan. Where changes in planned assignments are desired or new stations are to be established that were not included in the plan the administration will be involved in selecting carrier frequencies that will not cause interference to stations, domestic or foreign, which have planned assignments. A suitable frequency being found, the administration will also be involved in negotiating with other interested administrations the acceptance of an amendment to the plan, to provide for the new station. In these activities, the relevant plan, the associated procedures and the technical bases for spectrum management agreed in the CCIR provide a helpful framework to guide administrations (see below and pp. 174–178).

In Regions 2 and 3 the same functions have to be performed by the administrations but with neither the facilities nor the constraints of a formal plan. Where neighbouring countries have need for large numbers of assignments and their stations are liable to interfere with one another, it will often be helpful to set up a sub-Regional planning group to prepare a local plan. If this is done, or if the administration does its planning work independently with any necessary international co-ordination carried out as a separate process, the technical bases set out below will be found helpful.

In all Regions however, whatever use a country makes of these bands for broadcasting, an important and challenging task for the administration will be to optimise the national use of this range of the spectrum, on the one hand for broadcasting but on the other hand for fixed and mobile stations and indeed for the stations of any other service for which these bands are allocated and required. It is a task which is inextricably interlinked with the management of the same part of the spectrum for the MS; see the 'national spectrum management' discussion on pp. 225–230. It involves detailed co-ordination of plans with any other administration with radio stations within interference distance.

6.5.2 Technical bases of spectrum management for broadcasting at VHF/UHF

The optimisation of the use of the same frequency channels and adjacent channels at different transmitting stations without there being unacceptable interference is an essential part of good spectrum management. The radius of the coverage area of a VHF or UHF broadcasting transmitter is typically to be measured in tens of kilometres and the maximum range at which it is likely to cause significant interference is to be measured in hundreds of kilometres. Thus, except for small and isolated countries, efficient spectrum management involves rather similar principles, whether it is addressed on a national scale or a wider scale. The recommendations and reports of the CCIR and the experience of frequency assignment planning in Region 1 provide substantial assistance to spectrum managers in all Regions, and the most relevant material is outlined below. However, much of the detailed CCIR work on the various television parameters was done with frequency assignment planning for Region 1 in view; this material relates, in some cases, specifically to 625-line/50-field systems and further work may be needed to extend it for application to 525-line/60-field systems.

Ideal patterns of frequency re-use

The location, power etc. of broadcasting stations will depend primarily on the topography of a district and the distribution of the population which each station is to serve. These factors obviously have a substantial impact on the way in which the frequencies which are available for assignment can be re-used within a large geographical area. Nevertheless, there are lessons to be learned from studies of optimised distributions of assignments amongst an ideal geographical pattern of identical stations. CCIR Report 944[83] discusses the application of repeating lattices to idealised patterns of sound broadcasting stations. CCIR Report 1085[84] offers similar material on television broadcasting. A helpful analogy to the VHF or UHF broadcasting case arises with cellular UHF networks of the LM (see pp. 230–232).

System standards

CCIR Recommendation 450[64] gives the standards for maximum frequency deviation and the pre-emphasis characteristics which are currently recommended for monophonic and various kinds of stereophonic sound broadcasting signals. Corresponding material for the much more numerous parameters which define a television emission is to be found in CCIR Report 624[67].

The minimum usable field strength

The use of spectrum is likely to be inefficient if one of the objectives sought in choosing frequencies is that significant interference should not occur even when the wanted signals are too weak to be received satisfactorily. It is therefore necessary to set a lower limit to the level of a wanted signal that will be protected from interference. CCIR Recommendation 412[85] gives minimum usable field strengths estimated to be usable for various kinds of sound broadcast signal in the presence of man-made noise in various environments for use in frequency assignment planning. Corresponding information for use in planning television broadcasting is to be found in CCIR Recommendation 417[86]. It should however be noted that weaker field strengths may be satisfactorily receivable in electrically quiet locations, particularly if receiving antennas of high performance are used; see CCIR Recommendation 412 and Report 409[87].

Protection ratios

CCIR Recommendation 412[85] states protection ratios which should be used for determining the acceptibility of interference to an FM VHF sound broadcasting station from another FM sound broadcasting signal. The protection ratios are expressed as a function of the frequency difference between the wanted and interfering signals and can therefore be used for assessing both co-channel and adjacent channel interference; see also CCIR Report 1064[88]. CCIR Report 945[89] examines various methods for assessing the aggregate effect of interference from multiple sources.

CCIR Recommendation 655[62] provides corresponding protection ratio information for use in planning television assignments; see also CCIR Report 485[90]. CCIR Report 1084[91] discusses the signal-to-interference ratio required to protect digital data signals which are combined in TDM with video signals.

Television systems are very sensitive to interference from similar emissions at the same nominal frequency unless there is very precise control of the wanted and unwanted carrier frequencies. Such interference imposes an annoying, unstable moiré pattern on the picture even when the signal-to-interference ratio exceeds 50 dB. This sensitivity to interference becomes somewhat less if the carrier frequency tolerance is improved to ±500 Hz relative to the nominal frequency. A substantial further reduction in sensitivity to interference is obtained if the vision carrier frequencies are deliberately offset by an integral number of twelfths of the line frequency. Sensitivity can be still further reduced if the frequency offsetting is made very exact and is maintained with a tolerance of ±1 Hz; this is the so-called precision frequency offset technique. Relative to the basic co-channel condition without enhanced frequency tolerances, precision offsetting reduces the required protection ratio by as much as 25 dB; see Recommendation 655[62].

CCIR Report 947[92] provides some preliminary estimates of the signal-to-interference ratios required to protect sound broadcasting signals in Band II from television emissions using D/SECAM, the signal type used by the television stations operating in Band II; see also Chapter 5 of Annex 2 of the Geneva Agreement, 1984[79]. The ITU has also determined protection ratios applicable to interference to broadcasting signals from emissions typical of certain other services which share allocations with the BS and these are referenced on p. 165.

Radio wave polarisation

The CCIR has studied the advantages and disadvantages of using vertically or horizontally polarised waves for broadcasting at VHF and UHF; see

CCIR Reports 122[93] and 464[94]. Situations have been identified in which one polarisation mode is to be preferred to the other, although the advantage does not lie in all respects with the same mode. However, under most conditions depolarisation due to propagation or reflection from objects is not severe, and the use of orthogonal polarisations offers the prospect of significant reduction of interference or alternatively of a significant increase in the efficiency with which spectrum could be used.

Nevertheless, many sound broadcasting receivers are hand-portable and many others are installed in vehicles; for these receivers it has been found that the transmission of a signal of mixed polarisation, typically circular or a slanted-linear polarisation, gives better performance at the receiver. The widespread use of these polarisation modes would prevent interference rejection by orthogonal polarisations. The Geneva Agreement, 1984, gave each Region 1 administration freedom to use the polarisation mode of its choice for sound broadcasting in Band II.

The opportunity to use polarisation discrimination for the reduction of interference in television would be greater. The Stockholm Agreement of 1961 requires television transmissions in the European Broadcasting Area to have either horizontal or vertical polarisation. The option of reducing interference by means of polarisation discrimination was used little, if at all, in preparing the associated plan, but it may be found helpful in modifications of the plan to accommodate additional stations.

Propagation losses

Various basic CCIR texts on radio propagation at VHF are referenced in Appendix B.6.6, but a more convenient source of information, specially selected for use in Band II planning, may be found at Annex 2, Chapter 2 of the Geneva Agreement, 1984[79].

Other frequency planning considerations

Clearly the primary constraints on the frequencies that could be chosen for assignment to a broadcasting station are set by the level of interference from other stations transmitting in the same channel or an adjacent channel. However, satisfactory planning for VHF and UHF broadcasting also requires attention to be given to various other interference situations which are matters of electromagnetic compatibility (EMC). These situations include:

- interference from broadcasting channels at image frequencies of the broadcasting receiver,
- reception of the local oscillator output, or its harmonics, radiated by nearby receivers,
- powerful signals, far removed in frequency from the tuning point of the receiver, which may generate harmonics or intermodulation

products in the receiver input stages which interfere with the wanted signal,

and so on. CCIR Reports 946[95], 1086[96] and 655[62] list various interference mechanisms of this kind and identify means for avoiding the more important of them in planning frequency assignments. CCIR Report 625[97] offers data on the degree of protection that must be provided if typical domestic television receivers are not to be adversely affected through these interference modes.

Receiving antenna directivity

CCIR Recommendations 599[98] and 419[99] provide reference radiation patterns in the horizontal plane for typical domestic receiving antennas, for use in estimating the effect of interference.

6.5.3 The frequency assignment plans for Region 1

The Stockholm Conference of 1961 produced plans for television in the European Broadcasting Area which are still in force. The television plans for the African Broadcasting Area which were produced at the Geneva Conference of 1963 are also still in force, although they are to be replaced soon by new plans. The 1961 and 1963 Conferences also produced plans for sound broadcasting in their respective areas but, with the exception of a relatively small number of assignments in Band I, these were replaced by new Band II plans for the whole of Region 1 which were produced at the Geneva Conference of 1984. The 1961 television plan for Europe and the 1984 sound broadcasting plan are reviewed in this section.

In essence, these plans are schemes of assignments to meet stated requirements, it having been calculated, with a few exceptions which are identified, that if all of the assignments listed in the plans were put into operational use, the aggregate interference within any service area would not exceed the level that is considered acceptable. Each plan is part of an agreement which includes a set of administrative procedures for the flexible management of the plan throughout the foreseen lifetime of the agreement. The technical bases of the plans are also recorded in the agreements. Many of the technical aspects of these agreements and the methodology on which planning is based spring from studies carried out in the CCIR; the current state of these studies is reviewed briefly on pp. 170–174.

When an administration makes an assignment in accordance with one of these plans and notifies the IFRB accordingly, the assignment is automatically entered in the MIFR in accordance with RR 1245 and RR 1248. All such entries in the Register have the same status, irrespective of the date of bringing into service. Provision for co-ordinating desired departures from the Plan with other administrations with planned assignments that might be adversely affected is to be found in Article 4 of each Agreement.

The European Television Agreement, 1961

The Stockholm Agreement and the frequency assignment plans for television which are annexed to it are contained in the Final Acts of the conference[59]. All five of the VHF/UHF broadcasting bands are involved, as follows:

Band I: The plan for television was drawn up to cover the frequency range 41–68 MHz, although the band allocated has since been reduced to 47–68 MHz. The plan provides for a mixture of systems using 405, 625 and 819 lines per frame, and no clear pattern of carrier spacing emerges. This situation has recently been simplified by the withdrawal from use of 405-line and 819-line systems. The plan also provides for some FM sound broadcasting in Band I, virtually all of the assignments being between 65 and 68 MHz.
Band II: For most of Europe the 1961 plan fills the band 87·5–100 MHz with assignments for FM sound broadcasting. This element of the plan has since been replaced by the 1984 plan. However, a relatively small number of assignments for 625-line television were also made in this band, including an extension of the band downwards from 87·5 MHz.
Band III: The plan for television covers the band 174–230 MHz, with an extension downwards to 162 MHz to cover assignments to countries with appropriate footnote allocations for the BS. As in Band I, there is provision for systems using 405, 625 and 819 lines and a complex pattern of carrier spacing results. Here again, subsequent withdrawal from use of 405-line and 819-line systems has simplified the situation.
Bands IV and V: For television, the band between 470 and 958 MHz is divided into 61 channels, each 8 MHz wide, numbered from 21 (470–478 MHz) to 81 (950–958 MHz) although the plan does not contain any assignments to channels above number 69 (854–862 MHz), in keeping with RR 703. The plan provides for a variety of systems, with various required bandwidths but all using 625 lines per frame and none needing more bandwidth than 8 MHz. In every channel the nominal frequency of the vision carrier is 1.25 MHz above the lower limit of the channel.

Throughout the plan there is an unstated assumption that all vision/luminance carriers will be vestigial-sideband amplitude modulated (class of emission C3F). In all Bands the polarisation of television transmitting antennas has been specified as vertical or horizontal. Vision carrier frequency offsetting has been widely used to permit co-channel operation of stations, the geographical separation of which would otherwise be insufficient to limit interference. In general, an omnidirectional transmitting antenna and the power level specified by the responsible administration is assumed. In a relatively small number of cases, where interference would be excessive without further precautions, the use of a precision frequency offset, a limitation on the antenna gain over a specified arc in azimuth or a constraint on radiated power has been made a condition of the assignment.

The Region 1 sound broadcasting agreement, 1984

The Geneva Agreement, 1984 and the frequency assignment plans for sound broadcasting which are annexed to it are contained in the Final Acts of the Conference[79]. It covers all of Region 1, together with Afghanistan and Iran, which have land boundaries with Region 1 countries.

The Plan is limited to Band II, covering the frequency band 87·5–108 MHz. The basic structure of the plan is simple. All planned emissions are for FM sound broadcasting, class of émission F3E, although certain television emissions operating in Band II as agreed in the Stockholm 1961 plan are allowed for, using technical standards which are set out in Chapter 5 of Annex 2 of the Agreement. A uniform nominal carrier spacing of 100 kHz is used. Five transmission systems are provided for, as follows:

	Maximum frequency deviation
System 1: monophonic	± 75 kHz
System 2: monophonic	± 50 kHz
System 3: stereophonic, polar modulation system	± 50 kHz
System 4: stereophonic, pilot-tone system	± 75 kHz
System 5: stereophonic, pilot-tone system	± 50 kHz

The propagation data used in generating the plan are set out in detail in Chapter 2 of Annex 2 of the Agreement.

The procedure to be used when a contracting administration wishes to modify the plan is set out in Article 4 of the Agreement, and data which will be necessary or helpful in carrying out the co-ordination process that this involves are to be found in Annexes 4 and 5 to the Agreement.

It was not possible to decide with confidence, during the course of the 1984 Conference, that the interference that would occur between certain pairs of desired planned assignments would be acceptably low. Many of these possible assignments will have been discarded in the course of the preparation of the plan. Others, however, where perhaps the prospective interference was thought likely to be marginally unacceptable at worst, were retained for further investigation. These were listed for the period up to 1992–93, not in the main body of the Plan, but in an appendix to it. These assignments may be used for the time being. Meanwhile the administrations concerned are to study the interference levels actually experienced and seek acceptable ways of eliminating whatever problems may arise; see Article 6 of the Agreement.

The Plan was drawn up without regard for the various footnote allocations, time-limited or permitted, which share Band II with other services in Region 1. However, Resolutions 2 and 3 of the Final Acts of the Conference[79] indicate how these sharing problems are to be solved. In effect, the stations of the mobile service operating in the sub-band 87·5–88 MHz may continue to use existing assignments until the end of 1990, by which time their operation must be transferred, if necessary, to alternative frequencies which will not interfere with broadcasting stations operating in accordance with the Plan. The MS and FS stations operating in the sub-band 104–108 MHz may continue to operate up to 1995, or whatever earlier date may be specified in the relevant footnote, but the administrations concerned have agreed that this use will be flexible, so that it does not impede the gradual implementation of the Plan for broadcasting stations; details of this transition are to be co-ordinated between the administrations concerned.

Out-of-band interference to aeronautical receivers

It had been foreseen that the introduction of broadcasting with powerful transmitters in the band immediately below 108 MHz, previously used by the FS and MS with low power transmitters, might cause interference by various mechanisms to aeronautical radionavigation receivers operating in the band 108–117·975 MHz; see RR Recommendation 704[100], CCIR Report 929[101] and CCIR Recommendation 591[102]. The 1984 plan Agreement[79] contains, in particular in Chapter 7 of Annex 2, an analysis of the possibly troublesome mechanisms that were identified at the Conference. In brief it was concluded that the AeRN receivers at risk were airborne Instrument Landing System (ILS) or VHF Omnidirectional Range (VOR) system receivers and four possible interference mechanisms were identified:

Mechanism Type A1: Airborne AeRN receivers operating in the band 108–117·975 MHz might suffer interference from intermodulation products or other spurious emissions radiated by broadcasting stations that were assigned frequencies below 108 MHz.
Mechanism type A2: Out-of-band emissions radiated by broadcasting stations with assigned frequencies immediately below 108 MHz might interfere with airborne AeRN receivers operating immediately above 108 MHz.
Mechanism type B1: Intermodulation products in the band 108–117·975 MHz might be generated by non-linearities in the input stages of airborne AeRN receivers from strong broadcasting signals received at frequencies below 108 MHz.
Mechanism type B2: Airborne AeRN receivers might be desensitised by strong broadcasting signals received below 108 MHz.

RR 343 requires the frequency assignments to the stations of a service to:

> '... be separated from the limits of the band allocated to this service in such a way that . . . no harmful interference is caused to services to which frequency bands immediately adjoining are allocated.'

Some at least of these interference mechanisms may breach RR 343. These interference prospects raise concern in particular because human life might be endangered by false interpretation of navigational signals.

Various measures were incorporated into the broadcasting frequency planning process to reduce the incidence of interference problems of these kinds. Techniques were devised for identifying the kinds of situation which increased the likelihood of interference, to enable means to be found for averting interference in specific cases. Where the problem could arise wholly within the jurisdiction of one administration, it is a responsibility of that administration to identify and evaluate the risks and solve problems if they are found to be significant. Where interference to AeRN receivers could arise, in accordance with criteria given in the Plan, from broadcasting stations in another country, the administrations concerned are to co-operate to determine whether a problem exists and if so, to solve it; see Article 5 of the Agreement.

Resolution 4 of the Agreement seeks to ensure that these various precautions will be applied, not only by the parties to the Agreement but also by administrations that did not sign the agreement but might become exposed to the problem. The Conference proposed (Recommendation 7 of the Agreement) that RR Appendix 8 should be amended by some competent future WARC to put a more stringent limit on spurious emissions from broadcasting transmitters in the frequency bands in question. Finally, Recommendations 4, 5 and 6 of the Agreement call for further study of various aspects of this problem.

It was recognised that this interference problem arises, in part, from the use in aircraft of receivers, the performance of which is no longer to be considered adequate. Some easing of the constraints on the BS is foreseen, through the replacement of AeRN receivers, with effect from 1998; see Recommendation 5 of the Agreement.

6.6 References

1 'Terrestrial television broadcasting in the 12 GHz band (Band VI)', CCIR Report 961–1; Recommendations and Reports of the CCIR, 1986, Volume XI–1 (ITU, Geneva, 1986)

2 'Relating to the modification of carrier frequencies of LF broadcasting stations in Region 1'. Resolution 500; Radio Regulations, edition of 1982 (ITU, Geneva, 1982)

3 'Relating to the preparation of a broadcasting plan in the band 1605–1705 kHz in Region 2'. Recommendation 504; *ibid.*

4 Final Acts of the Regional Administrative LF/MF Broadcasting Conference (Regions 1 and 3), Geneva 1975 (ITU, Geneva 1975)

5 'Necessary bandwidth of emissions in LF, MF and HF broadcasting; CCIR Recommendation 639'. Recommendations and Reports of the CCIR, 1986, Volume X–1 (ITU, Geneva, 1986)

6 'Spectra and bandwidth of emissions'. CCIR Recommendation 328–6; *ibid.*, Volume I

7 'Terms and definitions used in frequency planning for sound broadcasting'. CCIR Recommendation 638; *ibid.*, Volume X–1

8 'Radio-frequency protection ratios in LF, MF and HF broadcasting'. CCIR Recommendation 560–2; *ibid.*, Volume X–1

9 'Objective measurement of radio-frequency protection ratios in LF, MF and HF broadcasting'. CCIR Recommendation 559–1; *ibid.*, Volume X–1

10 Final Acts of the Regional Administrative MF Broadcasting Conference (Region 2), Rio de Janeiro, 1981 (ITU, Geneva, 1982)
11 'Practical aspects of MF broadcasting coverage'. CCIR Report 616–3; Recommendations and Reports of the CCIR, 1986, Volume X–1 (ITU, Geneva, 1986)
12 'Factors influencing the limits of amplitude-modulation sound broadcasting coverage in band 6 (MF)'. CCIR Recommendation 598; *ibid.*, Volume X–1
13 'Transmitting antennas in LF and MF broadcasting'. CCIR Report 401–5; *ibid.*, Volume X–1
14 'Ionospheric cross-modulation in the LF and MF broadcasting bands'. CCIR Recommendation 498–1; *ibid.*, Volume X–1
15 'Frequency assignments to stations in the low-power channels'. Appendix 1 to the plan produced by the Regional Administrative LF/MF Broadcasting Conference (Regions 1 and 3), Geneva 1975 (ITU, Geneva, 1978)
16 'Radiocommunication vocabulary'. CCIR Recommendation 573–2; Recommendations and Reports of the CCIR, 1986, Volume XIII (ITU, Geneva, 1986)
17 'Maximum transmitter powers for broadcasting in the tropical zone'. CCIR Recommendation 215–2; *ibid.*, Volume X–1
18 'Fading characteristics for sound broadcasting in the tropical zone'. CCIR Report 304–2; *ibid.*, Volume X–1
19 'Choice of frequency for sound broadcasting in the tropical zone'. CCIR Recommendation 48–2; *ibid.*, Volume X–1
20 'Protection ratio for sound broadcasting in the tropical zone'. CCIR Recommendation 216–2; *ibid.*, Volume X–1
21 'Interference to sound broadcasting in the shared bands in the tropical zone'. CCIR Report 302–1; *ibid.*, Volume X–1
22 'Determination of the effects of atmospheric noise on the grade of reception in the tropical zone'. CCIR Report 303–3; *ibid.*, Volume X–1
23 'Relating to implementation of the changes in allocations in the bands between 4000 and 27 500 kHz'. Resolution 8; Radio Regulations, edition of 1982 (ITU, Geneva, 1982)
24 'Relating to the convening of a World Administrative Radio Conference for the planning of the HF bands allocated to the broadcasting service'. Resolution 508; *ibid.*
25 'Operation of HFBC transmitters in the extended bands above 10 MHz'. Resolution 512; Final Acts of WARC-HFBC-87; see[26] below
26 Final Acts of the World Administrative Radio Conference for the planning of the HF bands allocated to the broadcasting service (HFBC-87) (ITU, Geneva, 1987)
27 'Procedure in connection with the International Frequency Registration Board'. Article 11; Radio Regulations, edition of 1947 (ITU, Geneva, 1947)
28 'Relating to HF broadcasting'. Recommendation 503; Radio Regulations, edition of 1982 (ITU, Geneva, 1982)
29 'Characteristics of systems in LF, MF and HF broadcasting'. CCIR Report 458–4; Recommendations and Reports of the CCIR, 1986, Volume X–1 (ITU, Geneva, 1986)
30 'Transition from double-sideband (DSB) to single-sideband (SSB) emissions in the HF bands allocated exclusively to the broadcasting service'. RR Resolution 517; Final Acts of WARC-HRBC-87; see[26] above
31 'Single sideband (SSB) system for HF broadcasting'. CCIR Recommendation 640; Recommendations and Report of the CCIR, 1986, Volume X–1 (ITU, Geneva, 1986)
32 'Characteristics of single-sideband systems in HF broadcasting'. Report 1059; *ibid.*, Volume X–1
33 'Channel spacing for sound broadcasting in Band 7 (HF)'. CCIR Recommendation 597–1; *ibid.*, Volume X–1
34 'Propagation prediction method to be used in the HF bands allocated exclusively to the broadcasting service'. Recommendation 512; Final Acts of WARC-HFBC-87; see[26] above

35 'Fading allowances in HF broadcasting'. CCIR Recommendation 411–3; Recommendations and Reports of the CCIR, 1986, Volume X–1 (ITU, Geneva, 1986)
36 'Minimum AF and RF signal-to-noise ratio required for broadcasting in Band 7 (HF)'. CCIR Report 1058; *ibid.*, Volume X–1
37 'Relative RF protection ratio values for single-sideband (SSB) emissions in the HF bands allocated exclusively to the broadcasting service'. Recommendations 517; Final Acts of WARC-HFBC-87; see[26] above
38 'Multiple frequency use per programme in HF broadcasting'. CCIR Recommendation 410; Recommendations and Reports of the CCIR, 1986, Volume X–1 (ITU), Geneva, 1986)
39 'Directional antennas in HF broadcasting'. CCIR Recommendation 80–2; *ibid.*, Volume X–1
40 'Presentation of antenna diagrams'. CCIR Recommendation 414; *ibid.*,Volume X–1
41 'Transmitting antennas in HF broadcasting'. CCIR Report 32–5; *ibid.*, Volume X–1
42 'A set of simplified HF antenna patterns for planning purposes'. CCIR Report 1062; *ibid.*, Volume X–1
43 'Book of antenna diagrams, 1984' (ITU, Geneva, 1984)
44 'Synchronised transmitters in HF broadcasting'. CCIR Recommendation 205–2; Recommendations and Reports of the CCIR, 1986, Volume X–1 (ITU, Geneva, 1986)
45 'Use of synchronised transmitters in the HF bands allocated exclusively to the broadcasting service'. Recommendation 516; Final Acts of WARC-HFBC-87, see[26] above
46 'Tentative High Frequency Broadcasting Schedule'. Published four times per year (ITU, Geneva)
47 'High Frequency Broadcasting Schedule'. Published four times per year (ITU, Geneva)
48 'Annual High Frequency Broadcasting Frequency List'. Published annually on microfiche (ITU, Geneva)
49 'Improvements to the HFBC planning system and the consultation procedure'. Resolution 515; Final Acts of WARC-HFBC-87, see[26] above
50 'Programme of action relating to the improvement, testing, adoption and practical implementation of the planning system for the high frequency bands allocated exclusively to the broadcasting service, and associated provisions'. Resolution 511; *ibid.*
51 'Participation by administrations in the improvement of the planning system for the HF bands allocated exclusively to the broadcasting service'. Recommendation 509; *ibid.*
52 'Introduction of transmitters and receivers capable of both double sideband (DSB) and single sideband (SSB) modes of operation'. Recommendation 515; *ibid.*
53 'Improvement in the use of the HF bands allocated exlusively to the broadcasting service by avoiding harmful interference'. Resolution 513; *ibid.*
54 'Antennas to be used for the planning of the HF bands allocated exclusively to the broadcasting service'. Resolution 516; *ibid.*
55 'Improvements to the propagation prediction method to be used for the HF bands allocated exclusively to the broadcasting service'. Recommendation 514; *ibid.*
56 'HF broadcast receivers'. Recommendation 518; *ibid.*
57 'Possibility of extending the frequency spectrum allocated exclusively to HF broadcasting at a future competent world administrative radio conference'. Recommendation 511; *ibid.*
58 'Broadcasting for national coverage in the HF bands'. Recommendation 513; *ibid.*
59 Final Acts of the European VHF/UHF Broadcasting Conference, Stockholm, 1961 (ITU, Geneva, 1961)

60 'Protection ratios for 625-line television against radionavigation transmitters operating in the shared bands between 582 and 606 MHz'. CCIR Recommendation 565; Recommendations and Reports of the CCIR, 1986, Volume XI–1 (ITU, Geneva, 1986)
61 'Frequency sharing between the broadcasting service (television) and the fixed and mobile services below 1 GHz'. CCIR Report 1087; *ibid.*, Volume XI–1
62 'Radio-frequency protection ratios for AM vestigial sideband television systems'. CCIR Recommendations 655; *ibid.*, Volume XI–1
63 'Relating to the convening of a Regional Administrative Radio Conference to establish criteria for the shared use of the VHF and UHF bands allocated to fixed, broadcasting and mobile services in Region 3'. Resolution 702; The Radio Regulations, edition of 1982 (ITU, Geneva, 1982)
64 'Transmission standards for FM sound broadcasting at VHF'. CCIR Recommendation 450–1; Recommendations and Reports of the CCIR, 1986, Volume X–1 (ITU, Geneva, 1986)
65 'Stereophonic or multi-dimensional sound in frequency-modulation sound broadcasting'. CCIR Report 300–6; *ibid.*, Volume X–1
66 'Transmission of several sound programmes or other signals with a single transmitter in frequency-modulation sound broadcasting'. CCIR Report 463–4; *ibid.*, Volume X–1
67 'Characteristics of television systems'. CCIR Report 624–3; *ibid.*, Volume XI–1
68 'Video-frequency characteristics of a television system to be used for the international exchange of programmes between countries that have adopted 625-line colour or monochrome systems'. CCIR Recommendation 472–2; *ibid.*, Volume XI–1
69 'Television systems'. CCIR Recommendation 470–2; *ibid.*, Volume XI–1
70 'Teletext systems'. CCIR Recommendation 653; *ibid.*, Volume XI–1
71 'Additional broadcasting services using a television or narrowband channel'. CCIR Report 802–2; *ibid.*, Volume XI–1
72 'Data broadcasting systems, signal and service quality, field trials and theoretical studies'. CCIR Report 956–1; *ibid*, Volume XI–1
73 'Constitution of a system of stereoscopic television'. CCIR Report 312–4; *ibid.*, Volume XI–1
74 'The present state of high-definition television'. CCIR Report 801–2; *ibid.*, Volume XI–1
75 'Suitable sound systems to accompany high-definition and enhanced television systems'. CCIR Report 1072; *ibid.*, Volume X–1
76 'Enhanced quality television systems'. CCIR Report 1077; *ibid.*, Volume XI–1
77 Final Acts of the African VHF/UHF broadcasting conference, Geneva, 1963 (ITU, Geneva, 1963)
78 'Relating to the convening of a planning conference for sound broadcasting in the band 87·5–108 MHz for Region 1 and certain countries concerned in Region 3'. Resolution 510; Radio Regulations, edition of 1982 (ITU, Geneva, 1982)
79 Final Acts of the Regional Administrative Radio Conference for the planning of VHF sound broadcasting (Region 1 and part of Region 3), Geneva 1984 (ITU, Geneva, 1984)
80 Final Acts of the Regional Administrative Radio Conference of the Members of the Union in the European Broadcasting Area to revise certain parts of the Stockholm Agreement (1961), Geneva, 1985 (ITU, Geneva, 1985)
81 Final Acts of the Regional Administrative Radio Conference of the Members of the Union in the African Broadcasting Area to abrogate certain parts of the Geneva Agreement (1963), Geneva, 1985 (ITU, Geneva, 1985)
82 'Relating to the convening of a Regional Broadcasting Conference to review and revise the provisions of the Final Acts of the African VHF/UHF Broadcasting Conference, Geneva, 1963'. Resolution 509; Radio Regulations, edition of 1982 (ITU, Geneva, 1982)
83 'Theoretical network planning'. CCIR Report 944; Recommendations and Reports of the CCIR, 1986, Volume X–1 (ITU, Geneva, 1986)

84 'Planning methods for terrestrial television in VHF/UHF bands'. CCIR Report 1085; *ibid.*,Volume XI–1
85 'Planning standards for FM sound broadcasting at VHF'. CCIR Recommendation 412–4; *ibid.*, Volume XI–1
86 'Minimum field strengths for which protection may be sought in planning a television service'. CCIR Recommendation 417–3; *ibid.*, Volume XI–1
87 'Boundaries of the television service area in rural districts having a low population density'. CCIR Report 409–4; *ibid.*, Volume XI–1
88 'Protection ratios for FM sound broadcasting in the cases of substantial frequency differences and of interference from overmodulated transmitters'. CCIR Report 1064; *ibid.*, Volume X–1
89 'Methods for the assessment of multiple interference'. CCIR Report 945–1; *ibid.*, Volume X–1
90 'Contribution to the planning of broadcasting services'. CCIR Report 485–1; *ibid*, Volume XI–1
91 'Protection ratios for digital data signals multiplexed with television signals'. CCIR Report 1084, *ibid.*, Volume XI–1
92 'Radio-frequency protection ratio required by FM sound broadcasting in the band between 87·5 and 108 MHz against interference from D/SECAM television transmissions'. CCIR Report 947; *ibid.*, Volume X–1
93 'Advantages to be gained by using orthogonal wave polarizations in the planning of broadcasting services in bands 8 (VHF) and 9 (UHF)'. CCIR Report 122–3; *ibid.*, Volume XI–1
94 Polarisation of emissions in frequency-modulation broadcasting for Band 8 (VHF)'. CCIR Report 464–4; *ibid.*, Volume X–1
95 'Frequency planning constraints on FM sound broadcasting in Band 8 (VHF)'. CCIR Report 946; *ibid.*, Volume X–1
96 'Frequency planning constraints (625-line systems)'. CCIR Report 1086; *ibid.*, Volume XI–1
97 'Characteristics of television receivers and receiving antennas essential for frequency planning'. CCIR Report 625–3; *ibid.*, Volume XI–1
98 'Directivity of antennas for the reception of sound broadcasting in band 8 (VHF)'. CCIR Recommendation 599; *ibid.*, Volume X–1
99 'Directivity of antennas in the reception of television broadcasting'. CCIR Recommendation 419–1; *ibid.*, Volume XI–1
100 'Relating to the compatibility between the broadcasting service in the band 100–108 MHz and the aeronautical radionavigation service in the band 108–117·975 NHz'. Recommendation 704; Radio Regulations, edition of 1982 (ITU, Geneva, 1982)
101 'Compatibility between the broadcasting service in the band of about 87–108 MHz and the aeronautical services in the band 100–136 MHz'. CCIR Report 929–1; Recommendations and Reports of the CCIR, 1986, Volume VIII–3 (ITU, Geneva, 1986)
102 'Compatibility between the broadcasting service in the band of about 87–108 MHz and the aeronautical services in the band 108–136 MHz. CCIR Recommendation 591–1; *ibid.*, Volume VIII–3

Chapter 7

The mobile services

7.1 Introduction to the mobile services

Three definitions in RR Article 1 are intended to determine what is meant by 'the mobile service' (MS).

> RR 65 defines a mobile station as 'a station in the mobile service intended to be used while in motion or during halts at unspecified points.'
> RR 67 defines a land station as 'a station in the mobile service not intended to be used while in motion.'
> RR 26 defines the MS as 'a radiocommunication service between mobile and land stations or between mobile stations'.

Thus, the station at one end of a radio link in the MS is always mobile, being on a ship, an aircraft, or a land vehicle or being hand-portable. The station at the other end of the link may also be mobile but more usually it is at a specified location on land.

RR 65 acknowledges that a mobile station, which may be operated whilst the vehicle in which it is installed is in motion, is still properly classifiable as a mobile station if it is operated whilst the vehicle is at rest. It can be argued that a transportable station that is not designed to be operated whilst in motion, but may be set up for operation when the vehicles which carry it are at rest, is a mobile station within the terms of RR 65. The RR are ambiguous on this point and transportable microwave stations, such as those use to provide temporary television links in 'outside broadcasting' are commonly treated as mobile stations. However, it is probably more constructive to classify a transportable station, technically unsuitable for operation whilst in motion, as a station of the FS, the frequency assignments of which should be co-ordinated and notified for use at operating locations that can be foreseen.

The Radio Regulations also define three specialised mobile services, the maritime mobile service (MM), the aeronautical mobile service (AeM) and the land mobile service (LM) covering facilities for ships, aircraft and land vehicles respectively. Specific frequency bands have been allocated for each of these services in addition to those for the MS, the latter being available for the use of any kind of mobile station. Terms have been defined for the mobile and non-mobile stations of each of these services (see Table 7.1). The Table also indicates the paragraphs of the RR in which the formal definitions of these services and stations are to be found.

Table 7.1 Terms for stations in the mobile services (formal definitions for terms are in the RR paragraphs indicated)

Service	Stations which operate at permanent locations	Stations which operate when in motion or at unspecified halts
Mobile (RR 26)	land station (RR 67)	mobile station (RR 65)
Maritime mobile (RR 30)	coast station (RR 70)	ship station (RR 72)
Aeronautical mobile (RR 34)	aeronautical station (R 76)	aircraft station (RR 78)
Land mobile (RR 28)	base station (RR 68)	land mobile station (RR69)

Many frequency allocations for the AeM are designated for either the AeM (R) or the AeM (OR); the symbol R stands for 'route' and OR stands for 'off-route'. The AeM (R) bands are reserved for assignments for links between any aircraft and the aeronautical stations, such as air traffic control stations, which are concerned with safety and regularity of flight along civil air routes; see RR 34A[1] and RR 3630[1]. The AeM (OR) bands are reserved for links between aircraft and aeronautical stations used for other purposes; see RR 34B[1] and RR 3631[1].

There are frequency allocations for the MS and for the specialised mobile services in many parts of the radio spectrum. It is convenient, for regulatory purposes, to divide these allocations into two groups, those below and those above 28 MHz. The spectrum management of these two groups is considered respectively on pp. 188–213 and 214–238.

Two further definitions should be noted. The port operations service (defined in RR 32) is used in the vicinity of ports for messages relating to the operational handling of ships and the safety of ships and persons. The ship movement service (RR 33) has similar applications at a distance from port. No frequency allocations have been made specifically for these services, but channels are allotted to them in the channelling plan which has been agreed for the MM at 160 MHz (see p. 224). The port operations and ship movement services should be regarded, not as services in the sense in which the word is ordinarily used in spectrum management, but as specifically identified activities within the MM.

Distress and safety communication

The protection which radio communication could provide for human life in danger at sea did much to hasten the growth of the use of radio by ships in the early years of the twentieth century and to bring it under international regulation. The provision of an internationally agreed regulatory framework within which radio is used to bring aid effectively to ships and aircraft in distress or in other emergencies remains an important part of the activities of the ITU. This commitment has been carried out in recent years in collaboration with the International Maritime Organisation (IMO) and the International Civil Aviation Organisation (ICAO). Distress communication is largely an operational matter but it affects to some degree the regulation of the use of radio by mobile stations for other purposes.

Articles 25 and 31 of the new ITU Constitution give absolute priority to messages concerned with the safety of life and to distress messages. Article 32 of the Constitution binds administrations to take whatever steps are required to stop the transmission or circulation of false distress calls. RR Articles 37 to 42 set out the principles of distress and safety communication; for the spectrum manager, the main elements of these principles are as follows:

(*a*) Provisions for distress and safety communication are intended to apply without discrimination to all kinds of mobile station, whether ships, aircraft or manned spacecraft and the use of these facilities to serve disasters on land is not excluded. Obviously, the choice of operating frequencies will be affected by the nature of the stations involved and their equipment.

(*b*) A number of frequencies in bands allocated to one or other of the mobile services have been designated for use for distress and safety messages, and for communication between stations involved in rescue operations. The frequencies and their functions are identified on pp. 205–211 and 217–224; see also Chapter 10 for distress arrangements in the mobile-satellite services.

(*c*) RR 964 prohibits emissions capable of causing harmful interference to communication on the international distress and emergency frequencies and calls for adequate protection of supplementary distress frequencies.

(*d*) Emergency position-indicating radio beacons (EPIRBs) operate at several of the designated safety frequencies. Technical characteristics for beacons operating at 2182 kHz are defined in RR Appendix 37[2]; see also RR 2973[1] and CCIR Recommendation 439[3]. For information on beacons operating at 121·3, 156 and 243 MHz, see RR Appendix 37A[1] and CCIR Report 749 [4].

(*e*) Standard radiotelegraph and radiotelephone alarm and warning signals have been defined (RR 3268–RR 3286[1]); see also RR Appendix 36[1] and CCIR Recommendation 219[5]. The purpose of such signals is to draw attention to a distress message or some other emergency message (such as a urgent cyclone warning or a vital navigation warning) which is to follow the signal.

The regulations emphasise that these provisions are not intended to exclude the use of any other means, including other radio means, which may be available to help a mobile station in distress (RR 2933[2]–RR 2935[2]).

A major current programme, under the aegis of the IMO and the ITU, is bringing the system up-to-date; see Kent[6] and CCIR Report 747[7]. Many other texts of CCIR Study Group 8 report on progress in this programme, which is creating the Global Maritime Distress and Safety System (GMDSS). WARC-MOB-83 and WARC-MOB-87 established the regulatory basis for the GMDSS; see in particular RR Articles N37[1] to N41[1]. These regulations enter into effect on 1 July 1991; see section 8.1 of RR Article 69[1]. However, they will not fully displace the regulatory basis for the existing system which is provided by RR Articles 37 to 41, until a date, some years further into the future, which is to be determined by a future WARC in the light of decisions by IMO on the implementation of the GMDSS; see RR Resolution 331[8].

Operational regulations

For the mobile services, the functions of the ITU extend beyond the regulation of the use of the radio spectrum, covering many operational matters. The regulation of distress and safety communications is reviewed above. Many of the other operational matters have little relevance to radio spectrum management and they are not discussed here. However, some indication of the scope of these operational concerns of the ITU may be grasped from the fact that:

- RR Articles 42A[1], 43 to 51 and 51A[1] relate to manning, operation and operational procedures of stations in the aeronautical mobile service,
- RR Articles 54 to 66 relate to the corresponding aspects of the operation of stations in the maritime mobile service,
- RR Articles 67 and 68 relate to the operation of the land mobile service,
- RR Appendices 11 and 12 relate to the operation of ship and aircraft stations,
- RR Appendices 13 and 15 contain various signalling codes, used mainly in the maritime mobile service, and
- RR Appendix 41 relates to the procedure for using radio direction-finding facilities.

Many of these texts were revised at WARC-MOB-83[2], at WARC-MOB-87[1] or at both conferences.

Electromagnetic compatibility

Radio receivers of any service are susceptible to interference from unwanted signals received at frequencies which, being very close to the frequency of the wanted signal, are within the pass-band of the receiver filters. However, strong unwanted emissions outside the pass-band of the receiver may also cause interference, due to blocking or intermodulation in the early stages of the receiver. Interference from unwanted emissions and spurious emissions at other specific frequencies may also be troublesome, typically when those emissions are at image frequencies.

Mobile stations are particularly susceptible to interference of these latter kinds, because the receiver is physically close to the associated transmitter and it is therefore exposed to strong interference from its emissions, wanted and unwanted. There may also be strong interference from the transmitter of another mobile station which is nearby. Mobile stations are particularly susceptible to interference from the local transmitter if the system operates in the duplex mode (that is, the receiver may be receiving while the transmitter is transmitting).

When several or many transmitters and receivers are operating in a limited space, and especially on board a ship or an aircraft, signals and spurious emissions generated by the transmitters will be present on many frequencies. A large number of additional unwanted signals at the combination frequencies of the output signals of all of the transmitters and their harmonics will be generated by intermodulation. These intermodulation products are typically generated in the wide-band input stages of receivers and by the flow through non-ohmic junctions of multi-signal currents induced into metallic structures close to the transmitters. The reception of spurious emissions from other receivers may also cause interference. Thus, the risk that interference will arise from local sources is high whenever more than one transmitter is operating in a mobile station. Electromagnetic compatibility (EMC) analysis is required in the design of equipment and of assemblies of equipment to ensure that problems of this kind do not seriously degrade the performance of systems.

Electromagnetic compatibility can be achieved, to a greater or lesser degree, by the elimination of these various sources of interference, through:

- specifications for transmitters and receivers which place severe limits on spurious radiation and which require receivers to tolerate high levels of unwanted signals without degradation of the reception of the wanted signal,
- careful screening of transmitters,
- appropriate design and construction of buildings and vehicles, and

- attention to the critical factors in the installation and testing of radio equipment.

The level of spurious emissions from mobile transmitters may have to be made much lower than that required by RR 304 and RR Appendix 8 for transmitters in general (see Appendix A.5 below) and strict limits may have to be applied to the radiation of energy from receivers, for example at local oscillator frequencies.

Administrations ensure that EMC is given due attention by requiring stations to be equipped with type-approved equipment as a condition of licensing and by inspection of stations. There is guidance in this matter in CCIR Reports 739[9], 1019[10], 660[11] and 661[12] and CCIR Recommendation 218[13]. A review of related problems in the design of spacecraft structures is to be found in CIR Report 1049[14]. There is also an extensive literature on this subject.

Despite attention to the electromagnetic compatibility of installations, there may be residual interference from mechanisms of these kinds. It will sometimes be necessary to take such interference into account in selecting working frequencies for a station. However, this may deny to spectrum managers the option of making the frequency assignments that would be preferable for other reasons.

7.2 The mobile services below 28 MHz

7.2.1 Survey of MS regulation below 28 MHz

Between 4 MHz and 28 MHz, most of the allocations to mobile services in the international Table of Frequency Allocations (RR Article 8) have been made specifically to the MM, the AeM or the LM. There are further allocations below 4 MHz but these have mostly been made to the MS in general, or to the MS with the AeM specifically excluded. Numerous footnotes to the international Table indicate different frequency limits or sharing conditions applicable in specified sub-Regional areas, especially in frequency bands below 4 MHz, but the situation is not affected radically by this. It is convenient therefore to generalise, disregarding the footnotes, and to review the allocations in terms of the bands to which ship, aircraft and land vehicle stations respectively have substantially Region-wide or global access.

The frequency allocations for maritime use

(*a*) *The LF and VLF allocations:* Much of the spectrum belw 160 kHz is allocated for the MM, sharing with the FS, the RN or both. The use made of these allocations for the MM is declining.

(*b*) *The 500 kHz band:* The spectrum between 415 and about 525 kHz is allocated for the MM, some being shared with the AeRN. The Third

International Convention for the Safety of Life at Sea (1948) requires that all passenger and cargo ships of 1600 gross tonnage and upwards be equipped and manned to use this band. The international distress frequency for radiotelegraphy is 500 kHz and the band 495–505 kHz is allocated for the MS and is therefore available, not only to ships but also to other kinds of mobile station in distress.

(*c*) *The MF allocations:* Most of the bandwidth from 4063 kHz down to the upper limit of the medium wave broadcasting band (1606·5 kHz in Regions 1 and 3, 1705 kHz in Region 2) is allocated for the MM but the allocations are shared, in an intricate pattern of frequency limits and Regional variations, with various other services, such as the FS, AeM, LM, AmS, BS and various radiodetermination services. The mobile services are the dominant users of this part of the spectrum and use for MM is substantial, especially in Region 1.

(*d*) *The exclusive HF allocations:* Six long-established MM allocations, at 4, 6, 8, 12, 16 and 22 MHz, span the HF spectrum. All of these allocations have world-wide primary status and are exclusive to the MM. WARC-79 decided to increase the width of the bands at 12, 16 and 22 MHz by changing the allocation of adjacent bands from FS to MM. The allocation of narrow bands at 18·8 and 19·7 MHz was also changed from FS to MM at the same Conference. Also, narrow bands at 25·1 and 26·1 MHz, previously shared by MM and FS, were made exclusive to the MM at WARC-79. A phased transfer plan was devised to allow established FS assignments in the bands newly allocated to MM to be replaced by assignments in other bands before extensive use for ships of the new MM allocations began; see RR 532, RR 544 and RR Resolution 8[15]. This changeover period ended on 1 July 1989. All together these exclusive MM allocations now cover about 19% of the total HF spectrum. The frequency limits of all of these bands are given in Table 7.5. The use of these bands is subject to detailed frequency plans (see pp. 198–204).

(*e*) *The shared HF allocations:* Seventeen other HF bands are allocated for the MM, although the use of some of these bands by the MM is subject to geographical or operational constraints. All of these allocations are shared with other services, typically FS and LM, and in some bands the MM allocation has secondary status. The feasibility of planning channels for MM operation in some of these shared bands is under consideration (see p. 198).

The frequency allocations for aeronautical use

(*f*) *The 500 kHz band:* The band 495–505 kHz is available to the AeM, but only for distress and other safety traffic.
(*g*) *The MF allocations:* There are various AeM allocations between 1·6 and 2·5 MHz in all the three Regions. These allocations are limited to the AeM

(OR) in Region 1 and they are shared in all Regions with other services, typically FS, MM, LM, BS or one of the radiodetermination services.

(*h*) *The HF allocations:* About 30 relatively narrow bands, distributed throughout the spectrum from 2·5 to 24 MHz, are allocated for the AeM, all with primary status. Most of these allocations are world-wide and exclusive to AeM although a few are shared, typically with FS and LM, and there are some Regional variations in band frequency limits. Together they occupy about 10% of the total HF spectrum. These bands fall into two groups of approximately equal total bandwidth, one group designated for AeM (R) and the other for AeM (OR). The frequency limits of all of these bands are given in Table 7.6. The use of these bands is, in general, subject to frequency plans (see pp. 204–209).

(*i*) *Communication between aircraft and ships:* Aircraft stations may communicate with ship stations using MM frequency allocations but such use by aircraft must conform with the regulatory provisions which govern the use of frequencies by ships in those allocations; see RR 3571[1], RR 4143 and RR 4144.

(*j*) *The use of HF for public correspondence:* RR 3633[1] forbids the use of assignments in exclusive AeM allocations for circuits to be used by the public. See also RR Resolution 409[16]. RR 3571[1] and RR 4143–RR 4145 permit aircraft to use MM frequency allocations for this purpose but relatively little use is made of this option at HF. Regulatory provision for aeronautical public correspondence facilities is currently being developed at frequencies above 28 MHz (pp. 215–216).

The frequency allocations for land mobile use

(*k*) *The 500 kHz band:* The band 495–505 kHz is allocated for the MS and therefore is available to the LM, but for distress and other safety traffic only.

(*l*) *The MF and HF bands:* A number of bands below 28 MHz, mostly between 1·6 and 8·1 MHz and between 23 and 28 MHz, have been allocated for LM, mostly with secondary status. All of the allocations are shared with other services, typically other mobile services and the FS. The lower frequency bands show intricate Regional variations in band limits and sharing conditions.

Frequency band sharing

A few bands around 500 kHz and around 2 MHz which are allocated for the MM are the subject of frequency assignment plans and services sharing these bands have been given permitted or secondary allocation status.

Conversely, the AeRN is planned at 415–435 kHz in Region 1, as a result of which MM has permitted status in that band (see also pp. 396–398). There is information on technical aspects of sharing between these services in CCIR Report 910[17]. In general, however, sharing does not raise major issues of regulatory principle below 28 MHz because assignments to all services in shared bands, with the exception of the planned assignments mentioned above, are regulated by the same international procedures. However, the availability of these bands to two and often more services tends to reduce the spectrum capacity available to any one of them.

CCIR Recommendation 48[18] urges that the allocations for the FS and MS which share bands with the BS (limited to tropical broadcasting) at 2·4, 3·3, 4·9 and 5·0 MHz should be used for assignments to FS stations in preference to those of the MS, since the superior frequency stability to be expected of FS emissions should cause less interference to broadcast reception.

There is information in CCIR Recommendation 589[19] and CCIR Reports 915[20] and 911[21] on technical and operational aspects of sharing between various services at LF and HF.

Typical systems in operation

It is characteristic of systems of the MM below 28 MHz that they provide a temporary telephone or telegraph link, often quite brief, between a ship at sea and a coast station or between one ship and another. Often the link is between radio operators at the two ends, transmitting the text of a message or a telegram. Sometimes the link is extended to a crew member or to a passenger on the ship and to a telephone or telex subscriber ashore, through the public switched network or by private wire. For telegraphy, manual morse operation is common but the extent of use of direct printing systems with ARQ is growing.

For either telephony or telegraphy, it is usual for one spot frequency in each maritime frequency band to be designated as a common 'calling frequency'. Operators wishing to set up communication with a distant station use the calling frequency to make the initial contact; both stations then usually change to an agreed 'working frequency' to pass the traffic.

The MF MM allocations are suitable for radio links up to a few hundred kilometres in length. The 500 kHz band offers longer range, up to about 1000 km, but use is limited by international agreement to radiotelegraphy. The coverage offered by the HF allocations is potentially global, subject to the vagaries of ionospheric propagation.

The AeM resembles the MM in that brief links are established between operators, typically to carry air traffic control messages between aircraft in flight and aeronautical stations on the ground.

Land mobile stations operating below 28 MHz fall into two main groups:

(*i*) They may be installed in land vehicles and operated while the vehicle is in motion for telephone or more rarely, telegraphic, communication over

short distances with another mobile station or a base station, typically by ground wave propagation.
(*ii*) Various hand-portable low-power radio devices such as short-range telemetry and telecommand systems and cordless telephones are also part of the LM.

These short-distance LM systems, particularly (*a*), are liable to cause and suffer interference when ionospheric conditions favour long distance propagation at the frequency that is used. This can be avoided by a suitable choice of operating frequency; see p. 193.

International spectrum management

An extensive body of technical and operational regulations, recommendations and advice has been developed through the ITU to assist the mobile services and more especially the MM to use the radio spectrum below 28 MHz with the high efficiency that is made desirable by the importance of these services and the congestion of the medium.

Radio channel plans and associated frequency allotment plans have been implemented for three sets of MS HF frequency allocations, namely:

- the exclusive bands allocated to MM between 4 and 23 MHz,
- the exclusive bands allocated to AeM (R) between 2·85 and 22 MHz, and
- the exclusive bands allocated to AeM (OR) between 2·5 and 24 MHz.

A further 8 bands allocated to the AeM (OR) are shared with other services, and some degree of planning for AeM (OR) has been agreed for these bands also.

These plans are reviewed on pp. 198–204 and 204–209. The plan agreements incorporate various technical constraints on stations using planned channels and these are noted on pp. 204 and 207–208. Regulatory constraints which apply to all mobile systems are reviewed on pp. 194–198.

Various channelling schemes have been agreed for other MM allocations and these are reviewed on pp. 209–213. No frequency allotment or frequency assignment plans have been implemented yet on a global or Regional scale for other AeM or LM allocations below 28 MHz.

Special procedures, noted on pp. 201–208, are used in notifying to the IFRB assignments to the various kinds of land station in bands for which allotment plans have been agreed. Assignments of transmitting frequencies to land stations in unplanned bands are subject to the same considerations and procedures as are applied to assignments to fixed stations; (see pp. 84–86). The assignment of frequencies to land stations, to be used for receiving emissions from mobile stations, may also be notified for registration in the MIFR; see RR 1219. However, notification is not required for the assignment of certain specific frequencies (such as distress frequencies) which may be used by all stations in common; see RR 1220.

Frequency assignments to mobile stations are neither notified to the IFRB nor registered in the MIFR. However, the registration of both transmit and receive frequencies assigned to the associated land station provides some international recognition of the frequencies which mobile stations use.

National spectrum management

National spectrum management in the mobile service in the HF band and below has the following main aspects.

In the MM, the practice followed in assigning frequencies to coast stations must combine operational adequacy with efficiency in spectrum utilisation. In the exclusive HF maritime mobile bands, assignments for duplex telephony will be made in the context of the frequency allotment plan which is reviewed on pp. 201–202. Assignments to coast stations for some other purposes should conform to a channelling plan (see pp. 202–203 and 211) and it will often be desirable to co-ordinate the assignments informally with the administrations of neighbouring countries.

The policy to be followed in assigning frequencies to ship stations depends greatly on the nature of the facilities for which the assignments are to be used. In many cases a ship station is authorised to use, not one, but several or many radio channels, and the channel which is used on any particular occasion is often determined by the coast station with which the ship station wishes to communicate.

In the AeM (R), the assignment of frequencies to both aeronautical and aircraft stations is a joint responsibility of the administration and the national civil aviation regulating body, the latter having the more immediate responsibility. These responsibilities are discharged in collaboration with ICAO.

The LM at HF is not a well-characterised service and frequency assignment practice is likely to vary considerably from country to country according to national circumstances. Long distance interference to or from short distance links can be avoided by using a frequency by day (typically below 5 MHz) which is so low that sky wave signals are absorbed in the ionosphere and by using another frequency at night which is so high (typically above 20 MHz) that sky wave signals pass through the ionosphere without reflection.

An administration will wish to ensure that internationally agreed technical principles, devised to ensure efficient use of the radio spectrum, are implemented where appropriate to the land and mobile stations under its jurisdiction. This will, no doubt, be achieved mainly by making the issue of a licence to a station conditional on the installation of type-approved equipment, the requirement being supported by inspection of stations. The administration will also need to ensure that operating staff are adequately trained and that efficient operating practices are employed. Both of these concerns would probably be discharged jointly with other parties, such as government bodies responsible for the safety of life at sea and in the air.

Call signs or other means of identification must be assigned to stations; see RR Article 25[1], RR Appendix 42 and pp. 74–75 above. The principles used in the formation of identity numbers for use in multi-tone selective calling systems in the MM are defined in RR 2134–RR 2148; these principles and various aspects of their application are developed further in RR Article 62[1], RR Appendices 39 and 44 and CCIR Recommendation 257[22]. For the formulation of digital identity numbers and their use in digital selective calling systems, see RR 2149[2], Section III of RR Article 62[1], RR Appendix 43[1], RR Resolutions 320[23] and 336[24] and CCIR Recommendations 493[25], and 491[26].

Information on call signs for all mobile stations and additional data on maritime mobile stations is reported to the ITU for inclusion in the 'Service Documents' which the ITU publishes, in particular for the operational use of mobile stations; see RR Article 26[1].

7.2.2 Technical constraints on systems

There are carrier frequency tolerance limits and constraints on spurious emissions in the RR which apply to all stations of the mobile services, regardless of the frequency bands in which they operate. For frequencies below 28 MHz, these limits are reviewed below.

The Radio Regulations, the various current frequency plan agreements and the recommendations and reports of CCIR Study Group 8 contain a great deal of other material relating to the mobile services. Some of this material sets international standards of equipment design and interfaces, some establishes standard operational procedures and some presents technical information of value to system designers. Important though this material is, it is not directly relevant to spectrum management.

However, these texts also impose constraints on system parameters and operating practices which are intended to increase the efficiency with which spectrum is used, to reduce interference and to make possible a disciplined use of spectrum which facilitates the implementation of less labour-intensive operating techniques and procedures. Most of this latter material relevant to stations of the MM and the AeM is contained in the various frequency plan agreements. The measures which these agreements incorporate would, no doubt, contribute to good spectrum utilisation in any frequency band but they have regulatory force only for stations which operate in the frequency bands which are planned; they are therefore reviewed on pp. 198–209, along with the other provisions of those plan agreements.

The ITU has not, so far, established a general framework for enhancing the efficiency of the use that the LM makes of the spectrum below 28 MHz. Exceptionally, CCIR Recommendation 494[27] offers preferred technical characteristics for SSB radiotelephone systems of the LM operating at MF and HF.

Frequency tolerance

RR 303 requires that transmitting stations conform to the frequency tolerances specified in RR Appendix 7, as modified by WARC-MOB-87[1]. For MS transmitters operating below 28 MHz the list of tolerances applicable in the various services and frequency bands, for the various types of emission and to the various kinds of station, is quite intricate. This intricacy has arisen in two ways:

- There is a general need to reduce inefficiency of spectrum utilisation due to carrier frequency errors; this trend towards stringent tolerances is held back to some degree by a wish to ensure that existing equipment, perhaps old but still serviceable, which may not be capable of attaining the frequency stability standards of up-to-date equipment, does not have to be discarded.
- Secondly, however, new types of emission and new operating practices are coming into use in some mobile facilities and these are effective only if carrier frequencies are maintained with high precision.

The first of these motives has led to the application, from 1990, of somewhat more stringent tolerances than were applied before 1990 to transmitters used for basic, well-established facilities. These new tolerances are typically in the range of 20 to 200 parts per million, according to the service, the frequency band and the kind of station.

The second trend is leading towards much more stringent tolerances of between ±5 and ±50 Hz, regardless of the absolute frequency, for stations using SSB for telephony and for certain specified emissions used for telegraphy. The entry into force of the most severe of these tolerances, agreed at WARC-MOB-87, has been delayed until the beginning of 1992.

The frequency tolerances applied by the RR to MM stations are shown in Table 7.2. The corresponding figures for AeM and LM stations are given in Table 7.3 and Table 7.4.

Spurious emissions

RR 304 requires that transmitting stations do not radiate spurious emissions at levels exceeding those specified in RR Appendix 8 (see Appendix A.5 below). For transmitters with output power in the range typical of the MS below 28 MHz, this provision requires the mean power of any spurious component to be at least 40 dB below the output power of the transmitter or less than 50 mW, whichever is the more stringent requirement. Minor relaxations of these terms are provided for hand-portable transmitters of less than 5 W output and for mobile station transmitters.

In order that an adequate measure of EMC may be achieved, it will often be necessary for spurious emissions from transmitters at stations of the MS to be suppressed to a much greater degree than RR Appendix 8 requires and for radiation from receivers to be minimised also (see pp. 187–188 above).

Table 7.2 Transmitter carrier frequency tolerances below 28 MHz in the maritime mobile service from 1990 (RR Appendix 7 as amended at WARC-MOB-87[1])

Stations and emissions (note 1)	Tolerances in parts per million (p.p.m.) or absolute frequency (Hz) in stated frequency ranges and carrier power ranges (PEP for SSB telephony)		
	below 535 kHz	1·6 to 4·0 MHz	above 4 MHz
1 F1B emissions used for direct-printing telegraphy or data transmission			
(*a*) for narrow band PSK,			
coast stations	5 Hz	5 Hz	5 Hz
ship stations	5 Hz	5 Hz	5 Hz
(*b*) for FSK (note 2),			
coast stations	15 Hz	15 Hz	15 Hz
ship stations	40 Hz	40 Hz	40 Hz
2 For any transmitter used for digital selective calling installed after 1 January 1992 (note 3)	10 Hz	10 Hz	10 Hz
3 SBB telephony:			
coast stations		20 Hz	20 Hz
ship stations		40 Hz	50 Hz
4 A1A emissions			
coast stations	100 p.p.m.	< 200 W; 100 p.p.m. > 200 W; 50 p.p.m.	10 p.p.m.
ship stations	200 p.p.m.	50 p.p.m.	10 p.p.m.
5 Other classes of emission:			
coast stations	100 p.p.m.	< 200 W; 100 p.p.m. > 200 W; 50 p.p.m.	20 p.p.m.
ship stations	200 p.p.m.	40 Hz	50 Hz (note 4)

Notes

1 The tolerances applicable to ships' emergency transmitters and transmitters on survival craft are less stringent that those for ship stations.

2 The tolerance shown applies to transmitters installed before 1 January 1992. For later installations the tolerance for both coast stations and ship stations is 10 Hz.

3 The tolerance shown will apply to all transmitters after the date of full implementation of GMDSS.

4 A tolerance of 40 p.p.m. applies to mobile stations on board small craft operating in the band 26 175–27 500 kHz in or near coastal water utilising A3E, F3E or G3E emissions with a carrier power not exceeding 5 W.

Table 7.3 Transmitter carrier frequency tolerances below 28 MHz in the aeronautical mobile service from 1990 (RR Appendix 7 as amended at WARC-MOB-87[1])

	Tolerances in parts per million (p.p.m.) or absolute frequency (Hz) in stated frequency ranges and carrier power ranges (PEP for SSB telephony)		
Stations and emissions	below 535 kHz	1·6 to 4·0 MHz	above 4 MHz
1 SSB telephony, in AeM (R) exclusive bands			
(a) aeronautical stations		10 Hz	10 Hz
(b) aircraft stations on international routes		20 Hz	20 Hz
(c) aircraft stations on national routes		50 Hz	50 Hz
2 Other classes of emission			
(a) aeronautical stations	100 p.p.m.	< 200 W; 100 p.p.m. > 200 W; 50 p.p.m.	< 500 W; 100 p.p.m. > 500 W; 50 p.p.m.
(b) aircraft stations	100 p.p.m.	100 p.p.m.	100 p.p.m.

7.2.3 Planning in the HF exclusive MM frequency bands

Six frequency bands spread from 4 MHz to 23 MHz have been allocated exclusively for the MM worldwide since 1947. Most long distance traffic is transmitted in these bands, and plans for dividing them into channels for ships to use have been in existence since 1947, being revised from time to time as expanding requirements made it desirable to increase the number of channels and improvements in receiver selectivity and carrier frequency tolerances made it feasible to sub-divide them. WARC-79 extended three of these bands and allocated four more bands exclusively for the MM. WARC-MOB-87 made channelling plans to cover these extension bands,

Table 7.4 Transmitter carrier frequency tolerances below 28 MHz in the land mobile service from 1990 (RR Appendix 7 as amended at WARC-MOB-87[1])

Stations and emissions	Tolerances in parts per million (p.p.m.) or absolute frequency (Hz) in stated frequency ranges and carrier power ranges (PEP for SSB telephony)		
	below 535 kHz	1·6 to 4·0 MHz	above 4 MHz
1 SSB telephony			
(a) base stations		< 200 W; 50 Hz > 200 W; 20 Hz	< 500 W; 50 Hz > 500 W; 20 Hz
(b) land mobile stations		40 Hz	50 Hz (see note)
2 F1B telegraphy land mobile stations		40 Hz	40 p.p.m.
3 Other classes of emission			
(a) base stations		< 200 W; 100 p.p.m. > 200 W; 50 p.p.m.	20 p.p.m.
(b) land mobile stations		50 p.p.m.	40 p.p.m.

Note: A tolerance of 40 p.p.m. applies to SSB telephony transmitters at LM mobile stations operating in the band 26 175–27 500 kHz at a PEP not exceeding 15 W.

covering also a band allocated for the MM at 25 MHz which had previously been unplanned. These extension band arrangements enter into force on 1 July 1991. The frequency limits of these planned bands are given in Table 7.5. These bands, taken together, can conveniently be regarded as eight working bands, at 4, 6, 8, 12, 16, 18/19, 22 and 25/26 MHz, the 18/19 MHz and 25/26 MHz bands both being split.

The feasibility of including the bands 4000–4063 kHz and 8100–8195 kHz, which are shared with equal primary status by the FS and the MM, in the planned scheme is under consideration; see RR 517[1], RR Resolution 319[28], RR 4212A[1] and sections C-1 and C-2 of RR Appendix 16[1]. The other HF bands in which MM has a shared allocation are not currently being considered for planning, although three bands near 25 and 26 MHz were taken into account in the planning to some degree; see RR 4210[1].

Table 7.5 The exclusive MM HF bands

Band designation (MHz)	Band limits before the WARC-79 additions (kHz)	Additional allocations made at WARC-79 (kHz)
4	4063–4438	
6	6200–6525	
8	8195–8815	
12	12 330–13 200	12 230–12 330
16	16 460–17 360	16 360–16 460 and 17 360–17 410
18/19		18 780–18 900 and 19 680–19 800
22	22 000–22 720	22 720–22 855
25/26	25 070–25 110 (see note)	25 110–25 210 and 26 100–26175

Note: This band was not planned prior to WARC-MOB-87.

For operational purposes the eight planned working bands are each divided into three parts. Two parts of each working band are designated for duplex telephony. The third part is sub-divided into sub-bands to be used for each of the other facilities for which ships use the HF spectrum. The frequency limits of these various parts and sub-bands, applicable from 1991, are stated in RR 4195[1]–RR4207[1] and tabulated in RR Appendix 31[1]. The same paragraphs of the RR, in the versions agreed at WARC-MOB-83[2], give the frequency limits in force before that date. Fig 7.1 illustrates the detailed sub-banding structure of the 12 MHz band for the period up to 1991, by way of example.

A frequency allotment plan was agreed at a WARC in 1967 and revised at another WARC in 1974 for the parts of the six pre-WARC-79 bands which were identified for duplex telephony. Preparations for bringing the plan up-to-date were made at WARC-MOB-87; see below.

No frequency allotment or frequency assignment plans have been drawn up for the sub-bands that are used for MM facilities other than duplex telephony. However, the use of these sub-bands by ships has been made systematic by means of agreed channelling plans for ships' transmitters. The sub-bands used by coast station transmitters have not, in general, been formally structured. Various constraints and various system standards have been agreed for implementation in these bands. The channelling plans and their management, the constraints and the standards are discussed further below, in most cases in the terms of the situation that will come into force in 1991 as a result of the decisions taken at WARC-MOB-87. The situation prior to 1991 is not greatly different and is usually to be found in the RR text as it was before WARC-MOB-87 amended it.

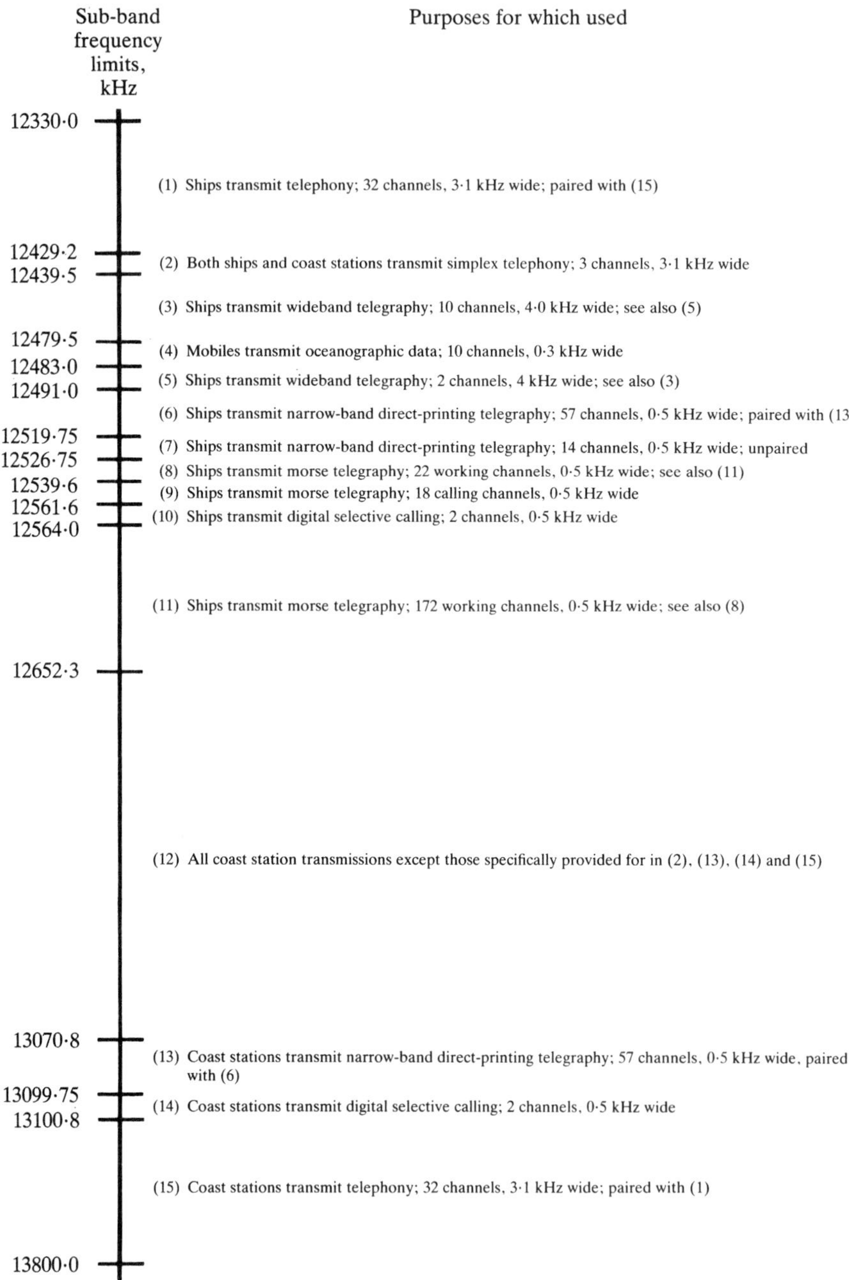

Fig. 7.1 Sub-band structure of the 12 MHz exclusive HF MM frequency band prior to 1991 (spectrum not precisely to scale). The structure of the other exclusive HF bands is similar, although the bandwidth available for each purpose differs from band to band

The duplex radiotelephone plan

Some of the eight working bands are wide but others are narrow; they range in total bandwidth from 1050 kHz (at 16 MHz) to 215 kHz (at 25/26 MHz). In consequence, the widths of the parts designated for duplex radiotelephony in the various working bands are also unequal. However, the two duplex radiotelephony parts in any one working band are equal in width; one is used by ships' transmitters and the other by coast station transmitters Each part is divided into channels just wide enough for an SSB emission. The channels are paired, a coast station channel being permanently associated with the corresponding ship channel. One channel-pair in each band is designated for calling; all the rest are working channel-pairs. RR Appendix 16[2] shows the channel frequencies in use up to 1 July 1991 and the frequencies to be used thereafter are in RR Appendix 16[1].

The coast station working channels then available were allotted by a WARC in 1974 to countries (or broadly defined regions within large countries) in a frequency allotment plan which is set out in RR Appendix 25. Each administration assigns allotted channels as appropriate to the coast stations within its jurisdiction. When a coast station and a ship station with business to transact have made initial contact over a calling channel, the coast station changes its transmitting frequency to a working channel which has been assigned to it and the ship's transmitter is retuned to the ship channel corresponding to the channel to which the coast station is being retuned.

The number of channel-pairs in the RR Appendix 25 plan in the various working bands ranges from 6 to 41. Even the larger number is greatly exceeded by the total of the allotments for which administrations have stated requirements. Thus each channel-pair is allotted in the plan to many different countries or different regions of a large country; in some working bands channel-pairs are allotted 40 times over. These allotments have been selected, having regard to typical radio propagation conditions and expected traffic flows, with the objective of ensuring that interference will be acceptably mild or infrequent, but clearly the minimisation of interference depends also on good operational practice.

Procedures for adding new allotments to the plan, to provide for new requirements or to replace existing allotments which have proved to be unsatisfactory, typically because of excessive interference, are to be found in RR Article 16. RR Article 16 also contains, at RR 1720, a procedure that is used for deleting from the plan allotments which have not been taken into use within a reasonable period of time after the making of the allotment. RR Appendix 25 is revised and reissued from time to time, in accordance with RR 1722, to carry into the published record the additions and deletions that arise from these procedures.

When an administration assigns a channel-pair to a coast station, the transmitting assignment and the receiving assignment are notified to the IFRB for registration in the MIFR in accordance with section IIB of RR Article 12 as amended by WARC-MOB-83[2] and WARC-MOB-87[1]. If the assignment is fully in accordance with the plan, the details of the assignment are recorded in the MIFR. If the assignment is in accordance with the technical constraints of the plan but there is no corresponding allotment, there is a procedure whereby an appropriate entry can be made in the MIFR.

The frequency allotment plan set out in RR Appendix 25, drawn up in 1974, implements a channel bandwidth of 3·1 kHz. WARC-MOB-87 decided that the channel bandwidth should be reduced to 3·0 kHz in order to provide additional channel-pairs. WARC-MOB-87 also extended channel planning to include additional spectrum which had become available as a result of new allocations for the MM made at WARC-79. Thus, RR Appendix 16[1] shows the channel frequencies that will apply from 1 July 1991, taking both of these changes into account. RR Resolution 326[29] provides a plan for the transfer of operations from the frequencies listed in RR Appendix 25 to the frequencies associated with the same channel-pair number at 0001 hours UTC on July 1991. RR Resolution 325[30] records an agreed procedure for the allotment to coast stations of the additional channel-pairs which the reduction of the channel bandwidth and access to additional planned spectrum will provide.

Only SSB suppressed carrier emissions (J3E) may be used for telephony in these bands; see RR 4371[1].The upper sideband is radiated. RR 4328[1] requires SSB transmitters to conform to the characteristics given in RR Appendix 17[1]. RR 4373[1] requires coast stations to use the minimum transmitter power that will cover their service area in these bands and sets an absolute maximum of 10 kW PEP per channel. RR 4374[1] limits to 1·5 kW PEP the power of ships' transmitters.

The sub-bands not used for duplex radiotelephony

The parts of the eight working bands which are not planned for duplex telephony have been divided into sub-bands and are used for other maritime facilities. The frequency limits of the sub-bands in the six pre-WARC-79 bands are set out in RR 4195[2]–RR 4209[2] and RR Appendix 31[2]; see for example Fig. 7.1. The new frequency limits agreed at WARC-MOB-87[1] which enter into force at 0001 hours UTC on 1 July 1991 are shown in RR 4195[1]–RR 4209[1].

These sub-bands are managed as follows:

(*a*) One sub-band in each band is used by both coast stations and ship stations for simplex telephony. These sub-bands are divided into channels which are just wide enough for an SSB emission and the technical constraints applicable to the use of the duplex telephony channels (see above) apply in the simplex channels also, except that the power of coast station transmitters is limited to 1 kW PEP. In the current channelling plan, to be found in section B of RR Appendix 16[2], some channels are 3·0 kHz wide but most are 3·1 kHz wide. In the plan that enters into force in 1991 (see RR Appendix 16[1]) all channels are 3·0 kHz wide. These channels are used for operator-to-operator communication, ship-to-shore or inter-ship; see also RR Recommendation 304[31].

(*b*) Two sub-bands in each band are used for digital selective calling, one sub-band for transmission from ships and the other for coast station transmitters. In the plan that enters into force in 1991 each ship sub-band will be either 2·0 or 1·75 kHz wide, each coast station sub-band will be 1·75 kHz wide and the sub-bands will be divided into 3 or 4 channels, each 0·5 kHz wide. Section III of RR Article 62, taking into account where appropriate the amendments incorporated in the Final Acts of WARC-MOB-87[1], gives the operational procedures for digital selective calling; see also RR 4323AB[1]–RR 4323AE[1].

(*c*) Two sub-bands in each band are used for narrow-band direct-printing telegraph and data transmission systems, one sub-band being used for ships' transmissions and the other for coast station transmissions. These systems operate at speeds below 100 bauds with FSK emissions or at 200 bauds with PSK emissions, The sub-bands are divided into radio channels 0·5 kHz wide, each ship channel being paired with the corresponding coast station channel. The channel frequencies are listed in RR Appendix 32, the frequencies that will be appropriate from 1991 being in RR Appendix 32[1]. The carrier frequencies, transmitting and receiving, corresponding to the various channels, are assigned by administrations to coast stations. Ship stations are authorised to transmit in whichever channel is assigned for reception at the coast station with which they are communicating. RR Resolution 300[32] qualifies the normal procedure for the registration of these coast station assignments. Some technical characteristics of equipment used on these channels are set out in RR Appendix 38[1] and in CCIR Recommendations 627[33] and 625[34].

(*d*) With a few exceptions, separate sub-bands have been defined in each band for transmissions from ships for five other groups of facilities. (The exceptions are that there is no provision for morse (items (i) and (ii) below) in the 18/19 MHz band, nor for oceanographic data transmission (item (v) below) in either the 18/19 MHz or the 25/26 MHz bands.) These sub-bands are used as follows:

(i) As calling bands for morse telegraphy, using A1A or A1B emissions at up to 40 bauds. These sub-bands are channelled in accordance with RR Appendix 34; see also RR Appendix 34[1]. Administrations assign one calling channel or more to each of their ship stations using these frequency bands and RR 4277[1]–RR 4285[1] and RR Resolution 312[35] indicate how this should be done to achieve efficient calling.

(ii) As working bands for morse operation using A1A or A1B emissions at up to 40 bauds. The basic channelling plan for these sub-bands is for channels 0·5 kHz wide and the plan itself is given in RR Appendix 35, the plan effective from 1991 being at RR Appendix 35[1]. Note (*e*) to RR Appendix 31 provides for the eventual sub-division of these basic channels, increasing the number of working frequencies available. Administrations assign working frequencies to their ships, seeking to spread traffic over the available bandwidth to reduce the incidence of interference; see RR 4306[1].

(iii) As working bands for narrow-band direct-printing telegraph and data transmissions operating at less than 100 bauds, not using paired channels (see (*c*) above). From 1991 these sub-bands may also be used for A1A and A1B morse. These sub-bands are divided into radio channels 0·5 kHz wide, as shown in RR Appendix 33, the plan effective from 1991 being at RR Appendix 33[1]. Administrations assign the frequencies of these channels to ships requiring to use them; see RR Resolution 335[36].

(iv) For wide-band telegraphy, facsimile and special transmission systems. These sub-bands are sub-divided into radio channels 4 kHz wide, as indicated in RR Appendix 31, the plan effective from 1991 being at RR Appendix 31[1]. The carrier frequencies are assigned by administrations to ships having need of them; see RR 4323AW[1]–RR 4323BC[1].

(v) For the transmission of oceanographic data. These sub-bands are sub-divided into radio channels 0·3 kHz wide, as indicated in RR Appendix 31, the plan effective from 1991 being at RR Appendix 31[1]; see RR 4323BE[1] and RR 4323BF[1].

(*e*) Finally, there is a sub-band in each working band which is used by coast station transmitters for the five groups of facilities for which ships use the sub-bands listed in (*d*) above. The bandwidth available for these purposes ranges from 22·5 kHz at 25/26 MHz to 418·5 kHz at 12 MHz. There is no channelling plan for these sub-bands; administrations assign frequencies to coast stations, and notify the IFRB accordingly for registration in the MIFR in the way that is conventional for the HF bands (see pp. 84–86). Exceptionally, special treatment is given to notifications of coast station receiving frequencies for narrow-band direct-printing telegraphy in unpaired channels; see RR Resolution 335[36].

Coast station transmitters are limited by various regulations in RR Article 60; typically to a mean power of 5 kW at 4 and 6 MHz, 10 kW at 8 MHz and 15 kW at higher frequencies. Exceptionally, a limit of 1 kW PEP is applied for simplex telephony.

7.2.4 Frequency planning for the AeM at HF

Twelve frequency bands between 2·85 and 22 MHz are allocated exclusively for the AeM (R) service and all but one of these allocations are world-wide. Ten frequency bands, also spanning the HF spectrum, are allocated exclusively for the AeM (OR) service, and all but one allocation are world-wide in this case also. There are eight further AeM (OR) allocations in one or more of the ITU Regions but all of these are shared. Table 7.6 shows the frequency limits for all of these bands. RR 3632[1] requires frequencies in these bands to be assigned in accordance with frequency allotment plans.

Table 7.6 The planned AeM HF bands

AeM (R) (kHz)	AeM (OR) Exclusive allocations (kHz)	AeM (OR) Shared allocations (kHz)
		2505–2850 (Region 2)
2850–3025	3025–3155	3155–3230
3400–3500		3800–3900 (Region 1)
	3900–3950 (Region 1)	3900–3950 (Region 3)
		4438–4650 (Region 2)
4650–4700	4700–4750	4750–4850 (Region 1)
5450–5480 (Region 2)		5450–5480 (Regions 1 and 3)
5480–5680	5680–5730	
6525–6685	6685–6765	
8815–8965	8965–9040	
10 005–10 100		
11 275–11 400	11 175–11 275	
13 260–13 360	13 200–13 260	
	15 010–15 100	
17 900–17 970	17 970–18 030	
21 924–22 000		23 200–23 350

Frequency allotment plans were drawn up for the HF bands allocated for the AeM (R) and AeM (OR) services at the International Administrative Aeronautical Radio Conference in 1949. These plans were adopted by the Extraordinary Administrative Radio Conference in 1951 and they were incorporated with some up-dating into the RR as Appendix 26[37] at WARC-59.

The AeM (R) Plan was revised at the WARC on the Aeronautical Mobile (R) Service in 1978, and incorporated into the RR, with the addition of a plan for the 22 MHz band, as Appendix 27[38] at WARC-79.

Relatively limited up-dating amendments to RR Appendix 26 with respect to the AeM (OR) Plan are to be found in the Final Acts of WARC-MOB-87[1]. PC-89 concluded that further up-dating of the AeM (OR) Plan was necessary, including revision of the bases of the plan to take advantage of advances in technology since 1949, and in particular the introduction of SSB emissions. Accordingly, PC-89 instructed the IFRB to prepare proposals of plan amendments, to be agreed bilaterally with the administrations concerned and to be implemented (subject to the approval of a WARC to be held in 1992) by 15 December 1992.

These two frequency allotment plans, as they now stand, and the arrangements for registering frequency assignments to aeronautical stations, are reviewed below.

The channels at 3023 and 5680 kHz may be used by stations of both the AeM (R) and the AeM (OR) services under conditions laid down in RR Appendix 27 (Part II, Section II, Article 3)[38] and also by any other station of the mobile services for the co-ordination of a search and rescue operation; see RR 505[1], RR 2980, RR 2984, RR N 2980[1] and RR N 2984[1] and also RR Resolution 403[39].

The allotment plan for the AeM (R) service

The plan in RR Appendix 27[38] is based on the assumption that A3J emissions will be used for radiotelephony or alternatively any of several specified kinds of radiotelegraph emission which occupy no more bandwidth. (For a discussion of the choice between various transmission techniques for digital data in the AeM, see CCIR Report 928[40]). The 12 exclusive AeM (R) allocations are divided into radio channels, the separation between channel centres being 3·0 kHz.

Unlike other ITU frequency plans, in which frequencies or channels are allotted to countries or to broad geographical areas within countries or are planned for assignment to specific stations, these channels are allotted for communication between aeronautical stations and aircraft stations on a broad geographical basis and according to the functions of the communication links which are to use the channel. There are four defined functions:

- Communication between any aeronautical station and an aircraft flying on any of a geographically defined group of long distance air routes, essentially international in character, extending through more than one country and requiring long distance communication facilities. Each of 15 groups of routes is designated a Major World Air Route Area (MWARA).
- Communication between any aeronautical station and an aircraft flying on any of a geographically defined group of routes which are not classified as a major world air route. Such a group of routes is designated a Regional and Domestic Air Route Area (RDARA). There are 14 RDARAs, subdivided into 73 sub-areas.
- VOLMET allotments are made to aeronautical stations within an VOLMET allotment area for use in broadcasts to aircraft within the corresponding VOLMET reception area. There are nine VOLMET areas.
- WORLD-WIDE allotments provide for communication between an aeronautical station and an aircraft operating anywhere in the world.

The MWARA, RDARA and VOLMET areas and sub-areas are defined in RR Appendix 27[38]. The MWARA, RDARA and WORLD-WIDE allotments are used bi-directionally in the simplex mode and the VOLMET function is uni-directional. Frequency allotments are provided in the plan in complements appropriate to the propagation conditions prevailing between the localities to be served. Inevitably most channels are allotted for several different complements within the plan, the repetition distances being calculated to be large enough to provide adequate protection from interference.

There is heavy pressure of traffic on these channels in some parts of the world and consideration is being given to various measures for relieving this pressure, by using the HF channels more effectively or by transferring traffic from HF channels to higher frequencies or to satellite systems; see RR Resolutions 405[41] and 406[42] and RR Recommendations 402[43], 403[44] and 405[45]. However, paragraph 27/21 of RR Appendix 27[38] also recognises that not all the allotment sharing possibilities have been exhausted in the HF allotment plan. Accordingly, administrations may assign frequencies from the HF AeM (R) bands in areas other than those to which they are allotted in the plan, provided that interference to planned allotments is not made unacceptable thereby.

In practice, the use of the MWARA, RDARA and VOLMET allotments is managed by co-operation between governments, mainly through ICAO. The management of the WORLD-WIDE allotments rests with the administrations but RR Recommendation 402[46] proposes that administrations should draw on the assistance of the IFRB to ensure that these allotments are used efficiently.

The procedure to be followed by the IFRB when an administration notifies an assignment in one of these AeM (R) exclusive bands to an aeronautical station is given in RR 1333–RR 1342 as amended by WARC-MOB-87[1]. In effect, all such notified assignments are to be included in the MIFR, but with an indication of whether the assignment is in accordance with the plan, and if not, with an appraisal of the effect of the unplanned assignment on the operation of planned assignments.

As indicated above, the AeM (R) plan is based on the assumption that radiotelephony will be used, with J3E emissions. Exceptionally, A3E or H3E emissions may be used in co-ordinated search and rescue operations in the channels at 3023 and 5680 kHz. However, the constraints within which stations using planned channels may operate are defined in RR Appendix 27[38], paragraphs 27/49 to 27/66C. A2A emmissions and various types of single sideband emission, such as J7B and J9D, are permitted for radiotelegraphy in the plan. A1A and F1B emissions are also permitted provided that they do not cause harmful interference to emissions of other classes. The power of single sideband emissions is limited to 6 kW PEP for aeronautical stations and 400 W PEP for aircraft stations, and templates are used in RR Appendix 27 to define an upper limit to the relative power level at various points across the spectrum of these emissions. The power of A1A and F1B emissions from aeronautical and aircraft stations is limited to 1·5 kW and 100 W respectively.

The allotment plan for the AeM (OR) service

The allotment plan in RR Appendix 26[37], Parts III and IV, is based on the assumption that A3E emissions will be used, although emissions of other classes are not excluded if they occupy no more bandwidth. Allotments are made to countries and for the aircraft which countries authorise to fly in their air-space.

The ten exclusive AeM (OR) bands listed in Table 7.6 have been divided into radio channels wide enough for A3E emissions; the frequency separation between channel centres is 7·0 kHz at the lower end of the HF spectrum, rising by stages to 10 kHz at the upper end. These channels are intended to be used in the simplex mode for communication between aeronautical stations and aircraft stations. Channels in each band are allotted to countries which have indicated requirements. Extensive requirements have been stated and, as a result, virtually every channel is allotted to a number of different countries. Typically 20 to 30 countries will have been allotted each channel in the lower frequency bands and rather fewer in the higher bands.

Inspection of the plan suggests that some channels in the plan have been allotted jointly to a group of countries for operational reasons. In other cases where a channel is shared, the geographical separation between the countries which have been allotted the same channel is great enough to ensure sufficient attenuation, by day if not by night, to reduce interference from one service area to the other to an acceptably low level. Where this is not a valid assumption, there may have been grounds for assuming that the activity on these channels will be so intermittent that interference will not impede shared use of the channel. In some cases the prospect of interference has been limited by the acceptance of specific sharing constraints, such as a transmitter power limitation (see below). Some allotments have been given a so-called 'secondary' status, inferior relative to other allotments of the same channel in ways which are defined in the text of the plan. Finally it may be supposed that administrations avoid interference in some channels by means of operating agreements or operators' procedures.

Among the shared frequency bands listed in Table 7.6, channelling plans and associated allotments are included in the plan for Region 3 at 3900–3950 kHz. Similar planning is included for Region 1 in part of the 3800–3900 kHz band. For most of the other shared bands the allotment plan identifies specified groups of countries which would share channels, but the determination of the frequencies at which these channels would be located is left open.

The procedure to be followed by the IFRB when an administration notifies an assignment in one of the AeM (OR) exclusive bands to an aeronautical station is given in RR 1343–RR 1349; amendments agreed at WARC-MOB-87[1] should be noted. In effect, all such notified assignments are to be included in the MIFR, but with an indication of whether the assignment is in keeping with the plan, and, if not, with an appraisal of the ways in which it departs from the plan. The process for registering a frequency assignment to an aeronautical station of the AeM (OR) service in a shared band follows the same course as for an HF FS assignment (pp. 84–86).

Except as indicated above, the allotment plan places few constraints on stations using it. For example there is no general constraint on transmitter power levels, although the transmitter power that may be used with some specific allotments is limited to reduce interference to other specific allotments in the same channel. However, it seems that the repeat distances used when the plan was drawn up were chosen so that interference would be acceptably low by day, and there is a general proposal at paragraph 5 of Section II of Part III of RR Appendix 26[37] 'that interested administrations should agree on a reduction in aeronautical station radiated power at night to the extent necessary to make possible night-time use of these frequencies'.

7.2.5 Schemes for using the MM bands below 4 MHz

Various schemes have been included in the Radio Regulations to assist in the systematic utilisation of spectrum allocated to the MM in certain frequency bands for which no frequency allotment plans have been drawn up. It is convenient to consider these schemes in two parts; those relating to the bands around 500 kHz and those relating to frequencies between about 2 and about 4 MHz which may, somewhat imprecisely, be called the MF band.

The 500 kHz band

Fig. 3.1 shows the international Table of Frequency Allocations (RR Article 8) in the neighbourhood of 500 kHz as it was in 1982. A few minor amendments were made to footnotes to this part of the Table at WARC-MOB-83 and WARC-MOB-87, but the only general impact that these changes have on the MM lies in the constraints which they place on the use of certain specified frequencies. These regulations, and others found mainly in RR Articles 38 and 60, have some of the features of a channelling plan.

Key features of the band betweeen 415 and 526·5 kHz (525 kHz in Region 2) are that:

- most of it is allocated exclusively to the MM,
- it may be used only for radiotelegraphy,
- the international distress frequency for radiotelegraphy is located within it at 500 kHz, and
- the basis of use of this band is global in scope.

It should also be noted however that the AeRN has a primary sharing allocation in Region 1 at 415–435 kHz; the MM allocation was given permitted status in this band in Region 1, pending the development of a frequency plan for the AeRN, and this has limited, to some degree, the opportunities for optimising the use of this band for the MM.

The pressure of MM traffic in the 500 kHz band is high, particularly in the European Maritime Area. (There is a formal definition of the European Maritime Area in RR 405.) A plan for the use of the band in Europe was agreed at a conference in Copenhagen in 1948 but it is no longer satisfactory. The use of channels is not as orderly as it could be if there were more channels to assign, and improvements in the frequency stability of transmitters and receivers have now made feasible the designation of more channels. There is also a need to prepare for growth in the use of direct-printing telegraphy.

WARC-MOB-83 concluded that a RARC was needed to develop plans for the use of this and other MM bands below 3·4 MHz in Region 1, including a frequency assignment plan for coast stations; see Resolution 704[47]. Another task of this RARC would be to draw up a frequency plan for the AeRN between 415 and 435 kHz in Region 1, one consequence of which would be to allow frequency planning in Region 1 for the MM to cover the whole bandwidth around 500 kHz for which the MM has allocations. Appendix 1 to Resolution 704 indicates the channelling plan that was foreseen for this band in Region 1.

The conference called for in RR Resolution 704 was convened in Geneva in 1985. The Final Acts[48] of that Conference contain a plan for the AeRN in Region 1 in this part of the spectrum (see p. 397 below) and, interlinked with it, a frequency assignment plan for Region 1 coast stations operating between 416 and 526 kHz. Planned coast station assignments fill the spectrum between 416 and 453 kHz and between 510·5 and 526 kHz, but with a gap around 425 kHz, the ships' working frequency. The nominal spacing between frequency assignments is 0·5 kHz and classes of emission A1A and F1B are assumed. The 1985 Agreement enters into force in April 1992. Frequency assignments made by administrations in accordance with this plan and notified to the IFRB will be registered in the MIFR without question of interference risks to assignments already registered for other administrations that are party to the Agreement, in accordance with RR 1245, RR 1248 and Article 5 of the Agreement.

In addition to its role as the distress frequency, 500 kHz is used as the calling frequency for both coast stations and ship stations operating in this band. Pressure of traffic would lead to congestion if all coast stations called on precisely the same frequency; so it has been agreed that administrations may assign calling frequencies to coast stations up to 2 kHz removed from 500 kHz; see RR 4225[2] and RR 4226[2].

It is, however, necessary to maintain a guard band around the distress frequency to ensure that distress calls shall not be masked by interference. Prior to 1979 the guard band was 490–510 kHz and no transmissions were permitted within that band except for those specifically authorised in the immediate vicinity of 500 kHz (see the previous paragraph). WARC-79 reduced the guard band, as shown in the international Table of Frequency Allocations, to 495–505 kHz but delayed the effective implementation of the change until a future conference should decide that the frequency stability of radio equipment in use was good enough to permit the change to be made without risk; see RR 471 and RR 3018. WARC-MOB-83 postponed that decision and WARC-MOB-87 decided that the 490–510 kHz guard band should be maintained, in practice, until the GMDSS was fully implemented; see RR Resolution 210[49].

In view of these various factors, a group of key operating frequencies have been identified in addition to 500 kHz. Thus:

- 518 kHz is to be used only for the transmission by coast stations to ships, using narrow-band direct-printing telegraphy, of meteorological and navigational warnings and other urgent information; this facility is called NAVTEX. Special provision has been made for co-ordinating the transmitting frequency assignments made to coast stations for this facility; see RR 474[1], RR Resolution 324[50] and RR Article 14A[1].
- 425, 454, 468 and 480 kHz are to be used by ships as working frequencies; coast stations may not transmit on these frequencies; see RR 4237[1] and RR 4238.
- 512 kHz is used by ships as a fifth working frequency except when 500 kHz is being used locally for distress traffic, when 512 kHz becomes a calling frequency and is available for coast stations as well as ships; see RR 4237[1], RR 4239[2], RR 4240 and RR 4241.

Frequencies in other parts of the band are assigned by administrations to coast stations as working frequencies.

The 'MF' band, 1606·5 kHz to 4000 kHz

The band 2173·5–2190·5 kHz is allocated world-wide as a guard band for distress and safety traffic at 2182 kHz, which is an international radiotelephony distress frequency (RR 500[1]). 2182 kHz is also the calling and reply frequency for MF radiotelephony (RR 4345). Within the guard band:

- 2174·5 kHz is used only for distress and safety traffic by narrow-band direct-printing telegraphy (RR 500B[1]),
- 2187·5 kHz is used only for distress and safety calls by digital selective calling techniques (RR 500A[1]), and
- 2177 kHz and 2189·5 kHz may be used for digital selective calling under the terms of RR 4682[1]–RR 4684[1], with effect from 1 July 1991.

With the exception of the uses indicated above, all transmissions within the guard band are prohibited to protect distress traffic (RR 3023[2] and RR 3023[1]). Narrow bands above and below the guard band are allocated world-wide for MM, with lower and upper limits of 2170 and 2194 kHz respectively, but these additional bands may not be used for narrow-band direct-printing telegraphy (RR 4319[1]). The structuring of these frequency band arrangements, having regard in particular to safety services, continues under active study; see in particular RR Resolution 330[51] and CCIR Report 1029[52].

In addition to those world-wide allocations around the radiotelephony distress frequency around 2182 kHz, there is a complex pattern of primary and permitted allocations for MM in the international Table of Frequency Allocations between 1606·5 and 4000 kHz. The MM is not the most important service with allocations in this part of the spectrum for many administration. Consequently there is no uniform global, or even Regional, pattern in the use which MM makes of radio at MF. Nevertheless, there is a fragmentary scheme for using the MF bands for ships in Regions 2 and 3, and a more substantial scheme for Region 1.

For Region 1, a scheme is set out in RR 4188[2] and RR 4188A[2] for the sub-division of MM allocations in the frequency range 1606·5–3800 kHz (excluding 2182 kHz and its guard band) between the various maritime facilities which are needed (see Fig. 7.2). In the bands 1606·5–1800 kHz and 2045–2160 kHz this scheme has mandatory force in Region 1 and a frequency assignment plan has been agreed; see the next paragraph. Above 2194 kHz this scheme is to be adopted wherever possible (RR 4188). No doubt administrations in Region 1 will follow this scheme in planning the use of bands between 2194 and 3800 kHz to the extent that the claims of the stations of other services permit. No comprehensive channelling scheme above 2194 kHz is contained in the RR, but RR 4189[2] sets a channel bandwidth of 0·5 kHz for narrow-band direct-printing telegraphy and digital selective calling. The channel bandwidth for SSB radiotelephony is to be 3·0 kHz. RR 4357–RR 4368 provide further rules and guidelines for the use of frequencies between 1605 and 2850 kHz for maritime radiotelephony; note should be taken of amendments made to these paragraphs at WARC-MOB-83[2] and WARC-MOB-87[1].

The Region 1 RARC, proposed in RR Resolution 704[47] and convened in 1985, planned the band between 1606·5 and 1800 kHz for coast stations, associated with the use of the band 2045–2169 kHz for ships' transmitters; see the Final Acts of the conference[48]. Assignments are planned with frequencies at 0·5 kHz intervals for F1B (narrow band direct printing telegraphy) emissions from 1607 to 1620·5 kHz and frequencies at 3·0 kHz intervals from 1635 to 1800 kHz for J3E emissions. As at 500 kHz, this plan enters into effect in April 1992 and assignments in accordance with the plan notified by administrations to the IFRB will be registered in the MIFR automatically.

For Region 2 (except Greenland), the essence of a channelling plan for the band 2067–2107 kHz is to be found in RR 497. With sub-Regional variations, 2068·5–2078·5 kHz is identified for use for wide-band telegraphy and similar systems and a series of preferred carrier frequencies for radiotelephony fills the remainder of the band.

For Regions 2 and 3, 2635 and 2638 kHz are designated for carrier frequencies for inter-ship links; see RR 4193[2] and RR 4369. In Region 3 the band 2634–2642 kHz is reserved for inter-ship links, providing small guard bands for these inter-ship channels.

For areas North of the Equator in Region 3, 2089·5–2092·5 kHz is designated as a calling and safety band, to be used with radiotelegraphy to serve ships in that area which use radiotelegraph assignments between 1605 and 2850 kHz; see RR 4246–RR 4251, taking into account the amendments made at WARC-MOB-87[1].

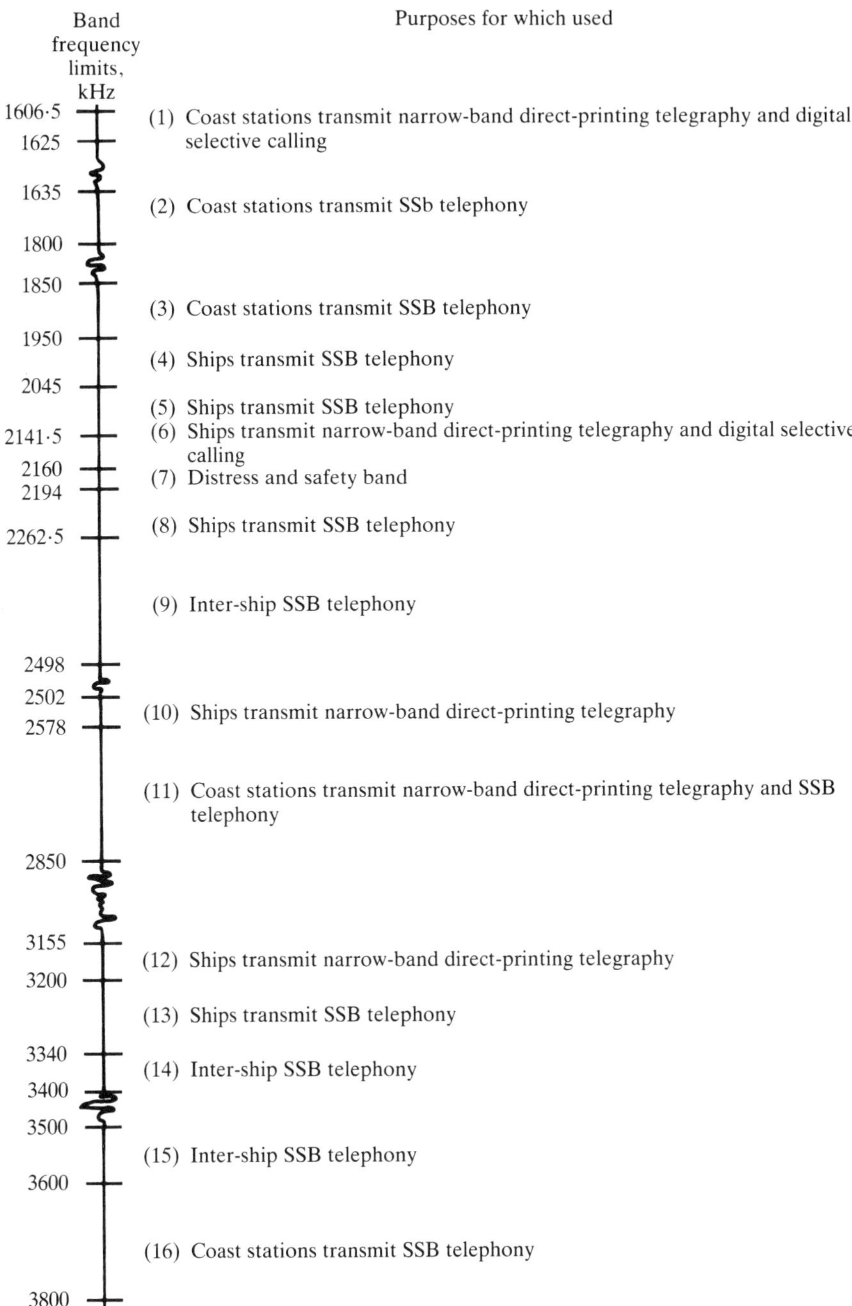

Fig. 7.2 Region 1 scheme for using MM allocations between 1606·5 kHz and 3800 kHz
Spectrum not precisely to scale

7.3 The mobile services above 28 MHz

7.3.1 Survey of MS regulation above 28 MHz

The frequency allocations and sharing

Frequency bands in all parts of the spectrum from 28 MHz to 275 GHz have been allocated to mobile services in the international Table of Frequency Allocations (RR Article 8), most of the allocations having primary status. Furthermore, many bands which are not allocated to mobile services globally or Regionally have a mobile service allocation by footnote for specified countries. In aggregate these allocations comprise a large percentage of the whole currently usable radio spectrum, far exceeding the needs of the mobile services. Most of these allocations have been made to the MS in general, but some specifically exclude the AeM and a few are limited to just one of the specialised services.

However, this appearance of ample frequency allocation is deceptive, owing to two factors, as follows:

- For a number of reasons there is a broad optimum frequency range for most mobile systems which is centred around a few hundred megahertz. These reasons include the variations with frequency of man-made noise, galactic noise, the transmission loss between the low-gain antennas which typical mobile stations use and the effect of obstacles on radio propagation. Frequencies well below this optimum range, down to the HF band, may be usable for the MS and have advantages for some purposes, although such other factors as long distance interference also arise. Frequencies well above this range can also be used by stations designed to be operated on moving vehicles, but the penalty in terms of the higher transmitter power required, the consequential increase in the size and cost of the batteries of hand-portable equipment and increased signal fading in an urban environment may become severe at about 2 GHz.
- With the exception of 117·975–136 MHz and 156·7625–156·8375 MHz, all of the bands allocated to mobile services above 28 MHz are also allocated to other terrestrial services with equal or higher status. Most MS bands are shared with the FS. Below 1 GHz, some bands are shared with the BS, for which the technical preference for access to this part of the spectrum applies with almost as much force as for the MS. In addition, much of the spectrum allocated to mobile services above 3 GHz is shared with the FSS and some is shared with the BSS.

These factors greatly reduce the opportunities which administrations have to make use of these seemingly ample MS frequency allocations.

Indeed in some countries there is now a severe shortage of spectrum allocated and suitable for mobile services. Since the late 1970s there has been a marked increase in the demand for assignments, preferably below 1 GHz, for LM facilities, typically for car-borne and hand-portable radio-telephones and cordless telephones. Another requirement, smaller but vigorously growing, is for video link equipment used for Electronic News Gathering (ENG), for which somewhat higher frequencies would be usable. Because of such needs for additional MS spectrum, not too far removed from the optimum frequency range for mobile systems, a considerable number of new LM allocations were made at WARC-79, many of them by footnotes to the international Table and subject to consultation in accordance with the procedures of RR Article 14. This trend continued at WARC-MOB-87. There has also been pressure for the transfer of spectrum from AeRN to LM at around 4 and 5 GHz; see RR Recommendations 606[53] and 607[54].

Specific allocations to mobile services have been taken into use globally in a few frequency bands to meet the needs of aircraft and ships that move from country to country (see p. 217). Some groups of administrations have agreed to implement a common policy in making assignments to land mobile systems, and in requiring harmonised system specifications to be used with them, so that radiotelephones installed in road vehicles which roam from country to country can be used in any participating country. International trade in radio equipment for mobile systems has also had some effect in standardising the bands which are used for specific MS facilities. However, most mobile stations do not move from country to country, and long-distance interference is uncommon at UHF and higher frequencies. Consequently, international standardisation of most mobile allocations is not essential and national allocations tend to be determined by national circumstances. There is wide variation from country to country among the national allocations to the mobile services above 28 MHz.

A need is perceived for the passengers of aircraft in flight to have access to the public telephone network. RR 3633[1] forbids the use of frequency assignments in exclusive AeM allocations for this purpose and RR Resolution 409[16] reinforces that prohibition. RR 4145 gives permission in general for aircraft stations to use MM allocations for public correspondence. There is reference on p. 190 to the occasional use of HF MM allocations for this purpose but RR 4148–RR 4153 effectively forbid the use for this purpose of the one frequency band above 28 MHz which is in common use for MM, namely 156–174 MHz. The AeM has many shared VHF and UHF allocations and some countries have assigned frequencies from bands around 900 MHz for national aeronautical public correspondence systems but there is no general international acceptance of this usage. New international regulatory provision is needed for this purpose, preferably in globally uniform frequency bands.

A start was made at WARC-MOB-87 in providing spectrum globally for aeronautical public correspondence. RR 731B[1] and RR 731C[1] make allocations to AeM specifically for this purpose at 1593–1594 MHz and 1625·5–1626·5 MHz. These allocations have primary status in much of Region 1 and certain countries elsewhere and secondary status throughout the rest of the world, although subject to various constraints; see RR 731A[1] and RR 731D[1]. If it is found that there is a substantial demand for this facility, it will no doubt be necessary to develop this regulatory provision further at some future WARC, and RR Recommendation 408[55] prepares the way for this. CCIR Report 1051[56] contains an initial discussion of the provision that would be required. The conference which PC-89 has planned for 1992 will, no doubt, be a convenient opportunity for progress in this matter.

Typical systems in operation

Vehicle-borne and hand-portable systems have many applications by land, at sea and in the air, providing for example:

- links allowing passengers of road vehicles, ships and trains and users of hand-portable radiotelephones to be connected to the public switched telephone network,
- links between the control office of a fleet of road vehicles and the drivers of the vehicles, which may be taxis, heavy transport vehicles, delivery vehicles, police patrol cars, ambulances etc,
- links between a doctor and his surgery, between a servicing technician and his workshop etc,
- links between an air traffic control centre or the operational control centre of an airline and the crews of aircraft in flight or between coastguards or harbour masters and ships approaching port,
- many kinds of military communication system,
- direct wireless communication between security guards in buildings, construction workers on building sites, sailors on ships at sea, production staff working on a film set or in a television studio, and so on,
- citizens' band facilities, and
- short distance connections between a telephone instrument and one specific telephone exchange connection ('cordless telephones'), and various kinds of unidirectional facilities such as radio-pagers, telecommand and telemetry links, cordless microphones and ENG links.

The links provided by these systems are typically intermittent, communication between two stations on a given occasion lasting a few minutes or less. Most links are between a mobile station and a land station but mobile-to-mobile links are also used.

Most vehicle-borne and hand-portable stations operating above 28 MHz are designed to carry a single analogue telephone channel, using A3E, F3E or G3E emissions, although a growing number can carry other types of signal, such as facsimile, digital data, and, more rarely, video. Combined transmitter/receivers ('transceivers') are usually employed where two-way communication is provided, with duplex or semi-duplex facilities. However, more complex systems which provide a better grade of service and use spectrum more efficiently are also coming into widespread use to meet a massive growth of demand, using for example multiplexed carriers, various kinds of multiple access system and cellular systems.

International spectrum management

Five frequency bands, or groups of bands, above 28 MHz have been made the subject of some degree of worldwide planning, as follows:

- 117·975–137·0 MHz; the AeM (R) channelling plan for this band is discussed on pp. 222–224.
- 156·0–157·45 MHz, 160·6–160·975 MHz and 161·475–162·05 MHz; the MM channelling plan for these band is discussed on pp. 224–225.
- The spot frequency 243 MHz is used for radio links with aeronautical survival craft, EPIRBs and for related purposes; see RR 501[1], RR 642 and RR Appendix 37A[1]. Any other use of this frequency is forbidden; see RR 3031B[2].
- Certain spot frequencies near 457 MHz and 467 MHz are specified for on-board communication on ships; see RR 669 and RR 670. Some characteristics of the equipment which is to be used for this purpose are set out in RR Appendix 20; note the amendments to this latter text which were agreed at WARC-MOB-87[1]; see also CCIR Recommendation 542[57].
- The new provision for an aeronautical public correspondence facility at 1593–1594 MHz and 1625·5–1626·5 MHz, referred to above.

Consideration is being given to the establishment of a new MM planned band at UHF for automatic, ship-shore telephony; see RR Recommendation 310[58]. Similarly, RR Recommendation 205[59] supports the work in progress in CCIR and CCITT on the definition of a world-wide standard LM system for public correspondence and recommends the identification at a future WARC of frequency bands in which it would be deployed.

The regulation of the use of bands shared with certain services is reviewed on pp. 220–222. In the absence of the complexities arising from sharing, a relatively simple procedure is followed in notifying frequencies assigned to land stations for transmitting and receiving to the IFRB for registration in the MIFR; the procedure is the same as is used for the FS in the same part of the spectrum and it is summarised on pp. 103–104. Where a number of land stations operate on the same frequency in a stated area, a single notification for a typical station may be provided in place of a separate notification for each station; see RR 1223. Assignments to mobile stations are not registered by the IFRB. There are generally applicable technical constraints on carrier frequency tolerance and spurious emissions (see p. 218). In other respects, responsibility for spectrum management rests with the administrations.

Nevertheless, the CCIR has performed valuable work on the standardisation of systems for use in these services and in gathering together and presenting technical advice, much of which has a bearing on spectrum utilisation. In particular:

- on automatic radiotelephone systems for maritime use at VHF or UHF, see CCIR Recommendations 586[60] and 587[61] and CCIR Reports 1033[62] and 1034[63],
- on cordless telephones, see CCIR Reports 1025[64],
- on radio-pagers, see CCIR Recommendations 539[65] and 584[66] and CCIR Reports 499[67] and 900[68].

There are references to texts on other aspects of LM systems on pp. 330–338.

Technical constraints on mobile systems

RR 303 requires that transmitting stations conform to the frequency tolerances specified in RR Appendix 7[1] (see Table 7.7). Rather less stringent tolerances applied in some cases before 1990.

RR 304 requires transmitting stations to conform to the spurious emission power limits specified in RR Appendix 8 (see Appendix A.5 below). However, the requirements of RR Appendix 8 are not stringent enough for application where mobile radio systems are used intensively. CCIR Recommendation 478[70] recommends for LM stations that no spurious emission component should exceed −70 dB or 2·5 μW (whichever is the less stringent) for any operating frequency between 25 and 1000 MHz. It may be desirable to limit spurious emissions to even lower values, perhaps 25 pW, for transmitters when they are in the standby condition, that is, when they are not required to be radiating a carrier wave.

National spectrum mangement

For the mobile services, perhaps more than for any other service, substantial responsibility for spectrum management rests upon the national administration. The frequency allocation aspects of this role are considered on pp. 225–210 and various aspects of LM system architecture are considered on pp. 230–238.

7.3.2. Frequency band sharing and the mobile services

The constraints which sharing imposes on mobile systems are sometimes severe and they take different forms, depending on whether the sharing is with terrestrial services or with space services and on whether or not the sharing service is subject to frequency planning.

Table 7.7 Transmitter carrier frequency tolerance above 28 MHz from 1990 (RR Appendix 7[1])

	Frequency tolerance in parts per million (p.p.m.)	
	Land stations	Mobile stations
MARITIME MOBILE SERVICE		
below 100 MHz	20	20 (1)
100–156 MHz	10	50
156–174 MHz	10	10 (2)
174–470 MHz	10	50 (3)
AERONAUTICAL MOBILE SERVICE		
below 100 MHz	20	20
100–470 MHz	20 (4)	30 (4)
LAND MOBILE SERVICE		
below 100 MHz	20	20 (1)
100–235 MHz	15 (5)	15 (5)
235–401 MHz	7 (5)	7 (5, 6)
401–470 MHz	5 (5)	5 (5, 6)
ALL MOBILE SERVICES		
470–2450 MHz	20 (7)	20 (7)
2·45–10·5 GHz	100	100
above 10·5 GHz	– (8)	– (8)

Notes

1 For hand-portable transmitters with power output not exceeding 5 W the tolerance is 40 p.p.m.

2 See also RR Appendix 19[1]

3 For transmitters used for on-board communication, the tolerance is 5 p.p.m.; see also RR Appendix 20.

4 Where the channel spacing is 50 kHz, the tolerance is relaxed to 50 p.p.m. Conversely, it may be noted that ICAO recommends that AeM (R) systems providing off-set carrier facilities in the 120 MHz band should achieve tolerances ranging from 15·3 p.p.m. (where two carriers are offset in a 25 kHz plan channel) to 0·3 p.p.m. (for five carriers); see p. 95 and pp. 231–232 of Reference [69].

5 RR Appendix 7[1] applies these tolerances where the channel spacing is 20 kHz or more; no tighter tolerance is laid down in the RR for narrower channel spacings. However, CCIR Recommendation 478[70] recommends that LM systems designed for channel-centres spaced by 12·5 kHz should be required to achieve a tolerance of 12 p.p.m. at 80 MHz, 8 p.p.m. at 160 MHz and 3 p.p.m. at 450 MHz.

6 For hand-portable transmitters with power output not exceeding 5 W the tolerance is 15 p.p.m.

7 RR Appendix 7 advises administrations to be guided by the latest CCIR recommendations in applying this tolerance.

8 No tolerances have been determined yet for stations of the MS operating above 10·5 GHz.

Sharing with terrestrial services

Sharing between mobile systems and stations of other terrestrial services raises two kinds of technical problem:

- MS transmitting stations which operate when in motion usually have low effective radiated power. Radio propagation between mobile stations and land stations is often subject to severe transmission losses due to obstruction of the ray-path, particularly for LM systems operating in towns. Thus land stations, particularly base stations, are usually designed and located so that their sensitivity for the wanted signal is high. Consequently land stations are very susceptible to interference from stations of sharing services.
- Frequency assignments to mobile stations cannot be co-ordinated effectively with assignments to other stations. Interference between a mobile station and a station of a sharing service could occur when the mobile station is at a considerable distance from the associated land station, particularly if the mobile station involved is on an aircraft.

Thus, in bands which are shared internationally between a mobile service and another terrestrial service, typically the BS, a large geographical separation is needed between stations of the two services. Useful information on required separation distances between areas using a band for the BS and other areas using the same band for a mobile service is to be found in CCIR Reports 358[71] and 1023[72] and in the various CCIR texts on propagation conditions; for references to the latter see Appendix B.6.11 below. Where the mobile service is the AeM, reference should be made to CCIR Recommendation 441[73], CCIR Reports 926[74] and 927[75] and relevant parts of the ICAO Convention[69].

When the sharing service is the FS, shorter separation distances are usually sufficient but the distance may be highly dependent on the geographical distribution of FS stations and on their antenna characteristics. In an area where a few FS stations have assignments in a band which is also used for mobile stations, administrations may find it convenient to use a technique like the one that is used for defining earth station co-ordination areas in bands shared with the FS; see RR Appendix 28 (noting in particular section 7) and pp. 230–238. Techniques of this kind could be used, for example, to define an area surrounding an FS station outside which mobile stations with defined characteristics would neither cause interference to the FS station nor suffer interference from it.

Ultimately, interference between mobile systems and the stations of other terrestrial services sharing a frequency band is controlled by the registration of assignments in the MIFR. Below 1 GHz, the preferred part of the spectrum for communication with moving vehicles, much of the spectrum allocated to the MS is shared with the BS, and BS assignments may have two regulatory advantages over those of the MS. Firstly, the status of the BS allocation in RR Article 8 has usually been made superior to that of the MS, sometimes by means of a footnote to the Table; see for example RR 621 and RR 674. Secondly, the MS is the expanding service, competing for spectrum with a service which is mature. The BS is often already established and probably has the advantage of priority of registration. In addition, in Region 1, the BS is usually regulated by means of a frequency assignment plan. Consequently the BS often presents a substantial obstacle to the expansion of the MS.

In bands which the MS shares with unplanned terrestrial services, such as the FS, a local agreement between neighbouring administrations may determine where and on what frequencies mobile stations may operate. Failing that, and given equal allocation status, the use of the band may be decided piecemeal by priority of date of entry of assignments in the MIFR.

Sharing with co-ordinated space services

Frequency band sharing between mobile services and space services such as the FSS involves the limitation of interference between MS stations and space stations and between MS stations and earth stations. Different means are used to solve these two problems.

The sharing constraints which have been devised to limit interference between space stations of the FSS and stations of the FS apply also for stations of the MS (see pp. 110–118). In the bands allocated to the FSS for space-to-Earth links, limits placed on the PFD from satellite emissions by RR Article 28, Section IV, may be expected to provide ample protection from interference for MS receivers. In the bands allocated to the FSS for Earth-to-space links, the power fed to the antenna of a MS transmitter is limited by Section II of RR Article 27 to 20 W between 1625·6 MHz and 10 GHz and to 10 watts between 10 GHz and 29·5 GHz. This power constraint may deny access to shared bands for some kinds of mobile system, particularly those requiring relatively powerful transmitters.

Land stations are co-ordinated with earth stations in the same way as are stations of the FS before frequency assignments are notified for registration in the MIFR; see RR Articles 11, 12 and 13, RR Appendix 28 and pp. 230–238 above.

In bands allocated for FSS space-to-Earth links and shared with mobile services, the most difficult problem would be posed by interference from aircraft transmitters to earth station receivers. Accordingly, in most such bands the mobile service allocation either excludes the AeM or gives it secondary status.

There is no formal procedure by means of which aircraft or ship stations of the MS can be co-ordinated with foreign earth stations and, thereby, establish a right to operate. Nevertheless, administrations may co-operate to determine how stations of the two services can make the best use of spectrum. The technique for defining a co-ordination area which is set out in RR Appendix 28 can be adapted to define the area around an earth station within which a mobile station should not be operated, because it may cause interference to the earth station or suffer interference from the earth station.

Sharing with planned space services

When a mobile service shares a frequency band with the BSS (or an FSS band designated for feeder links to broadcasting satellites) and a frequency assignment plan has been agreed for the BSS, the situation for the mobile service is similar in principle to that which arises in a shared, planned terrestrial broadcasting band (see above). This situation arises at present only in certain 12 GHz BSS bands and in corresponding feeder link bands at 17·7–18·1 GHz and 14·5–14·8 GHz (Regions 1 and 3 only). In the 12 GHz bands the status of the MS allocation has been made inferior to the status of the BSS allocation, through a secondary allocation in the international Table or by a footnote (see RR 838). However, some protection is given to the MS allocation by a special co-ordination procedure. This procedure applies also to sharing between the FS and the BSS and its feeder links in the same bands, although there are differences of detail (pp. 131–133).

WARC-ORB-88 agreed a frequency allotment plan for the FSS is various frequency bands between 4 and 14 GHz (see pp. 297–302). However, this agreement did not change the status of the frequency allocations of other services sharing these bands. The co-ordination procedures and sharing constaints used to limit interference between stations of space and terrestrial services in bands in which spectrum use by the space service is managed by co-ordination (see above) apply in these bands also.

7.3.3 The AeM (R) and MM radio channel plans at VHF

The Aeronautical Mobile (R) plan

The frequency band 117·975–136 MHz is allocated to AeM (R) in the international Table of Frequency Allocations. RR 501[2], RR 592[2] and RR 593[2] make provision for the use of 121·5 MHz as the aeronautical emergency frequency. RR 591 provides for the use of this band by mobile-satellite stations, subject to conditions which protect the AeM (R) from interference and RR 594 adds a permitted allocation to AeM (OR) in part of the band in certain countries. For most practical purposes however the band can be regarded as an exclusive, worldwide AeM (R) allocation.

The AeM (R) allocation was extended upwards to 137 MHz at WARC-79, the extension to take effect in 1990. The band 136–137 MHz is allocated to various other services, some with with primary status, up to 1990 and from that date, when the primary AeM (R) allocation took effect, the allocations for these other services became secondary (see RR 595[1]) except for a permitted allocation to AeM (OR) in certain countries (see RR 594A[1]). There is some concern that the AeM (R), a 'safety service', may suffer interference after 1 January 1990 from stations of services which become secondary at that time and RR Resolution 408[76] calls for action to prevent this.

Some concern about interference to the AeM (R) also arises from recent increases in the use for sound broadcasting of spectrum immediately below 108 MHz (see pp. 177–178). RR Recommendation 714[77] calls for attention to this problem from administrations and for further study of it in CCIR and ICAO.

The AeM has had the use of 118–132 MHz since 1947 and the aircraft receivers then available would be considered now to have poor selectivity, requiring a minimum separation between adjacent carriers of 100 kHz. Successive improvements in receiver performance have since made carrier separations of 50 kHz and 25 kHz readily feasible, and the demand for channels to carry the traffic arising in some parts of the world requires these options to be exercised. However, not all receivers in service have this improved selectivity. The channelling plan which has been drawn up by ICAO defining the carrier frequencies which may be assigned to aircraft and aeronautical stations (see pages 160–161 of[69]) takes account of this situation.

The structure of the ICAO channelling plan assumes the general use of A3E emissions for telephony, operated in the single channel simplex mode, although other types of signal are also used. Between 118 and 132 MHz the plan is based on 100 kHz carrier separation, the channels centred on these frequencies being called Group A. An additional set of carrier frequencies, interpolated between the Group A set, is defined as Group B, providing for channels with carrier separations of 50 kHz. A third set of carrier frequencies, called Group E, interpolated between the Group A and Group B frequencies, provides for channels with carrier separations of 25 kHz.

The use of the band above 132 MHz started more recently and it has been assumed that no receivers operating in this band need carriers to be separated by more than 50 kHz. Accordingly, the basic set of carrier frequencies give separations of 50 kHz (Group C), the interpolated set (Group D) providing for 25 kHz separation.

As with other AeM (R) bands, the prime purpose for which these channels are used is communication between aircraft and air traffic control stations in order to maintain safe and regular flight on national and international air routes; see RR 3630[2]. The plan allots frequency sub-bands for the various operational functions which are needed. The use of channels is determined regionally and nationally, and is co-ordinated through ICAO. When and where Group A does not provide enough channels, additional frequencies are assigned from Groups B and C, introducing some carriers spaced at 50 kHz. This may be followed in due course by drawing on Groups D and E for 25 kHz spacing. Frequencies assigned to aeronautical stations can be notified by administrations to the IFRB and registered in the MIFR, as indicated for FS stations (see pp. 103–104). Where an extensive area, covered by several aeronautical stations, is controlled by a single air traffic controller, it is sometimes found convenient to use the same channel at all of these aeronautical stations; if so, the aeronautical station carrier frequencies are offset from one another by a few kilohertz to limit the effects of wave interference.

The aeronautical emergency frequency, 121·500 MHz, is used for distress and urgency traffic and it may be used also by survival craft. 123.100 MHz is auxiliary to the emergency frequency and it is used by aircraft and ships engaged in search and rescue operations. No other use is permitted for these two channels; see RR 593[2], RR 2989[2]–RR 2991[2] and RR 3031A[2]–RR 3031C[1]. There is a specification of certain characteristics of survival craft radio equipment on page 142 of Reference[69]; see also RR Appendix 37A[1] and RR Resolution 601[78].

The maritime mobile plan

The frequency band 156·7625–156·8375 MHz is allocated worldwide and exclusively to the MM as a distress and calling band; see RR 501[1] and RR 613[1]. This narrow exclusive band lies within a wider band, 150·05–174 MHz, which is allocated to the MS worldwide, shared with various other mobile services and the FS.

A radio channel plan has been agreed for MM covering the narrow exclusive band and parts of the shared band, namely 156·0–157·45 MHz, 160·60–160·975 MHz and 161·475–162·05 MHz. In accordance with RR 613, administrations undertake to give priority to the MM within the shared bands, but covering only the channels in the plan which the administration in question assigns for the use of its own MM stations. A less binding protection is conceded for other channels.

Within the distress and calling band, the spot frequency 156·800 MHz is the distress frequency for ships using this frequency band, although it is also used as a calling frequency, and 156·825 MHz is used with direct-printing telegraphy for distress and safety traffic, subject to causing no harmful interference to the traffic for which 156·800 MHz is reserved; see RR 613, RR 2994[1], RR 2995, and RR 3033[1]. Apart from these specified uses of 156·800 and 156·825 MHz, there are to be no emissions within the guard band 156·7625–156·8375 MHz. A simplex channel in the channelling plan, at 156.525 MHz, is reserved for digital selective calling, to be used exclusively for distress and safety purposes; see RR Resolution 323[79] and RR 613A[1].

The radio channel plan has been drawn up assuming that phase-modulated telephony emissions, class G3E, will be used, although in practice other classes of emission are also used. The basic technical characteristics of transmitters and receivers used for telephony in the planned channels are stipulated in RR Appendix 19[1]; CCIR Recommendation 489[80] offers additional advice on these matters.

The plan itself, with 25 kHz separation between the carriers of adjacent channels, is given in RR Appendix 18[1]. It provides 19 channels for simplex use and 35 channel-pairs for duplex use, all outside the distress and calling band. The channels are designated for:

- intership communication (simplex channels only),
- the port operations service (simplex or paired channels),
- the ship movement service (simplex or paired channels),
- public correspondence (paired channels only), and
- digital selective calling for distress and safety purposes (see above).

Most channels are designated for more than one of these purposes, to increase flexibility of use. Some use is made of frequencies in this band for on-board communication: but see RR Recommendation 305[81].

Administrations assign some or all of these frequencies to ship stations and coast stations as appropriate. In many cases the assignments are first co-ordinated between the administrations of neighbouring countries to minimise interference. The output carrier power of ship station transmitters is limited to 25 W.

The frequencies in this plan may be used by aircraft having a requirement for direct communication with a ship, subject to restrictive conditions set out in RR 4148–RR 4153, RR 4154[1] and RR 4155[1].

7.3.4 National spectrum management for mobile services

An administration has responsibilities towards national users of the bands planned internationally for the AeM and MM. Oversight of the use of the 120 MHz aeronautical band will probably be done jointly by the administration and the government department responsible for air transport, such use being co-ordinated with other governments through ICAO. Maritime use of the 160 MHz band is likely to involve the administration more directly in licensing ships, coast stations and harbour authorities to use channels in accordance with the plan, and it will often be desirable for administrations to co-operate with neighbouring administrations in deciding which channels each coast station should use.

A more onerous task, however, will be to develop the national frequency allocation table to accommodate the MS facilities which operate above 28 MHz and are not internationally planned, and to manage the bands so provided.

Some of these facilities are long-established, such as:

- aeronautical channels which are used for airline company business,
- many kinds of military application for radio involving land vehicles, aircraft and ships, and
- land mobile facilities for government civilian services, such as the police, firefighting and ambulance services.

Substantial bandwidth is needed for these facilities, mostly between 28 and 1000 MHz. However, these facilities are mature and the magnitude of the spectrum requirement, once established, is relatively stable.

A second group of facilities consists of short distance devices, such as cordless telephones, paging systems, personal radio-telephones, 'walkie-talkies' etc. and a wide range of security devices and simple telecommand systems. All these kinds of system need spectrum, preferably a sub-band set aside for the particular use of each product, and the number of users of such systems is growing rapidly in some countries. However, by limiting the power of the transmitters to the minimum that is required to achieve limited range objectives, it becomes possible to satisfy these requirements with quite narrow allocated bands. If the transmitter power can be limited in practice to a very low value, typically a small fraction of one watt, the administration may not require such stations to be licensed, although unlicensed use would carry no guarantee of freedom from interference.

However, there is a third group of MS facilities, namely those providing radiotelephone links between commercial road vehicles, taxis, railway trains and private motor cars on the one hand and either a controlling office or the public telephone network on the other hand. Public correspondence facilities for the passengers on aircraft will soon be added to this list. In some countries the use of these products is growing rapidly and this growth is likely to be sustained or even accelerated by the addition of data transmission facilities, by the growing availability of hand-portable radiotelephones suitable for carrying in a briefcase or a pocket and by falling equipment prices.

Given the necessary spectrum resources, a large fraction of all road vehicles will probably be equipped with some kind of radiotelephone facility by the end of the present century, and a substantial fraction of all telephone calls may be connected into the public switched network by a radio link by early in the next century. The provision of frequency assignments for hundreds of thousands of mobile stations, perhaps millions of them, in metropolitan areas in a part of the spectrum which is technically and economically suitable is the greatest challenge that administrations face today.

There seems to be no doubt that the MS spectrum requirements now foreseeable could be provided for in all countries. For some countries they could be supplied with ease. Elsewhere, however, it will be more difficult and a good solution will require careful spectrum management by the administration and decisions and constraints which will raise the cost of equipment. Much will depend on the amount of bandwidth nationally allocated for broadcasting and on the geographical pattern of its use. Each administration's task, therefore, is to find and implement the programme of decisions which gives the best solution to its own country's needs, as they develop, taking into account the effect that the decisions of neighbouring administrations will have on its own situation.

Depending on the present level and foreseen growth rate of the demand for spectrum for these mobile service facilities in each country, and the demand for other services which compete for the same part of the spectrum, an administration will need to consider many factors in developing its national frequency allocation table, of which the following are likely to be the most important.

Preferred frequency range: A broad frequency range centred around 300 MHz is technically best for most LM systems. For much lower frequencies, interference propagated by various long distance mechanisms may be troublesome. For much higher frequencies, wave diffraction round minor obstructions and penetration into buildings are less and the rate of decline in external noise (man-made and galactic) with rising frequency is insufficient to compensate for increased link transmission losses, due to the decline in the effective aperture of substantially omnidirectional receiving antennas (see Appendices B.6.6 and B.6.7). However, frequencies from 30 to 500 MHz are already in very heavy use for mobile systems and use of bands around 900 MHz is growing. To meet the preessure for further growth, higher frequencies will no doubt be taken into use within the forseeable future, particularly for short-distance systems. The disadvantages of higher frequencies would have less force for aeronautical links.

The planning period: Effective frequency allocation planning for the MS may require re-organisation of the use of frequency bands already allocated nationally for mobile systems in order to increase the efficiency of spectrum utilisation. Alternatively or additionally, new national MS allocations may have to be established in frequency bands already allocated to, and used for, other services. If these processes of re-organisation and re-allocation are to be carried out at an acceptably low cost, planning must be far-sighted. Changes in national allocations should desirably be planned ten years or more before they are implemented.

Level of demand: In estimating, from a variety of indicators, the way in which the demand for new mobile systems will grow within the planning period, and in converting that demand estimate into an estimate of the amount of bandwidth required for mobile services, it is essential to recognise the co-existence of several different telecommunications products, each of which may need different system architecture if it is to be provided at least cost. These various kinds of system can each be designed and managed to use spectrum efficiently, but the efficiency-promoting measures which can best be applied are likely to be different in the several cases. It is also important that demand forecasts take geographical factors into account. The demand for most, perhaps all, of these MS products will be heaviest in locations where high population density and intense business activity extend over an area of hundreds or even thousands of square kilometres, that is in so-called metropolitan areas. The extent of the demand in metropolitan areas is likely to determine the spectrum management policy for the country as a whole. The various facilities and the optimisation of systems are discussed further on pp. 230–238.

Policy towards new systems: If it is clear that the demand for LM facilities will grow sufficiently to raise severe problems of spectrum availability, the use of alternatives to LM radiotelephone systems may be encouraged. For example:

(i) Wide-area radio paging satisfies the needs of many users and can serve tens of thousands of users through a single narrow radio channel; see, for example CCIR Report 499[67].

(ii) New local area systems using freely radiated radio waves might not be licensed if their function could be performed by other systems with little or no radiation and with no substantial increase in cost; in time it may also be feasible to have existing radiating systems replaced by non-radiating ones. CCIR Report 902[82] discusses radiating cable systems and induction-loop; leaky-waveguide and infra-red systems also find application.

Existing mobile systems: The frequency band which an administration would prefer to use to accommodate a major expansion of mobile systems may already be in use for the MS. If so, it may be possible to change the pattern and intensity of this existing use to make room for the new facility. Administrations should give attention to the efficiency with which the existing systems use spectrum and to the way carrier frequencies are assigned in these bands. It may be feasible to bring about considerable improvements, in the course of time and at an acceptably low cost, by careful attention to the terms under which licences are renewed and by more economical planning of frequency assignments. Some aspects of the technical specifications for equipment are considered further below, and others on pp. 230–238.
System specification: Some elements of the equipment specification which have a major impact on the efficiency with which spectrum is used can be identified as follows:

(i) Carrier separation. Some MS radiotelephone systems, typically using angle modulation, operate with a separation of 50 kHz between adjacent assignable frequencies but most use 25 or 20 kHz carrier separation and considerable use is now being made of 12·5 kHz systems. There are prospects of even narrower channels with SSB amplitude modulation or digital modulation using speech signals compressed by voice coding. There is an opportunity here to trade off occupied bandwidth against susceptibility to interference and noise and the optimum technique for any given network depends on the network configuration. A requirement of the administration that the channel density provided by systems should be high involves some equipment cost penalty, but where the demand for LM products is heavy, high channel density is cost effective and conceptually simple.

(ii) Adjacent channel interference. Given close carrier spacing, it is necessary to ensure that a transmitter will not interfere with a well-designed receiver tuned to the adjacent channel, and similarly that a receiver will not suffer interference from a well-designed transmitter operating in the adjacent channel. CCIR Recommendation 478[70] recommends that the adjacent channel response should be, in each case, of the order of −65 to −70 dB relative to the wanted channel response, or better.

(iii) Spurious emissions. The receivers of mobile systems, and in particular those of land mobile stations, must function satisfactorily under conditions of exceptional difficulty, often having to receive weak signals from distant land stations whilst closely surrounded by other transmitters and receivers operating in the same frequency band. Interference from spurious emissions must be controlled. CCIR Recommendation 478[70] recommends that the strongest spurious radiation from the antenna of a receiver should be less than 2 nW. Spurious emissions from transmitters should be far below the value required by RR Appendix 8; 25 pW might be taken as a

suitable value in land mobile systems.
(iv) The specifications of mobile system receivers should also call for:

- Adequate protection from signals remote from the wanted frequency, for example at image frequencies.
- Adequate freedom from interference due to intermodulation of unwanted signals in the receiver input stages.
- Adequate freedom from receiver de-sensitisation by powerful signals which are not at the wanted frequency.
- For angle-modulated systems, sufficient suppression of weak interfering signals in the presence of a stronger signal at the wanted frequency.

(v) Transmitter power: The transmitter power which a licensee is permitted to use is one of the terms of the licence and it should be related to the size of the area to be served and the height of the land station antenna. Transmitters should therefore be designed so that they can be preset to deliver, sufficiently accurately, the permitted carrier power.

Re-allocation of spectrum: If changes in the use of frequency bands which are already used for the MS does not produce enough bandwidth for foreseen new mobile systems, it will be necessary to consider changing to LM the national allocations of bands, within the preferred frequency range, which are already allocated for other services. Most of the suitable bands are likely to have been allocated internationally for BS or FS, sharing with MS in some cases, and the environment established by these international allocations affects the problem of domestic re-allocation.

(i) In Region 1, although not in the other Regions, broadcasting stations operating at VHF and UHF use frequency assignments which have been approved in formal international frequency assignment plans (see pp. 174–178). These plans may make it impossible to use the whole of wide bands for LM, even if there is no need for broadcasting locally. Nevertheless it will usually be possible to find sub-bands within the broadcasting band which can be used for LM, given the goodwill of the administrations of neighbouring countries.
(ii) Some fixed links have such adverse propagation paths that, while a VHF and UHF assignment provides satisfactory service, an SHF assignment would not be satisfactory, but such links are few. It may be necessary to cease making assignments below 1 GHz to FS stations within interference range of a metropolitan area and to adopt as a long term policy the systematic transfer of existing FS assignments below 1 GHz to higher bands, in order to conserve this part of the spectrum for mobile systems. The need for international collaboration in changing the use of an FS band is much less than when the BS is involved, being limited, perhaps, to cases where the FS use is for tropospheric scatter systems.

Harmonisation of plans: Administrations may wish to harmonise their national allocation planning with that of other administrations, and especially neighbouring countries, to minimise sharing problems, to enable the same mobile station equipment to be used in different countries and to facilitate international trade in radio equipment.

Frequency band pairing: Full duplex or two-frequency simplex MS systems are assigned two working frequencies, typically used for transmission from the land station and from the mobile station respectively. It is advantageous to make land station and mobile station assignments from different sub-bands, in particular to avoid the need for a mobile station to receive in a channel adjacent to the transmission from another mobile station, which might be physically very close to it. Thus administrations make systematic use of frequency bands allocated to the MS by dividing them into one or more pairs of sub-bands, one sub-band of each pair being used for land station transmissions and the other for mobile station transmissions. For the LM these sub-bands are defined by administrations acting either individually or in sub-Regional groups, but the design of available equipment may also exert a strong influence on this choice.

The frequency separation of paired sub-bands has little direct effect on the efficiency with which a system uses spectrum. However, many mobile stations operating in the full duplex mode use a single antenna for transmitting and receiving, the antenna being coupled to the transmitter and the receiver by means of duplexer. The cost of effective duplexers is raised if the frequency difference is made too small. Too small a frequency difference may also make more complex the design of high-capacity integrated systems such as cellular networks. There is guidance on the choice of the frequency difference in CCIR Recommendation 478[70].

7.3.5 Spectrally-efficient LM radiotelephone systems

The demand for radiotelephone facilities for road vehicles and more recently for personal hand-portable radiotelephones is growing rapidly. In some countries the use of these products in metropolitan areas is already so heavy that the achievement of efficiency in the use of the spectrum is a matter of acute concern. The same situation seems likely to develop, in some degree in many other countries in the foreseeable future, particularly where the use of spectrum below 1 GHz for other purposes, typically broadcasting, is also extensive.

There are three basic principles for achieving spectral efficiency in the LM and they are reviewed briefly below. Their applicability varies from product to product and their application has a considerable effect on system architecture and cost. A consideration of the application of these principles to two major LM products follows.

Principles for achieving efficient spectrum utilisation

The three principles for achieving efficient spectrum utilisation are effective geographical frequency re-use, minimisation of channel bandwidth and the intensive use of channels within a given geographical area. In more detail:

Frequency re-usability: The necessary service area of a base station having been defined, the area which the base station could effectively serve (that is, the coverage area), in particular in the transmit mode, should not avoidably extend beyond the limits of that service area. The same frequencies can then be re-used at other base stations located some distance beyond the edge of the coverage area. With good spectrum management, it may be feasible to use a frequency many times over within the same country. The main factors determining the degree of frequency re-use that is feasible within a given large geographical area are the transmitter power, the height and gain in relevant directions of the base station antennas, the nature of the terrain, the frequency band used, the type of modulation and the wanted field strength that is to be protected against unacceptable interference.

In estimating how effectively frequencies can be re-used, attention must be given to the statistical distribution of field strengths of the wanted and unwanted signals near the edge of the service area, due to obstructions and reflections in the propagation path. The sketch at Fig. 7.3 illustrates the situation arising where the right choice of transmitter power and of the height and the gain of the base station transmitting antenna has ensured that the quasi-minimum field strength of the base station signal falls to level F_1 at distance D_1 where F_1 is the minimum usable field strength for a mobile station receiver (allowing for the presence of interfering signals from other base stations using the same frequency) and D_1, is the distance from the base station to the edge of the service area. At a greater distance D_2, the quasi-maximum field strength falls to a value F_2, which is low enough to be tolerated as interference by mobile stations of another system receiving signals at a field strength of not less than F_1 from their own base station.

The difference between F_1, and F_2 is the protection ratio which the mobile stations require, disregarding fading. CCIR Report 358[71] shows that this is around 10 dB when the wanted signal is angle-modulated, as most LM signals are. See also CCIR Reports 319[83] and 1018[84]. For references to relevant propagation data, see Appendix B.6.6. The distance $D_2 + D_1$ is a measure of the 'repetition distance', that is, the minimum distance between base stations at which it is feasible to re-use a frequency assignment; $D_2 + D_1$ may typically equal 4 or 5 times D_1 but the ratio varies with the height of the base station antenna, the frequency band used and the nature of the terrain.

It has been assumed above that the maximum length of links between a base station and a mobile station is determined by the performance in the base-to-mobile direction. This is usually the case, but it may be necessary to consider conditions in the other direction of transmission if, for example, the base station antennas are rather low.

At various millimetre-wave frequencies, in the bands where the absorption of radio signals in the gases of the atmosphere is very severe, the

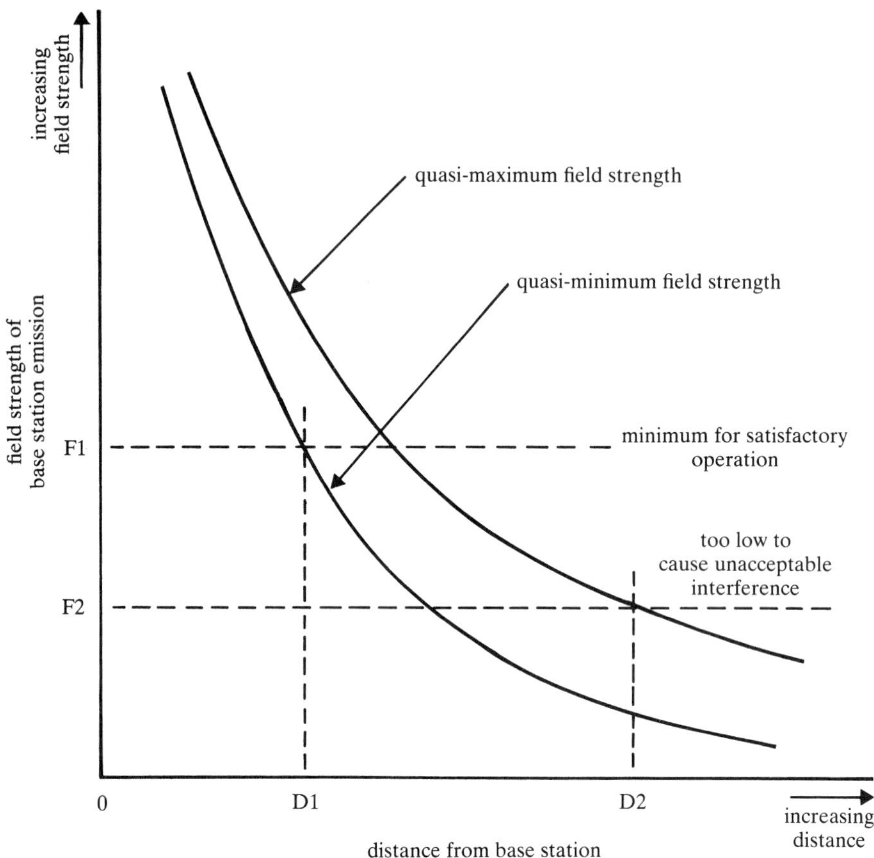

Fig. 7.3 Field strength conditions determining the maximum operating range and the minimum co-channel repetition distance in a cellular network

distance represented by $D_2 - D_1$ is only a few kilometres, regardless of D_1 (see Appendix B.4.4). This may, one day, provide a medium of very great available bandwidth and high frequency re-usability, capable of carrying an almost unlimited amount of traffic for short-range mobile services. It is not used in this way at present mainly because low-cost active devices operating at these frequencies are not yet available and economical ways have not yet been found for overcoming severe fading in the service area due to obstructions.

Channel bandwidth: The signal processing, modulation methods, modulation parameters and band-limiting filter characteristics of systems and the carrier frequency tolerance of radio transmitters and receivers should all be such as to permit the use of narrow-band channels. Most present-day LM systems use narrow-deviation FM, with careful limitation of the amplitude and the bandwidth of the modulating signal; a separation between adjacent carriers in the range 12·5 kHz to 25 kHz is typical. A number of trials have been made of SSB emissions, some using compandors or similar devices to compress the dynamic range of the speech signal; good test results have been obtained in RF bandwidths of 5 kHz or less and some limited operational use is reported; see CCIR Report 899[85]. Commercial use of digital transmission for telephony is beginning, low-cost reduced-bit-rate codecs having become available; see also CCIR Reports 903[86] and 1021[87].

It should be noted that the required protection ratio depends to some degree on the type of modulation of both the wanted and the unwanted signals. Some types of modulation which provide narrower-band channels require higher protection ratios. Accordingly there may be a trade-off to be optimised between the objectives of maximum geographical frequency re-use and minimum per-channel bandwidth occupation.

Channel occupancy: Channels should be effectively loaded. It is typical of telecommunications facilities that the traffic that is offered varies greatly throughout the day. Regular busy hours occur, alternating with periods when little traffic is offered. An effectively loaded channel will be carrying traffic for most of the time during the busy hours; it will carry perhaps 0·7 or 0·8 'busy-hour Erlangs'. However, it is important to the user that a channel should usually be available very quickly after the need for it has arisen, that is, the 'grade of service' should be acceptable. There will often be conflict between the needs for both high traffic loading and an acceptable grade of service unless the system has appropriate architecture.

The grade of service may be defined with sufficient precision for the present purpose as the probability that a channel will not be available for a randomly-arising call before the end of an arbitrarily chosen waiting time. Thus, if the mean holding time of calls on a single channel system is 20 seconds and the total holding time during the busy hour is 30 minutes (that is, a busy-hour traffic flow of 0·5 Erlangs) then Erlang's delayed-call formula indicates a probability of 30% of that single channel will not be available for a new call within 20 seconds of the call being required. This is said to be a grade of service of 30% for a waiting time of 20 seconds. For further information on the statistics of traffic flow in telecommunications systems, see, for example, Bear[88].

Consider a LM user with a heavy traffic demand, perhaps a dispatcher at the control office of a taxi company. Such a user might be assigned the exclusive use, within a given service area, of one radio channel. The peaks of demand for the facility tend to be automatically smoothed out by the way that the operator regulates the flow of traffic over the channel under such conditions. There may be enough traffic to load the channel for half the time in the busy hour, or more, in these circumstances and the user may well be satisfied with the grade of service that the exclusive use of one frequency channel provides under these conditions. A busy-hour traffic flow of 0·5 Erlangs or more may be regarded as efficient use of spectrum in a low-cost system.

More usually the traffic flow generated by one user is quite small, and much less than a channel could be carrying. If so, the administration may seek to increase the loading on the channel by assigning it to two or more users in the same geographical area, depending on operator discipline or an interlock system to limit interference between the assignees. However, with several users having access to the channel, the grade of service will probably be bad even when traffic load carried by the channel in the busy hour is low. Call delays on shared channels are a source of great irritation to users.

Trunked systems, in which a group of channels is made available to a group of users, are more complex but they combine higher potential traffic flows with a much better grade of service, since any free channel in the group can be seized for a call by any user. For the same holding time (that is, 20 seconds) and the same waiting time (also 20 seconds) as was assumed in the single channel example cited above, a trunked system of 10 channels could carry 0·793 Erlangs per channel with a grade of service of 5%.

Spectrum economy in private mobile radio systems

Private mobile radio (PMR) systems typically provide a direct connection between a base station operator and a mobile station operator, usually the driver of a road vehicle which remains within the service area of a single base station. Obviously it may be technically feasible for the base station operator to be given access to a mobile station that is outside the local service area by routing the call via another base station. Also a base station may be equipped to connect a call from a mobile station into the public telephone network. However, neither of these modes of operation changes in principle the characteristics of the system.

Most PMR systems at present are technically and organisationally simple. The user owns or leases the base station and mobile station equipment and the administration assigns a channel, shared or exclusive, for his use. The base station may be located at the user's business premises if the service area is small and local to the premises, but a remote installation at the top of a high building or a hill may be needed for larger service areas, up to perhaps 40 km radius. The administration will include in the licence a condition that some appropriate narrow-band modulation system be used and will probably ensure that the equipment is capable of operating satisfactorily in the channel assigned, through type-approval of the equipment that is used. The licence should also include a limit on transmitter power in order to prevent avoidable overspill beyond the edge of the service area. On channels shared by networks with overlapping coverage areas there should be some means of avoiding simultaneous use of the channel by two systems. A widely-used, elementary solution is to require each user to listen on a channel, to make sure that it is not already in use, before transmitting on it, delaying his own call as long as necessary if the channel is engaged. Systems are available with electrical means for locking out a would-be caller if the channel is already in use. Either solution tends to be irksome to users.

The spectrum management process, as applied to systems of this kind, includes assessment of the feasibility of channel sharing within a service area, assessment of the probability of interference between systems that cannot share a channel and verification that channels are being used in the licensed way and to the degree claimed by licensees; see CCIR Report 319[83]. In countries where tens of thousands of such systems and hundreds of thousands of mobile stations are licensed to operate, this is a major burden on the administration. Some administrations devolve parts of the management function to bodies representing users or equipment suppliers, perhaps retaining only an administrative oversight.

Trunked PMR systems are coming into use in countries where pressure on the spectrum requires higher spectrum utilisation efficiency than simple single channel systems can provide. Some trunked systems have a single user with a very heavy traffic load. More usually a trunked system base station is owned by a service-providing company, and users obtain radio access to their mobile stations through the base station under service agreements; see CCIR Reports 741[89] and 901[90].

Spectrum management for trunked PMR systems is much less onerous for the administration, partly because the number of base stations is relatively small and partly because the base station service provider has both the means and an incentive to ensure that the channels of the trunked system combine intensive use with a good grade of service.

Spectrum economy in public LM radiotelephone systems

Public LM radiotelephone systems connect mobile stations on road vehicles with telephone subscribers through the public switched telephone network (PSTN). Personal mobile stations, battery-powered and small enough and light enough to be carried in a pocket, are coming into use as part of a great expansion of the use of such facilities.

These systems are always trunked. Some established systems do not differ greatly in principle from trunked PMR systems, and they are available in many countries in metropolitan areas; see CCIR Report 742[91]. However, the combination of systems with wide geographical coverage and a need for high spectrum occupancy has created requirements for new system capabilities. It has become necessary for the radio connection to a moving vehicle to be transferred quickly and automatically from the base station of a coverge area being left to the base station of the coverge area being entered and for a wanted mobile station to be located and contacted wherever it may be within the area covered by the system. Systems with the complexity that these requirements entail have only recently become economically viable through the introduction of cellular networks.

To provide wide-area coverage in a large country it may be necessary to set up many base stations. To facilitate consideration of frequency re-use in such a system, the geographical distribution of base stations may be idealised in the form of a hexagonal lattice plan, typically with distances of about 50 km between stations. A mobile station should always be within the service area of one of the base stations.

In an idealised lattice such as is shown in Fig. 7.4, it will be impossible to use a channel at base station K for a call to one mobile station and to use the same channel simultaneously at the base station of any of the six adjacent service areas, L to Q, for a call to another mobile station, since interference levels will obviously be much too high in large parts of the service areas of both the base stations involved. Thus, for operation in service area K and the six service areas contiguous with it, the available channels must be split into seven packets, one packet of channels being assigned to each base station. Provided that the necessary protection ratio is obtained when $D_2 + D_1$ is not more than 4·6 times D_1 (using the symbols of Fig. 7.3) it will be possible to cover an indefinitely large service area by repeatedly re-using these same seven packets of channels in a systematic way (see Fig. 7.4*b*).

If the minimum repetition distance were greater than 4·6 times D_1, the optimum ideal lattice configuration and the optimum pattern of frequency re-use would depart from the hexagonal pattern shown in Fig. 7.4 and the channels would have to be distributed between more than seven packets. The attainable density of channel utilisation would be reduced accordingly.

In real systems the physical geography of the terrain will lead to departures from the idealised hexagonal lattice of base stations and base station antenna characteristics and transmitter power will be optimised to serve local concentrations of mobile stations. Large-scale non-uniformities in the distribution of mobile stations will make it desirable to assign more channels to some base stations than to others. Nevertheless the concept of the uniform hexagonal lattice provides a useful measure of the geographical re-usability of the spectrum for cellular systems.

To meet growth in the traffic offered, it may be possible to assign more bandwidth to the system, increasing the number of channels that can be assigned in each packet. When this easy solution is no longer practicable, the alternative is to scale down the size of the coverage area of each base station, allowing frequencies to be re-used at much smaller geographical intervals. Using the idealised hexagonal lattice model for base stations, the theoretical capacity of a given bandwidth is inversely proportional to the square of the radius of the service area of a unit base station. Present-day cellular systems are using cells of 2 to 20 km radius; they do not adhere to a regular lattice or a uniform distribution of channels between base stations, but even so the application of the small-cell principle increases the efficiency of spectrum utilisation by many times. For information on cellular systems, see CCIR Reports 742[91], 740[92] and 1020[93].

It would seem that the principle of cellular operation could extend indefinitely the traffic-carrying capacity of a given bandwidth through further reduction of cell size, but other factors arise as the number of cells increases and their size is reduced. For example:

- The difficulty of locating a given mobile station, so that a telephone

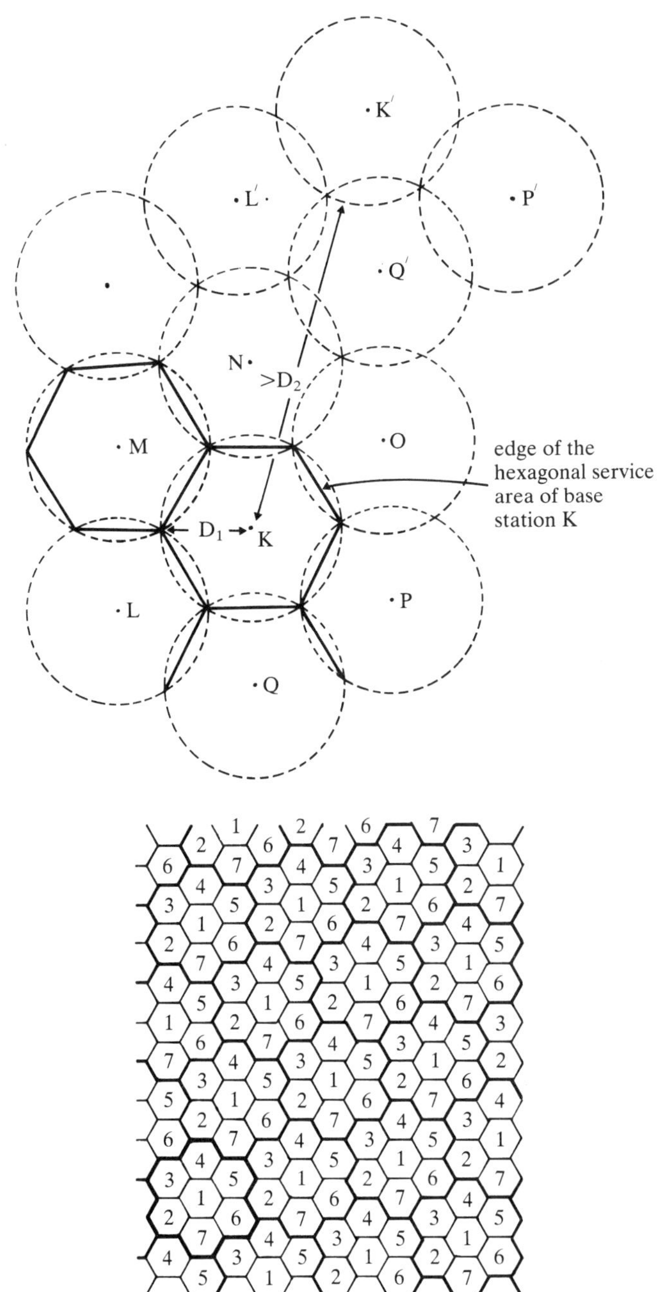

Fig. 7.4 An idealised regular hexagonal lattice of service areas for a cellular system, based on a seven-cell repeating pattern of frequency re-use, which is feasible if D_2 (as shown in Fig. 7.3) exceeds $4{\cdot}6\,D_1$

call from the telephone network can be routed to it, increases.

- In cellular systems a moving vehicle must be handed over automatically from the cell which it is leaving to the cell which it is entering without a perceptible break in the connection. As the cells get smaller, a vehicle spends less time within any one cell and the management burden of carrying out several transfers during the course of a call becomes more expensive.
- As the size of cells becomes comparable with the size of the radio propagation shadows cast by obstructions, the boundaries of base station service areas become less determinate and this may lead to a further increase in the frequency of handovers,
- the capital cost of systems, per unit of traffic capacity, rises.

It is not yet possible to say where the technical and economic limits of geographical frequency re-use will lie.

Public LM radiotelephone systems are likely to be large and usually there will be few of them within the jurisdiction of any one administration. As with trunked PMR systems, they will be operated by expert organisations with an interest in securing efficient use of the spectrum. Thus they do not create a heavy spectrum management burden for the administration.

7.4 References

1 Final Acts of the World Administrative Radio Conference for the Mobile Services (MOB-87) (ITU, Geneva, 1987)

2 Final Acts of the World Administrative Radio Conference for the Mobile Services (MOB-83) (ITU, Geneva, 1983)

3 'Emergency position-indicating radio beacons operating at the frequency 2182 kHz'. CCIR Recommendation 439–3; Recommendations and Reports of the CCIR, 1986, Volume VIII–2 (ITU, Geneva, 1986)

4 'Future use and characteristics of emergency position-indicating radiobeacons in the mobile service and the mobile-satellite service'. CCIR Report 749–2; *ibid,* Volume VIII–2

5 'Alarm signal for use on the maritime radiotelephony distress frequency of 2182 kHz'. CCIR Recommendation 219–1; *ibid,* Volume VIII–2

6 KENT, P.E. 'The future global maritime distress and safety system', *Telecom. J,* 1986, **53,** pp. 339–342

7 'Technical and operating considerations for a future global maritime distress and safety system'. CCIR Report 747–1; Recommendations and Reports of the CCIR, 1986, Volume VIII–2 (ITU, Geneva, 1986)

8 'Introduction of provisions for the Global Maritime Distress and Safety System (GMDSS) and the continuation of the existing distress and safety provisions'. Resolution 331; Final Acts of WARC-MOB-87; see[1] above

9 'Interference due to intermodulation products in the land mobile service between 25 and 1000 MHz'. CCIR Report 739–1; Recommendations and Reports of the CCIR, 1986, Volume VIII–1 (ITU, Geneva, 1986)

10 'Sources of unwanted signals in multiple base station sites in the land mobile service'. CCIR Report 1019; *ibid*, Volume VIII–1

11 'An electromagnetic compatibility figure of merit for single channel voice communications systems'. CCIR Report 660; *ibid*, Volume I

12 'Measurement procedures for an electromagnetic compatibility figure of merit for single channel voice communications systems'. CCIR Report 661; *ibid*, Volume I

13 'Prevention of interference to radio reception on board ships'. CCIR Recommendation 218–1; *ibid*, Volume VIII–2

14 'Control of passive intermodulation products'. CCIR Report 1049; *ibid*, Volume VIII–3

15 'Relating to implementation of the changes in allocations in the bands between 4000 kHz and 27 500 kHz'. Resolution 8; Radio Regulations (ITU, Geneva, 1982)

16 'Use of frequency bands allocated exclusively to the aeronautical mobile service for various forms of public correspondence'. Resolution 409; Final Acts of WARC-MOB-87; see[1] above

17 'Sharing between the maritime mobile service and the aeronautical radio-navigation service in the band 415–526·5 kHz'. CCIR Report 910–1. Recommendations and Reports of the CCIR, 1986, Volume VIII–2 (ITU, Geneva, 1986)

18 'Choice of frequency for sound broadcasting in the tropical zone'. CCIR Recommendation 48–2; *ibid*, Volume X–1.

19 'Interference to radionavigation services from other services in the bands between 70 kHz and 130 kHz'. CCIR Recommendation 589–1; *ibid*, Volume VIII–2

20 'Interference between fixed and maritime mobile and radionavigation services in the bands between 70 and 130 kHz'. CCIR Report 915–1; *ibid*, Volume VIII–2

21 'Frequency sharing between services in the band 4 to 30 MHz'. CCIR Report 911; *ibid*, Volume VIII–2

22 'Selective calling system for use in the maritime mobile service'. CCIR Recommendation 257–2; *ibid*, Volume VIII–2.

23 'Relating to the allocation of Maritime Identification Digits (MIDs), and the formation and assignment of identities in the maritime mobile and maritime mobile-satellite services (Maritime mobile service identities)'. Resolution 320; Final Acts of WARC-MOB-83 (see[2] above)

24 'Early implementation of the use of digital selective calling on maritime HF radiotelephone channels'. Resolution 336; Final Acts of WARC-MOB-87 (see[1] above)

25 'Digital selective-calling system for use in the maritime mobile service'. CCIR Recommendation 493–3; Recommendations and Reports of the CCIR, 1986, Volume VIII–2 (ITU, Geneva, 1986)

26 'Translation between an identity number and identities for direct printing telegraphy in the maritime mobile service'. CCIR Recommendation 491–1; *ibid*, Volume VIII–2

27 'Technical characteristics of single sideband equipment in the MF and HF land mobile radiotelephone'. CCIR Recommendation 494; *ibid*, Volume VIII–1

28 'General review of the bands 4000–4063 kHz and 8100–8195 kHz allocated on a shared basis to the maritime mobile service'. Resolution 319; Final Acts of WARC-MOB-87; see[1] above

29 'Transfer of frequency assignments of radiotelephone stations operating in accordance with Appendix 25 (Rev)'. Resolution 326; *ibid*

30 'Use of additional channels reserved for use for duplex radiotelephony in the HF bands allocated to the maritime mobile service'. Resolution 325; *ibid*

31 'Relating to the frequencies in Appendix 16, Section B of the Radio Regulations, provided for worldwide use by ships of all categories and by coast stations'. Recommendation 304; Radio Regulations, edition of 1982 (ITU, Geneva 1982)

32 'Use and notification of the paired frequencies reserved for narrow-band direct-printing telegraph and data transmission systems in the HF bands allocated on an exclusive basis to the maritime mobile service'. Resolution 300; Final Acts of WARC-MOB-87; see[1] above

33 'Technical characteristics for HF maritime radio equipment using narrowband phase shift keying (NBPSK) telegraphy'. CCIR Recommendation 627; Recommendations and Reports of the CCIR, 1986, Volume VIII–2 (ITU, Geneva, 1986)

34 'Direct printing telegraph equipment employing automatic identification in the maritime mobile service'. CCIR Recommendation 625; *ibid*, Volume VIII–2

35 'Calling procedures for HF A1A morse telegraphy'. Resolution 312; Final Acts of WARC-MOB-87; see[1] above

36 'Use of non-paired ship station frequencies for narrow-band direct-printing telegraph and data transmission systems'. Resolution 335; *ibid*

37 'Frequency allotment plan for the aeronautical mobile service and related information'. Appendix 26 to the Radio Regulations, 1959, published separately (ITU, Geneva, 1959; reprinted 1967)

38 'Frequency allotment plan for the aeronautical mobile (R) service and related information'. Appendix 27 to the Radio Regulations, published separately (ITU, Geneva, 1983)

39 'Relating to the use of frequencies 3023 kHz and 5680 kHz common to the aeronautical mobile (R) and (OR) services'. Resolution 403; Radio Regulations, edition of 1982 (ITU, Geneva, 1982)

40 'Optimum coding and modulation method(s) for the transmission of digital data between terrestrial stations and aircraft stations operating in the aeronautical mobile (R) service'. CCIR Report 928; Recommendations and Reports of the CCIR, 1986, Volume VIII–3 (ITU, Geneva, 1986)

41 'Relating to the use of frequencies in the aeronautical mobile (R) service'. Resolution 405; Radio Regulations, edition of 1982; (ITU, Geneva, 1982)

42 'Relating to the use of frequency bands higher than the HF bands in the aeronautical mobile (R) service and the aeronautical mobile-satellite (R) service for communication and for meteorological broadcasts'. Resolution 406; *ibid*

43 'Relating to co-operation in the efficient use of worldwide frequencies in the aeronautical mobile (R) service'. Recommendation 402; *ibid*

44 'Relating to the development of techniques which would help to reduce congestion in the high frequency bands allocated to the aeronautical mobile (R) service'. Recommendation 403; *ibid*

45 'Relating to a study of the utilisation of the aeronautical mobile-satellite (R) service'. Recommendation 405: *ibid*

46 'Relating to co-operation in the efficient use of worldwide frequencies in the aeronautical mobile (R) service' Recommendation 402; *ibid*

47 'Relating to the holding of a Regional Administrative Radio Conference to prepare frequency assignment plans for the maritime mobile service in the bands between 435 kHz and 526·5 kHz and in parts of the band between 1606·5 kHz and 3400 kHz in Region 1 and to plan for the aeronautical radionavigation service in the band 415–435 kHz in Region 1; Resolution 704; Final Acts of WARC-MOB-83; see [2] above

48 Final Acts of the Regional Administrative Radio Conference for the planning of the MF maritime mobile and aeronautical radionavigation services (Region 1), Geneva 1985 (ITU, Geneva, 1986)

49 'Date of entry into force of the 10 kHz guardband for the frequency 500 kHz in the mobile service (Distress and Calling)'. Resolution 210; The Final Acts of WARC-MOB-87; see[1] above

50 'Procedure to be applied for the co-ordination of the use of the frequency 518 kHz for the International NAVTEX System'. Resolution 324; *ibid*

51 'Frequencies for routine (non-distress) calling in the bands between 1605 kHz and 4000 kHz'. Resolution 330; Final Acts of WARC-MOB-87; see[1] above

52 'The future use of the band 2170–2194 kHz'. CCIR Report 1029; Recommendations and Reports of the CCIR, 1986, Volume VIII–2 (ITU, Geneva, 1986)

53 'The possibility of reducing the band 4200–4400 MHz used by radio altimeters in the aeronautical radionavigation service'. Recommendation 606; Final Acts of WARC-MOB-87; see[1] above

54 'Future requirements of the band 5000–5250 MHz for the aeronautical radionavigation service'. Recommendation 607; *ibid*

55 'Development of a world-wide system for public correspondence with aircraft'. Recommendation 408; *ibid*

56 'Public mobile telephone service with aircraft'. CCIR Report 1051; Recommendations and Reports of the CCIR, 1986, Volume VIII–3 (ITU, Geneva, 1986)

57 'On-board communications by means of portable radiotelephone equipment'. CCIR Recommendation 542–1; *ibid*, Volume VIII–2

58 'Relating to an automated UHF maritime mobile radiocommunication system'. Recommendation 310; Radio Regulations, edition of 1982 (ITU, Geneva, 1982)

59 'Future public land mobile telecommunication system'. Recommendation 205; Final Acts of WARC-MOB-87; see[1] above

60 'Automated VHF/UHF maritime mobile telephone system'. CCIR Recommendation 586–1; Recommendations and Reports of the CCIR, 1986, Volume VIII–2; (ITU, Geneva, 1986)

61 'Coast station identities and initiation of location registration in an automated VHF/UHF maritime mobile telephone system'. CCIR Recommendation 587–1; *ibid*, Volume VIII–2

62 'VHF radiotelephone system for the maritime mobile service with automatic facilities using sequential single frequency coding'. CCIR Report 1033; *ibid*, Volume VIII–2

63 'Operational procedures for an international VHF radiotelephone system with automatic facilities based on DSC signalling format'. CCIR Report 1034; *ibid*, Volume VIII–2

64 'Technical and operating characteristics of cordless telephones'. CCIR Report 1025; *ibid*, Volume VIII–1

65 'Technical and operational characteristics of future international radio-paging systems'. CCIR Recommendation 539–2; *ibid*, Volume VIII–1

66 'Standard codes and formats for international radio paging'. CCIR Recommendation 584–1; *ibid*, Volume VIII–1

67 'Radio-paging systems'. CCIR Report 499–4; *ibid*, Volume VIII–1

68 'Radio-paging systems, standardisation of code and format'. CCIR Report 900–1; *ibid*, Volume VIII–1

69 'Aeronautical Telecommunication'. Annex 10, Volume I, fourth edition of The Convention on International Civil Aviation, ICAO, Montreal, 1985

70 'Technical characteristics of equipment and principles governing the allocation of frequency channels between 25 and 1000 MHz for the land mobile service'. CCIR Recommendation 478–3; Recommendations and Reports of the CCIR, 1986, Volume VIII–1 (ITU, Geneva, 1986)

71 'Protection ratios and minimum field strengths required in the mobile services'. CCIR Report 358–5; *ibid*, Volume VIII–1

72 'Frequency sharing between the land mobile service and the broadcasting service (television) below 1 GHz'. CCIR Report 1023; *ibid*, Volume VIII–1

73 'Signal-to-interference ratios and minimum field strengths required in the aeronautical mobile (R) service above 30 MHz'. CCIR Recommendation 441–1; *ibid*, Volume VIII–3

74 'Factors that should be considered when establishing protection criteria for aeronautical safety services'. CCIR Report 926; *ibid*, Volume VIII–3

75 'General considerations relative to harmful interference from the viewpoint of the aeronautical mobile services'. CCIR Report 927–1; *ibid*, Volume VIII–3

76 'The use of the band 136–137 MHz by services other than the aeronautical mobile (R) service'. Resolution 408; Final Acts of WARC-MOB-87; see[1] above

77 'Compatibility between the aeronautical mobile (R) service in the band 117·975–137 MHz and sound broadcasting stations in the band 87·5–108 MHz; Recommendation 714; *ibid*

78 'Recommendations and standards for emergency position-indicating radiobeacons operating on the frequencies 121·5 and 243 MHz'. Resolution 601; *ibid*

79 'Implementation and use of frequency 156·525 MHz for digital selective calling for distress, safety and calling'. Resolution 323; Final Acts of WARC-MOB-87; see[1] above

80 'Technical characteristics of VHF radiotelephone equipment operating in the maritime mobile service in channels spaced by 25 kHz'. CCIR Recommendation 489–1; Recommendations and Reports of the CCIR, 1986, Volume VIII–2 (ITU, Geneva, 1986)

81 'Relating to the use of channels 15 and 17 of Appendix 18 by on-board communication stations'. Recommendation 305; Radio Regulations, edition of 1982 (ITU, Geneva, 1982)

82 'Radiating/leaky cable systems in the land mobile services'. CCIR Report 902; Recommendations and Reports of the CCIR, 1986, Volume VIII–1 (ITU, Geneva, 1986)

83 'Characteristics of equipment and principles governing the assignment of frequency channels between 25 and 1000 MHz for land mobile services'. CCIR Report 319–6; *ibid*, Volume VIII–1

84 'Co-channel and adjacent-channel co-ordination criteria for simultaneous use of different modulation techniques in the mobile service'. CCIR Report 1018; *ibid.*, Volume VIII–1

85 'Systems of modulation with high spectrum efficiency for the land mobile service'. CCIR Report 899; *ibid*, Volume VIII–1

86 'Digital transmission in the land mobile service'. CCIR Report 903–1; *ibid*, Volume VIII–1

87 'Equipment characteristics for digital transmission in the land mobile service'. CCIR Report 1021; *ibid*, Volume VIII–1

88 BEAR, D. 'Telecommunication traffic engineering' revised edition, (Peter Peregrinus, Stevenage, England, 1988)

89 'Multi-channel land mobile systems for dispatch traffic (with or without PSTN interconnection)'. CCIR Report 741–2; Recommendations and Reports of the CCIR, 1986, Volume VIII–1 (ITU, Geneva, 1986)

90 'Frequency assignment methods of trunked mobile radio systems'. CCIR Report 901–1; *ibid*, Volume VIII–1

91 'Public land mobile telephone systems'. CCIR Report 742–2; *ibid*, Volume VIII–1

92 'General aspects of cellular systems'. CCIR Report 740–2; *ibid*, Volume VIII–1

93 'Adaptation of system specification to ease the practical implementation of Radio equipment'. CCIR Report 1020; *ibid*, Volume VIII–1

Chapter 8

The fixed-satellite service

8.1 Survey of FSS Regulation

8.1.1 Allocations, sharing and systems

Radio stations on satellites (and sub-orbital spacecraft) are called space stations and radio stations on Earth which communicate directly with space stations are called earth stations; see RR 61 and RR 60. 'Satellite network' is another key term in space radio; it is defined in RR 106 as one satellite, together with the co-operating earth stations.

RR 22, as amended in the Final Acts of WARC-ORB-88[1], defines the fixed-satellite service (FSS) as:

> a radiocommunication service between earth stations at given positions, when one or more satellites are used; the given position may be a specified fixed point or any fixed point within specified areas; . . .

Thus, transportable earth stations which are operated only whilst at rest are unambiguously included within the FSS definition, but frequency assignments made to them and registered in the MIFR have no international status if they are used outside the areas which have been specified and, if necessary, co-ordinated.

RR 22 states that the FSS includes satellite-to-satellite links. However, all of the frequency bands allocated to the FSS at present are allocated specifically for Earth-to-space links, for space-to-Earth links or, in a few cases, for both; no allocation allows for satellite-to-satellite links. The bands allocated to the ISS are available for these satellite-to-satellite links (see Chapter 13)

Feeder links

The third clause of RR 22 states that the FSS may also include feeder links for other space radiocommunications services. RR 109[1] defines a feeder link as:

> a radio link from an earth station at a given location to a space

station, or vice versa, conveying information for a space radiocommunication service other than for the fixed-satellite service. The given location may be at a specified fixed point, or at any fixed point within specified areas.

This definition includes, for example, the up-links carrying programme signals to broadcasting satellites, the down-links carrying data from Earth exploration satellites and the up-links and down-links between a satellite of the MMSS and a coast earth station.

The definitions of some space radio services include associated feeder links, allowing assignments to feeder links to be made in the bands allocated for those services, although the option for systems of these services to use FSS allocations for their feeder links remains open. This is true, for example, of the MSS and EESS. One very important exception is the BSS, the definition of which, at RR 37, makes no reference to feeder links.

The option for feeder link assignments to be made in FSS allocations is, in general, a good one. Most space services requiring feeder links are unlikely to involve many satellites at any one time and the bandwidth required for the feeder links of most satellites of these services is comparatively small. Feeder link earth stations, being located at specified fixed points or within specified areas, can have their frequency assignments co-ordinated with those of terrestrial stations operating in the same frequency band. Flexibility and efficient spectrum utilisation is generally to be achieved by accommodating these feeder links in the broad bands allocated to the FSS and shared with terrestrial services, rather than by allocating bands specifically for these various groups of feeder links.

However, the feeder links required for broadcasting satellites are an exception to this generalisation. The access of BSS feeder links to the frequency bands used by FSS networks could raise major problems, technical and regulatory. The technical problems arise from the very extensive implementation of systems foreseen for satellite broadcasting and the low tolerance of interference which current broadcasting satellite system designs allow for their feeder links; see CCIR Reports 561[2] and 952[3]. The regulatory problems arise from the application of frequency assignment planning to satellite broadcasting and in consequence to feeder links for broadcasting satellites. These plans give higher status to feeder link assignments than any assignment to an FSS network in the same band can have. Consequently, as the use of satellite broadcasting grows, FSS networks would be ejected from their Earth-to-space allocations to make room for feeder links.

The feeder links to satellites broadcasting in planned BSS bands have not, in general, been denied access to the Earth-to-space bands allocated for the use of FSS networks, although exceptionally RR 858[1] specifically excludes feeder links of satellites broadcasting to European countries from the band 14·0–14·5 GHz. However, a feeder link assignment made in one of these bands must be co-ordinated on equal terms with assignments made to FSS networks. To provide for the majority of feeder links to satellites broadcasting in BSS bands in accordance with agreed frequency assignment plans, additional Earth-to-space frequency allocations have been made to the FSS and their use is limited to this specific purpose. Feeder link frequency allocation plans are drawn up as required in these bands. The bands are identified on p. 312. In principle these feeder links are part of the FSS but it is more convenient to treat them in this book as a different service and they are considered in Chapter 9.

The frequency allocations

A WARC in 1963 made the first allocations for what was then called the communication-satellite service, a designation which covered what is now both the FSS and the MSS. These allocations were all between 3 and 9 GHz.

Another WARC in 1971 defined the FSS and MSS as separate services. It re-allocated for the FSS all of the bands which WARC-63 had allocated for the communication-satellite service and it made allocations for the MSS, the MMSS and AeMSS, some sharing with the FSS allocations. WARC-71 also made allocations for the FSS between 10 and 15 GHz, around 20 and 30 GHz and at various frequencies between 40 and 230 GHz. A third WARC, in 1979, made minor changes to the FSS allocations that were already in existence and made additional allocations between 4 and 14 GHz.

The allocations above 31 GHz are little used at present. Figure 8.1 shows the allocations below 31 GHz as they have been since WARC-79. They make up five sets of bands, each set consisting of one or more Earth-to-space bands and one or more space-to-Earth bands, as follows:

- a pair of narrow bands at 2·6 GHz in Regions 2 and 3 only, the use of which is governed by RR 761,
- a pair of wide bands at 6 and 4 GHz, the space-to-Earth band being split, allocated world-wide although with minor variations between Regions in the Earth-to-space band,
- a pair of bands at 8 and 7 GHz, world-wide,
- a pair of bands at 13/14 GHz and 11/12 GHz, both bands being split and there being considerable Regional variation at 11/12 GHz. The conditions of use for the FSS of the band 11·7–12·7 GHz in Region 2 are set out in RR 836[4] and RR 839[1]; likewise, for the use of the band 12·2–12·5 GHz for the FSS in Region 3, see RR 845, and
- a pair of wide bands at 30 and 20 GHz, world-wide.

All of these allocations are primary but all are, to some degree, shared; see below.

Frequency band sharing with terrestrial services

As indicated on pp. 92–93, there is extensive sharing between the FSS and

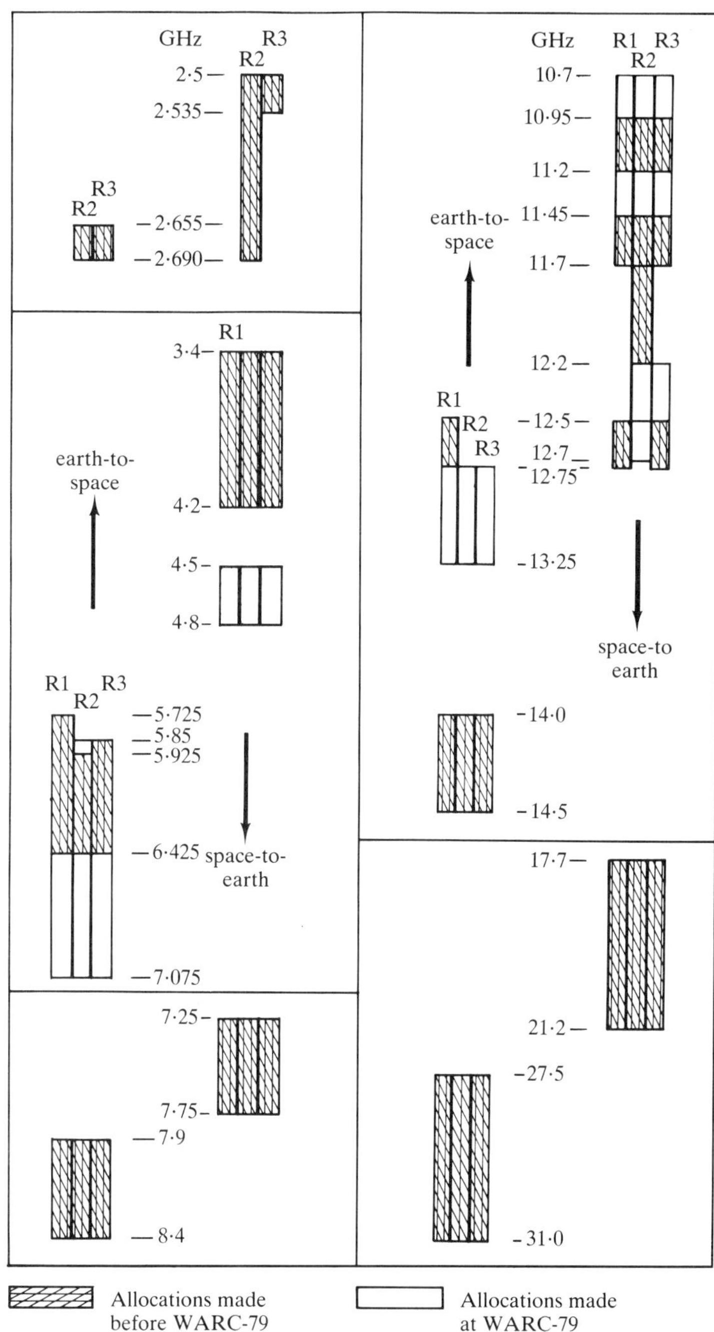

Fig. 8.1 The current frequency allocations for the FSS below 31 GHz in ITU Region 1 (R1), Region 2 (R2) and Region 3 (R3)

the fixed and mobile services. In fact, with few exceptions, all FSS allocations below 31 GHz are shared on equal terms with the FS, and many bands are shared with the MS also. The exceptions are 12·1–12·2 GHz (Region 2 only), 12·5–12·75 GHz (Region 1 only), 14·0–14·3 GHz, 19·7–21·2 GHz and 29·5–31·0 GHz. There are footnotes to the international Table of Frequency Allocations (RR Article 8) adding primary allocations to the FS in large numbers of specified countries for most of these latter bands also and MS allocations are included in some of these footnotes. RR 850 and RR 873 mitigate to some degree the impact on the FSS of this sharing in the 12·5 GHz and 20 GHz bands. The sharing constraints and co-ordination procedures which are used to limit interference between stations of the FSS and stations of these terrestrial services are discussed on pp. 110–135.

The FSS (Earth-to-space) shares 5·725–5·85 GHz with RL in Region 1. This band is allocated to RL but not to FSS in Regions 2 and 3. Depending upon the location of stations and the characteristics of systems, RL transmitters in any Region might interfere with FSS space station receivers. Also, RL receivers might suffer interference in Region 1 in the vicinity of FSS earth stations. No sharing constraints have been determined by the ITU to facilitate such sharing. In countries where this frequency band is also allocated with primary status to the FS and MS (see RR 803, RR 805 and RR 807) the constraints applied to earth station emissions radiated at low angles of elevation by RR 2540, RR 2541 and RR 2547 to facilitate sharing with the FS and the MS may give limited protection to RL receivers also (see p. 266). CCIR Reports 827[5], 828[6] and 837[7] may contribute to the resolution of interference problems involving RL transmitters and analogue FSS networks; see also Annex 5, Section 5.3.3 of the Report of the CCIR CPM to WARC-ORB-85[8].

The FSS (Earth-to-space) shares the band 14·0–14·3 GHz with the RN. However, RR 856 requires RN stations using this band to provide sufficient protection to space station receivers of the FSS. Measures to provide such protection are discussed in CCIR Report 560[9] and agreed limits on the power of RN signals are contained in CCIR Recommendation 496[10].

Frequency band sharing with other space services

The FSS shares some of its allocated frequency bands with other space radio services with equal allocation status. The sharing situations arising below 31 GHz fall conveniently into four groups:

(*a*) In the following frequency bands the sharing service operates in the same direction of transmission (upwards or downwards) as the FSS:

BSS: 2·5–2·535 GHz (Regions 2 and 3)
BSS: 2·535–2·69 GHz (Region 2 only)
MetS: 7·45–7·55 GHz, 8·175–8·215 GHz and 18·1–18·3 GHz
MSS: 7·25–7·375 GHz, 7·9–8·025 GHz, 20·2–21·2 GHz and 30–31 GHz
BSS: 12·5–12·75 GHz (Region 3 only)

Frequency assignments in these bands to stations of all of these services are subject to the sharing constraints, designed to facilitate sharing with terrestrial services, which apply also to the FSS (see p. 111). Partly as a consequence of this, the parameters of the emissions used by networks of these various services are likely to be of the same order of magnitude as those of typical networks of the FSS. Also the same international spectrum management procedures are used for all space services in these bands. It is therefore feasible to use the same basic processes of frequency co-ordination to resolve potential problems of interference between the networks of all of these services (see pp. 256–271 and 281–294). However, MSS assignments in the 7 and 8 GHz bands are, in addition, made subject to agreements reached under the procedures of RR Article 14; see RR 812.

It may be noted that the MetS allocation at 18·1–18·3 GHz is limited by RR 870 to geostationary satellites but the allocations at 7 and 8 GHz are not so limited and some meteorological satellites operating in those bands use other orbits.

Finally, there is an allocation for the BSS in Region 2 in the band 11·7–12·2 GHz under conditions that relate its use closely to the use of that band for the FSS; see RR 836[4].

(*b*) In the following frequency bands the sharing service and the FSS operate in opposite directions of transmission:

BSS: 2·655–2·69 GHz (Regions 2 and 3)
EESS: 8·025–8·4 GHz (Region 2 only)

These sharing situations are similar in principle to those in (*a*), but the interference geometry is different. Here again, the level of interference between networks of the different services is kept acceptably low by frequency co-ordination, augmented at 2·6 GHz by contraints on the BSS set out in RR 757 and by a requirement for agreement to assignments in accordance with RR Article 14 (see RR 761).

(*c*) Frequency assignment plans have been agreed for the BSS in the following bands:

11·7–12·5 GHz (Region 1)
12·2–12·7 GHz (Region 2)
11·7–12·2 GHz (Region 3)

Related frequency assignment plans have been agreed for feeder links to broadcasting satellites using those assignments in the following bands:

17·7–17·8 GHz (Region 2);
17·7–18·1 GHz (Regions 1 and 3)

All of these bands are also allocated to the FSS for assignments to links in FSS networks, although at 12 GHz the sharing does not, in general, take place in the same Region. The status of assignments made under the agreed plans is higher than any FSS assignment can attain in these bands, and this creates technical and regulatory problems which are discussed on pp. 294–297.

It may also be expected that broadcasting satellites will radiate unwanted spectral products, typically due to the non-linearities of high power amplifiers, outside their wanted transmission bands. These distortion products might well be powerful enough to cause significant interference to FSS networks operating down-links at lower power levels in adjacent frequency bands, contrary to the principles of RR 343. See CCIR Report 712[11]. However, the frequency plans drawn up for the BSS at 12 GHz include guard bands at the edges of the allocated frequency bands to limit the occurrence of such interference; see pp. 339–340.

(*d*) The band 18·6–18·8 GHz is allocated to the FSS for space-to-Earth links and to the EESS and the SRS for use with passive sensors. In Regions 1 and 3 the passive sensor allocations have secondary status but in Region 2 they are primary. The problem which this creates in Region 2 is discussed on p. 372.

Typical systems in operation

Most present-day commercial FSS networks operate in a transmission bandwidth of 500 MHz, within the bands allocated to the FSS before 1979, around 6 and 4 GHz or around 14 and 11/12 GHz. Some networks are equipped to operate in both pairs of bands. Most government FSS networks, mostly military systems, operate in the bands at 7 and 8 GHz. There is no basis in the international regulations for this separation of commerciál and government systems, but system characteristics and the pattern of sharing with other services in these bands make the arrangement a convenient one.

Almost all operating FSS satellites are in quasi-geostationary orbits. The footprint of their antenna beams (that is, the area on the Earth's surface illuminated by a beam) varies from the whole Earth disc as seen from the orbit to spot beams with apex angles as small as 1°. Many networks obtain double use of the spectrum (called 'frequency re-use') by means of dual polarisation, and some networks, especially those with a very extensive service area, are able to further re-use the spectrum by means of high-gain satellite antenna beams which illuminate different parts of the service area.

A typical space station has a wide-band receiver and frequency changer for each receiving beam, the output from which is split by filter networks between a number of transmitters, each of which provides power amplification for the signals within a band of 36 MHz, 72 MHz or some such bandwidth before feeding them to a transmitting beam. This receiver/transmitter combination is called a transponder.

Space stations differ in their antenna beam configurations depending on their intended mission. The maximum output power of typical transponders ranges from around 1 W to 20 W and some recent satellites have transponders rated at 50 W. Most transponders are 'transparent', relaying the signals received at the space station unchanged except in frequency band and power level, although some process the signals in various ways before retransmitting them. The satellite vehicles that carry the space stations differ in the technologies that they use, for example for the fine adjustment of orbital parameters and the control of attitude. In other significant respects however they tend to be very similar to one another, most providing wide-band quasi-linear amplification which can be used to transmit any kind of signal for a variety of applications.

With appropriate earth station equipment, FSS networks are providing such facilities as:

- International connections between national public telephone networks.
- Domestic trunk telephone connections between provincial centres.
- The transfer of sound broadcasting and television material from the point of origin to locations where they will be incorporated into broadcast programmes; these transmissions may be long distance links between national gateways or short distance links used for satellite news gathering (SNG).
- The distribution of broadcasting programme signals, particularly television, to earth stations which feed the signals to terrestrial radio broadcasting stations and cable broadcasting centres. This facility is called 'direct distribution by satellite'.
- Private communications networks linking the premises of large commercial organisations.
- Multi-destination distribution networks carrying market information and news.
- Point-to-multi-point low-capacity data networks.
- All kinds of telecommunications facilities between earth stations serving isolated communities.

For each of these applications, appropriate types of radio carrier are used. The FSS is unusual among radio services in that most of its networks serve not one but several or many purposes. All FSS space stations relay a substantial number of carriers and some relay very large numbers of them. These carriers may be used for a variety of types of modulation and emission parameters. A satellite may emit signals with a wide range of power density levels, according to the needs of the various earth stations it serves. Emissions may include:

- FM carriers for analogue systems with baseband capacities of one video signal or telephone systems with capacities ranging from one channel (so-called single channel per carrier (SCPC) emissions) to FDM systems with basebands continuing many supergroups of telephone channels.
- PSK carriers for digital systems having bit rates ranging from a few

hundred bits per second to time division multiplex (TDM) and time division multiple access (TDMA) systems operating at up to 120 Mbit/s.
- Spread spectrum emissions, used to provide channels of low information rate to earth stations with very low figures of merit.

Where more than one carrier is amplified simultaneously by a transponder, the carriers being separable after retransmission because of differences of carrier frequency, frequency division multiple access (FDMA) is said to be used.

Depending on a range of factors, including the frequency band, the gain of the satellite antennas, the power rating of the transponder, the information capacity of the link and the number of earth stations receiving a carrier, the economic optimum size for the earth station antenna will range from a diameter of 30 metres, typical of many INTELSAT standard A earth stations used at 6/4 GHz, to a diameter of 1·2 metres which is typical of the very small aperture terminals (VSAT) used at 14/12 GHz for multi-point low capacity data networks. The spectral power flux density of satellite emissions, measured at the Earth's surface, varies over a correspondingly wide range, to accommodate earth stations with figures of merit ranging from above 40 dB/K to below 20 dB/K.

In addition to the up-links and down-links which carry the signals which the satellite relays, each satellite transmits a down-link to carry telemetry signals to a control station on Earth and each satellite receives an up-link carrying telecommand signals. Additional links will be needed from time to time for use in determining the orbital parameters of the satellite. Frequencies are usually assigned for these purposes in bands allocated to the SpO and these frequencies are used, in particular, during the satellite launch phase. However frequencies are also assigned for these links in bands allocated to the service to which the satellite belongs, and these are the assignments that are used for telemetry and telecommand in normal network operation (see also pp. 373–374). The telemetry emission from the satellite, operated in the FSS, is commonly also used as a beacon by earth stations that have active satellite-tracking antennas.

8.1.2 International spectrum management

With good system engineering and effective radio spectrum management, the circuits that FSS networks can provide are excellent in quality in almost all respects and they cost less than terrestrial systems in a number of applications, national and international. Thus these networks are a resource of considerable economic importance.

However, FSS networks suffer interference, potentially significant in level, mostly from other FSS networks and interference to a national network is likely to come from a network serving another country. Furthermore, the total transmission capacity that is feasible with this medium, though very large, is not limitless, being determined mainly by inter-network interference and the cost of the high-performance earth station and space station equipment that would limit that interference. Consequently the capacity available from the FSS for any one country, and its cost, depend to a considerable degree on the extent of the use made of the medium by other countries.

For these reasons, radio spectrum management in the FSS is a matter of major international concern. Up to the present time the basis of international spectrum management for the FSS has been as follows:

- A number of regulations and technical constraints are applied to FSS networks by RR and by CCIR recommendations. Most of these measures relate only to networks using geostationary satellites. Some have the purpose of limiting interference from FSS networks to the stations of terrestrial services but most are intended to increase the number of satellites that can operate in the GSO in the same part of the spectrum without unacceptable mutual interference. (see pp. 265–267 and 271–281).
- A two-stage procedure is used to harmonise, as well as possible, the configurations of networks which could interfere with one another and then to determine whether the remaining interference will be acceptable. The first stage, called Advance Publication, requires an administration that proposes to set up a new satellite network to publish at least an outline of its intentions whilst the project is still in the planning stage. This provides an opportunity for informal discussions with other administrations with an interest. The second stage, called Frequency Co-ordination, applies only to networks that are to use geostationary satellites. It consists of negotiations, in a framework of conditions which are laid down in detail in the RR, to complete the harmonisation of networks and to determine whether interference between the planned network and other networks will be acceptable (see pp. 285–291).
- Harmonisation having been completed and interference having been found acceptable in level, the frequency assignments made to the new network are notified by the responsible administration to the IFRB for registration (see pp. 291–294). If unacceptable interference arises after the entry of the new network into operation, the problem would usually be resolved in accordance with priority of registration, as indicated on pp. 64–65.

An administration wishing to secure access to the GSO for a new network must often obtain the co-operation of the administrations responsible for networks which are already in operation or which are more advanced in their preparations for operation. This is an unequal contest, advantage lying with the networks which are already set up, the first-comers. The cost of late-comer networks may be made high by the need to use more expensive equipment to minimise inter-network interference, or the need to operate in parts of the spectrum where radio propagation is less favourable. It may ultimately be impossible to find room in orbit for some new networks. It is arguable that these disadvantages will fall, in particular, on developing countries.

There is a general reference on p. 63 to a tide of opinion among the administrations of developing countries in favour of using international frequency assignment planning instead of priority of registration of frequency assignments as the basis for international recognition of the right to use a radio frequency. Strong pressure for a change of procedures arose at WARC-79, aiming to ensure that no country would ever be prevented from obtaining an equitable share of access to space for radio services and especially for the FSS. The matter was not on the agenda of WARC-79 and there was no opportunity to consider the matter in depth at that conference, but interim action was agreed in the form of three resolutions:

RR Resolution 2[12] is a re-affirmation of a resolution which had been agreed eight years before, at WARC-71, and it proclaims basic principles of the established regulatory method. It emphasises that the registration of frequency assignments in the MIFR 'should not provide any permanent priority for any individual country or group of countries'. Similarly, such registration 'should not create an obstacle to the establishment of space systems by other countries' and 'that, accordingly, a country or group of countries having registered with the IFRB frequencies for their space radiocommunication services should take all practicable measures to realise the possibility of the use of new space systems by other countries or groups of countries so desiring'.
RR Resolution 4[13] seeks to give regulatory force to the first of the points contained in RR Resolution 2 by clarifying the circumstances in which assignments registered in the MIFR should be assumed to be void at the end of the operational life of the network. Minor amendments were made to this resolution at WARC-ORB-88.
RR Resolution 3[14] determined that another WARC should be convened 'to guarantee in practice for all countries equitable access to the geostationary-satellite orbit and the frequency bands allocated to space services'. The conference was to be held in two sessions, the first to decide what should be done and the second to implement the decisions.

The first session called for in RR Resolution 3 became a major part of the agenda of WARC-ORB-85 and the decisions that were reached are to be found in the report of that conference to the second session[15]. The second session was held in 1988, forming a large part of the work of WARC-ORB-88. The outcome is to be found in the Final Acts of the latter conference[1]; in brief:

(i) The FSS allotment plan: A frequency allotment plan was drawn up for two groups of frequency bands allocated to the FSS, namely:

Earth-to-space: 6725–7025 MHz and 12·75–13·25 GHz
Space-to-Earth: 4500–4800 MHz, 10·7–10·95 GHz and 11·2–11·45 GHz

For more details of the plan, see pp. 297–302.

(ii) Multilateral co-ordination in the FSS: The normal basis of frequency co-ordination is bilateral negotiation between the administration planning a new network and each of the other administrations concerned. This is expected to continue as the normal mode. However provision was made by WARC-ORB-88 for simultaneous co-ordination between all concerned administrations at Multilateral Planning Meetings (MPMs) for FSS networks in specified frequency bands if the parties so wish. The bands specified include most of the FSS allocations below 15 GHz which are not covered by the allotment plan (see p. 287).

(iii) Simplified procedures: A large number of amendments of detail were made to the RR which concern the procedures for co-ordination and registration of frequency assignments for all space radio services, in order to make the procedures more effective and easier to implement. These amendments are taken into account on pp. 281–294.

8.1.3 National spectrum management

The FSS differs from other services in that the initiative for selecting frequencies for assignment often lies, not with the administration or the earth station operating organisations but with the satellite operating organisation. An administration will play a major part in the implementation of the procedures which are used for the international management of frequency assignments to earth stations within its jurisdiction and to space stations for which it is responsible. The administration will also wish to ensure that interference between networks which are wholly within its own jurisdiction is kept within appropriate limits. The administration of a country which is host to an international organisation owning a satellite system usually provides the formal interface for regulatory matters between the organisation and the ITU; it is called a notifying administration.

The internationally agreed technical constraints reviewed on pp. 271–281 are fairly comprehensive. However, if an administration authorises a large number of networks to serve its own territory, it may need to require those networks to implement more stringent constraints. This might involve, for example, the provision of enhanced standards of earth station antenna sidelobe gain suppression, the acceptance of higher levels of interference from other networks than the ITU recommends or the acceptance of a more limited freedom of choice in the assignment of carrier frequencies to emissions. Such additional constraints are not, however, likely to be found to be necessary in most countries.

There is, however, one other important domestic matter for an administration to consider. For an earth station which is used primarily for international links it is usual to choose a site in an area where relatively little use is made of the same frequency bands for terrestrial systems. It should not be difficult to arrange that the frequencies assigned to such an earth station are not also used by national terrestrial systems in ways that raise serious interference problems, and most well-chosen sites should be free from significant interference to and from foreign terrestrial stations also.

However, there is less freedom to choose locations for the earth station of a domestic FSS network so as to avoid interference problems in shared bands and freedom is even more limited in the case of VSAT and similar earth stations, which are often located on the users' premises in city centres. Radio relay stations are often near such stations. Recent growth in the reception on domestic antennas of satellite television emissions, the primary purpose of which is the distribution of programme signals to terrestrial broadcasting stations and cable network head ends, has brought a new dimension to this problem.

It is important that these domestic FSS systems should operate in frequency bands which are not used in the same area for terrestrial systems. Unfortunately the international Table of Frequency Allocations (RR Article 8) provides very few opportunities for this, although for example it may be noted that the FSS allocation at 12·5–12·75 GHz for Region 1 is exclusive, apart for secondary allocations, except for the countries listed in RR 848 and RR 850 and that RR 837[4] makes similar provision for much of North America in the band 11·7–12·1 GHz. However, any administration is free to make a suitable FSS allocation exclusive to that service within its own jurisdiction, preferably in agreement on corresponding action on the part of the administrations of neighbouring countries, and this action should be considered where domestic FSS networks are contemplated.

8.2 Principles of frequency sharing and co-ordination for space systems

8.2.1 Introduction

Space radio networks may cause interference to terrestrial radio systems operating in the same frequency bands, or suffer interference from them, by four modes, as follows:

Mode 1: Earth station transmitters may cause interference at the receivers of nearby terrestrial stations.
Mode 2: Terrestrial station transmitters may cause interference at the receivers of nearby earth stations.
Mode 3: Space station transmitters may cause interference at the receivers of terrestrial stations spread over a wide area.
Mode 4: Terrestrial station transmitters spread over a wide area may cause interference at the receivers of space stations.

These modes are illustrated in Fig 5.1. The principles of the methods which are used to limit such interference are summarised on pp. 265–268 and are reviewed on pp. 93 and 259–263.

Space radio networks may interfere with other space radio networks operating in the same frequency bands by four modes also, as follows:

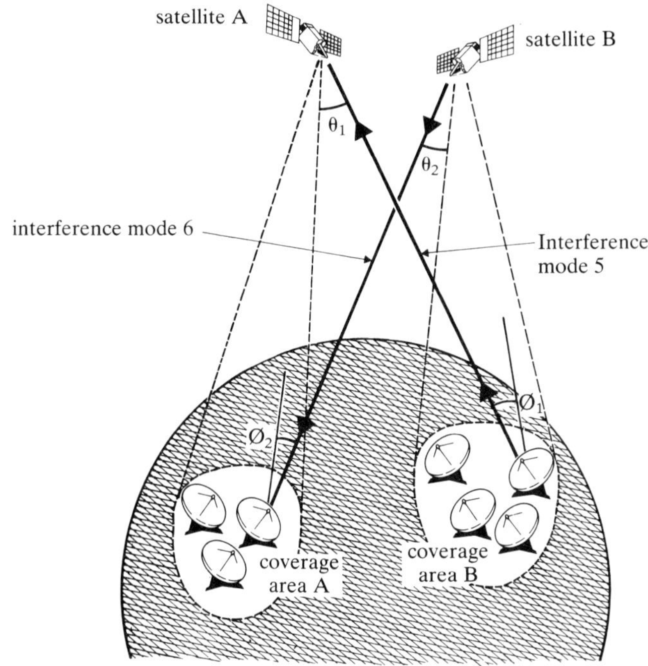

Fig. 8.2 Interference modes between satellite networks which have the same pattern of frequency band utilisation (typical of FBW)

Mode 5: When both networks use the same band for up-links, earth station transmitters may cause interference to space station receivers (see Fig. 8.2).
Mode 6: When both networks use the same band for down-links, space station transmitters may cause interference to earth station receivers (See Fig. 8.2).
Mode 7: When network A uses band X for up-links and network B uses the same band for down-links, the earth station transmitters of network A may interfere with the receivers of nearby earth stations of network B. If in addition network A uses band Y for down-links and network B uses the same band for up-links, this earth station to earth station interference would be mutual (see Fig. 8.3).
Mode 8: Given the same frequency utilisation situation as mode 7, interference may also occur between the space station transmitters of one network and the space station receivers of the other network. The level of interference may be significant when the satellites are close together in orbit. Significant interference may also arise when the satellites are widely separated, such that a straight line from one to the other passes close to the Earth; this is called the quasi-antipodal case (see Fig, 8.3).

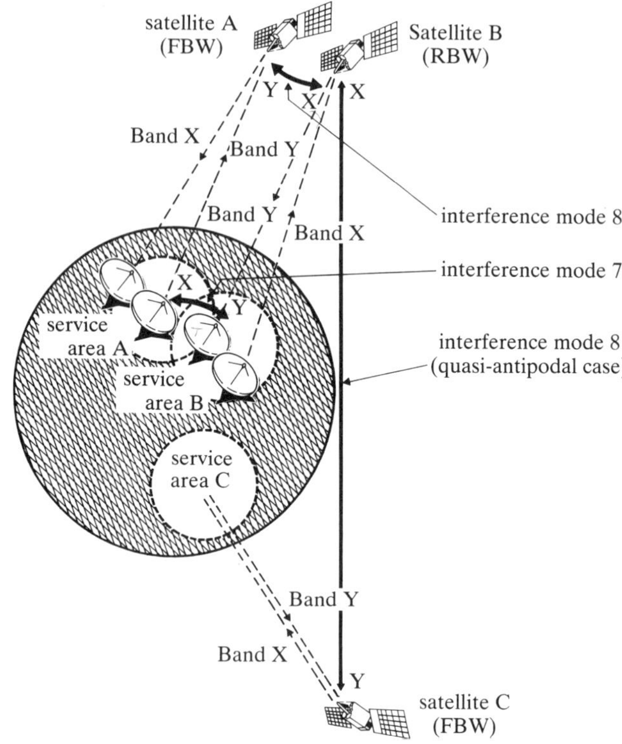

Fig. 8.3 Interference modes between satellite networks which have contrary frequency band utilisation patterns, FBW and RBW

The severity of these various interference entries depends on many factors. When both satellites are geostationary, the most significant of these factors are the angular separation between the satellites as seen from the earth stations, the geographical separation (if any) between the coverage areas on the Earth's surface of the space station antennas, the side-lobe characteristics of the earth station and space station antennas and the distribution of each network's signals in the frequency domain. If either of the satellites is non-geostationary, the relative positions and the orbital motion of the satellites have a large effect on interference levels.

Typical networks of the FSS use the FSS frequency allocations in conventional ways. Most currently operating networks use the 6 GHz band for up-links and the 4 GHz band for down-links, and/or the 14 GHz band for up-links and the 11 GHz or 12 GHz band for down-links. Such networks cause and suffer mutual interference by modes 5 and 6. They are said to use these bands in the 'forward band working' (FBW) mode, to distinguish them from networks that use bands allocated to the FSS in the unconventional direction of transmission, that is, for example, using the 11 or 12 GHz band for up-links and the 14 GHz band for down-links.

At the present time the international Table of Frequency Allocations (RR Article 8) does not allow FSS networks to operate in this latter 'reverse band working' (RBW) mode although the interference problems that this could cause have been studied; see CCIR Report 557[16]. However, several situations arise in which feeder links to broadcasting satellites and links for space services which share frequency bands with the FSS are permitted by the Table to operate in the RBW mode relative to FBW FSS networks operating in the same frequency band. These situations are listed on pp. 249–250. Such conflicts between FBW and RBW networks permit interference modes 7 and 8 to arise, although interference is not necessarily mutual.

Intricate methods have been adopted for managing the assignment of orbital locations and frequencies to FSS networks so that interference will not be unacceptably high. The same methods are applicable in principle within other space services and in sharing between different space services allocated the same frequency band, although interference criteria have not been agreed internationally in all cases. The methods are similar in principle to and interlinked with those that are used for managing sharing between space and terrestrial services.

The principles of this interference management method are as follows. The CCITT and CCIR have agreed performance and interference objectives for channels which are to form part of the international public telecommunications network (see below). A system of technical constraints is being established to ensure that the medium is used efficiently, although with due regard to system costs (see pp. 271–281). Finally, co-ordination procedures have been devised to identify situations in which the interference levels between an existing network and a proposed new network would be so high that channel performance objectives would not be achieved, and to delay registration in the MIFR of assignments to the new network until ways have been found for ensuring that such interference will be reduced to an acceptable level (pp. 268–271 and 281–294).

Exceptionally, the management of sharing between the FSS and the BSS and its feeder links involves additional regulatory factors, and these are considered on pp. 295–298.

8.2.2 Channel performance objectives and interference criteria

The principles of the methods which are used in the ITU to define performance standards and interference criteria for channels in FS and FSS systems which form part of the international public telecommunications network are discussed on pp. 105–110. The application of these principles in the FSS is considered in more detail below. Standards and criteria intended for spectrum management and applicable to other kinds of FSS traffic have not been developed yet.

Satellite HRCs and HRDPs

The concept of defining a standard long distance channel for use in setting noise, bit error ratio and interference objectives, in particular for FS systems, is discussed in detail on pp. 105–110. These standard channels are called hypothetical reference circuits (HRCs) (if analogue) or hypothetical reference digital paths (HRDPs) (if digital). The same principles and the same terms are applied also in the FSS although whereas in the FS the HRC or the HRDP is truly hypothetical, for the FSS the standard channels are in essence the same as the main part of any conventional actual channel, analogue or digital, routed through an FSS network.

For analogue transmission for the public telecommunications network in the FSS, the HRC is defined in CCIR Recommendation 352[17] (see also Fig. 8.4). The HRC, applicable to both telephone and video channels, consists of a single satellite hop from the input of the modulator of the transmitting earth station to the output of the demodulator of the receiving earth station, regardless of the distance between the two stations. If an inter-satellite link (ISL) were used, then those links and the additional space station equipment involved would be included within the HRC. Likewise, if earth station diversity were used, typically to mitigate the severity of fading at the higher frequencies due to heavy rainstorms, then the HRC would be considered to include the diversity earth stations and the terrestrial links between earth stations and the diversity switching point.

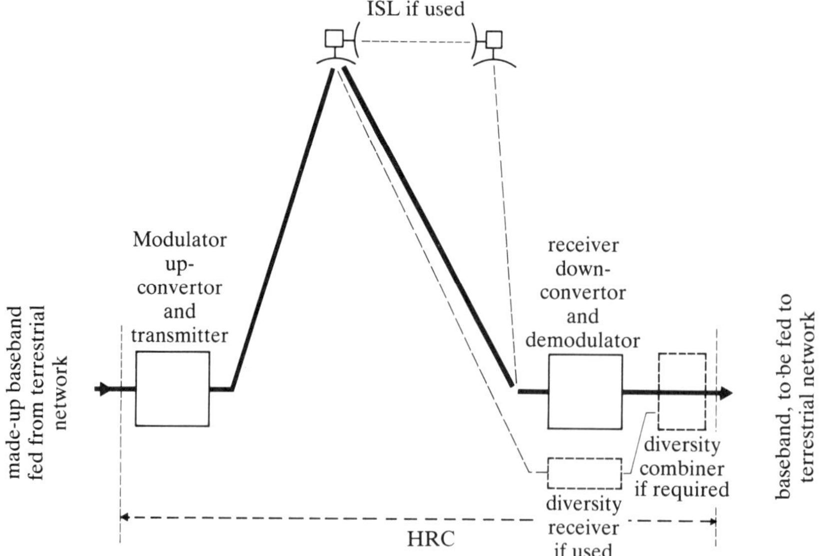

Fig. 8.4 CCIR HRC for multiplexed analogue telephone channels in the FSS

For digital transmission for the public telecommunications network in the FSS, the HRDP is defined in CCIR Recommendation 521[18]. In principle the concept is the same as for the HRC but the definition of the interface with the terrestrial network differs in detail.

Transmission performance objectives for an FSS channel

The channel noise objectives, including interference, for analogue telephone channels carried by the FSS and conforming to its HRC are given in CCIR Recommendation 353[19]. The key figures are that the mean channel noise power, averaged over one minute, should not exceed 10 000 pW0p for more than 20% of the time in the worst month, nor 50 000 pW0p for more than 0·3% of the time in the worst month. Thus, for example, if the time incidence of the noise-plus-interference level were plotted on Fig. 8.5 for the worst month, the objective would have been achieved if the curve remained below and to the left of the relevant benchmarks.

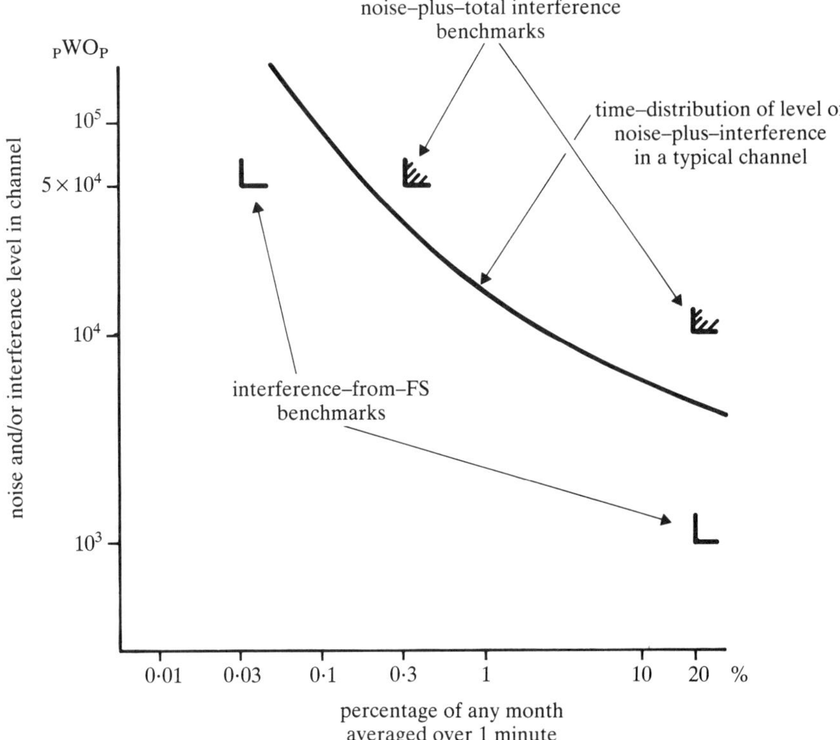

Fig. 8.5 CCIR recommended limits, defined by benchmarks, for noise-plus-total-interference and interference-from-FS in analogue telephone channels conforming to FSS HRC (CCIR Recommendations 353[19] and 356[26])

The noise objectives for international analogue video channels transmitted by an FSS network are given by CCIR Recommendations 354[20] and 567[21]. These recommendations say, for example, that with the received noise weighted by the low-pass filter defined in Annex II to Part C of Recommendation 567, the signal-to-noise ratio at the receiver should not fall below 53 dB for more than 1·0% of any month, nor below 45 dB for more than 0·1% of any month. There is a discussion of the issues arising in defining the noise objectives for telephony and video channels in CCIR Report 208[22].

The performance objectives for the HRDP for channels used for 8-bit encoded PCM telephony are given in CCIR Recommendation 522[23]. The key figures of the recommendation are that the bit error ratio (BER) at the receiver should not exceed 1 per million averaged over a 10-minute period for more than 20% of any month nor 100 per million averaged over a 1-minute period for more than 0·3% of any month.

The performance objectives for the HRDP for channels which may form part of an international connection in an ISDN are given in CCIR Recommendation 614[24]. The key figures are that a BER of 0·1 per million should not be exceeded for more than 10% of any month, nor 1·0 per million for more than 2% of any month. The period of time over which these latter BERs should be averaged is left somewhat open in the current version of the recommendation, since the optimum value may depend, to some degree, on the circumstances; the choice of averaging time is one of the factors bearing on Recommendation 614 which are discussed in CCIR Report 997[25].

It should be emphasised that these standards are set, in general, for international channels. They are high standards, required for channels which will be used for signals which must if possible, reach their destination in an acceptable condition even though they may have been transmitted in tandem through various other links, each of which may introduce degradations of various kinds, before or after transmission through the international link. Significantly inferior performance would often be fully acceptable for purposes that could be more narrowly defined. However, it is not the ITU's role to develop such standards.

Interference from FS transmitters

Maximum recommended levels of interference, reckoned as noise, into an FSS analogue telephone channel conforming to the HRC, due to transmitters of the FS, are given in CCIR Recommendation 356[26]. These levels are 1000 pW0p for not more than 20% of any month and 50 000 pW0p for not more than 0·03% of any month. These maximum interference levels are shown by benchmarks on Figure 8.5. They are directly related to the noise-plus-interference levels recommended in Recommendation 353[19], the 20%-of-the-month figure being 10% of 10 000 pW0p, the total long term noise level and the 50 000 pW0p level being allotted to interference for 10% of the time for which that degree of channel degradation is accepted due to all causes. The 1000 pW0p long term interference entry must, in turn, be divided into two parts, one allotted to mode 2 interference (as defined on p. 256) and the other allotted to mode 4 interference; these allotments are usually taken to be equal.

The same principle has been followed in CCIR Recommendation 558[27] in allotting to interference from FS transmitters a share of the total bit error degradations of a FSS HRDP used for PCM telephone channels; see also CCIR Report 793[28]. It is recommended there that the power of the interference from FS systems, averaged over 10 minutes, should not exceed for more than 20% of any month 10% of the noise-plus-interference power at the input to the earth station demodulator which would produce a BER of 1 per million. Similarly the FS interference power, averaged over 1 minute, should not exceed for more than 0·03% of any month the level of noise-plus-interference which would cause a BER of 100 per million.

No recommendations have been arrived at yet for interference from transmitters of the FS into an FSS analogue video channel or an HRDP which is used for ISDN channels.

Interference between networks of the FSS

Interference may arise within a satellite network. For example in a satellite network that re-uses the spectrum by means of orthogonal polarisations or multiple satellite antenna beams, the isolation between the two polarisation modes or the various antenna beams will not be perfect and signals from one transmission mode may leak through into another, causing interference. Keeping this interference to an acceptable level is a problem for the system designer; the ITU treats it, in most circumstances, as part of the in-system transmission degradations.

Interference between different satellite networks is of concern, not only to system designers but also to spectrum managers. The inter-network frequency co-ordination process requires each FSS network to accept interference from other FSS networks, up to a limit recommended by the CCIR. The co-ordination process is usually a bilateral one, so the limit is set in terms of the interference entry from one single network into another network, the so-called maximum single-entry permissible interference criterion. The chief purpose of co-ordination is to ensure that the level of interference entering a network from any one other network will not exceed that permissible single-entry level, unless a higher level is specifically accepted,

However, the system designer will need an estimate of the total interference that will be entering the worst-affected channel of the network from all other FSS networks, in order that the network may be designed so that channel noise objectives are achieved, despite the interference. If this total interference figure is under-estimated, channels may not achieve their performance objectives, but if it is over-estimated, the potential maximum traffic capacity of the satellite may not be realised, Therefore, the CCIR recommendations on permissible interference also contain an estimate of the sum of all the interference entries from other FSS networks that is not likely to be exceeded in the worst-affected channel of a network.

The value chosen for the single-entry interference criterion is important. If the criterion were set too low, network managers could justify seeking wide separations between satellites or severe constraints on emission parameters or carrier frequency plans in the frequency co-ordination process, making it more difficult for other satellites to achieve a satisfactory outcome in subsequent co-ordination negotiations. If however the single entry interference criterion were made too large, the total interference figure on which networks were designed may subsequently be found to be an optimistic under-estimate and channels may fail to achieve their performance objectives.

CCIR Recommendation 466[29] contains recommendations for the single-entry interference level that should be permitted in the worst-affected channel of a geostationary satellite network that is used for telephony with analogue transmission. It also recommends a value to be assumed for the total entry of interference into the worst-affected channel:

- The version of this recommendation that was approved at the CCIR Plenary Assembly in 1974 set these figures at 400 pW0p and 1000 pW0p respectively.
- At the 1978 Plenary Assembly these figures were revised for new networks, becoming 600 pW0p and 2000 pW0p respectively, although the 1974 figures remained in force for networks which were already in being. As a concession for new networks that might have significant internal interference due to frequency re-use, the total interference figure was reduced to 1500 pW0p in such cases.
- At the Plenary Assembly in 1986, the 1978 figures of 600, 2000 and 1500 pW0p were changed to 800, 2500 and 2000 pW0p respectively for new networks, but the 1974 and 1978 figures remained applicable, as appropriate, for networks which were already in being.

These successive increases in the permissible interference levels have substantially increased the number of satellite networks which will be able to operate, using the GSO.

CCIR Recommendations 483[30] and 523[31] give corresponding values of permissible interference for video channels carried by analogue transmission and for 8-bit encoded telephone channels carried by digital transmission.

Recommendations 466 and 523 are limited, at present, to systems operating in frequency bands below 15 GHz. At higher frequencies propagation conditions will be more severely affected by rain in most earth station locations, and this may significantly change the desirable relationship between in-system noise and permissible interference. No conclusions have been reached on this point in CCIR yet, but see Report 710[32].

The actual value of interference from any one network into the worst-affected channel of another network depends on the orbital separation between the satellites, the frequency assignment plans, the carrier parameters and the antenna characteristics of both networks, among other factors. After any necesary frequency co-ordination, these single entries of interference will usually be less, often much less, than the maximum single entry interference criterion. Also, in many FSS satellite networks, especially those using FDMA, the total interference in the worst-affected channel may be considerably less than the sum of all of the single entries of interference, because the various single entries of interference are not likely to coincide in frequency. There is a discussion of this matter in CCIR Report 455[33] and studies continue. With geostationary satellites in a heavily loaded arc of the orbit, the interference level in the worst-affected channel in an FDMA system is likely to be between 2 and 4 times the single-entry limit.

Criteria for FSS networks using non-geostationary satellites

The HRCs, the HRDPs and the transmission performance objectives reviewed above apply to all FSS networks, regardless of the satellite orbit. The recommended limit on interference from FS transmitters also applies to all FSS networks, although it is evident that a channel in a non-geostationary network which suffers interference which never exceeds the recommended limit will have a much lower interference entry from FS transmitters for most of the time. The recommended levels of permissible interference between networks of the FSS relate specifically to interference which enters a geostationary network.

The CCIR makes no recommendations as to the level of interference that should be permitted between FSS networks using non-geostationary satellites, nor for interference entering a non-geostationary network from a geostationary network. These are matters which are left to the discretion of the administrations concerned. Interference from non-geostationary networks to geostationary networks is discussed on pp. 271–272.

8.2.3 Sharing between the FSS and terrestrial services

Four interference modes between space services like the FSS and terrestrial services like the FS are identified on p. 256 and illustrated in Fig, 5.1. The measures that are used to limit the interference via these modes are reviewed in this section. These measures assume that all FSS networks use geostationary satellites. There is a brief discussion on interference to FS receivers from transmitters on non-geostationary satellites in CCIR Report 387[34].

Modes 1 and 2: Interference between earth stations and terrestrial stations

The basis of interference containment via modes 1 and 2 within the limits stated on pp. 262–263 is the frequency co-ordination process discussed on pp. 118–131 (see also pp. 131–133). However, in addition, sharing constraints are imposed upon earth stations to prevent the occurrence of extremely extensive co-ordination areas.

(*a*) RR 2540–RR 2548 place a limit on the spectral EIRP density which an earth station may radiate at an angle of elevation, θ degrees relative to the horizon as seen from the earth station, in bands shared with terrestrial services. The value of this constraint for frequencies between 1 and 15 GHz is:

> + 40 dBW in any 4 kHz sampling bandwidth in the direction of the horizon or below it
>
> + 40 + 3 θ dBW per 4 kHz when θ is between 0 and + 5°.

The corresponding figures for frequencies above 15 GHz are + 64 dBW and + 64 + 3 θ dBW, both measured in a 1 MHz sampling bandwidth. No limit is applied at angles above 5° relative to the horizon. The case for this constraint is made in CCIR Report 386[35]. These limits may be relaxed by 10 dB, given the agreement of all affected administrations; see RR 2546.
(*b*) RR 2549–RR 2551 forbid the use of any FSS earth station antenna for transmission at an angle of elevation less than 3° above the horizontal plane, regardless of the parameters of the emission, without the consent of other administrations that would be affected. There is no corresponding ban on the use of earth station antennas in the receive mode at very low angles but RR 2550 requires the angle of elevation to be treated in co-ordination negotiations as 3° if the actual angle of elevation is less than that figure.

In addition, RR 2501 and RR 2539 draw attention to the need to locate stations of both the FS and the FSS having regard to the needs of the stations of the other service. CCIR Report 385[36] provides general advice on the subject and CCIR Report 390[37] contains information on the protection from interference that can be obtained from obstacles, such as hills and buildings, in the interfering ray-path. Locating earth stations in large pits in the ground and in disused quarries may considerably reduce the interference level. CCIR Report 709[38] considers the reduction of interference that can be obtained by the careful positioning of FS and earth station antennas which are unavoidably in close proximity to one another.

Mode 3: Interference from space stations to terrestrial stations

The spectral PFD limits, as measured at the Earth's surface, to which space station emissions in shared bands are constrained, in order to contain interference to receivers at terrestrial stations within the limits stated on pp. 105–110, are discussed on pp. 110–115. The values of the PFD limits are shown graphically in Fig. 5.4 and they are set out in detail in RR 2552–RR 2584; see also CCIR Report 387[34]. RR 2585 permits these limits to be exceeded over the territory of administrations that agree to it.

If no action were taken to disperse the spectral energy in the immediate vicinity of the carrier wave of a typical radio emission, earth stations would require high figures of merit (G/T) in the receiving mode in order for networks to function satisfactorily within these constraints. The modulation of the carrier by traffic signals etc. provides some dispersal of the carrier energy over a significant proportion of the bandwidth which the emission requires for its satisfactory transmission. CCIR Report 792[39] shows how the peak spectral power density of typical emissions, modulated by traffic-carrying basebands, may be calculated. Techniques have been developed for complementing the dispersal effect of traffic modulation with artificial carrier energy dispersal waveforms when necessary. CCIR Report 384[40] describes these techniques and CCIR Recommendation 446[41] and RR Recommendation 103[42] call for their use. Artificial carrier energy dispersal is, indeed, in widespread use in the systems of the FSS. See also p. 000.

In FSS networks which operate in frequency bands where spectral PFD limits are applicable and which serve earth stations with a very low figure of merit, it may be necessary to use spread spectrum modulation techniques to obtain sufficient dispersal of the energy of the emission. CCIR studies of the use of these techniques in the FSS have not been completed yet, but CCIR Reports 651[43], 826[44] and 974[45] provide some guidance.

Mode 4: Interference from terrestrial stations to space stations

RR 2502–RR 2511 place limits on the carrier power and EIRP of terrestrial transmitting stations operating in frequency bands shared with the FSS and these regulations also seek to ensure that radio relay beams will not be directed towards the points where the GSO intersects the horizon. These constraints are discussed on pp. 115–118. In general they are sufficient to contain interference entering FSS networks via this Mode within the objectives stated on pp. 261–263; see for example CCIR Report 790[46].

The impact on FSS networks of sharing with terrestrial services

Sharing limits the space station transmitter power that the FSS can use and this might be expected to prevent FSS networks from using earth stations with small antennas and cheap low noise receiver input amplifiers. So far, this does not seem to have become a major problem. It is otherwise with the second consequence of sharing.

When the concepts of frequency co-ordination between the FSS and the terrestrial services were first agreed at WARC-63 and when the procedures were revised in detail at WARC–71, the primary application foreseen for the FSS was international communication. Use of the FSS on a large scale for domestic communication was not clearly foreseen then. The earth stations used for international communication were few and they could be located in thinly-populated districts where the use by the FS of the same frequency bands could often be dispensed with. Where such isolated sites were not feasible, advantage might be taken of site screening by hills etc. The avoidance of severe interference between earth stations and terrestrial stations presented no great problems then.

Nowadays many countries use the FSS for domestic communication. The earth stations of domestic satellite networks need to be located in or near to the towns which they serve. Some networks involve large numbers of small-antenna earth stations, many of them located within cities. Some degree of site screening is sometimes feasible in these situations. However, many earth stations antennas are mounted on the roofs of buildings. Typical FS stations also have to be located in city centres, their antennas also being mounted at or above the roof-line. Consequently earth station antennas are often very close to the terrestrial station antennas and there is sometimes nothing to screen the one from the other. Thus, the conditions on which the concept of frequency band sharing between the FS and the FSS was based are often absent now.

When large numbers of earth stations are required and these must be located in or near towns, co-ordination is labour-intensive and success is difficult to achieve. It is better then to use different frequency bands for these two services. In this way potential limitations on the operation of each service, due to the presence of the other, will be avoided and the amount of frequency co-ordination required will be greatly reduced. Failing that, it will be necessary to exercise great care in selecting radio station sites, taking advantage of natural topographical shielding and the shielding provided by large buildings, etc, Alternatively it may be necessary to limit the assignment of frequencies for one service to gaps in the interference spectrum set up by the stations of the other service.

8.2.4 Harmonisation of FSS networks by frequency co-ordination

Interference between two networks of the FSS, typically via modes 5 and 6 as defined on pp. 256–257 and illustrated in Figure 8.2, is likely to be small or negligible if there is a large angular separation between the satellites as seen from the earth stations (angle ø in the figure) or between the coverage areas as seen from the satellites (approximately represented by angle θ). With smaller angular separations, interference may be significant.

If interference can arise between networks, some operating in the FBW mode and others in the RBW mode, interference modes 7 and 8 (see Fig. 8.3) will also have to be taken into account.,

If unacceptable interference arises from time to time between two networks, both using non-geostationary satellites, it will usually be possible to reduce the interference to within acceptable limits by adjusting the timing of the hand-over of links from one satellite to another. A minor modification of orbital parameters may also be desirable. However, the

ITU leaves the solution of such problems to the discretion of the administrations involved.

The ITU, however, has given close attention to the situation which arises when both of the satellite networks are geostationary. One way to reduce an entry of interference, so as to meet the permissible interference recommendation, would be to increase the separation between the satellites, since the off-beam gain of earth station antennas can be expected to fall as ø, the off-beam angle, increases. However, this may not be feasible in frequency bands which are already in heavy use by the FSS, since a satellite, in moving away from another satellite, is likely to be moving towards a third satellite; in diminishing the interference from one network it may be increasing the interference from another. Furthermore, the use of this tactic will lead to the ineffective occupation of the limited capacity of the GSO in frequency bands which are not already in heavy use. Effective means are required for limiting interference without sacrificing efficiency of orbit utilisation.

The size of the entry of interference from one satellite network into another depends on many factors in addition to ø and θ. Most of these factors are peculiar to a given pair of networks. Some involve aspects of the hardware design of the network which cannot readily be changed once operation has begun but others can be changed at any time. There is thus a need for organisations intending to set up and operate a new satellite network to meet organisations that have networks that are already in operation, if interference problems can be foreseen, to determine how the designs of the various networks can be harmonised. In this way it should be possible to find ways for all the operating organisations to achieve all or most of their objectives at an acceptably low cost, using the orbit and spectrum resources that are available. The solutions emerging from these discussions should also take into account the need of future networks to obtain access to the GSO.

The more important of the factors which can be varied in order to harmonise the use which different satellites make of the GSO are as follows:

(*a*) *Orbital location:* At the start of the harmonisation process the proposer of a new geostationary network will announce the longitude of the location in the GSO which is wanted for the new satellite. However, any such statement must be provisional and RR Appendices 3 and 4 require an Administration to declare the longitudinal limits of the arc of the GSO over which all the earth stations of the proposed network would be able to see the satellite at an angle of elevation not less than 10° (the 'visible arc') and likewise the limits of the arc within which the proposed satellite could function efficiently (the 'service arc'). The feasibility of solving harmonisation problems by moving a satellite which is already co-ordinated and built and may be already in service should also be considered. However, this raises various technical problems, which are examined in CCIR Report 1002[47] and may require the re-negotiation of existing co-ordination agreements for that satellite.

(*b*) *The reduction of satellite antenna overspill:* At a sufficiently early stage in the construction of the proposed satellite, it will usually be feasible to redesign satellite antennas so that prospective interference to or from earth stations in the service area of other networks is made less (see pp. 278–279). Opportunities for modifying the footprints and the rate of roll-off of gain of the beams of satellites which are no longer under construction are very limited in most cases.

(*c*) *Earth station antenna off-axis radiation:* The spectral EIRP density radiated off the axis of the earth station antennas, in the direction of other satellites, is determined for each type of emission used in a network by a complex of parameters, the most important of which are:

- earth station antenna on-axis gain,
- sidelobe radiation pattern of the earth station antenna in the transmitting mode,
- gain of the satellite transponder and the associated antennas,
- figure of merit (G/T) of the receiving earth station, and
- power spectrum of the emission.

The use of earth station antennas which have lower off-beam radiation levels, typically because their sidelobe gain is less (see pp. 275–278) or their on-axis gain is greater (thereby reducing the required antenna input power level), is a very effective way of reducing prospective interference. Carrier energy dispersal may reduce interference between satellite networks, although there are circumstances in which conventional energy dispersal techniques make interference between satellite networks worse (see pp. 273–274). The CCIR recommends a limit for earth station off-axis radiation (see p. 280).

(*d*) *Harmonisation of frequency assignment plans:* It will often be found that conformity with the recommended constraint on earth station off-axis radiation density is not enough to solve harmonisation problems. When the feasible ways of reducing earth station off-axis radiation have been exhausted, consideration should be given to optimising the distribution in the frequency domain of the various kinds of emissions which the networks carry. In any point in the frequency bands that the networks use, one network should not set up an emission which has a high spectral power density if the same frequency is assigned in another network to an emission which is particularly susceptible to interference. A systematic approach to this interference minimisation practice is called spectrum segmentation; see CCIR Report 1000[48]. A substantial further reduction in interference levels can be obtained by detailed co-ordination of carrier frequencies.

(*e*) *The pre-demodulator interference level:* The interference energy density which is acceptable for each carrier is determined mainly by the carrier-to-noise-density (C/T) ratio needed for the proper operation of the link and the post-demodulator channel interference level that is acceptable. The latter will usually be limited to the single-entry value recommended by the CCIR, but it may be feasible to negotiate a higher level of interference in specific cases. If the carrier suffering the interference does not have other strong single interference entries, the acceptance of more than the recommended level of permissible interference from a new single entry will not necessarily cause the recommended level of total inter-network interference to be exceeded. If the recommended level of total interference is exceeded, it will often be possible to make a compensating adjustment to the level of noise from other sources by means of a small change in carrier parameters at virtually no cost. For FSS networks which are not used to provide channels to CCITT standards of performance, the channel performance objectives reviewed on pp. 261–262 may be unnecessarily stringent; a higher level of noise-plus-interference might be acceptable without any compensating adjustment.

The frequency co-ordination procedure reviewed on pp. 285–288 is designed to provide an opportunity for this harmonisation process to be carried out.

8.3 Rules and technical constraints for enhancing the FSS

8.3.1 General rules and constraints

Almost all satellites of the FSS are nominally geostationary and this predominance seems likely to continue, but other orbits have advantages for special applications and a few non-geostationary FSS satellites are in operation. However, all non-geostationary satellites must pass through the equatorial plane twice in every revolution, that is, several times per day, and at around these times there could be severe interference between networks using geostationary and non-geostationary satellites. The geometry of such interference paths is considered in CCIR Report 455[33]. RR 2613 forbids such interference in the following terms:

> Non-geostationary space stations shall cease or reduce to a negligible level their emissions, and their associated earth stations shall not transmit to them, whenever there is insufficient angular separation between non-geostationary satellites and geostationary satellites, and whenever there is unacceptable interference to geostationary-satellite space systems in the fixed-satellite service operating in accordance with these Regulations.

This rule gives a clear advantage to geostationary satellites. However, the GSO is a special case of the geosynchronous orbit; a geostationary satellite is a geosynchronous satellite with a direct and equatorial orbit. Furthermore, both the GSO and a geosynchronous orbit are virtually unattainable ideals in that they involve an orbital period which is exactly the same as the Earth's sidereal period of rotation. The criteria by which a satellite is judged to be geostationary are therefore a matter of some importance.

These criteria are not defined precisely in the RR. RR 181 defines a geostationary satellite as:

> a geosynchronous satellite whose circular and direct orbit lies in the plane of the Earth's equator and which thus remains fixed relative to the Earth; by extension, a satellite which remains approximately fixed relative to the Earth.

Some quantitative significance is given, elsewhere in the RR, to the word 'approximately' in the last line of RR 181 (see p. 275) This vagueness does not, at present, lead to many problems.

Great importance is attached to the efficient use of the GSO for FSS networks and a number of technical constraints have been agreed in the ITU to help in the attainment of that objective. Most of these constraints have been applied only to geostationary satellite networks and they are reviewed on pp. 274–281. However, other rules and constraints apply to all networks of the FSS, regardless of the orbit that their satellites use, and these are reviewed below.

Uncontrolled satellites and debris in space

Working satellites, and in particular geostationary satellites, are at risk from other satellites, the orbits of which are no longer being controlled and from pieces of debris in or passing through their orbit. The risk arises in two ways;

(*a*) Radio interference to a working network may occur if a satellite, which is no longer in service but may perhaps be emitting its telemetry carriers and may have transponders switched on, drifts close to the working satellite.
(*b*) If a physical collision occurred, the working satellite would be damaged or destroyed. Indeed destruction or major damage might be caused by collision with quite small pieces of debris.

Clearly the risk is increasing. The probability of radio interference arising as in (*a*) is already significant and RR 2612 seeks to limit its incidence, thus:

> Space stations shall be fitted with devices to ensure immediate cessation of their radio emissions by telecommand, whenever such cessation is required under the provisions of these Regulations.

The likelihoood of a physical collision occurring at present is held to be remote, but the amount of debris in space increases with every new launch and collision may become a significant hazard in time. CCIR Question

34/4[49] calls for study of this matter, and CCIR Report 1004[50] contains some early results. It is usual to arrange for fuel for the orbit adjustment thrusters to be available at the end of a satellite's operating lifetime, so that the satellite can then be propelled into a higher orbit, obviating the risk of collision with operating satellites.

Frequency tolerance

Appendix 7 of the Radio Regulations imposes a tolerance on the carrier frequencies of all earth stations and space stations of the FSS operating below 40 GHz. The tolerance is 50 parts per million below 10·5 GHz and 100 parts per million above 10·5 GHz, and it is applicable to all new equipment from the beginning of 1985 and to all equipment, regardless of age, from the beginning of 1990. However, much better frequency stability is necessary for the satisfactory operation of many satellite links.

Spurious emissions

The Radio Regulations do not impose specific limits on spurious emissions from earth stations or space stations of the FSS, although footnote 13 to the Table in RR Appendix 8 requires that 'the levels of their spurious emissions should be reduced to the lowest possible values compatible with the technical and economic constraints to which the equipment is subject'. RR Recommendation 66, agreed at WARC-79[51], recommends urgent study by the CCIR of the specific limits that should be applied to spurious emissions in the FSS, and CCIR Report 713[52] shows the present state of these studies.

Carrier energy dispersal

If the distribution of energy across the spectrum of a radio emission is very non-uniform, and in particular if there is a strong residual carrier spike, the emission is liable to cause disproportionately severe interference, especially to a narrow-band emission. These spikes of spectral energy can be dispersed by relatively simple methods in most cases; see CCIR Report 384[40]. Such methods are used to enable relatively powerful satellite emissions to conform to the down-link PFD constraints which make frequency band sharing with terrestrial services feasible.

CCIR Recommendation 446[41] recommends that these methods should also be available and used where necessary to reduce interference between FSS networks. See also RR Recommendation 103[42]. Considerable benefit may be obtained in this way. However, two issues which do not arise in connection with interference between space systems and the wide-band systems which are typical of the terrestrial services need to be taken into account in the context of interference between satellite networks:

- Whereas carrier energy dispersal by, for example, the addition of a large-amplitude low-frequency triangular waveform to the baseband of a frequency modulated carrier can greatly reduce the spectral density in the immediate vicinity of the carrier frequency, it may increase the average spectral density over a considerable bandwidth above and below the carrier frequency. Thus, the inter-network interference situation may, in some specific circumstances, be made, not better, but worse. If a satellite network is not required to use carrier energy dispersal to protect terrestrial services, careful co-ordination of the carrier frequency plans of satellite networks which interfere may sometimes be a more satisfactory solution to inter-network interference problems than carrier energy dispersal.
- The triangular waveforms, having a period of a few milliseconds, which are commonly added to the modulating signal to disperse the residual carrier energy of frequency modulated emissions carrying multiplexed analogue telephone channels, are effective for reducing interference to all types of signal in common use in the FSS. However, for FM emissions carrying video signals, a triangular waveform with a period of 20 or 16·7 ms is commonly used, synchronised with the video field. This does not reduce very much the interference of an FM video emission to a narrow-band emission, typically an SCPC telephony emission, in the immediate vicinity of the video carrier frequency and it broadens the frequency band over which severe interference to SCPC emissions can occur; see CCIR Reports 384[40] and 867[53].

Various other techniques for dispersing the residual energy of an FM video carrier are under study; see Annex II of CCIR Report 384.

8.3.2 Constraints on specific parameters of geostationary FSS networks

There is a general survey of the technical factors which influence the efficiency with which the GSO is used for FSS networks in CCIR Report 453[54]. The outcome of the studies reflected in that CCIR report, and others, in the form of constraints and prospective constraints which are applicable to specific parameters of geostationary satellite networks, is reviewed in this section. An alternative approach to orbit utilisation optimisation, which is also applicable, mainly, to geostationary satellite networks, is examined on pp. 280–281.

Satellite station-keeping

CCIR Report 556[55] discussses the principal perturbations of a nominally geostationary satellite's orbit which occur after the satellite has been set accurately on station in the GSO. These perturbations cause apparent satellite motions, as seen from Earth, both east–west and north–south. The perturbations can be corrected by the operation of thrusters on-board the satellite as necessary, but they recur.

In the course of the co-ordination process which is reviewed on pp. 283–288, a nominal orbital longitude is determined for each geostationary satellite, a margin being left in the nominal angular separation between pairs of satellites to allow for residual errors in the orbital period and for diurnal east–west movements due mainly to orbital ellipticity. If it were necessary to allow a large margin of arc for this purpose, the efficiency of use of the GSO would be significantly reduced. Accordingly, stringent limits have been agreed for the residual error in east–west location.

RR 2615–RR 2619 require the east–west movement of all geostationary satellites using frequencies in bands allocated to the FSS to be controllable within close limits. The standard tolerance which is to be achievable is ±0·1°, this figure being relaxed to ±0·5° for experimental satellites. However, these standards were agreed at WARC–79 and they were not intended to apply to satellites that were then already in service or under construction. For satellites which were put into service before the beginning of 1987 and for which advance publication information had been published before the beginning of 1982, the requirement is ±1·0° (see RR 2624–RR 2627). Furthermore, whilst satellites are required to be maintainable within the stated range of positions, there is no regulatory requirement to implement these tolerances except to avoid causing unacceptable interference to the network of another satellite; see also CCIR Recommendtion 484[56].

Thus, despite the apparent imprecision given to the ITU definition of a geostationary satellite by the inclusion of the word ‘approximately’ in RR 181 (quoted on p. 272) RR 2615–RR2619 and RR 2624–RR 2627 ensure that the orbital period of an operational satellite which is not to be constrained by RR 2613 (quoted on p. 271) is maintained very exactly equal to one sidereal day.

There are no direct regulatory limits at present on the inclination of the plane of the orbit of a nominally-geostationary FSS satellite, and thus on its apparent north–south movement. However, here again it should be noted that indirect constraints apply; a geosynchronous satellite with an orbital inclination exceeding 5° would make an apparent daily excursion of more than 0·1° to the east and to the west of its mean longitude for that reason alone, and thus might be in breach of RR 2617. The IFRB does not regard as geostationary a satellite which has an orbital inclination exceeding 5°. The desirability of applying a regulatory limit to the orbital inclination of a satellite which is to enjoy the privileges given to geostationary satellites is currently under study in CCIR.

Earth station antenna sidelobe gain

The gain of the sidelobes of practical reflector antennas, outside the immediate vicinity of the main lobe, is considerably higher than basic theory indicates. The sidelobe gain of many axi-symmetrical Cassegrain antennas, a type very commonly used at earth stations, is particularly high, due mainly to sub-reflector spill-over, diffraction at the edge of the sub-reflector, scattering at the sub-reflector supports and large-scale imperfections in the profile of the reflectors. CCIR Reports 391[57] and 998[58] contain details of studies of the sidelobe gain characteristics of existing earth station antennas. Careful design, still using axi-symmetrical Cassegrain geometry, can reduce these defects, although sometimes at the cost of a small reduction in forward gain. Off-set feed geometry can provide a further improvement.

CCIR Recommendation 465[59] provides a 'reference radiation pattern' for the sidelobes of earth station antennas. This pattern is intended to be an approximation to the peak gain, as a function of off-axis angle, of the sidelobes which are not in the immediate vicinity of the mainlobe. Its use is recommended by the CCIR for estimating the levels of interference to and from an antenna in the absence of specific information about the polar diagram of that antenna (see Fig. 8.6). The basic pattern contained in that recommendation, namely the radiation pattern for big antennas, the diameter of the main reflectors of which are more than one hundred times the wavelength, has not been changed since 1970.

High sidelobe gain within 20° or 30° of the principal axis of the earth

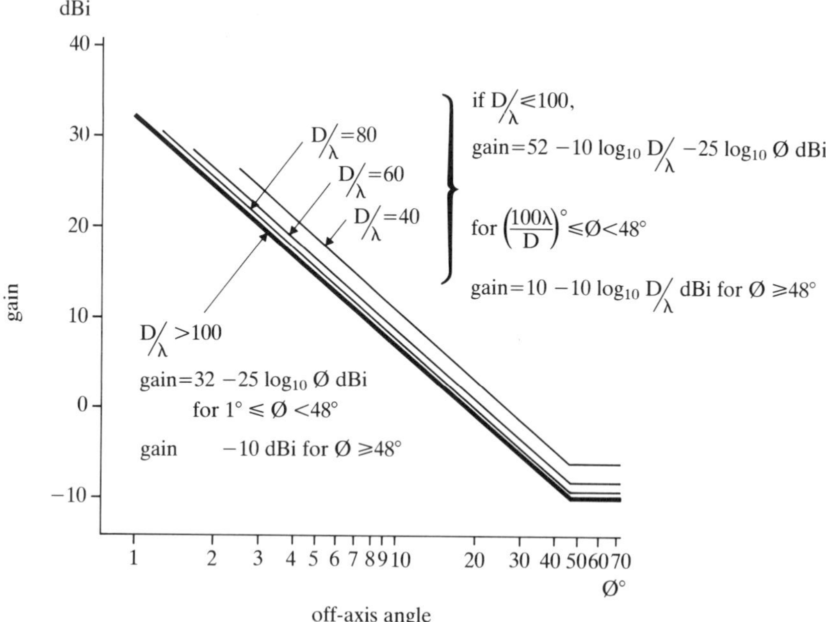

Fig. 8.6 Reference radiation pattern (CCIR Recommendation 465[59]) for an earth station antenna, for use in calculations of interference levels in the absence of gain data specific to the antenna in question, as a function of main reflector diameter D, wavelength λ, and off-axis angle Φ (degrees)

station antenna substantially increases the level of interference transmitted to and received from satellites nearby in orbit. This is a principal cause of requirements for wide orbital separation between satellites. It is therefore important for low sidelobe gain to be specified in the procurement of earth station antennas.

Accordingly the 1982 CCIR Plenary Assembly set a design objective for big new earth station antennas procured for use with geostationary satellites. It recommended that the maximum gain of at least 90% of the sidelobe peaks of such antennas in defined directions of the greatest concern should be at least 3 dB less than the gain given by the reference radiation pattern of CCIR Recommendation 465[59]. Fig. 8.7 illustrates the angular directions under normal operating conditions in which that design objective would apply, namely a solid angle extended 3° north and south of the GSO and 20° east and west of the satellite, but excluding a cone with a semi-apex angle of 1° surrounding the principal axis. The CCIR Plenary Assembly in 1986 extended this recommendation to cover small antennas, over a somewhat smaller angular range and with appropriate dates of implementation. For the smallest antennas, implementation is to be staged; see CCIR Recommendation 580[60].

It must, of course, be emphasised that CCIR Recommendation 580 provides design objectives. The CCIR has made no recommendation yet that an antenna which does not achieve these objectives should not be used. Nevertheless, the texts which recommend the levels of interference which should be permitted to enter an FSS network from another FSS network

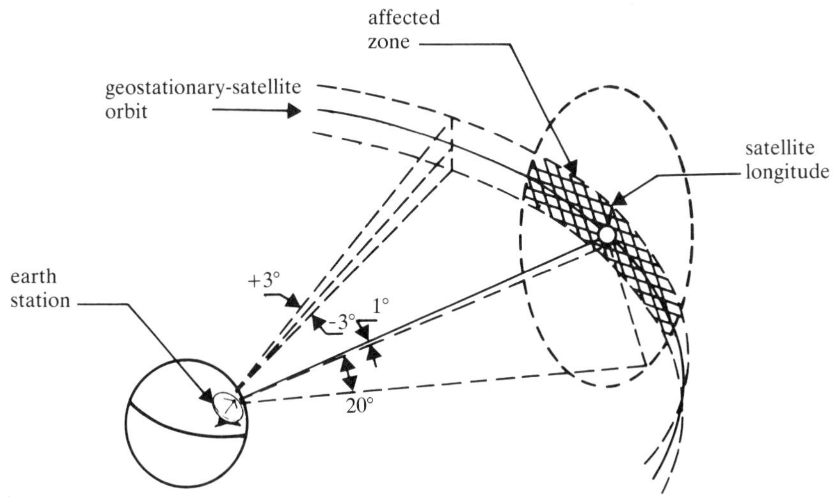

Fig. 8.7 Angular range, relative to the off-axis angle, over which CCIR Recommendation 580[60] offers sidelobe gain design objectives for new earth station antennas with a main reflector diameter exceeding 100 wavelengths

call for the interference levels to be calculated using the assumption that the receiving antenna sidelobe gain conforms to the reference radiation pattern of Recommendation 465[59] unless the gain of the antenna in the relevant direction is known to be less, in which case the actual value is to be used; see for example CCIR Recommendation 466[29]. Thus, the owner of an antenna which does not achieve the recommended performance standard is expected to accept the consequential high level of interference in the channels that the antenna receives.

Satellite antenna spillover

Some geostationary satellites serve much or all of the Earth's surface visible from the satellite and these typically use antennas with beams which cover the whole disc of the Earth; the minimum apex angle of such a beam is about 17°. Other geostationary satellite antennas serve relatively small areas. The apex angle of the beams of such antennas, out to the −3 dB contour may, at present, be as small as 1°, depending on the size of the service area and the frequency. Outside the −3 dB contour, the gain of such an antenna may roll off rapidly.

If there is overlap of the geographical areas which two satellites serve in the same frequency bands, it is likely that the directivity of the satellite antennas will provide no assistance in reducing the interference between the two networks. If however the service areas are well separated, the directional properties of the satellite antennas may provide a very considerable amount of protection against interference; that is, the angles θ_1 and θ_2 in Fig. 8.2. may be large enough to ensure that the gains of the satellite antennas in the direction of unwanted earth stations are much less than the gains of the same antennas in the direction of their service areas.

The limitation of inter-network interference by means of satellite antenna directivity may, however, fall short of what is feasible for two reasons:

- A satellite antenna beam may not always be directed accurately towards its wanted service area, and this mis-direction may increase the gain of the antenna in the direction of earth stations of a different network. Furthermore, the network designer may broaden the antenna beam in order that beam pointing errors do not result in failure to provide an adequate standard of service to the outermost parts of the wanted service area, thereby further increasing potential interference outside the service area.
- Antennas capable of generating beams with footprints which cover service areas relatively precisely, with a rapid roll-off of gain outside the service area, are feasible but they tend to be heavier, larger and costlier than a simpler antenna which provides comparable in-beam gain but less satisfactory out-of-beam characteristics. The antenna giving better protection against interference may therefore be economically unattractive to the satellite owner.

RR 2628–RR 2630 require the pointing directions of the antenna beams of geostationary satellites to be maintainable within 10% of the −3 dB beamwidth or within ±0·3° (whichever is the less stringent requirement) of the nominal pointing direction. As with satellite station-keeping, this tolerance is not required to be maintained unless excessive interference to another network would otherwise arise.

The devising of design objectives or practicable constraints to limit the coverage area of satellite antenna beams to the required service area and to ensure a rapid roll-off of the gain of the antenna in directions outside the coverage area is technically difficult. Nevertheless, CCIR studies in this area are progressing and they are now approaching the stage where a reference radiation pattern might be defined for use in estimating interference potential in the absence of specific performance data. A design objective may also be developed; see CCIR Report 558[61].

Polarisation discrimination

For most of the time the propagation path between a satellite and an earth station causes relatively little depolarisation of microwave radio signals (although see the references to Faraday Rotation in Appendix B.3.6 and to cross-polarisation and depolarisation in the troposphere in Appendices B.4.2 and B.4.3 respectively).

Some FSS networks which require high transmission capacity transmit in two orthogonal polarisation modes. If the purity of the polarisation of the transmitted signals and the discrimination of the receiving antennas between the two polarisation conditions is sufficiently high, coupling between them may be quite small and the spectrum can be used twice over within the same satellite antenna beam. If the isolation of the two polarisation modes is less but still substantial, polarisation discrimination may be augmented by satellite antenna directivity or by selection of carrier frequencies, producing a significant if smaller degree of frequency re-use.

The possibility arises therefore that two satellite networks with a lower transmission capacity requirement, each using a single polarisation mode and with their polarisation modes orthogonal, might be able to operate in the same frequency bands without excessive interference, with overlapping service areas and with their satellites very close together in orbit, or even with the same nominal satellite location. There are substantial difficulties in realising this option by regulation, particularly if linear polarisation is used, but they are under study; see CCIR Reports 555[62] and 453[54].

There are, at present, neither standard conventions nor performance objectives for polarisation in FSS networks. There is, however, provision in RR Appendix 29 for taking polarisation discrimination into account when calculating potential interference levels between geostationary networks, given the consent of both parties (see also p. 291).

8.3.3 GSO utilisation efficiency through generalised parameters

It is feasible to improve significantly the efficiency with which the FSS uses the GSO by ensuring that certain specific parameters of networks conform to good standards of design. The progress that has been made so far in agreeing technical constraints of this kind is indicated on pp. 271–279. However, the observance of single-parameter constraints may not be the only way to increase the number of satellite networks that can share a pair of frequency bands without unacceptable mutual interference, and other ways may indeed be preferable.

Thus, earth-station antenna sidelobe gain is obviously an important parameter in determining the power of the signal which an earth station antenna radiates off-beam, causing interference at the receiving antennas of satellites belonging to other networks. Accordingly, benefit is to be obtained from reducing sidelobe gain levels through agreed antenna design targets or recommended antenna performance standards. However, earth station antenna sidelobe gain is by no means the only factor determining the interference potential of energy radiated through these sidelobes. The level of power that is fed by the earth station transmitter to the antenna and the way in which that power is distributed across the spectrum of the interfering emission are also important. The earth station transmitter power will be affected, in turn, by such other parameters of the network as the earth station antenna on-axis gain, the satellite receiving antenna gain, the satellite transponder gain and the required down-link signal power.

In a similar way the level of the interference which satellite emissions present to the antennas of earth stations which are served by other satellites is determined, partly by the out-of-beam gain of the satellite antenna but also by many other network parameters. The same is true of the susceptibility of emissions to interference.

It may be possible to ensure that the GSO is used efficiently by placing limits on the susceptibility of networks to interference and on the liability of networks to cause interference, instead of using constraints on specific parameters. Such limits are called generalised parameters. Network designers would then be free to optimise network design within those generalised constraints. This would probably allow most networks to be designed using less costly components than the present methods allow.

One generalised constraint has already been agreed. CCIR Recommendation 524[63] sets limits on the spectral density of the off-beam EIRP of earth stations transmitting at 6 GHz. The formulation of the constraint is rather complicated but, for example, for most types of emission it is recommended that the maximum EIRP in any 4 kHz bandwidth should not exceed $35-25\log\phi$ dBW at an off-beam angle ϕ between 2·5° and 48°. Some relaxation of this limit is allowed for SCPC emissions, the spectral density of which cannot readily be dispersed. Studies now in progress will produce a recommendation for application at 14 GHz; see Report 1001[64]. RR 2636 emphasises the importance of limiting sidelobe radiation in this way.

There is a discussion of generalised constraints and generalised parameters, in particular for application to orbit management, in Annex III to CCIR Report 453[54]. A comprehensive set of generalised parameters, valid for all kinds of FSS network, which might replace some or all of the constraints on specific parameters which are reviewed on pp. 271–279, has yet to be devised and evaluated. However, a set of generalised parameters is to be used as one important element in the management of the FSS Allotment Plan (see pp. 301–302).

8.4 Procedures for frequency co-ordination and registration of assignments

8.4.1 Introduction

Special procedures are used for the international management of frequency assignments to FSS networks in two groups of frequency bands. These are:

- The bands around 12 and 17 GHz which have been planned for the BSS or for associated feeder links (the band limits vary from Region to Region) but are also allocated to the FSS for assignments to its networks. These bands are managed by means of the procedures set out in RR Appendices 30[4] and 30A[1], not only for assignments for satellite broadcasting systems but also, wholly or in part, for the FSS networks which operate in these bands (see pp. 294–297).
- The Earth-to-space bands 6725–7025 MHz and 12·75–13·25 GHz and the space-to-Earth bands 4500–4800 MHz, 10·7–10·95 GHz and 11·2–11·45 GHz. These bands are managed by means of the procedures set out in RR Appendix 30B[1] for the FSS Allotment Plan (see pp. 298–302).

However, frequency co-ordination followed by the registration of frequency assignments in the MIFR is the normal way of establishing a recognised right to use assigned frequencies in FSS networks and it is applied in all other bands. The procedures used for these purposes are set out in RR Articles 11 and 13 and they are reviewed in this section. Both Articles were revised in detail at WARC-ORB-88; see the Final Acts of that conference[1].

In a few frequency bands an allocation for the FSS is made subject to agreements reached in accordance with RR Article 14; see, for example, RR 761 and the FSS assignments in Regions 2 and 3 between 2500 and 2690 MHz. This requirement is, in principle, additional to co-ordination in accordance with RR Article 11. However, a footnote added to RR 1060 at WARC-ORB-88 affirms that successful completion of the RR Article 14 procedures in such cases shall be deemed to constitute successful completion of the corresponding part of the RR Article 11 procedure; see RR 1060.1[1].

It is assumed that an administration which is responsible for more than one satellite network will do all that is necessary to ensure that the frequency assignments which it makes to one of its networks will not cause unacceptable interference to another of them. The technical principles which are used in international frequency co-ordination are available for calculating interference levels in this case also, but the administrative procedures do not take the domestic function into account.

In outline, the international co-ordination procedure is as follows:

- An administration which is planning a new satellite network makes an announcement of its intentions, including an outline of the technical characteristics intended. This is called 'advance publication'. It provides an opportunity for discussions with administrations responsible for other networks, already existing or in a more advanced state of planning, which might suffer excessive interference from the new network. The purpose of these discussions is to identify and perhaps to implement changes to the new network and to these other networks that would reduce the interference.
- For an administration planning a network using non-geostationary satellites, the discussions which follow advance publication help to determine the precautions that must be taken, once operation begins, in fixing hand-over times to avoid causing interference to geostationary satellites, in conformity with RR 2613 (see p. 271 above). The discussions also provide an opportunity for the adjustment of proposed orbits, the definition of frequency assignment plans and the formulation of operating practices so as to prevent one non-geostationary satellite network from interfering with another.
- If the planned new network is to use a geostationary satellite, a second set of consultations must usually be carried out when the proposals for the new network can be stated in full detail. This, the formal frequency co-ordination process, involves the administrations responsible for all geostationary satellite networks that might be affected by the new network. The right to participate in this formal co-ordination process is precisely defined in the RR, the basic criteria of the co-ordination process are internationally agreed and the IFRB has some degree of oversight over the conduct of the co-ordination process. The purpose of this co-ordination process is to determine whether interference in excess of the single-entry permissible interference level recommended by the CCIR will arise from the operation of the new network in the configuration which is proposed and, if so, to find and agree ways of modifying any or all of the networks involved so as to reduce the interference to an acceptable level.

When these discussions and co-ordinations have been successfully completed, the administration planning the new network, and those with responsibilities for other networks that have been involved, are free to notify full details of the new network, including frequency assignments, and the agreed changes to existing networks, to the IFRB for registration in the MIFR. Any necessary co-ordination of earth station assignments with terrestrial stations will have been going on in parallel. When the IFRB has been advised that the whole co-ordination process, with both space and terrestrial systems, has been completed, the registration process can be completed in accordance with RR Article 13.

This summary of the procedures applies to the introduction of a new satellite network. Substantially the same procedure has also to be followed when changes are made to an existing network that increases prospective interference to another network.

Advance publication and the associated discussions are considered further below. Formal co-ordination as between geostationary networks and the criterion that is used to determine which administrations have a right to participate in the co-ordination process are considered on pp. 285–288 and 288–291 respectively. The procedures used for the subsequent notification and registration of frequency assignments are considered on 292–294.

8.4.2 Advance publication and associated discussions

The procedures associated with advance publication are set out in Section I of RR Article 11. Substantial changes of detail were made to these regulations at WARC-ORB-88, mainly to clarify the previous text where necessary and to provide more specifically for IFRB assistance to be given to administrations needing it; see the Final Acts of the conference[1]. The main elements of the procedure, as it applies to a proposed new satellite, are as follows:

Notification: Not more than six years, and preferably not less than two years, before any new FSS satellite network is to brought into operation, the administration responsible for the proposed network sends to the IFRB the details called for in RR Appendix 4[1]. It is convenient to designate the notifying administration as 'Administration A'. The information which RR Appendix 4[1] calls for is rather general, consisting mainly of:

- the orbital elements, if the satellite is to be non-geostationary. If the orbit is to be geostationary, the range of feasible orbital locations is to be given,
- the frequency bands to be used,
- the characteristics of earth station and space station antennas,
- the noise temperature of earth station and space station receivers,
- the gain of space station transponders in the various configurations that are to be available,
- the maximum spectral power density of the up-link and down-link emissions, and
- the planned date for the start of network operation.

More specific information could not usually be supplied at an early stage in the planning of a new network.

Publication and risk estimation: The IFRB publishes the information in its weekly circular. This information, although somewhat unspecific, is sufficient to enable other administrations to decide whether there is any significant risk that interference from the proposed new network in excess of the permissible level would enter a network, in operation or planned, for which those other administrations are severally responsible. If any such risk is detected, the administrations concerned report their concern to Administration A and to the IFRB. It is convenient for these concerned administrations to be designed 'Administration B1', Administration B2', Administration B3' etc. Likewise Administration A determines whether other networks, which are in operation or which have already been announced by the advance publication procedure, might cause interference to its proposed new network; if so, it also notifies the administrations concerned and the IFRB.

The Radio Regulations do not specify how an administration should decide whether a significant risk of excessive interference will arise. However, the method set out in RR Appendix 29, which is specified for use in the formal co-ordination of geostationary networks, is technically suitable for use for geostationary networks at this preliminary stage also.

Resolution of problems: Administrations which become aware in these ways of potential difficulties arising from the proposal for the new network are required to supply any additional information on their networks that would help to clarify the prospect of interference and they are to endeavour to resolve the difficulties; see RR 1049[1]. Perhaps inevitably, the pressure to make changes falls most heavily on the intending owner of a proposed new network if the spacecraft that is to become its satellite is still being designed and constructed. It is for that reason that the advance publication can be made up to six years before the new network begins operation, when there would still be ample opportunities for changing proposed orbital locations, modifying the design of antennas etc. However, the RR also press for changes to be made to existing networks if this is necessary and feasible; see RR 1051[1]–RR 1053[1]. Agreed changes to network characteristics are notified to the IFRB and published for the information of all concerned. The IFRB is also kept informed of the progress made in solving interference problems and the Board's technical help is available if required.

Of course, not all advance publications relate to spacecraft which are still being designed and constructed. For example, it is sometimes proposed that a satellite that is already in orbit shall be operated in a new orbital location. In such cases there is less opportunity to optimise equipment for the new location. On the other hand, more complete information on the parameters of antennas and emission is likely to be available in these circumstances, permitting more precise calculations of prospective interference levels to be made.

The period that must be kept available for this informal search for solutions to foreseen interference problems extends for at least six months from the date of the original notification of information to the IFRB for publication; see RR 1058E[1]. Discussions may continue beyond six months if necessary unless Administration A prefers to proceed to the next stage; that is:

- for new non-geostationary satellite networks, the notification to the IFRB of frequency assignments to earth stations and space stations, and
- for new geostationary satellite networks, the formal frequency co-ordination process.

If the proposed new network is to use non-geostationary satellites, it is important that solutions be found to interference problems. If unresolved foreseen interference problems remain, or if unforeseen problems emerge when the new network starts operation, B administrations suffering interference will initiate the procedure for the elimination of harmful interference in due course, and their networks will presumably have the advantage of priority of registration of frequency assignments. Thus, the opportunity for an untroubled entry into operation will have been lost for the new network.

Success in finding solutions to interference problems between geostationary networks at the advance publication stage is obviously desirable. Success will facilitate the formal frequency co-ordination process which is to follow, perhaps by suggesting modifications to the new network at a time when such changes can be readily incorporated into the design, producing the most effective solution at the least cost. However, even if such evidently constructive results are not obtained, advance publication provides useful information for administrations which have plans to prepare for other networks.

8.4.3 Frequency co-ordination for geostationary satellite networks

When the administration responsible for a proposed new geostationary satellite network (Administration A) is ready to declare all the details of the network, including the frequency assignments that are proposed for the space station and the earth stations, one of two courses of action is followed:

- If the potential interference entry, calculated in a specified way, from the new network to any other network, in operation or already co-ordinated and awaiting the start of operation, does not exceed a specified criterion, and if Administration A is prepared to disregard whatever interference may enter the new network from existing networks, the details can be notified to the IFRB for registration of the frequency assignments in the MIFR; see RR 1060[1] and RR 1067[1].

- If calculated potential interference levels do exceed the criterion, the details must be made available to initiate the frequency co-ordination procedure (see RR 1060[1]) because only when co-ordination has been completed will the frequency assignments be registered.

Corresponding rules apply when a new earth station or a new frequency assignment is added to an existing geostationary satellite network, or if the parameters of existing stations or emissions are significantly changed. If potential interference levels are increased thereby so as to exceed the criterion, or if an interference entry that has already been co-ordinated is made worse, then co-ordination is required before the new assignment details are notified to the IFRB for registration.

Section II of RR Article 11 sets out the regulations governing these procedures. To initiate co-ordination, Administration A is required by RR 1073[1] to prepare the 'co-ordination data' for the new network (or, for example, the new assignment). The elements of these data are listed in RR Appendix 3[1]. For a new satellite network the co-ordination data consist mainly of:

- the limits of the arc of the GSO within which the new satellite could satisfactorily operate.
- all of the information on the stations (earth and space) and emissions of the new network needed to allow the determination of the extent to which the earth station transmitters and space station transmitters might interfere with other satellite networks and the space station receivers and earth stations receivers will be susceptible to interference from other satellite networks.

The statement of data includes a complete set of proposed frequency assignments, transmitting and receiving, for the space station together with sufficient information on earth station assignments, transmitting and receiving. Important changes were made to the information required on earth station assignments at WARC-ORB-88. Previously the RR required full particulars of all earth station frequency assignments. With effect from 16 March 1990, the date of entry into force of the amendments to the RR that were agreed at that conference, it is sufficient to disclose the characteristics of typical earth stations and the geographical areas within which they will be found.

Administration A calculates the maximum potential level of interference which the new network will cause to other networks which are responsibilities of other administrations, and vice versa. The methods for calculating, for this purpose, the potential maximum level of an inter-network interference entry and the criterion which determines whether co-ordination is required, are set out in RR Appendix 29; see pp. 288–291. It is assumed that the interference in the worst-affected channel of a network receiving pre-demodulator interference below that criterion will not exceed the permissible single-entry level recommended by the CCIR.

Administration A sends the co-ordination data to the administrations which are responsible for other satellite networks which it has identified for co-ordination in this way and asks for co-ordination. Administration A also sends the co-ordination data to the IFRB, together with a list of the other networks for which co-ordination has been requested. The IFRB examines the data, checks the interference entry calculations which Administration A has done and publishes the data and the Board's own list of networks with which co-ordination is required in the IFRB weekly circular. Any other administration that considers that it should have been asked to co-ordinate has a right to demand co-ordination. Thus, a list of the B administrations, with which Administration A is to co-ordinate, is built up; see RR 1073–RR1081 as amended at WARC-ORB-88[1].

No procedure for the conduct of frequency co-ordination negotiations is prescribed by the ITU. Usually administrations and system operators meet bilaterally and informally to identify the problems, find solutions and undertake to implement them, although no doubt informal multilateral meetings are found to be useful on occasion. However, WARC-ORB-88 decided to establish a new option whereby an Administration A might initiate action for the convening of a Multilateral Planning Meeting (MPM) for resolving the co-ordination problems of a network which was to operate in the following heavily loaded frequency bands;

3700–4200 MHz	11·7–12·2 GHz (Region 2 only)
5850–6425 MHz	12·5–12·75 GHz (Regions 1 and 3)
10·95–11·2 GHz	14·0–14·5 GHz
11·45–11·7 GHz	

see RR 1085C[1], RR 1085D[1], RR Resolution 110[65] and CCIR Report 1003[66]. RR Resolution 110 invites the Administrative Council to monitor the use of this procedure.

The method used to determine the need to co-ordinate, given in RR Appendix 29[1], is cautious. It identifies for scrutiny all situations in which the single-entry permissible interference level recommended by the CCIR would be exceeded, provided that care is also taken to determine whether powerful high-density carriers with a slow-sweeping energy dispersal waveform would interfere with narrow-band carriers. However it is often pessimistic, the criterion being exceeded when the worst-case post-demodulator interference level would be substantially below the level recommended by the CCIR. Thus, the first stage of co-ordination consists of more precise calculations of interference levels for the networks as they are and as they are planned to be, carried out in full knowledge of network characteristics; see CCIR Reports 455[33], 449[67] and 867[53] for information on the calculation of interference levels.

It will probably be found that the single-entry interference limit will not be exceeded between the proposed new network and several, perhaps many, of the networks for which B administrations are responsible. This permits some networks to be eliminated from the second stage of co-ordination, in which the situations where the recommended single-entry level of permissible interference would indeed be exceeded are identified and ways are found for eliminating the excess. The main techniques that are available for reducing interference levels between networks are reviewed on pp. 268–271.

The same basic processes are used to secure agreement, if necessary, to the addition of new stations or new assignments to an existing network or to significant changes to existing assignments.

The administrations involved in co-ordination report the state of progress to the IFRB from time to time, and agreements to change networks in significant ways which are reached in the course of the negotiations, which may be of interest to other administrations, are published in the IFRB weekly circular. When all B Administrations confirm that they accept the proposals of Administration A, in whatever state of amendment the proposals have reached during the co-ordination discussions, Administration A is free to notify the frequency assignments to the IFRB for registration.

In the event that Administration A does not succeed in getting the agreement of one or more of the B Administrations, the help of the IFRB in finding a basis for agreement may be requested; see RR 1088–RR 1094[1]. The Board will try to resolve the difficulties that remain and has authority to disregard a breakdown of co-ordination if this arises from a refusal to co-operate; see RR 1095–RR 1103[1]. Ultimately, if disagreements remain on matters of substance, Administration A may notify the frequency assignments to the Board without the unanimous acceptance of the B Administrations, the Board being asked to register the assignments in the MIFR, with an appropriate endorsement, in accordance with provisions in the Radio Regulations.

The process of co-ordination is neither quick nor easy, and it becomes more laborious as the number of networks in operation in a frequency band or in some stage of planning increases. Nevertheless, and perhaps surprisingly, the record of success is good.

8.4.4 The criterion for frequency co-ordination

RR Article 11 gives an administration which publishes the RR Appendix 3 information on a proposed new geostationary satellite network the right to require co-ordination of administrations which are responsible for networks, already in operation, which might interfere with the new network or which might later suffer harmful interference from the new network. This is an important right, providing a means for optimising the use of the GSO, thereby enabling new networks to be brought into service in frequency bands which are becoming heavily loaded. However, frequency co-ordination is burdensome to administrations and its use must be limited to situations where there is indeed a risk of significant interference. Thus there is a need for an agreed objective criterion which will determine whether frequency co-ordination must be granted in any specific case. That criterion, and the method by which it is applied, are contained in RR Appendix 29. The appendix was amended in significant details at WARC-ORB-88; see the Final Acts of the conference[1].

The same criterion can be used, somewhat more tentatively, to identify the possibility of significant interference to and from a new network at the advance publication stage, even though the information available on the new network may be limited to that required by RR Appendix 4.

The method set out in RR Appendix 29 is considered in more detail in CCIR Reports 454[68] and 871[69]; see also CCIR Report 455[33]. It is basically as follows. The worst-case spectral power density of the interference from a single source, entering the wanted up-link or down-link from another satellite network, is treated as thermal noise and is expressed in the form of the equivalent increase, ΔT_s or ΔT_e, in the system noise temperature of the receiver of the link, this being T_s for an up-link or T_e for a down-link. Thus

$$\text{For the up-link, the proportionate increase} = \Delta T_s/T_s \qquad (8.1)$$

$$\text{For the down-link, the proportionate increase} = \Delta T_e/T_e \qquad (8.2)$$

For a satellite link consisting of an up-link and a down-link in tandem, and in particular in the FSS in a network using transparent satellite transponders (that is, without on-board signal processing), it is usually convenient to refer both T_s and ΔT_s to the input to the earth station receiver. This involves calculating the transmission gain γ and the equivalent satellite link noise temperature T.

γ is the gain that would be measured from the output of the receiving antenna of the wanted space station to the output of the receiving antenna of the worst-affected wanted earth station. This gain is expressed as a numerical power ratio, and it is usually less than unity; see CCIR Report 871[69].

T is the noise level, expressed as a noise temperature referred to a receiving port of the feed of the earth station antenna, due to all of the thermal noise, transmission impairments and interference arising within the network but disregarding interference entering from outside the network; see RR 168 and CCIR Report 871[69].

Then
$$T = T_e + \gamma T_s + T_a \text{ kelvins} \qquad (8.3)$$
where T_a = the thermal noise equivalent of other transmission impairments in the system (kelvins)

and
$$\Delta T = \gamma \Delta T_s + \Delta T_e \text{ kelvins} \qquad (8.4)$$

The method of RR Appendix 29 assumes, in effect, that the performance of the worst affected channel of a wanted carrier received with this interference will not be degraded by more than the maximum value of the single entry of permissible interference recommended by the CCIR (see p. 264) if the apparent increase of thermal noise does not exceed 6%, but that recommended limit might be exceeded if there were a larger interference entry.

Whether interference from an unwanted satellite enters the wanted satellite through an up-link, a down-link or both, RR Appendix 29 sets the criterion for co-ordination as

$$\Delta T/T > 6\% \qquad (8.5)$$

RR Appendix 29 provides all the necessary geometrical data for calculating pre-demodulator interference levels involving interference modes 5, 6 and 8 (see Figs. 8.2 and 8.3). The appendix advises on the assumptions that should be made as to antenna off-beam gain in the absence of specific information and on the identification of probable worst-case interference situations.

Appendix 29 recommends that the methods of RR Appendix 28 should be used to identify earth stations for co-ordination where interference mode 7 (Fig. 8.3) arises (see pp. 118–131). Dealing with co-ordination between BSS feeder link earth stations transmitting in the band 17·7–18·1 GHz and FSS earth stations receiving in that band, Section 3 of Annex 4 of RR Appendix 30A[4] defines departures in detail from RR Appendix 28 which are specific to this case. This text was amended in detail at WARC-ORB-88; see the Final Acts of that conference[1]. The CCIR is seeking a more general, less labour-intensive, treatment of this problem; the results so far are to be found in CCIR Report 999[70].

Interference to narrow-band carriers

For almost all types of emission the entry of interference into communication channels can be estimated from the mean interfering spectral power density averaged over a period of a few tens of milliseconds. Consequently, allowance may be made for dispersal of carrier power by traffic signals (if a high level of modulation is continuously present due to the traffic signal) or by means of an artifical energy dispersal waveform, typically a triangular waveform.

Exceptionally among emissions that are commonly used in the FSS, this is not true when the bandwidth occupied by the wanted carrier is small, typically an SCPC emission, and the interfering carrier is frequency modulated by an analogue video signal with an artifical carrier energy dispersal waveform having a period equal to the frame or the field period of the video signal (see p. 274). In this case the impact of the interference is much more severe than consideration of the mean interfering spectral power density would indicate and the interference potential must be estimated from a comparison of peak power ratios.

RR Appendices 3 and 4 call for the declaration of the information necessary for the early identification of such situations and section 4 of RR Appendix 29 draws attention to the need for special consideration of them. An amendment to Appendix 29 made at WARC-ORB-88 directs the attention of administrations to CCIR texts on this subject. However, it is best to harmonise the frequency assignment plans of satellite networks so that these situations do not arise; see p. 000.

Polarisation discrimination

Some reduction of interference between networks arises from polarisation discrimination, but the cross-polar rejection that can be obtained from antennas in directions away from the principal axis and especially outside the main lobe is not well established yet. Accordingly, no account is taken of polarisation in the mandatory provisions of RR Appendix 29. However, the administrations involved may agree to include a limited allowance for polarisation discrimination in calculations of the need to co-ordinate that affect them both; see section 2.2.3 of RR Appendix 29.

The effectiveness of the Appendix 29 method

The method set out in RR Appendix 29 provides a dependable means for detecting the risk of a single entry of interference above the permissible level but in many instances a positive result is produced when more precise calculations show that interference will be below the permissible level, perhaps far below it.

These false positive indications are particularly frequent, and are perhaps inevitable, when the method is applied using RR Appendix 4 information at the advance publication stage. At the formal co-ordination stage, when the more detailed information on emissions required by RR Appendix 3 is available, there are fewer false positive indications. Nevertheless the method takes no account of differences in the interference reduction factors (factor B as discussed in CCIR Report 388[71]) of wanted carriers with different types of modulation and different modulation parameters; see also Annex II of CCIR Report 455[33]. Consequently co-ordination discussions may be entered into unnecessarily even when the Appendix 3 information has been made available.

The CCIR has the method of Appendix 29 under continuous review and the present state of the work is to be seen in CCIR Report 454[68]. The most important objective of this work at present is to reduce the likelihood of false positive indications without creating a risk that a need to co-ordinate will remain undetected. From time to time competent world administrative radio conferences consider the incorporation of proposed improvements into Appendix 29.

A more radical extension of the $\Delta T/T$ method for detecting with greater discrimination whether the limit for a single entry of interference was likely to be exceeded would involve using values for the co-ordination criterion which are normalised according to the types of emissions which are causing and suffering interference; see section 2.6 of Annex III of CCIR Report 453[54].

Some methods that might be used for multi-lateral co-ordination of several proposed new networks with the existing networks are considered in outline in CCIR Report 1003[66]. The specific concepts suggested, including the use of generalised parameters for determining an optimum scheme of satellite locations are considered in more detail in CCIR Report 870[72] and Annex III of Report 453[54].

Resolution 69[73] of WARC-ORB-88 invites the CCIR to do further work in this area.

8.4.5 Notification and registration of frequency assignments

Any FSS frequency assignment, transmitting or receiving, must be notified to the IFRB if its use is capable of causing harmful interference at the stations of another country or if it is to be used for international communication. Any other FSS assignment may be notified if the assigning administration wishes it to receive international recognition; see RR 1488–RR 1491. One or other of these conditions will apply to almost any FSS assignment. The IFRB processes notices of assignments in accordance with RR Article 13. Substantial amendments to this article were made at WARC-ORB-88, the amended regulations entering into force on 16 March 1990; see the Final Acts of that Conference[1]. The main provisions of these regulations are as follows.

Notices of frequency assignments should provide the information called for in Section II (for space stations) or Section III (for earth stations) of RR Appendix 3[1]. Space station assignments, transmitting and receiving, may be defined in terms of associated typical earth stations, the geographical areas in which they operate being identified. Earth station assignments, transmitting and receiving, may also be notified for typical earth stations, one notice serving for any number of earth stations within the stated area, unless either:

- the co-ordination area calculated in accordance with RR Appendix 28 (see pp. 128–129 above) overlaps the territory of another administration in which the frequency band is allocated with equal rights to terrestrial services, or
- the characteristics of the earth station are such that the interference caused or suffered would be greater than for any of the typical earth stations which have already been co-ordinated.

Details of notifications received are published in the IFRB weekly circular. Then all notices are examined to verify that the assignments are in accordance with the RR; see 1503[1]. This regulatory examination includes, for example, verification that the assignments are in keeping with the international Table of Frequency Allocations (RR Article 8), that mandatory technical constraints are observed and that the procedure laid down by RR Article 14 has been carried out successfully where necessary.

If the result of the regulatory examination is unsatisfactory but the administration undertakes to ensure that the emission will be discontinued immediately if there is a complaint of harmful interference, in accordance with RR 342 and RR 1560, the assignment is registered in the MIFR, suitably annotated. Without that undertaking, the notice is rejected by the Board.

However, if the result of the regulatory examination is satisfactory:

(*a*) Assignments to space stations on non-geostationary satellites are registered in the MIFR (RR 1515 and RR 1516).

(*b*) Assignments to space stations on geostationary satellites are registered in the MIFR, provided that any necessary frequency co-ordination with other geostationary satellite networks has been completed satisfactorily (RR 1525, RR 1526 and RR 1504).

(*c*) Assignments to earth stations are registered in the MIFR, provided that any necessary frequency co-ordination with terrestrial stations (see pp. 118–131) has been completed satisfactorily (RR 1525, RR 1526 and RR 1505).

If the notifying administration has not attempted necessary co-ordination procedures, or if co-ordination has been attempted but has not been brought to a successful conclusion, the Board is authorised to assist by determining whether there are significant prospective interference problems. If not, the assignments will be registered in the MIFR. If there are problems, the Board will if possible suggest ways of changing the assignments so as to avoid the problem (RR 1527–RR 1543). If, despite the IFRB's conclusion that interference would be harmful, the administration resubmits the assignment notices unchanged, after having had the assignments in use for four months without complaint of harmful interference, and insists on further consideration of the notices by the Board, there is provision in the Regulations for the assignments to be registered in the MIFR, suitably annotated (RR 1544).

RR 1573 requires administrations to have an MIFR entry deleted when the assignment to which it refers is no longer to be used.

A similarity will be noted between these procedures and those used for the fixed service operating below 28 MHz (see pp. 84–86). As in all such cases, the MIFR provides evidence of declared use of the spectrum and of foreseen interference problems. Conflicts of use, leading to actual interference, are for administrations to resolve amongst themselves with the help of that evidence (see pp. 64–65).

The time-frame for satellite network registration

A procedural time-frame has been set for satellite networks to prevent the co-ordination process being made unworkable by the inclusion in it of large numbers of tentative projects which are unlikely to reach maturity. A second objective has been to help to ensure that all nations have equitable access to the GSO by denying that prior use of an orbital location gives permanent rights of its use. The elements of this time-frame are as follows.

(i) When an administration makes advance publication of proposals for a new satellite network, the information published includes an indication of the date by which the network is to be brought initially into use. The notification of this information may not be made more than six years before the forecast start of operation (RR 1042[1]).

(ii) When an administration notifies frequency assignments to the IFRB for registration, the notice includes a statement of the date (actual or foreseen) when that assignment is to be brought into use. The notification of the assignment may not be made more than three years before that date (RR 1496).

(iii) The notified date for bringing into use the first frequency assignment for a satellite network shall be not later than six years after the date of issue of the IFRB circular which contained the advance publication of the proposal to set up the network. This period will be extended by not more than three years on request by the notifying administration (RR 1550[1]).

Thus, if the completion of a project for a new satellite network becomes delayed by several years for any reason, it may become necessary to return to the beginning of the co-ordination process.

(iv) Resolution 4 of WARC-79, amended at WARC-ORB-88[13], requires the notice of a frequency assignment to a space station on a geostationary satellite to include a declaration of the lifetime of the network. This declared lifetime is not to exceed the design lifetime of the network. When that declared lifetime has expired, the Resolution lays down rules for discussion between the IFRB and the notifying administration to determine when all of the frequency assignments for a network should be cancelled, it being assumed that the network has reached the end of its life.

RR Resolution 4 is, at present, an experimental procedure, to be reviewed at some future competent WARC.

Finally, Section VI of RR Article 13 requires the IFRB to ensure that the MIFR remains a realistic record of operating systems, not retaining assignments which are no longer in use.

8.5 The FSS in bands planned for the BSS and its feeder links

Planned frequency assignments for satellite broadcasting in Region 2 in the bands 12·2–12·7 GHz may be used for down-links for FSS networks, in accordance with RR 846[4]. provided that such FSS emissions neither cause more interference nor require more protection from interference than BSS facilities operating in accordance with the BSS plan (RR Appendix 30A[4]). Such FSS assignments can be treated from the international standpoint as a particular use of the planned BSS assignments and they do not need to be considered further here. In Regions 1 and 3 there are no internationally agreed allocations to the FSS in frequency bands which have been used for a BSS frequency assignment plan for the same Region.

However, the BSS frequency assignment plans for the three Regions contained in RR Appendix 30A[4] cover various parts of the band 11·7–12·7 GHz and at any frequency within this range where there is a BSS plan in one Region, there is an allocation for the FSS in another Region. Also, the band 17·3–18·1 GHz is allocated world-wide to the FSS (Earth-to-space) for use for feeder links for the BSS and, of this band, 17·7–18·1 GHz is also allocated to the FSS (space-to-Earth) for use for down-links for networks of the FSS. In Region 2, part of this shared band has been planned for feeder links for broadcasting satellites transmitting on planned assignments in the

12 GHz band and in Regions 1 and 3 the feeder link plan covers the whole of the shared band.

These situations produce problems, technical and regulatory. The problems and the means that have been agreed to deal with them are reviewed below.

Inter-Regional sharing at 12 GHz

The BSS frequency assignment plans at 12 GHz, drawn up in 1977 and 1983, have been approved by competent WARCs. Thus, subject to ratification, all ITU Members have agreed that the assignees have an unqualified right to use the planned assignments, and to use them without unacceptable interference. This includes freedom from interference from FSS networks operating in primary allocations in other Regions, even though the FSS frequency assignments that might cause interference may have priority of registration in the MIFR.

The frequency bands where this inter-Regional sharing situation arises are as follows:

In Region 1: The FSS allocation at 12·5–12·7 GHz is affected by the Region 2 BSS plan.
In Region 2: The FSS allocation at 11·7–12·2 GHz is affected by the BSS plans in Regions 1 and 3.
In Region 3: The FSS allocation at 12·5–12·7 GHz is affected by the BSS plan in Region 2. In addition, the FSS allocation at 12·2–12·5 GHz, made by RR 845, is affected by the BSS plans in Regions 1 and 2.

CCIR Report 873[74] shows that the powerful carrier component of BSS signals could cause significant interference to FSS networks in adjacent Regions. This interference could be reduced in various ways, one of which is the application of artificial carrier energy dispersal to the BSS signal in the absence of programme modulation; see also CCIR Report 809[75]. Accordingly, Annex 6 of RR Appendix 30 as amended at WARC-85[4] discusses these matters and section 3.18 of Annex 5 to RR Appendix 30[4] indicates that, in planning for the BSS, a measure of carrier energy dispersal has been assumed.

The use of the 12 GHz band by the FSS is growing rapidly, whereas it may be many years before most of the BSS assignments are put into service. Some planned BSS assignments may never be used. Clearly, it would be undesirable for the use of the spectrum by the FSS to be needlessly inhibited. Thus, an administration wishing to establish an FSS network in one of these bands needs the right to insist that other administrations provide information on their intention to implement planned BSS assignments that would affect or be affected by the FSS proposals. The rules for co-ordinating and registering FSS frequency assignments in the MIFR, set out in RR Article 11 and 13 and reviewed on pp. 283–294, are not designed to protect a planned BSS assignment which may be dormant.

Accordingly, FSS frequency assignments in these bands must be treated in accordance with RR Article 15 and Article 7 of RR Appendix 30. Both of these texts were amended at WARC-85[4]. RR Article 15 provides a basis for co-ordination with the assignees of BSS assignments. Information on required protection ratios is to be found in Annex 6 of RR Appendix 30[4]; see also CCIR Report 631[76].

The essence of Article 7 of Appendix 30 is to be found in its section 7.2.1, which requires co-ordination between an administration proposing an assignment to an FSS network and an administration with a planned BSS assignment if the power flux density of the FSS signal within the service area and within the necessary bandwidth of the BSS assignment would exceed criteria given in Annex 4 of RR Appendix 30[4]. These criteria are such that the need for co-ordination seems likely to be limited to a proposed FSS network having its satellite within 20° of longitude of the location planned for a broadcasting satellite of another Region.

This RR Article 15 requirement is, of course, additional to the procedures set out in RR Article 11 for co-ordination with other FSS networks.

RR Appendix 30[4] incorporates procedures whereby an administration may propose a change to the parameters of a BSS frequency assignment which has been planned for it. It might be greatly to the collective disadvantage of systems of the fixed-satellite service if an administration could choose to locate its broadcasting satellite in an arc of the orbit where the predominant use of the frequency band in which it was to operate was the FSS. Accordingly, limits have been placed on the range of orbital locations to which a planned assignment may be moved (see p. 339).

Sharing between feeder links and FSS space-to-Earth links at 18 GHz

The high status given to planned frequency assignments applies to feeder link assignments as well as BSS down-link assignments, and two very different modes of interference have to be considered, namely:

- interference from a feeder link earth station entering the receiver of an FSS earth station nearby, and
- interference from an FSS satellite entering a BSS satellite feeder link receiver.

See interference modes 7 and 8 in Fig. 8.3

Assignments to the FSS in these frequency bands must be treated in accordance with the requirements of RR Article 15A[1] and RR Appendix 30A[4], in addition to the procedure under RR Articles 11 and 13 required for co-ordination with other FSS networks.

When a country brings its 12 GHz BSS assignments into use, an earth station will be required for the feeder links. This earth station may cause interference to earth stations used for FSS networks in the 18 GHz band in neighbouring countries. The prospect of interference from a feeder link earth station with planned frequency assignments at an unknown place in a neighbouring country may inhibit the development of networks of the FSS, operating at 18 GHz, over wide areas near national frontiers. A procedure has been agreed to assist in the resolution of this problem; see sections 7.2–7.7 of Article 7 and section 3 of Annex 4, both to be found in RR Appendix 30A[4].

Section 1 of Article 7 of RR Appendix 30A[4] and Section 1 of Annex 4 of the same appendix provide for the co-ordination of FSS satellites operating at 18 GHz in the same band as planned feeder link assignments. Co-ordination is required if the proposed orbital location of the FSS satellite is within 10° of a planned broadcasting satellite, subject to the potential single-entry interference level at the BSS space station feeder link receiver, as calculated by the method of RR Appendix 29, exceeding a threshold value of $\Delta T/T = 10\%$. Co-ordination is also required, subject to the same $\Delta T/T$ criterion, if the location of the proposed FSS satellite is more than 150° away from a broadcasting satellite (the quasi-antipodal case) unless the spectral power density to be radiated by the FSS satellite in the affected band in the direction of the Earth's limb, measured as the PFD at the Earth's surface at the equator, is less than −123 dBW per square metre in a bandwidth of 24 MHz.

8.6 The FSS Frequency Allotment plan

WARC-ORB-85 agreed that a frequency allotment plan should be drawn up for geostationary satellite networks of the FSS in specified frequency bands, and WARC-ORB-88 drew up and agreed the plan and the procedures for its implementation. The bands planned were:

Earth-to-space: 6725–7025 MHz and 12·75–13·25 GHz
Space-to-Earth: 4500–4800 MHz, 10·7–10·95 GHz and 11·2–11·45 GHz

The plan and the associated procedures are to be found in RR Appendix 30B[1].

Objectives of the Allotment Plan

The primary objective in drawing up the allotment plan, in the words of RR Appendix 30B[1], was 'to guarantee in practice, for all countries, equitable access to be geostationary-satellite orbit ...'. To this end the plan allots to each ITU member-country:

- A transmission bandwidth of 800 MHz (300 MHz at 6 and 4 GHz plus 500 MHz at 13 and 11 GHz) to be used for domestic facilities.
- The opportunity to cover once the whole of its territory.

- An orbital location (more than one location if the territory to be covered is so widespread that a single location in the GSO cannot serve all of it).

These are the 183 allotments which form Part A of the plan. The plan is intended to remain in force for at least 20 years and there is provision for the addition of new allotments from time to time for new ITU members as they join the Union. The plan is intended to accommodate as wide a range as possible of the facilities that systems of the FSS provide. It is also intended that two or more neighbouring countries should be able to exchange some or all of their national allotments for an allotment that could be used for a joint network, called a 'sub-Regional system'.

To combine this range of options with a minimum of costly constraints on equipment performance, it has been necessary to design the plan and the procedures that go with it with several sorts of flexibility, reviewed briefly below. In this respect the FSS plan differs considerably from the frequency assignment plans that have been drawn up for the BSS (see pp. 320–341), which provide for a single kind of facility, namely a colour television broadcast with the same picture standards as are currently used for terrestrial broadcasting and with good signal quality. The FSS plan is more closely analogous to the terrestrial frequency allotment plans that have been drawn up, for example, for maritime radio telephony at HF (see pp. 199–202) and it is from this analogy that the FSS Allotment Plan takes its name.

The Allotment Plan

The plan consists essentially of a list of allotments, one for each satellite in Part A of the plan, stating essentially the nominal orbital position allotted to the satellite, the boresight pointing direction and the half-power beamwidth of the satellite antenna beam and a measure of the maximum up-link and down-link spectral EIRP density that may be used. The nominal orbital positions have been chosen so that the interference entering each network from all other networks when all allotments have been activated will be acceptably small, provided that certain key network characteristics and parameters are observed.

The network parameters used in drawing up the plan are set out in section A of Annex 1 of RR Appendix 30B[1]. The principal assumptions are as follows:

- Earth station antennas will be 7 m in diamter at 6/4 GHz and 3 m in diameter at 13/11 GHz. The receiver system noise temperatures will be 140 K and 200 K at 4 and 11 GHz respectively. The sidelobe reference radiation pattern will be

$$G = 32\text{–}25\log\phi \text{ dBi} \quad (8.6)$$

for values of ϕ, the off-axis angle, between 1° and 48° and

$$G = -10\text{dBi} \quad (8.7)$$

where $\phi > 48°$

- Space station antennas will have beams of elliptical or circular cross-section. The half-power beamwidth will be just sufficient to cover all geographical test-points listed by an administration as a means of defining the required national coverage area, after allowance has been made for beam pointing errors not exceeding 0·1°, the minimum beamwidths assumed being 1·6° at 6/4 GHz and 0·8° at 13/11 GHz. A reference radiation pattern, in-beam and out-of-beam, is specified, virtually identical with curve B of Fig. 9.3.
- The nominal satellite location will be within an orbital arc, any part of which can be seen from all of the geographical test-points defining the required national coverage area at a minimum angle of elevation between 10° and 40°, depending on climatic conditions in the service area. Special rules apply to countries which are at high latitudes or are very large.
- No assumptions are made as to carrier frequencies or modulation techniques. The carrier powers, averaged over the occupied bandwidths of emissions, are sufficient to provide a total carrier-to-noise ratio of 16 dB for 99·95% of the year at 4 GHz and 99·9% of the year at 11 GHz. The aggregate carrier-to-interference ratio under free space propagation conditions will be not worse than 26 dB.

It should, however, be noted that these parameters do not have to be implemented in their entirety; proposals for implementation can depart from them provided that interference to other allotments, or susceptibility to interference from other allotments, is not increased.

Allotments for new ITU members

An article in RR Appendix 30B[1] provides a procedure for the creation of an allotment for a new ITU member. It is very simple in principle. The administration of a new member country states what is required, and in particular the locations of the test-points which define national coverage. The IFRB then identifies a suitable nominal orbital position and enters the allotment in Part A of the plan.

However, it will often happen that a suitable orbital location cannot be found without changing the orbital location of existing allotments. For this reason, among others, every allotment is associated with a pre-determined arc (PDA) within which, in circumstances such as these, the nominal location for an existing allotment may be changed.

Before an administration begins the process of implementing an allotment, that is in the pre-design stage, the PDA is ±10° relative to the nominal position, unless a smaller range is necessary to ensure that a satisfactory angle of satellite elevation is maintained at all test-points; 20 years after the plan enters into force (that is, in 2010), the PDA is to become ±20° for allotments which are still at the pre-design stage. When a network is in the design stage and it becomes less feasible to accommodate the changes in network design, and in particular the design of satellite antennas, which a change of satellite location may demand, the PDA is reduced to ±5° about the nominal position. At the final stage, when a network is ready for operation, the PDA shrinks to zero.

Existing systems

The frequency bands selected for the FSS allotment plan were chosen at WARC-ORB-85 partly because they were not so high in the spectrum that radio propagation would be adverse. However, these bands had been newly allocated to the FSS at WARC-79 and another important reason for choosing them was that, being newly allocated, there were relatively few networks operating in them. There would therefore be fewer impediments to planning than would arise in other frequency bands below 15 GHz.

Nevertheless the number of networks using these bands that had been notified to the IFRB for advance publication of information by the first day of WARC-ORB-85 (8 August 1985) was far from negligible. Some networks are equipped to use the whole of the bandwidth available for the allotment plan at 6/4 GHz or 13/11 GHz or in both band-pairs; others occupied no more than a small part of this spectrum. They had been, or were being, co-ordinated and brought into use in accordance with RR Articles 11 and 13, the regulations that were in force at the relevant time. If these procedures are successfully completed, the ITU has no authority to deny to these networks the right to operate in these bands. They are called 'existing systems' whatever stage of implementation they have reached, and they are listed, 36 in number, as Part B of the plan.

RR Resolution 108[77] prevents any initiation of the RR Article 11 procedure for additional networks using the allotment plan frequency bands with effect from the first day of WARC-ORB-88 (5 October 1988). RR Resolution 107[78] provides a limited compromise status for networks for which the procedure was initiated at dates between the two conference sessions. An article of RR Appendix 30B[1] requires existing systems, listed in Part B of the plan, to cease operation by the year 2010 at the latest.

In drawing up the plan for Part A allotments, no account has been taken of the orbital location of the satellites of existing systems. There are flexible procedures in RR Appendix 30B[1] for harmonising existing systems which are in the course of coming into use with existing systems which are already in operation and with allotments which are brought into use. This flexibility will, no doubt, enable the use of the planned frequency bands to be developed for both communities of satellite networks until the Part B systems are phased out and the whole capacity of the bands becomes available for the implementation of allotments.

Implementation of allotments

RR Appendix 30B[1] provides a regulatory framework for the conversion of an allotment in the plan into assignments for a working network. The principal elements of the process are as follows:

(i) When the network enters the design stage, but not more than five years before the planned start of operation, the information listed in Annex 2 of RR Appendix 30B[1] is notified to the IFRB by the responsible administration. Annex 2 calls for precise information on antenna characteristics but information on emissions is generalised, being limited mainly to statements of maximum spectral power density radiated.

(ii) The IFRB makes a technical examination of this information. Using the method of generalised parameters (see pp. 280–281), it is examined to determine whether the proposed network would cause more interference than the set of parameters on which the plan was based, or require better protection from interference. The information is examined for adherence to a concept of spectrum segmentation set out in Annex 3B of RR Appendix 30B[1], whereby emissions of high spectral density would be segregated from emissions of low spectral density of other conforming networks. And the Board ascertains whether the proposed network would be compatible with the existing systems listed in Part B of the Plan.

(iii) Appropriate solutions are sought if the results of this technical examination are unfavourable, including possible changes of satellite location within its PDA. When any such problems have been solved, the network under design is included, as a planned assignment, in a list to be maintained as part of the plan.

(iv) The administration is then free to notify frequency assignments to the IFRB in accordance with RR Article 13, for registration in the MIFR. Where appropriate, it will of course also be necessary to coordinate earth station frequency assignments with foreign terrestrial station assignments, in accordance with Section III of RR Article 11, before notification for registration.

Implementation of sub-Regional systems

When two or more neighbouring countries decide to set up a sub-Regional system, they follow a procedure set out in RR Appendix 30B[1]. An orbital position is selected for the network, desirably from amongst the nominal positions of the national allotments of the countries involved, and an appropriate set of information is assembled in accordance with the requirements of Annex 2 of RR Appendix 30B[1]. If necessary, the national allotments of countries in the group will be suspended while the sub-Regional system is in operation, but the suspended allotments will continue to be protected from interference. The methods used to find a suitable orbital location for a sub-Regional satellite are the same as those that are used when the implementation of a network intended to serve a single national area encounters difficulties (see above).

If, subsequently, countries decide to leave the sub-Regional network, the registered details of the latter are cancelled and the countries revert to their original national allotments.

Additional systems

RR Appendix 30B[1] also defines conditions under which a country, or a group of countries, may operate in the planned bands a satellite network which is additional to those listed in Parts A and B of the plan and may notify assignments to the network to the IFRB for registration. However, the conditions are such that little use of this option is likely to be feasible over most arcs of the GSO.

8.7 References

1 Final Acts adopted by the second session of the World Administrative Radio Conference on the use of the geostationary-satellite orbit and the planning of space services using it (WARC-ORB-88) (ITU, Geneva, 1988).

2 'Feeder links to space stations in the broadcasting-satellite service'. CCIR Report 561–3; Recommendations and Reports of the CCIR, 1986, Volume IV–1 (ITU, Geneva, 1986)

3 'Technical characteristics of feeder links to broadcasting satellites'. CCIR Report 952–1; *ibid.*, Volume X/XI–2

4 'Final Acts of the World Administrative Radio Conference, 1985 (ITU, Geneva, 1985)

5 'Radar spectrum utilisation'. CCIR Report 827; Recommendations and Reports of the CCIR, 1986, Volume I (ITU, Geneva, 1986)

6 'Theoretical and experimental results for spectrum sharing between FDM-FM and radar systems using pulse blanking'. CCIR Report 828–1; *ibid.*, Volume I

7 'Methods for calculating pulsed radar emission spectrum bandwidth'. CCIR Report 837; *ibid.*, Volume I

8 'Technical bases for the World Administrative Radio Conference on the use of the geostationary satellite orbit and the planning of the space services utilising it (WARC-ORB(1)), Part II'. Report of the CCIR Conference Preparatory Meeting, 1984 (ITU, Geneva, 1984)

9 'Sharing criteria for the protection of space stations in the fixed-satellite service receiving in the band 14 to 14·4 GHz'. CCIR Report 560–2; Recommendations and Reports of the CCIR, 1986, Volume IV–1 (ITU, Geneva, 1986)

10 'Limits of power flux density of radionavigation transmitters to protect space station receivers in the fixed-satellite service in the 14 GHz band'. CCIR Recommendation 496–2; *ibid.*, Volume VIII–3

11 'Factors concerning the protection of fixed-satellite earth stations operating in adjacent frequency band allocations against unwanted emission from broadcasting satellites operating in frequency bands around 12 GHz'. CCIR Report 712–1; *ibid.*, Volume IV–1

12 'Relating to the equitable use, by all countries, with equal rights, of the geostationary satellite orbit and of frequency bands for space radiocommunication services'. Resolution 2; Radio Regulations; edition of 1982 (ITU, Geneva, 1982)

13 'Period of validity of frequency assignments to space stations using the geostationary satellite orbit'. Resolution 4; Final Acts of WARC-ORB-88, item [1] above

14 'Relating to the use of the geostationary satellite orbit and to the planning of space services utilising it'. Resolution 3; Radio Regulations, edition of 1982 (ITU, Geneva, 1982)

15 'World Administrative Radio Conference on the use of the geostationary satellite orbit and the planning of space services utilising it, First session, Geneva, 1985; Report to the Second Session of the Conference'. (ITU, Geneva, 1985)

16 'The use of frequency bands allocated to the fixed-satellite service for both the up link and the down link of geostationary satellite systems'. CCIR Report 557–2; Recommendations and Reports of the CCIR, 1986, Volume IV–1 (ITU, Geneva, 1986)

17 'Hypothetical reference circuit for systems using analogue transmission in the fixed-satellite service'. CCIR Recommendation 352–4; *ibid.*, Volume IV–1

18 'Hypothetical reference digital path for systems using digital transmission in the fixed-satellite service'. CCIR Recommendation 521–2; *ibid.*, Volume IV–1

19 'Allowable noise power in the hypothetical reference circuit for frequency division multiplex telephony in the fixed-satellite service'. CCIR Recommendation 353–5; *ibid.*, Volume IV–1

20 'Video bandwidth and permissible noise level in the hypothetical reference circuit for the fixed-satellite service'. CCIR Recommendation 354–2; *ibid.*, Volume IV–1

21 'Transmission performance of television circuits designed for use in international connections'. CCIR Recommendation 567–2; *ibid.*, Volume XII

22 'Form of the hypothetical reference circuit and allowable noise standard for frequency division multiplex telephony and television in the fixed-satellite service'. CCIR Report 208–6; *ibid.*, Volume IV–1

23 'Allowable bit error ratios at the output of the hypothetical reference digital path for systems of the fixed-satellite service using pulse-code modulation for telephony'. CCIR Recommendation 522–2; *ibid.*, Volume IV–1

24 'Allowable error performance for a hypothetical reference digital path in the fixed-satellite service operating below 15 GHz when forming part of an international connection in an integrated services digital network'. CCIR Recommendation 614; *ibid.*, Volume IV–1

25 'Characteristics of a fixed-satellite service hypothetical reference digital path forming part of an integrated services digital network'. CCIR Report 997; *ibid.*

26 'Maximum allowable values of interference from line-of-sight radio systems in a telephone channel of a system in the fixed-satellite service employing frequency modulation when the same frequency bands are shared by both systems'. CCIR Recommendation 356–4; *ibid.*, Volume IV/IX–2

27 'Maximum allowable values of interference from terrestrial radio links to systems in the fixed-satellite service employing 8-bit PCM encoded telephony

and sharing the same frequency bands'. CCIR Recommendation 558–2; *ibid.*, Volume IV/IX–2

28 'Derivation of interference criteria for digital systems in the fixed-satellite service sharing bands with terrestrial systems'. CCIR Report 793–1; *ibid.*, Volume IV/IX–2

29 'Maximum permissible level of interference in a telephone channel of a geostationary satellite network in the fixed-satellite service employing frequency modulation with frequency-division multiplex, caused by other networks of this service'. CCIR Recommendation 466–4; *ibid.*, Volume IV–1

30 'Maximum permissible level of interference in a television channel of a geostationary satellite network in the fixed-satellite service employing frequency modulation, caused by other networks of this service'. CCIR Recommendation 483–1; *ibid.*, Volume IV–1

31 'Maximum permissible levels of interference in a geostationary-satellite network in the fixed-satellite service using 8-bit PCM encoded telephony, caused by other networks of this service'. CCIR Recommendation 523–2; *ibid.*, Volume IV–1

32 'Interference allocations in systems operating at frequencies greater than 10 GHz in the fixed-satellite service'. CCIR Report 710–2; *ibid.*, Volume IV–1

33 'Frequency sharing between networks of the fixed-satellite service'. CCIR Report 455–4; *ibid.*, Volume IV–1

34 'Protection of terrestrial line-of-sight radio-relay systems against interference due to emissions from space stations in the fixed-satellite service in shared frequency bands between 1 and 23 GHz'. CCIR Report 387–5; *ibid.*, Volume IV/IX–2

35 'Determination of the power in any 4 kHz band radiated towards the horizon by earth stations of the fixed-satellite service sharing frequency bands below 15 GHz with terrestrial services'. CCIR Report 386–3; *ibid.*, Volume IV/IX–2

36 'Feasibility of frequency sharing between systems in the fixed-satellite service and terrestrial radio services. Criteria for the selection of sites for earth stations in the fixed-satellite service'. CCIR Report 385–1; *ibid.*, Volume IV–1

37 'Earth station antennas for the fixed-satellite service'. CCIR Report 390–5; *ibid.*, Volume IV–1

38 'Consideration of the coupling between an earth station antenna and a terrestrial link antenna'. CCIR Report 709–1; *ibid.*, Volume IV/IX–2

39 'Calculation of the maximum power density averaged over 4 kHz of an angle-modulated carrier'. CCIR Report 792–2; *ibid.*, Volume IV/IX–1

40 'Energy dispersal in the fixed-satellite service'. CCIR Report 384–5; *ibid.*, Volume IV–1

41 'Carrier energy dispersal for systems employing angle modulation by analogue signals or digital modulation in the fixed-satellite service'. CCIR Recommendation 446–2; *ibid.*, Volume IV–1

42 'Relating to carrier energy dispersal in systems in the fixed-satellite service'. Recommendation 103; Radio Regulations, edition of 1982, (ITU, Geneva, 1982)

43 'Spread spectrum techniques'. CCIR Report 651–2; Recommendations and Reports of the CCIR, 1986, Volume I (ITU, Geneva, 1986)

44 'Examples of band sharing by employing spread-spectrum techniques'. CCIR Report 826–1; *ibid.,* Volume I

45 'Procedures for analysing spread spectrum interference to conventional receivers'. CCIR Report 974; *ibid.,* Volume I

46 'EIRP and power limits for terrestrial radio-relay transmitters sharing with digital satellite systems in bands between 11 and 14 GHz and around 30 GHz'. CCIR Report 790–1; *ibid.,* Volume IV/IX–2

47 'Flexibility in the positioning of satellites'. CCIR Report 1002; *ibid.,* Volume IV–1

48 'Spectrum utilisation methodologies'. CCIR Report 1000; *ibid.,* Volume IV–1

49 'Physical interference in the geostationary satellite orbit'. CCIR Question 34/4; *ibid.,* Volume IV–1

50 'Physical interference in the geostationary satellite orbit'. CCIR Report 1004; *ibid.,* Volume IV–1

51 'Relating to studies of the maximum permitted levels of spurious emissions'. Recommendation 66; Radio Regulations, edition of 1982, (ITU, Geneva, 1982)

52 'Spurious emissions from earth stations and space stations of the fixed-satellite service'. CCIR Report 713–1; Recommendations and Reports of the CCIR, 1986, Volume IV–1 (ITU, Geneva, 1986)

53 'Maximum permissible interference in single-channel-per-carrier transmissions in networks of the fixed-satellite service'. CCIR Report 867–1; *ibid.,* Volume IV–1

54 'Technical factors influencing the efficiency of use of the geostationary satellite orbit by radiocommunication satellites sharing the same frequency bands. General summary'. CCIR Report 453–4; *ibid.,* Volume IV–1

55 'Factors affecting station-keeping of geostationary satellites of the fixed-satellite service'. CCIR Report 556–3; *ibid.,* Volume IV–1

56 'Station-keeping in longitude of geostationary satellites using frequency bands allocated to the fixed-satellite service'. CCIR Recommendation 484–2; *ibid.,* Volume IV–1

57 'Radiation diagrams of antennas for earth stations in the fixed-satellite service for use in interference studies and for the determination of a design objective'. CCIR Report 391–5; *ibid.,* Volume IV–1

58 'Performance of small earth station antennas for the fixed-satellite service'. CCIR Report 998; *ibid.,* Volume IV–1

59 'Reference earth-station radiation pattern for use in co-ordination and interference assessment in the frequency range from 2 to about 30 GHz'. CCIR Recommendation 465–2; *ibid.,* Volume IV–1

60 'Radiation diagrams for use as design objectives for antennas for earth stations operating with geostationary satellites'. CCIR Recommendation 580–1; *ibid.,* Volume IV–1

61 'Satellite antenna patterns in the fixed-satellite service'. CCIR Report 558–3, *ibid.,* Volume IV/–1

62 'Discrimination by means of orthogonal circular and linear polarisations'. CCIR Report 555–3; *ibid.,* Volume IV–1

63 'Maximum permissible levels of off-axis EIRP density from earth stations in the fixed-satellite service transmitting in the 6 GHz frequency band'. CCIR Recommendation 524–2; *ibid.*, Volume IV–1

64 'Off-axis EIRP density limits for fixed-satellite service earth stations'. CCIR Report 1001; *ibid.*, Volume IV–1

65 'Improved procedures for certain bands of the fixed-satellite service'. Resolution 110; Final Acts of WARC-ORB-88; item [1] above

66 'Methods for multilateral co-ordination among satellite networks'. CCIR Report 1003; Recommendations and Reports of the CCIR, 1986, Volume IV–1 (ITU, Geneva, 1986)

67 'Measured interference into frequency modulation television systems using frequencies shared within systems in the fixed-satellite service or between these systems and terrestrial systems'. CCIR Report 449–1; *ibid.*, Volume IV/IX–2

68 'Method of calculation for determining if co-ordination is required between geostationary satellite networks sharing the same frequency bands'. CCIR Report 454–4; *ibid.*, Volume IV–1 (ITU, Geneva, 1986)

69 'Calculation of the equivalent satellite link noise temperature and the transmission gain'. CCIR Report 871–1; *ibid.*, Volume IV–1

70 'Determination of the bidirectional co-ordination area.' CCIR Report 999; *ibid.*, Volume IV–1

71 'Methods for determining interference in terrestrial radio-relay systems and systems in the fixed-satellite service'. CCIR Report 388–4; *ibid.*, Volume IV/IX–2

72 'Technical co-ordination methods for communication-satellite systems'. CCIR Report 870–1; *ibid.*, Volume IV–1

73 'Estimation of interference between satellite networks using simplified methods'. Resolution 69; Final Acts of WARC-ORB-88; item [1] above

74 'An analysis of the interference from the broadcasting-satellite service of one region into the fixed-satellite service of another region around 12 GHz'. CCIR Report 873–1; Recommendations and Reports of the CCIR, 1986, Volume IV–1 (ITU, Geneva, 1986)

75 'Inter-regional sharing of the 11·7 to 12·75 Ghz frequency band between the broadcasting-satellite service and the fixed-satellite service'. CCIR Report 809–2; *ibid.*, Volume X/XI–2

76 'Frequency sharing between the broadcasting-satellite service (sound and television) and terrestrial services'. CCIR Report 631–3; *ibid.*, Volume X/XI–2

77 'Use of the bands 4500–4800 MHz, 6725–7025 MHz, 10·70–10·95 GHz, 11·2–11·45 GHz and 12·75–13·25 GHz prior to the date of entry into force of Appendix 30B'. Resolution 108; Final Acts of WARC-ORB-88; item [1] above

78 'Satellite networks intended for use in the frequency bands of the plan in Appendix 30B for which information was communicated to the IFRB between 8 August 1985 and 5 October 1988'. Resolution 107; *ibid.*

Chapter 9

The broadcasting-satellite service

9.1 Survey of BSS regulation

In defining the (terrestrial) broadcasting service, RR 36 limits the service to 'transmissions intended for direct reception by the general public'. The definition of the broadcasting-satellite service (BSS) in RR 37 is rather broader, as follows:

> a radiocommunication service in which signals transmitted by space stations are intended for the direct reception of the general public. In the broadcasting-satellite service the term 'direct reception' shall encompass both individual reception and community reception.

RR 123 and RR 124 define individual and community reception as follows:

> *Individual reception:* 'The reception of emissions from a space station in the broadcasting-satellite service by simple domestic installations and in particular those possessing small antennas.'
> *Community reception:* 'The reception of emissions from a space station in the broadcasting-satellite service by receiving equipment, which in some cases may be complex, and have antennae larger than those used for individual reception, and intended for use by a group of the general public at one location or through a distribution system covering a limited area.'

The term 'direct broadcasting by satellite' (DBS) is widely used for satellite transmissions intended for individual reception.

The definition of the BSS in RR 37 covers the link from the space station to Earth but not the link which delivers the programme signals to the space station for re-transmission. This upward link is called a 'feeder link' which RR 109, as revised at WARC-ORB-88[1], defines as:

> 'a radio link from an earth station at a given location to a space station, or vice versa, conveying information for a space

radiocommunication service other than for the fixed-satellite service. The given location may be at a specified fixed point, or at any fixed point within specified areas.'

Feeder links for broadcasting satellites are not permitted to operate in frequency bands allocated for the BSS. Instead, RR 22 permits frequencies in frequency bands allocated for the FSS for Earth-to-space links to be assigned to feeder links, so these links are, in that respect at least, to be regarded as a part of the fixed-satellite service. However, in most other respects they are treated as if they were part of the BSS, and it is convenient to consider them in this Chapter.

Many television programme signals intended ultimately for broadcasting are transmitted to terrestrial broadcasting stations and the head-ends of cabled distribution systems through networks of the FSS, with both up-links and down-links operating in frequency bands allocated for the FSS. In recent years, large numbers of receive-only earth stations have been set up by members of the public to intercept these signals for domestic enjoyment. In some countries such stations are licensed by the administration. These receiving installations are similar to the stations used for domestic reception of BSS signals, although the receiving antenna is usually somewhat larger and the transmission margins are likely to be small. Such facilities are often regarded as satellite broadcasting. However, from the standpoint of international radio regulation they are pseudo-DBS, part of the FSS, escaping from the tight planning constraints of the BSS but becoming involved instead in the technical constraints and sharing problems of the FSS. Interference from stations of terrestrial services, in particular the FS, may be significant in countries where there is a terrestrial allocation.

The BSS frequency allocations

There are the following frequency allocations for the BSS in the international Table of Frequency Allocations (RR Article 8) as revised at WARC-ORB-85:

620–790 MHz, worldwide
2500–2690 MHz, worldwide
11·7–12·2 GHz (Region 2)
*11·7–12·2 GHz (Region 3)
*11·7–12·5 GHz (Region 1)
*12·2–12·7 GHz (Region 2)
12·5–12·75 GHz (Region 3)
22·5–23·0 GHz (Regions 2 and 3)
40.5–42.5 GHz, worldwide
84–86 GHz, worldwide

This would seem to be ample provision for a newly-emerging service. However, all of these allocations are shared with other services and the allocations are qualified by many conditions and constraints, regulatory and technical, which are reviewed on pp. 317–320 and 338–341. When these qualifications are taken into account, the following assessment of the opportunities currently open to the BSS emerges:

- The three 12 GHz allocations, one per Region, marked * above, are available for early use for satellite broadcasting with individual or community reception. Frequency assignment plans for television broadcasting in these bands were drawn up in 1977 and 1983 (see pp. 320–338).
- The allocations at 41 and 85 GHz provide for expansion of the service at some time in the future, when equipment capable of operating at these high frequencies is available at an acceptable cost. The conditions under which these bands will be shared with other services will, no doubt, be finalised when widespread initiation of operational use for the BS can be more clearly foreseen.
- The other allocations, at 700 and 2600 MHz and at 12 and 23 GHz, have been made for specific applications, perhaps in limited geographical areas. These bands are shared with other services and these other services are, in general, protected from the BSS by sharing constraints and various procedural measures.

In recent years several administrations have begun to develop plans for high definition television (HDTV) broadcasting facilities. Most HDTV systems currently in development require much more baseband bandwidth than the 625- and 525-line systems for which frequency assignments in the VHF and UHF terrestrial broadcasting allocations have been planned (see pp. 315–316). Attention has therefore been concentrated on HDTV broadcasting by satellite.

The 12 GHz BSS frequency assignment plans also have frequency channels optimised for 625- and 525-line signals, carriers being frequency modulated by the television signal. Some proposed HDTV systems have been devised for transmission over one such channel or two in parallel, but these systems have not found general acceptance. Other systems would need considerably wider channels or much higher space station transmitter power for reception using small, low-cost antennas. The 12 GHz plan assignments could not be used economically for these latter systems. Thus, the constraints on and limitations of the other BSS allocations are a major impediment to the introduction of HDTV. Resolution 521[2] of WARC-ORB-88 calls for action at an early WARC for the provision of appropriate BSS allocations for this purpose and for further study of the technical issues arising; the WARC which PC-89 planned for 1992 may provide an opportunity for such action.

Frequency allocations for sound broadcasting by satellite

All of the frequency allocations listed above were made in the expectation that they would be used for broadcasting to receivers with fixed antennas of substantial gain, although this expectation is not made explicit in the RR. Assignments planned for the 12 GHz band, although optimised for television, may be used for sound broadcasting if the assignee administration so chooses, provided that interference to other assignments is not increased thereby, but this latter requirement ensures that the down-link EIRP will not be high enough to allow fixed receiving antennas of substantial aperture to be dispensed with. No frequency allocation made so far for the BSS could be used with present-day technology for broadcasting sound programmes by satellite to personal portable receivers or car radios. There is growing interest in establishing such facilities in some countries.

Consideration has been given to the technical feasibility of using frequencies at the upper end of the HF part of the spectrum, around 25 MHz, for satellite sound broadcasting; see CCIR Report 955[3]. These studies have not reached final conclusions yet but it is clear that very powerful satellite transmitters and extremely large transmitting antennas would be required and good audio frequency quality seems unlikely to be achievable. This option is not being pursued at the present time within the ITU.

Other studies, also reported on in Report 955, have shown that high quality satellite sound broadcasting would be feasible using analogue-FM or digital signals at UHF, preferably between about 500 and 2000 MHz. Frequencies below 500 MHz are likely to be unsatisfactory because of high man-made noise levels on Earth and because space station transmitting antennas of practicable size would not give sufficient gain. Frequencies above 2000 MHz may be unsatisfactory because the antenna of a hand-portable receiver or car radio capable of operating at such frequencies would have a very small effective aperture, increasing considerably the space station transmitter power that is needed, and consequently increasing both the potential interference problems and the cost. Also, with increasing frequency, obstacles in the transmission path cause more signal attenuation. Even within the preferred frequency band, the PFD required for broadcasting would be very high, making sharing with terrestrial services very difficult; see CCIR Report 941[4].

WARC-79, WARC-ORB-85 and WARC-ORB-88 have each in turn discussed the feasibility of allocating a frequency band for satellite sound broadcasting. So far, no agreeable allocation has been found. The UHF frequency range is intensively used in many countries already, and current growth in the demand for land mobile facilities, for example, is placing additional demands upon it. The latest statement of proposals for action is to be found in Resolution 520[5] of WARC-ORB-88, which extended the range of search to 500–3000 MHz, called for the making of an allocation to be considered again at another WARC in the near future and invited the CCIR to study in particular the possible solutions to the interference problems that such an allocation is likely to create. Further action may be taken at the WARC planned for 1992.

Feeder link frequency allocations

BSS feeder links can be assigned frequencies in any FSS up-link band, subject to co-ordination of the assigned frequency with networks of the FSS operating in the band. Exceptionally it may be noted that RR 858[1] excludes the feeder links of European countries from the band 14·0–14·5 GHz. However, the use for BSS feeder links of frequency bands which have been allocated for use for networks of the FSS raises regulatory and technical problems for both satellite broadcasting and the FSS.

The regulatory problem arises from the decision, taken at WARC-71 (see p. 313), that the frequency bands allocated for the BSS should, in general, be managed by frequency assignment planning. The plans that have been agreed so far occupy virtually the whole of the band affected. If such down-link planning is not to be frustrated, the associated feeder links must be given a privileged regulatory status relative to the up-links of FSS networks operating in the same frequency band. Thus, as the use of satellite broadcasting increases, new networks of the FSS would be denied access to these bands and existing FSS networks would be required to cease operation.

The basic technical problem arises from differences between the design and performance objectives of FSS networks and BSS systems. The BSS frequency assignment plans drawn up for the 12 GHz band in 1977 and 1983 have been optimised to permit the delivery of as many programmes per country as possible, with high quality signals, to receiving installations of low cost with small antennas. As part of this optimisation, the feeder link parameters have been chosen so that very little thermal noise is fed forward from the feeder link into the broadcasting down-link and one of the planning objectives is to ensure that very little interference enters the space station receiver. It may be assumed that these principles will be applied to other BSS systems also.

Systems of the FSS are designed for a more equal division of the noise and interference budgets between the up-link and the down-link. Consequently, if BSS feeder links are compared with up-links of the FSS, it is generally found that the feeder links are transmitted with a higher spectral EIRP density than the up-links to FSS satellites, and they require interfering signals to be kept at lower levels. If a frequency band is used for both BSS feeder links and FSS up-links, these differences in parameters and degradation budgets lead to a need for wider spacings between the satellites than would be necessary if the band were used for either purpose alone. Thus the geostationary satellite orbit would be used inefficiently; see CCIR Report 561[6].

To limit the impact of BSS feeders on FSS systems, and to allow for feeder link frequency assignment planning, a number of frequency bands have been allocated to the FSS (Earth-to-space) specifically for use for BSS feeder links.

- The most immediately important of these are:

  10·7–11·7 GHz (Region 1 only; see RR 835)
  *14·5–14·8 GHz (worldwide, but excluding Europe; see RR 863[1])
  *17·3–18·1 GHz (worldwide; see RR 869)

  Frequency assignment plans for feeder links, corresponding to the BSS down-link plans at 12 GHz, have been agreed for the bands marked * above (see pp. 320–341).
- Also, the band 27·0–27·5 GHz has been allocated for the FSS (Earth-to-space) in Regions 2 and 3 and there are BSS allocations in those two Regions at 22·5–23·0 GHz, although the intention that this FSS allocation should be used for feeder links is not made explicit in the RR.
- There is an imbalance in the bandwidths of the Earth-to-space and space-to-Earth allocations for the FSS between 37·5 and 51·4 GHz and RR 901 urges that the surplus Earth-to-space bandwidth, identified as 47·2–49·2 GHz, be reserved for feeder links to broadcasting satellites operating at 40·5–42·5 GHz.
- Similarly there is more bandwidth allocated to the FSS for up-links than for down-links between 71 and 84 GHz and it may be supposed that the surplus was intended for use for feeder links to broadcasting satellites operating at 84–86 GHz.

All of these feeder link frequency allocations are shared with other services. These sharing situations and the means that have been adopted for limiting interference between services in these bands are reviewed on pp. 316–320 and 338–341.

International spectrum management

Spectrum management techniques like frequency co-ordination can be technically efficient and are operationally feasible in multifarious services such as the FSS, but for full effectiveness they require flexibility in satellite location, frequency plan, emission parameters and so on. For the BSS, and in particular where 'individual reception' is intended, the convenience of the public requires that such system parameters be changed very seldom, if ever. Furthermore, satellite broadcasting has few end-products and its basic technical characteristics can be expected to change only very slowly with time. Thus, some of the potential advantages of frequency co-ordination could not be realised in the BSS, whereas frequency assignment planning, despite potential disadvantages, meets well some of the basic requirements of the service.

Frequency assignment planning has other characteristics, when used for satellite broadcasting, which are attractive to many administrations. Thus:

- with planning, administrations can be sure that the early implementation of the service by some countries will not exhaust the transmission medium, denying other countries an opportunity to set up systems later.

- planning may facilitate frequency allocation sharing, gaps between the frequency channels planned for the reception of satellite broadcasting in a given geographical area being potentially available on a long term basis for assignment to stations of services with sharing allocations.

Furthermore, factors which are not directly related to radio regulation may not be disregarded. Satellite broadcasting of television is an effective medium for commercial advertising and for influencing public opinion. Many governments wish to ensure that neither the influence nor the profit arising from satellite broadcasting addressed to their citizens should pass avoidably into foreign hands.

These last two concerns gave rise to discussions extending over a number of years in the United Nations and in particular in its Committee on the Peaceful Uses of Outer Space. The discussions led in 1982 to Resolution 37/82[7] of the UN General Assembly which, amongst other provisions, requires a state which intends to authorise international satellite broadcasting to first secure the agreement of the other states affected and requires unavoidable overspill of satellite signals to be minimised.

WARC-71, which made a number of frequency allocations for the BSS, also resolved that BSS stations should, in general, operate in accordance with frequency plans which had been internationally agreed. The substance of the resolution is to be found, little changed in substance, in RR Resolution 507[8]. The 12 GHz frequency assignment plans drawn up at administrative radio conferences in 1977, 1983 and 1988 relate, with almost no exceptions, to national service areas and the required rate of roll-off of satellite antenna gain at the edge of the nominal coverage area is, in most circumstances, probably as rapid as is currently feasible, minimising overspill onto the territory of other countries. However, in these planned bands the BSS is the dominant service and stations of sharing services are severely constrained. The procedures used in the international management of assignments in these bands are reviewed on pp. 337–341.

A second resolution of WARC-71 provided procedures for the co-ordination of BSS frequency assignments with those to other stations operating in frequency bands, allocated for the BSS, for which no frequency plan had been drawn up. This resolution is to be found, little changed, as RR Resolution 33[9]. According to this resolution, BSS frequency assignments would be harmonised with assignments to terrestrial stations in a way that is similar to consultation under RR Article 14, while co-ordination with other space networks would be based on the corresponding procedure for networks of the FSS, which are set out in RR Article 11. Thus, in the absence of an agreed frequency assignment plan, BSS systems have no regulatory advantage over systems of other services in shared frequency bands. Indeed, these procedures and the sharing constraints which accompany them in some frequency bands place the BSS at a disadvantage relative to sharing services.

International technical constraints

Many technical constraints are applied to satellite systems broadcasting at 12 GHz in accordance with the 1977 and 1983 frequency assignment plans and the associated feeder link plans (see pp. 325–336). The sharing constraints applicable in certain other bands are noted on pp. 317–320. The following additional requirements and constraints apply to broadcasting satellites in general:

Cessation of emissions: RR 2612 requires all space stations to be fitted with telecommand devices to ensure that their radio emissions can be switched off without delay if this should become necessary.

Satellite station-keeping and antenna pointing accuracy: The standards of station-keeping and antenna beam pointing accuracy required of geostationary satellites of the FSS, set out in RR 2615–RR 2627 and RR 2628–RR 2630 and reviewed on pp. 274–275 and 279, apply also to geostationary broadcasting satellites in the absence of more stringent requirements imposed by frequency assignment plans.

Carrier frequency tolerance: RR 303 and RR Appendix 7 apply the following tolerance to the carrier frequencies emitted by the transmitters at BSS space stations and the associated feeder link earth stations:

20 parts per million (p.p.m.) for carrier frequencies below 2450 MHz
50 p.p.m. for carrier frequencies between 2450 MHz and 10·5 GHz
100 p.p.m. for carrier frequencies between 10·5 and 40 GHz

No tolerance has been set for carrier frequencies above 40 GHz.

Spurious emissions: The BSS uses space station transmitters which are powerful, in comparison with those of other space services, and there is particular concern that their spurious emissions may cause interference to the earth station receivers of other space radio services operating at lower power levels and above all to radio astronomy stations.

RR 304 and RR Appendix 8 require broadcasting satellites and feeder link earth stations operating below 960 MHz to limit the power of any spurious component of their emission to 20 mW, or −60 dB relative to the mean power of the signal, whichever is less. Lower limits apply if the transmitter output power is less than 25 W. Note 11 to RR Appendix 8 draws attention to the possibility that these limits may not give adequate protection to stations of other services and indicates that more stringent limits may need to be applied in individual cases, depending on the positions of the stations involved. However, few BSS systems will operate at such low frequencies.

As with other space radio services, the Radio Regulations impose no specific limits on spurious emissions for BSS stations operating above 960 MHz, although Note 13 to RR Appendix 8 calls for spurious emission levels 'to be reduced to the lowest possible values compatible with the technical and economic constraints to which the equipment is subject'. However, RR Recommendtion 66[10] calls on the CCIR to develop recommendations for maximum permitted levels of spurious emission from stations of space services. RR Recommendation 507[11] emphasises the need for CCIR studies in this area in the specific context of interference to radio astronomy in the band 23·6–24·0 GHz, arising from the radiation of the second harmonic of BSS space station emissions in the 12 GHz band. RR Recommendation 506[12] asks for limits to be placed on second harmonic radiation when specifying 12 GHz broadcasting satellites. CCIR Report 807[13] indicates that good progress is being made in basic studies of spurious emissions from BSS space station transmitters and this may lead soon to the formulation of recommended limits. See also CCIR Reports 712[14], 789[15] and 980[16].

National spectrum management

It is likely that the management of BSS allocations will be dominated by the procedures associated with the internationally agreed frequency assignment plans, leaving little scope for independent national management action.

Satellite broadcasting systems in prospect

The implementation of satellite broadcasting for individual reception, using antennas substantially less than one metre in diameter, is only just beginning, but studies and plans for the new service are well advanced. There are general discussions of satellite broadcasting system principles in CCIR Reports 215[17] and 632[18].

Interest in satellite broadcasting for early implementation has been concentrated almost entirely on analogue colour television with good signal quality and picture standards that are basically the same as those used in terrestrial broadcasting. Geostationary satellites would be used. To obtain an adequate post-demodulator signal-to-noise ratio with an acceptably low space station transmitter output power requirement, frequency modulation of the main carrier is used.

The colour television systems, PAL, NTSC and SECAM, which are used with VSB/AM for terrestrial broadcasting at VHF and UHF, could be used with FM for satellite broadcasting. This would have the great advantage of a high degree of compatibility with terrestrial broadcasting. However, when these established systems were developed, compatibility with monochrome systems and minimum bandwidth using amplitude modulation were prime requirements. In these systems there is spectrum overlap between the luminance and chrominance signals. Imperfections in the filters at the receiver which separate the luminance and chrominance signals lead to degradation of picture quality.

Also, demands have emerged in recent years for new facilities which these systems cannot readily be modified to supply, such as enhanced sound channel performance, multiple sound channels and substantial videotext capacity. Furthermore, when transmitted using an FM main carrier, signals with high level sub-carriers for the chrominance and sound channels, located near the top of the baseband, do not provide adequate post-demodulator sub-carrier signal-to-noise ratios without undesirably high transmitted carrier power.

Variants of the terrestrial systems have been developed for use in satellite broadcasting; see CCIR Reports 1073[19] and 1074[20]. A variant of the M/NTSC system, in which the analogue sound sub-carrier is replaced by a digitally modulated sub-carrier carrying sound and data signals in time division multiplex, has been taken into use in one country. Most countries that have declared their intentions, however, have chosen to implement one of the Multiplexed Analogue Components (MAC) family of systems. MAC sysems have been developed for both 525- and 625-line picture standards. All of these systems use the principle of time division multiplex to combine in sequence time-compressed analogue signals carrying luminance and colour-difference information and bursts of digital signals carrying sound channels and data signals. The various MAC systems differ from one another mainly in the nature of the digital bursts and the facilities which the digital bursts can transmit.

In the course of the 1990s it is expected that HDTV will be introduced, using satellite broadcasting. Basic HDTV system principles are discussed in CCIR Report 801[21]. At the present time a major effort is being made in the CCIR, seeking to ensure that the basic picture standards adopted for HDTV are uniform worldwide. Various different ways of implementing these principles into systems for satellite transmission are discussed in CCIR Report 1075[22].

Up to 20 channels of sound broadcasting, typically frequency division multiplexed onto a single carrier, can be transmitted by satellite in place of a television signal for reception by means of a fixed domestic antenna of moderate aperture; see CCIR Report 215[17]. Single channel and multi-channel systems, both analogue FM and digital PSK, have also been studied for implementation in satellite sound broadcasting systems for reception on vehicle-borne and person-portable radio receivers with antennas of negligible gain, in the expectation that a suitable frequency band will be allocated for this purpose; see CCIR Report 955[3], and also CCIR Report 954[23].

9.2 The BS and frequency allocation sharing

9.2.1 Sharing in planned bands

Frequency assignment plans have been agreed for the BSS and feeder links in the following bands:

Region 1: 11·7–12·5 GHz (BSS) and 14·5–14·8 GHz and 17·3–18·1 GHz (feeder links)
Region 2: 12·2–12·7 GHz (BSS) and 17·3–17·8 GHz (feeder links)
Region 3: 11·7–12·2 GHz (BSS) and 14·5–14·8 GHz and 17·3–18·1 GHz (feeder links)

All of these bands are also allocated to other services. Sharing between these various services presents technical problems, but these problems have been dealt with by regulatory provisions which have been incorporated into the BSS frequency assignment plans. It is convenient to consider these technical aspects (see pp. 338–341) in the context of a review of the provisions of the plans.

9.2.2 Sharing in unplanned bands

The regulatory basis for the BSS and its feeder links in frequency bands that have not been made the subject of frequency assignment plans is set out in RR resolution 33[9]. In general, sharing with terrestrial services involves consultation between administrations similar to the provisions of RR Article 14 and sharing with space services is similar to that required for networks of the FSS (see pp. 58–60 and 281–294 respectively).

However, the sharing situations in the various bands differ and specific constraints qualify most of the unplanned allocations. Taking in order of frequency the BSS allocations and the potential feeder link bands which may reasonably be associated with them, the situation is as follows:

620–790 MHz

The footnote allocation made to the BSS by RR 693 in the band 620–790 MHz is worldwide. The primary allocation for this frequency band in the framed part of the ITU Table of Frequency Allocations (RR Article 8) is to the BS in all three Regions, allocations to FS and MS being added in Region 3. Other footnotes make sub-Regional allocations to the FS and AeRN with primary or permitted status in Regions 1 and 2 and there are widespread secondary allocations to FS and MS. There is a frequency assignment plan for BS (television) in Region 1 and the band is widely and intensively used for terrestrial television broadcasting in all Regions.

In order to protect the operation of these terrestrial services from BSS emissions, RR 693 limits the BSS to frequency modulated television signals and places a constraint on the PFD of the BSS signals which an administration may allow to fall onto the territory of another country without the consent of the administrations affected. RR 693 also requires an administration proposing to assign frequencies in this band to a BSS space station to co-ordinate as necessary with other administrations, using the RR Resolution 33[9] procedure; see also RR Recommendation 705[24].

CCIR Report 631[25] contains a technical examination of the PFD constraint which is needed to protect terrestrial broadcasting from the

BSS emission, coming to a tentative conclusion that the constraint in RR 693 is more severe than would be necessary to protect terrestrial broadcasting in some countries but it may be insufficient for other countries. Much depends on the way in which receiving stations are distributed in the outer parts of the service areas of terrestrial broadcasting stations, the performance of antennas in general use at these remote receiving stations and consequently on the minimum terrestrial station field strength which is to be protected.

Report 631 also draws attention to other terrestrial services with allocations in this frequency band which would be highly susceptible to interference from the BSS, and in particular the LM and tropospheric scatter stations of the FS.

There is no specific feeder link band associated with the 700 MHz BSS allocation. The FSS (Earth-to-space) band at 6 or 14 GHz would probably be used for this purpose where the need arose, and the feeder link would be co-ordinated with FSS networks and with stations of the terrestrial services which share these bands in the same way as a conventional FSS up-link

2500–2690 MHz

The BSS shares 2500–2690 MHz with the FS and the MS (excluding the AeM) with equal primary allocations in all three Regions. RR 757 limits BSS systems using this allocation to national and regional networks for community reception. This latter limitation is given substance by the application of a sharing constraint in the form of a limit on the PFD at the Earth's surface at the same level as would apply to an FSS down-link in this part of the spectrum. Administrations proposing to assign frequencies in this band to broadcasting satellites are also required to consult and co-ordinate with other administrations using the procedures of RR Article 14 and RR Resolution 33[9].

In Regions 2 and 3 there are also FSS (space-to-Earth) allocations in this frequency band, subject to the same use and PFD limitations as the BSS and also subject to RR Article 14 consultation. In Region 2 part of this band is allocated to the FSS for both directions of transmission. It may be supposed that these BSS and FSS allocations would be used together for some specialised purpose, perhaps for educational or limited-access broadcasting with feedback.

The feeder link situation is the same as for the 700 MHz BSS band.

The constrained 12 GHz allocations

At 12 GHz two Regional allocations have been made for the BSS, additional to the bands which have been planned. These are:

Region 2; 11·7–12·2 GHz
Region 3; 12·5–12·75 GHz

The Region 2 allocation is by footnote RR 836[26], sharing with the FS and FSS, and subject to an EIRP limit and other constraints. This allocation may have been made in order to regularise the possible use for the BSS of transponders on space stations primarily intended for the FSS and the footnote indicates an expectation that the FSS will remain the principal space service using this band.

In Region 3 the BSS allocation is in the framed part of the Table of Frequency Allocations (RR Article 8), sharing with the FS, the FSS (space-to-Earth) and the MS (except AeM) with equal status. However, RR 847[26] applies a PFD constraint to the BSS and limits BSS use of this band to community reception; see also RR Resolution 34[27].

The feeder link situation is the same as for the 700 MHz BSS band.

22·5–23·0 GHz

This band has been allocated to the BSS in Regions 2 and 3, sharing with the FS and MS and (except for 22·50–22·55 GHz) the ISS. Use of the BSS allocation is subject to the RR Article 14 consultation procedure (see RR 877) as well as those of RR Resolution 33[9]. RR 879 draws attention to radio astronomy observations which are made in the band 22·81–22·86 GHz and urges that consideration should be given to the protection of these observations, in particular when assigning frequencies for transmission from space stations.

No specific sharing constraints have been adopted for this band. Studies reported in CCIR Reports 874[28] and 951[29] indicate that it would be feasible for the BSS and ISS to share it, the ISS allocation being used as it might be used for inter-satellite links in FSS networks, with co-ordination of satellite locations, given adequate orbital separations. However, such sharing would become impracticable if intensive use were made of the band for the BSS. The CCIR studies also show that interference from BSS signals to certain kinds of FS and MS system might be excessive if the BSS signal were strong enough for individual reception.

The use of inter-satellite links has scarcely begun and the development in the ITU of arrangements for their spectrum management is in a correspondingly early stage. Fortunately the procedures contained in RR Articles 11 and 13 are appropriate for both BSS and ISS in this band, providing a forum for co-ordination. The technical support available for the co-ordination process is also quite limited. However, section 2.2.2 of RR Appendix 29 is relevant, see also CCIR Report 451[30].

It may be assumed that the band 27·0–27·5 GHz was allocated to the FSS (Earth-to-space) in Regions 2 and 3 for use for feeder links to satellites broadcasting at 22·5–23·0 GHz, although this intention is not recorded in the RR. This band is shared with the FS and MS with equal primary status. The arrangements for the co-ordination and registration of frequency assignments for the FSS would be appropriate for feeder links in this band also.

40·5–42·5 GHz and 84–86 GHz

There is a primary allocation for BSS in the 41 GHz band; a sharing allocation for BS has permitted status and there are secondary allocations to FS and MS. In the 85 GHz band the BSS, FS, MS and BS all have primary allocations but RR 913 require stations of the FS, MS and BS to not cause harmful interference to BSS stations operating in accordance with an agreed BSS frequency assignment plan. All of these allocations are worldwide.

RR 901 prepares the way for the FSS (Earth-to-space) allocation at 47·2–49·2 GHz to be used for feeder links to satellites broadcasting at 41 GHz; this allocation is shared with FS and MS. There is no corresponding footnote, preparing the way for feeder link assignments for the 85 GHz BSS band, although this might be one use for the FSS (Earth-to-space) allocation at 74–75·5 GHz.

Thus, the way is clear for the BSS frequency assignment planning process to be initiated in both of these BSS bands when required. Some amendments to RR Article 8 to prepare the way for feeder link planning would be needed. However, propagation conditions between space and Earth are adverse at these frequencies (see Appendices B.4.3, B.4.4 and B.6.10) and it may be some considerable time before the large-scale use of these bands for satellite broadcasting begins.

9.3. The 12 GHz BSS frequency assignment plans

9.3.1 Introduction

A resolution of WARC-71 asked for an ARC to be convened to draw up frequency assignment plans for the BSS, together with the necessary regulatory provisions. Various footnotes were added by WARC-71 to the international Table of Frequency Allocations (RR Article 8) to protect the operation of the BSS in some of its allocated bands from interference from stations of any other service, even where the latter had priority of registration in the MIFR and its service had primary status.

A resolution of the Plenipotentiary Conference held in 1973 identifed frequency bands at 12 GHz for such planning, namely 11·7–12.2 GHz for Regions 2 and 3 and 11·7–12·5 GHz for Region 1.

A WARC met in 1977 to prepare the plans. The decision at WARC-71 to manage the use of spectrum and orbit by the BSS by means of frequency assignment planning had not been unanimous and the argument was reopened when WARC-BS-77 met. It was generally believed that substantial implementation of satellite broadcasting was not imminent. Some administrations argued that, since the technology of satellite broadcasting was also immature and the optimum parameters for satellite broadcasting systems could not therefore be determined at that time, planning should be postponed. There was considerable support for delaying frequency assignment planning amongst Region 2 administrations.

Other administrations at WARC-71, however, supported planning for the BSS because the 12 GHz BSS allocations were shared with terrestrial services, and it was foreseen that constructive use could be made of the BSS bands for these sharing services, using sub-bands which would not be needed at the same location for BSS reception, once BSS planning had been completed. A third view was that immediate planning was desirable because planning was going to enable all countries to obtain access to an equitable share of the medium for BSS, even though some countries might not be ready to implement satellite broadcasting systems for many years.

The outcome was that the administrative basis for the BSS under a planned regime was drawn up at WARC-BS-77 and frequency assignment plans were prepared for Regions 1 and 3; see the Final Acts of the conference[31]. The drawing up of a plan for Region 2 was postponed to a later conference, foreseen to take place in 1982; see Resolution Sat-9 of WARC-BS-77[32].

One important decision taken at WARC-BS-77 was that space stations, regardless of their service, operating in frequency bands that had been planned for the BSS, should be geostationary. This text was brought up to date in minor details at WARC-79 and WARC-ORB-88; see RR Resolution 506[33].

WARC-BS-77 foresaw that it would be necessary to prepare and agree a frequency assignment plan for BSS feeder links in a suitable FSS Earth-to-space frequency allocation. However, it was noted that there was already an imbalance around 12 GHz in the frequency allocations to the FSS for Earth-to-space and space-to-Earth links, there being more bandwidth allocated for down-links than for up-links. It was recognised that the difficulties for the FSS arising from this imbalance would be made worse if frequency assignments for BSS feeder links were planned in one of its existing up-link bands. Accordingly, Recommendation Sat-1[34] of the Conference asked for the spectrum requirements for feeder links to be studied by administrations and the CCIR and for the outcome of these studies to be drawn to the attention of WARC-79.

At WARC-79 the WARC-BS-77 plan for Regions 1 and 3 and the associated regulatory texts were incorporated into the RR as Article 15 and Appendix 30. Various other minor changes were made to the RR to give effect to the new procedures. RR Resolution 701[35] prepared the way for a Region 2 planning conference, to be held not later than 1983. WARC-79 also made changes to the Region 2 frequency allocations around 12 GHz, in consequence of which the band intended for the BSS plan became 12·3–12·7 GHz. The band, 12·1–12·3 GHz was provisionally allocated to both the BSS and the FSS (space-to-Earth) in Region 2, it becoming a responsibility of the Region 2 planning conference to determine how that band should be divided between the two services in the future.

WARC-79 responded to the feeder link allocation problem by making new allocations to the FSS (Earth-to-space) at 18 GHz, 14·5 GHz and 11 GHz, the last-mentioned being allocated in Region 1 only. RR Resolution 101[36] determined that these bands should be used for planned assignments for feeder links to 12 GHz broadcasting satellites serving Regions 1 and 3 and provided an interim procedure for co-ordinating feeder link assignments pending the agreement of such plans. The preparation of a feeder link plan for Region 2 in the band 17·3–18·1 GHz was included by RR Resolution 701 amongst the tasks of the prospective Region 2 conference.

The Regional Administrative Conference for the planning of the Broadcasting-Satellite Service in Region 2 (RARC-SAT-83) met in 1983. It decided to divide the 12·1–12·3 GHz band into two equal parts, the lower half to be allocated for the FSS and the upper half to be absorbed into the BSS plan. Frequency assignment plans for Region 2 were drawn up, the BSS assignments being at 12·2–12·7 GHz and those for the feeder links being at 17·3–17·8 GHz. The Conference drafted management procedures required to support the feeder link plan and prepared a set of management procedures for the BSS (down-link) plan which differed in detail from the set that had been agreed at WARC-BS-77. These plans and the management texts form the Final Acts of the Conference[37].

WARC-ORB-85 took note of the decision of RARC-SAT-83 on the re-allocation of the band 12·1–12·3 GHz and approved this change, and various consequential changes, to the international Table of Frequency Allocations (RR Article 8). RR Appendix 30 was amended to harmonise the management procedures proposed for BSS assignments by the two planning conferences and the Region 2 plan was incorporated into the appendix. The Region 2 feeder link frequency assignment plan and the associated management texts became RR Appendix 30A, linked into the main body of the regulations by a new RR Article 15A. All of these new RR texts are to be found in the Final Acts of the Conference[26]. The Conference also decided that a feeder link frequency assignment plan should be prepared for Regions 1 and 3 in the bands 17·3–18·1 GHz and 14·5–14·8 GHz at a subsequent WARC (namely WARC-ORB-88); see Chapter 6 of the Report of WARC-ORB-85[38].

WARC-ORB-88 produced and agreed a feeder link frequency assignment plan for Regions 1 and 3 and amended RR Article 15A and RR Appendix 30A to accommodate the plans and management texts for the feeder links for all three Regions; see the Final Acts of the Conference[1].

The principal technical provisions of these various plans are reviewed on pp. 323–336. The regulations for managing BSS and feeder link assignments in the planned bands are reviewed on pp. 337–338. The regulatory provisions which control interference between systems of the BSS and those of other services sharing these frequency bands are reviewed on pp. 338–341. Provision has also been made for 'interim systems' in Region 2, outside the Region 2 plan but harmonised with it (see p. 341).

The various provisions for BSS planning have entered into force at various times since 1979, the most recent on 16 March 1990. Interim arrangements were made for any systems which might begin operation before all of the necessary permanent regulations were in force, but there have been few such systems. The plans and the associated regulations were prepared to meet requirements up to January 1994 or later and they remain in force until revised by a competent ARC; see Article 14 of RR Appendix 30[26] and Article 11 of RR Appendix 30A[26 and 1].

Radio links are needed to and from broadcasting satellites, additional to the feeder links and the broadcasting down-links, to carry signals for telemetry, tracking and command (TTC) functions. For use during the launch phase, these links will usually be assigned frequencies in bands allocated to the SpO. However, the need for these functions continues throughout the working life of the satellite, and additional facilities may be needed, such as the reception at the satellite of a beacon signal, continuously transmitted at a fixed point on Earth, to assist in the maintenance of precise satellite antenna beam pointing. The frequency allocations made to the SpO are narrow, heavily loaded and their allocation status is often precarious. For these reasons they are unsuitable for continuous use for operational satellites. Accordingly, RR 25 indicates that carriers for TTC functions, for use during normal operation, should be accommodated within the frequency bands allocated to the service to which the satellite belongs. The technical requirements for TTC in the BSS under conditions of normal operation are discussed in CCIR Reports 633[39] and 1076[40].

No specific provision was made for TTC facilities in the frequency assignment plans that were drawn up in 1977 and 1983. However, Report 1076 foresees the possibility that frequencies for TTC down-links might be assigned in the guard bands near the upper and lower limits of the 12 GHz BSS bands, from which the powerful BSS assignments have been excluded in order to limit interference to FSS networks operating in adjacent bands. There are similar unassigned sub-bands at the edges of the feeder link bands and these might be used for TTC up-link assignments.

9.3.2 The technical provisions of the plans

A planned BSS assignment in one of the 12 GHz plans takes the form of a nominated carrier frequency, a satellite orbital location and a set of equipment and emission parameters and constraints. Some of the latter are mandatory limits required to protect other assignments from excessive interference and others indicate what must be achieved if the wanted channel is to meet the common performance objectives. If the administration for which an assignment had been planned makes that assignment to a space station transmitter, all of the various constraints being observed and all the assumed parameters being achieved, then the down-link noise performance objectives should also be achieved at a receiving station that also meets the performance assumed in planning. Even if all of the planned assignments were simultaneously active, the down-link noise-plus-inteference level objective should also be achieved if all of these other systems were also achieving the performance objectives on which the plan is based.

The specific parameters and constraints associated with a planned feeder link assignment are analogous to those of a planned BSS (down-link) assignment. Finally, when planned feeder link and BSS assignments are assigned and implemented in a satellite broadcasting system, the end-to-end performance should attain the end-to-end performance objectives even if all of the assignments in the feeder link and BSS plans have also been implemented and are operating.

There are administrative procedures which allow the plan to be amended, to meet the changed needs that an assignee country may have at the time when implementation is in prospect, subject to the agreement of all of the administrations with interests in the plan that would be affected by the amendment.

Finally, if the planned assignment is brought into use with parameters which differ from those set out in the plan but only in ways that would cause no increase in the interference caused to another planned assignment, the terms of the agreements permit the assignee administration to authorise the system to operate in this mode, although, of course, the performance objectives may not be achieved.

The planning objectives

At WARC-BS-77, at the start of the planning process for Regions 1 and 3, administrations were invited to state how many simultaneous programmes their countries would require to transmit, and with what geographical coverage. Requirements were rather indeterminate initially, but after some exploration of what was technically feasible for Europe and Africa, where the pressure on spectrum and orbital resources would be greatest, most administrations asked for the option of transmitting as many programmes as was feasible, with national coverage, the same number of programmes being available for every country. Only a few groups of countries with close affinities asked for assignments for one or more programmes serving their countries jointly. Other factors to be taken into account included the need for separate programme coverage for the different time zones of very big countries, and the sub-division of beams serving very large areas, in particular where rainfall is sometimes high, in order to avoid a need for space station transmitters of impracticably high power output.

A more flexible approach to the number of channels assigned per country became feasible after 1977 and was adopted for Region 2.

An important planning objective was to ensure that the television picture standards already in use for terrestrial broadcasting in each country would be available for satellite broadcasting also. This was, in fact, not difficult to provide, since the only significant factor was the picture line rate. In Regions 1 and 3 almost every country uses a 625-line/25-frames-per-second standard, in Region 2 almost every country uses a 525-line/30-frames-per-second standard and in each Region it was found feasible to base the plan on the majority requirement and to make adjustments, if necessary, to meet a minority requirement.

The plans were designed for individual reception. A high standard of overall circuit performance was aimed at corresponding approximately to a 45 dB video signal-to-weighted-noise ratio for most individual receiving stations in the service area for more than 99% of the worst month in any year. Such a signal would, of course give an even better quality of reception at a community receiving station with enhanced antenna performance. Some rather limited consideration was given, in the plans prepared for Region 1 and 3, to low-power systems intended only for community reception.

Factors which would make the domestic receiving installation more costly or more difficult to use effectively were to be avoided. Thus, where feasible, all programmes for one service area were to be transmitted from a single orbital location and with the same sense of polarisation. Satellite orbital locations were to be chosen so that the angle of elevation of the satellite would be high, as seen from the area which it would serve, to minimise the effect on reception of obstructions in the vicinity of the receiving antenna. In Region 1, where the allocated bandwidth was greater than elsewhere, the design of receivers and receiving antennas was made easier by concentrating all of the assignments planned for any one service area into the upper half or the lower half of the allocated band.

Finally, it was recognised that the cost of broadcasting satellites and of launching them would be high, primarily because of the high power output required from the space station transmitter to enable the cost of domestic receiving installations to be kept low, leading to high satellite mass. Planning principles which tend to reduce the required mass of a satellite were to be favoured.

The plan configurations and constraints

The various means that can be used to achieve planning objectives such as those set out above are examined in detail in CCIR Reports 215[17], 633[39], 811[41], 812[42], 814[43] and 952[44]. The means and the criteria actually used in producing the 1977 and 1983 Plans, and the results of the planning, are set out in detail in RR Appendices 30[26] and 30A[26 and 1], and in particular in Annex 5 and Annex 3 of these appendices, respectively. The main elements are summarised below. This summary considers only individual reception:

(a) General measures

To maximise the number of assignments that could be planned for each country, the technical constraints on systems were made almost as stringent as the technological means in prospect would permit. For the same reason, systems were required to be as homogeneous as possible, differing only in geographical respects. The parameters and signal levels in each channel were required to be virtually the same as those of every other channel. A mechanism was, however, provided in the Articles 4 of RR Appendices 30[26] and 30A[26 and 1] whereby the use of planned assignments with non-standard parameters can be given the status of a planned assignment if any administration that would be adversely affected by a departure from the standard parameters agreed to it.

The space station transmitter power required was reduced, to the extent possible, by basing the plan on the assumption that frequency modulation with optimum deviation would be used for main carriers and by allocating most of the total budget of noise and interference to the down link, the feeder link performance objectives being pitched high.

One other substantial opportunity for a system designer to reduce spacecraft mass lies in providing a satellite battery with no more than enough capacity for control purposes during solar eclipse. If this option were implemented, programmes would be interrupted during an eclipse. Accordingly, most orbital locations were selected so that the satellite would not enter the Earth's shadow until after local midnight in the service area, when the loss of programme transmission might not be of great importance.

Finally, the plans were drawn up with the assumption that the spectral energy density of a BSS emission, averaged over any 4 kHz band, would nowhere exceed a level 22 dB below the power level of the emission as a whole; see section 3.18 of Annex 5 of RR Appendix 30[26]. Some of this dispersal of carrier energy would be provided by programme signals but there are various means for artificially increasing the dispersal and for maintaining dispersal during breaks in programmes; see, for example CCIR Report 384[45]. Through limiting the peaks of spectral energy in this way, the interference from BSS emissions of many kinds of systems of services with sharing allocations is substantially reduced. Article 4 of RR Appendix 30[26] requires an intention not to implement this measure operationally to be treated as a departure from the planned parameters.

(b) Satellite antenna beam

The transmitting antenna is a key element in the design of a broadcasting satellite, determining the shape and gain of the transmitting beam. The technology of satellite antenna design is in rapid development at present, offering a variety of options. Thus it was not possible to know at the frequency assignment planning stage what the antenna characteristics will be when systems start operating. Nevertheless, the radiation properties of the transmitting antenna are also a key element in developing an efficient frequency assignment plan, in particular because they determine the level of down-link interference outside the service area. It was therefore necessary for planning purposes to define a hypothetical antenna beam for each planned frequency assignment, to provide a basis for constraints on the radiation properties of the antenna which will ultimately be used. There is a corresponding requirement to constrain the characteristics of the satellite antenna which receives the feeder links.

The constraints on the down-link beam involve three elements, namely the service area on the Earth, the area which the hypothetical antenna beam would illuminate at the standard PFD level (the coverage area) and a reference antenna radiation pattern constraining antenna gain outside the minus 3 dB gain contour of the hypothetical antenna beam. In planning it has been assumed that the space station transmitting antenna will have the smallest feasible circular or elliptical beam which is capable of covering the whole of the required service area from the planned satellite orbital location. The out-of-beam reference radiation pattern is considered in (*d*) below.

The term 'service area' is defined in Annex 5 of RR Appendix 30[26] as:

> The area on the surface of the Earth in which the administration responsible for the service has the right to demand that the agreed protection conditions be provided.

In the context of the plans drawn up at WARC-BS-77 and RARC-SAT-83, the service area is almost always the land area under the jurisdiction of the administration for which an assignment is planned, or some defined part of that area.

'Coverage area' is defined in Annex 5 of RR Appendix 30[26] as follows:

> The area on the surface of the Earth delineated by a contour of a constant given value of power flux-density which would permit the wanted quality of reception in the absence of interference.

Fig. 9.1 illustrates how these concepts are applied to BSS planning, using two imaginary countries, a large country called Rectanguland and a small country called Trianguland. Note that Fig. 9.1 has been drawn as a map projected from the location of the satellite onto a plane surface; the shape of geographical features is distorted in this projection but the shape of the cross-section of a beam from the satellite is undistorted.

To a first approximation the contour which delineates the coverage area associated with the hypothetical elliptical beam serving Rectanguland is the line of intersection with the Earth's surface of the −3 dB gain contour of the beam. However, this approximate definition of the coverage area disregards differences in the path length and therefore the path loss between the satellite and various segments of this contour of constant gain. Departures from the −3 dB constant-gain contour to compensate for differences in path loss produce the true constant-PFD contour which delineates the coverage area. The difference between the constant-gain contour and the constant-PFD contour is exaggerated in the Figure to make the difference more evident. The PFD at the centre of this coverage area will, hypothetically, be 3 dB higher than the PFD at the edge.

Thus the hypothetical antenna beam used for planning down-link assignments to Rectanguland is the beam of elliptical cross-section which, radiating from the assigned orbital location, generates the coverage area contour which most closely circumscribes the service area, allowance being made for differences in the path loss to the various parts of the service area. This beam can be defined, for example, by the orbital location, the point where the axis of the beam cuts the Earth's surface, the angular

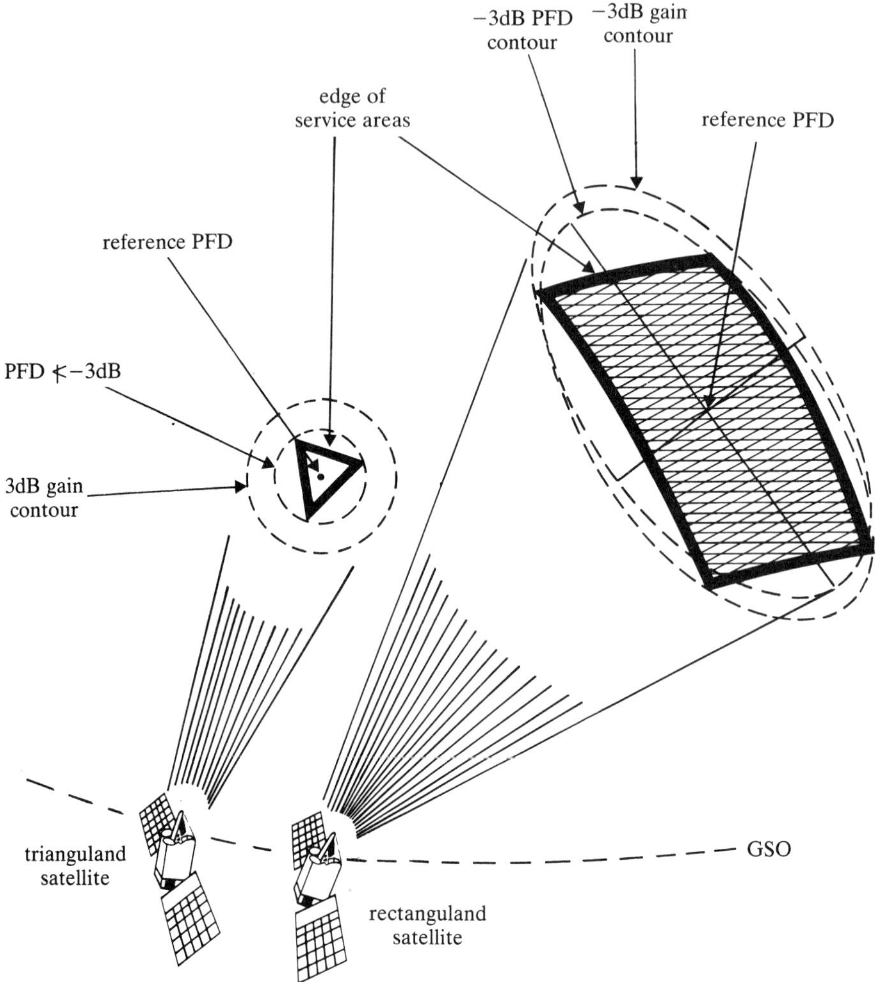

Fig. 9.1 'Rectanguland', a hypothetical big country, is shown, served by an elliptical beam which fits the service area with minimum overspill. 'Trianguland' is small and there is considerable overspill from the circular beam of minimum feasible diameter that serves it

width of the major and minor axes to the −3 dB gain contour and the orientation of the major axis as projected onto the Earth's surface.

However, Rectanguland is a large country. It is not feasible to define the hypothetical antenna beam for, for example, little Trianguland such that the contour of constant PFD, −3 dB below the maximum PFD of the beam, closely circumscribes the service area, because a satellite antenna with such a high gain could not readily be stowed within the shroud of a launch vehicle. WARC-BS-77 decided that 0·6° should be used as the minimum apex angle for the −3 dB gain contour of these hypothetical antenna beams. RARC-SAT-83 used a minimum of 0·8° for Region 2 planning. However, with the orientation of the beam optimised, a small service area may be circumscribed by a contour of constant PFD which lies within the −3 dB PFD contour and it is this inner contour that delineates the coverage area for a small country.

In determining whether carrier-to-interference objectives are achieved by any given arrangement of assignments, no account is taken of conditions within the coverage area but outside the service area. In practice, each administration defines the geographical co-ordinates of a number of 'test points', typically 10, which serve to define the service area and the interference level is computed for each test point.

The feeder link beam area is defined in Annex 3 of RR Appendix 30A[26] as:

> the area delineated by the intersection of the half-power beam of the satellite receiving antenna on the surface of the Earth.

Thus there are close similarities between the concepts of feeder link beam area and (down-link) coverage area. The feeder link beam area is also based upon hypothetical beams of elliptical or circular cross-section, a minimum of 0·6° being assumed for all three Regions for the apex angle of the minor axis at −3 dB relative to the maximum gain. A feeder link service area is defined, corresponding to the down-link service area concept, and the liability of a feeder link earth station to interfere at the space station receiver of an unwanted satellite is assessed with the aid of a set of test points.

(c) The transmission plan

The necessary bandwidth assumed in the plan drawn up at WARC-BS-77 for Regions 1 and 3, based on 625-line/25-frames-per-second picture standards and FM, was 27 MHz. The corresponding figure for 525-line/30-frames-per-second pictures in the RARC-SAT-83 plan from Region 2 was 24 MHz.

There were minor differences of principle between the feeder link plans produced for Region 2 at RARC-SAT-83 and for Regions 1 and 3 at WARC-ORB-88 in the determination of the feeder link EIRP. Somewhat simplified, the two approaches were as follows:

For Region 2, it was assumed that a feeder link earth station antenna would have a diameter of 5 m and an aperture efficiency of 65%, and thus, at the mid-band frequency, an on-axis gain of 57·4 dBi. With this standard antenna the transmitter power into the antenna may not exceed 1000 W, providing the standard EIRP of 87·4 dbW. Smaller antennas, not less than 2·5 m in diameter, may be used but if so, the off-beam EIRP may not exceed that of a 5 m antenna, fed with 1000 W and achieving the high standard of sidelobe suppression specified in Appendix 30A[26] (see item (*d*)

below). Antennas bigger than 5 m may be used, subject to various conditions and in particular provided that the on-axis EIRP does not exceed 87·4 dBW. Automatic up-link power control, rather limited in range and related in magnitude to the angle of elevation of the feeder link, may be used to increase the EIRP relative to 87·4 dBW when rain attenuation on the up-links exceeds 4 dB.
For Regions 1 and 3, the maximum permitted EIRP for each feeder link is specified in the plan. The values range between 82 and 89 dBW at 18 GHz, providing some compensation for differences in the gains of space station receiving antennas; at 15 GHz the maximum EIRP is uniform at 82 dBW. As for Region 2, these EIRPs are associated with a standard 5 m antenna (6 m at 14·5 GHz) and a very high standard of sidelobe suppression is specified in RR Appendix 30A[1]. Antennas smaller than 5 m, but not smaller than 2·5 m, in diameter may be used, but the level of off-beam radiation which the standard 5 m antenna would exhibit when providing the permitted EIRP may not be exceeded. Automatic up-link power control, without a threshold as in Region 2 but subject to safeguards to protect adjacent cross-polar channels, is permitted, to compensate for rain attentuation in the up-link, up to a maximum value specified for each feeder link in the plan.

Despite up-link power control, the feeder link carrier level at the space station receiver will vary considerably, mainly because of up-link transmission loss variations and earth station antenna pointing errors. It was assumed that these variations would be removed by an automatic level control system.

The maximum down-link EIRP for each planned assignment is one of the parameters of the plan, being calculated so that the PFD at the edge of the coverage area that will be attained for at least 99% of the worst month of the year, despite predicted down-link rain loss, will be:

−103 dBW per square metre in Regions 1 and 3
−105 dBW per square metre in Region 2

For individual receiving stations having the performance assumed in planning, these values of PFD should provide a received carrier-to-noise ratio of 14 dB. The PFD will be 3 dB higher at the centre of a large coverage area, but somewhat less if the service area is small.

The plans do not specify the diameter or the on-axis gain of the domestic receiving antenna used for individual reception. However, for the plans prepared for Regions 1 and 3 it was assumed that the figure of merit (G/T) of the antenna/input-amplifier combination would be 6 dB (K^{-1}) and that the antenna would be big enough to achieve a half-power beamwidth of 2·0°. The corresponding figures for the Region 2 plan were 10 dB (K^{-1}) and 1·7°.

(d) Antenna out-of-beam gain characteristics

The feeder link earth station antenna is required by Annex 3 of RR Appendix 30A[26 and 1] to attain a stringent reference radiation pattern. RARC-SAT-83 and WARC-ORB-88 presented the requirements for earth stations in the Regions for which they were planning in slightly different ways, but the intentions of the two Conferences can be represented approximately by the curves in Fig. 9.2. If a feeder link earth station antenna were put into service but the off-beam EIRP exceeded these levels when the input to the antenna from the transmitter was sufficient to produce the on-axis EIRP permitted by the plan, then the power transmitted would have to be reduced.

Expressions for the reference radiation patterns which the satellite antennas, receiving and transmitting, are required to attain are given in Annex 3 of RR Appendix 30A[26 and 1] and Annex 5 of RR Appendix 30[26] respectively but they can be expressed approximately in graphical form (see Fig. 9.3). There constraints are not intended to extend below a gain value of 0 dBi. The Region 2 plan offered the option to administrations of securing additional planned assignments by undertaking to use satellite antennas with particularly rapid gain roll-off characteristics in the outer skirt of the main lobe (not shown in Fig. 9.3) and a similar option was available for space station receiving antennas, but not space station transmitting antennas, in Regions 1 and 3. A number of administrations accepted this commitment for some of their planned assignments.

Expressions for the reference gain patterns assumed for antennas used for individual reception are to be found in Annex 5 of RR Appendix 30[24]. The assumptions are indicated graphically in Fig. 9.4.

(e) Polarisation

The plans for all three Regions were drawn up on the assumption that all emissions, whether feeder links or broadcasting links, would be circularly polarised. The planned sense of circular polarisation was chosen, assignment by assignment, to reduce interference by means of polarisation discrimination.

Transmitting antennas, at the feeder link earth station and at the space station, are required to radiate signals of high polarisation purity and receiving antennas at the satellite and the domestic receiving station are assumed to show good rejection of signals of the unwanted sense of polarisation. These characteristics are to be exhibited in all parts of the main lobe (see Fig. 9.2–9.4). Indeed the reference radiation pattern for the feeder link earth station in Annex 3 of RR Appendix 30A[26] calls for a low level of cross-polar radiation in the side-lobes and the domestic receiving antenna is expected to provide some considerable discrimination against the unwanted polarisation sense in the first sidelobe region. These requirements are technically demanding.

The planning conferences recognised that some administrations might wish to use linear polarisation instead of circular polarisation; an option to

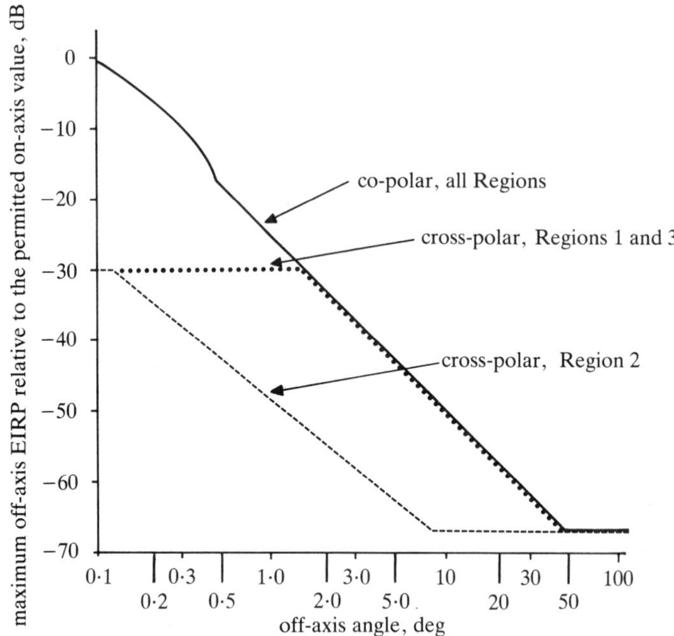

Fig. 9.2 Maximum off-axis EIRP, co-polar and cross-polar, which may be radiated by the antenna of a feeder link earth station, relative to the assigned on-axis EIRP with the assigned polarisation, in accordance with RR Appendix 30A, Annex 3[26 and 1]

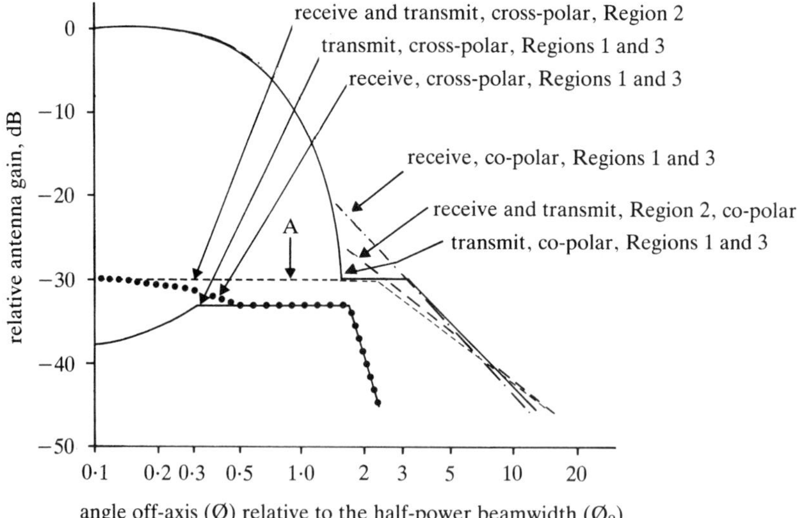

Fig. 9.3 Reference radiation patterns for space station antennas,. showing the gain which is not to be exceeded, co-polar or cross-polar, relative to the maximum gain in the assigned polarisation; see Annex 5 of RR Appendix 30[26] and Annex 3 of RR Appendix 30A[26 and 1]

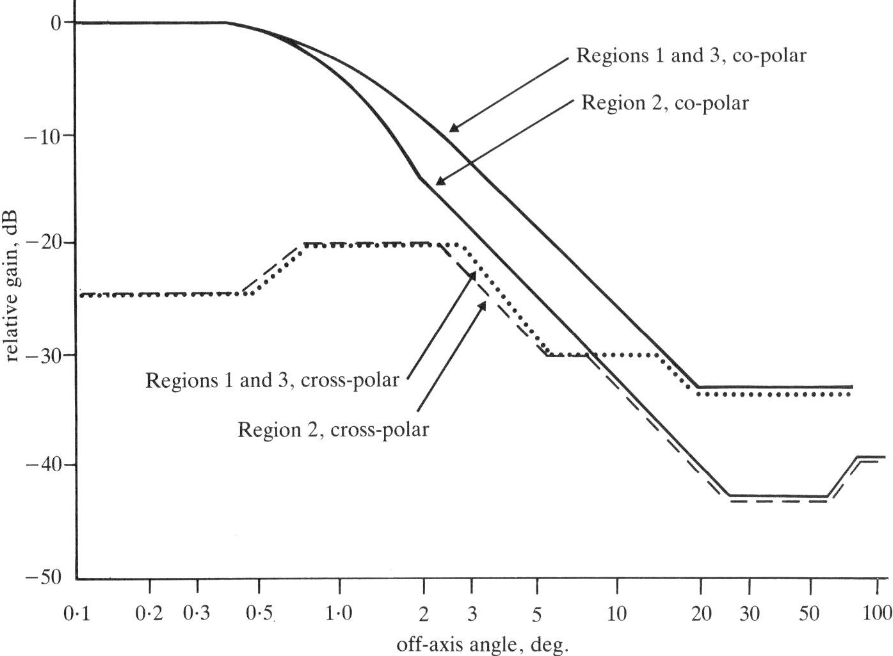

Fig. 9.4 Reference antenna gain patterns assumed for planning purposes, showing gain relative to the maximum gain in the assigned polarisation mode, for receiving antennas used by the public for individual reception

do so was left open, provided that all other administrations that would be affected agreed.

(f) Control of satellite location and beam pointing direction

The frequency assignment plans for all Regions require satellites to be maintained within 0·1° of the longitude of their nominal orbital positions. The deviation of satellite antenna beams from their nominal pointing directions may not exceed 0·1° in any direction.

In Regions 1 and 3 the inclination of the orbital plane of satellites may not exceed 0·1°. In Region 2 the maintenance of this limit on orbital inclination is recommended to administrations but it has not been made mandatory.

Angular rotation of an elliptical transmitting beam about its axis may not exceed 2° in regions 1 and 3; in Region 2 the limit is 1°.

(g) The arrangement of satellites in orbit

The 1977 plan for Region 1 placed satellites in clusters at intervals of 6° at nominal locations from 37° West longitude to 29° East longitude. If necessary, all assignable frequencies were available for transmission from every cluster. With minor irregularities the Region 3 satellites, together with a few for Region 1, extend in a similar series of clusters at 6° intervals from 38° to 170° East longitude.

Through careful selection of coverage areas for inclusion in each satellite cluster, it is readily possible for a domestic receiving antenna meeting the performance expectations referred to in (*d*) and (*e*) above to reject interference from satellite clusters which are 6° to either side of the wanted satellite.

By 1983, when the Region 2 plan was drawn up, computer techniques for optimising planning were available and these permitted the orbital separation angle between satellites to be made a variable of the planning process. The concept of satellite clusters was used again but, subject to there being sufficient protection from interference for the specific assignments being planned, the angles between cluster centres were made much smaller in parts of the orbit where the demand for assignments was high.

Furthermore, the 1983 conference was particularly concerned to safeguard adequately the protection of feeder links from interference from other feeder links, and above all from other feeder links serving satellites of the same cluster, close in frequency but cross-polarised. For these reasons the nominal orbital position given in the plan for an assignment which was to use right-hand circular polarisation was 0.2° west of the nominal cluster centre and assignments using left-hand circular polarisation were given nominal positions 0.2° east of the nominal cluster centre. This division of the cluster into two sub-clusters with a nominal orbital separation of 0.4° allows the high directivity of the feeder link earth station antenna to augment the protection against inteference provided by polarisation discrimination (see Fig. 9.5).

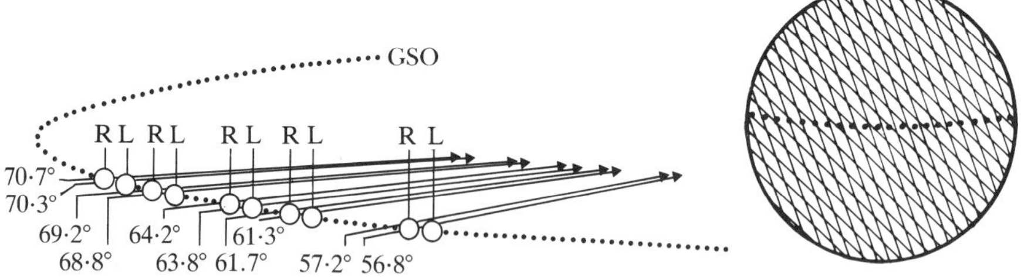

Fig. 9.5 An arc of the GSO showing a typical arrangement of satellites planned for Region 2, satellites of each orbital group being segregated into two sub-groups according to the polarisation which has been assigned to them

(h) Frequency channelling plan

In Region 1 the frequency band, 800 MHz wide, allocated to the BSS was divided into 40 channels, the centre frequencies of the channels being spaced at intervals of 19·18 MHz from 11 727·48 MHz to 12 475·5 MHz.

The necessary bandwidth of a 625-line/25-frames-per-second signal using frequency modulation was taken to be 27 MHz, and this combination of necessary bandwidth and choice of centre frequencies leads to substantial overlap of the spectra of the emissions occupying adjacent channels (see Fig. 9.6).

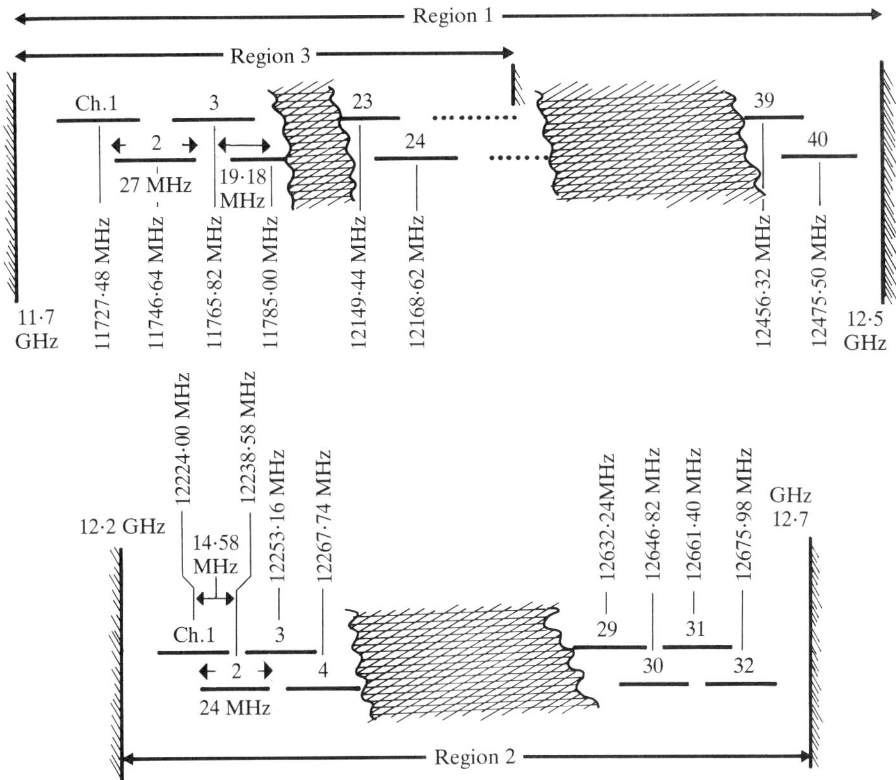

Fig. 9.6 The frequency plan for the BSS at 12 GHz in the three Regions

This Region 1 channelling plan creates sub-bands at the lower and upper edges of the allocated band which are free from powerful broadcasting signals, providing guard bands to protect sensitive FSS down-links in adjacent bands but potentially usable for the low-power TTC and ancillary signals which BSS systems will need.

Region 3 uses the same basic frequency plan as Region 1 but, the allocated bandwidth being less, the highest usable channel is number 24, centred at 12 168·62 MHz.

In Region 2 the 500 MHz allocated bandwidth is divided into 32 channels, with centre frequencies spaced by 14·58 MHz from 12 224·0 MHz to 12 675·98 MHz. The necessary bandwidth of a 525-line/30-frames-per-second FM signal was taken to be 24 MHz (see Fig. 9.6). Thus, in the

Region 2 plan there is a large degree of spectrum overlap between adjacent channels and interference from the so-called 'second adjacent channels' was taken into account in planning.

The Region 2 feeder link carrier frequency plan is an exact counterpart to the BSS frequency plan, with every frequency increased by 5100 MHz. In a similar way the feeder link frequency plan at 18 GHz for Regions 1 and 3 has the same form as the BSS frequency plan, the translation frequency being 5600 MHz, but with the assignments omitted for which feeder links at 14·5 GHz had been assigned. However, the bandwidth allocated for feeder links at 14·5 GHz is less than the BSS bandwidth at 12 GHz and there is therefore no simple relationship between down-link frequency assignments and feeder link frequency assignments at 14·5 GHz.

The outcome of planning

A few assignments were planned with multi-national coverage, with the agreement of the groups of countries concerned. However, almost all administrations had made their first priority the acquisition of as many planned assignments as possible with national coverage, and that is what the plans chiefly contain.

Despite the availability of a bandwidth of 800 MHz allocated for BSS in Region 1, compared with 500 MHz in other Regions, the difficulty of satisfying the demand for the option for transmitting a large number of programmes was greatest in Europe and Africa, where a large number of countries share a relatively limited orbital arc. A call for equitable treatment of all countries were strong at WARC-BS-77 and 'equitable' was interpreted as equality in the number of assignments per country. In the plan, virtually every country was assigned five channels and this exhausted the capacity which the methods used at WARC-BS-77 made available. Undoubtedly, some administrations foresaw a need for more channels.

The situation was more open in Asia and Australasia. In this part of the world, a relatively small number of countries share a long arc of the geostationary satellite orbit. The bandwidth allocated to the BSS is less in Region 3 than in Region 1 but not all of the channels that the plan could have made available were in fact used. It may be assumed that all or most administrations in Asia and Australasia received assignments for all of their stated requirements.

In Region 2 the situation was again different. Here a relatively small number of countries share a long orbital arc with a wide expanse of ocean to the west. Administrations stated large requirements, but planning techniques had become more sophisticated by 1983. A plan with a substantial number of assignments was produced and various measures were devised to give flexibility to its implementation. It may be supposed that, in Region 2 also, administrations received as many assignments as they wanted.

9.3.3 Management procedures for the BSS plans

RR Articles 15[26] and 15A[26 and 1] require the administrative procedures contained in RR Appendices 30[26] and 30A[26 and 1] to be applied in the bands planned for the BSS and its feeder links.

When an administration decides to bring into use planned assignments for satellite broadcasting and the associated feeder links, it notifies the IFRB in order that the assignments may be included in the MIFR. Articles 5 of RR Appendices 30[26] and 30A[26 and 1] state the procedures which the administration and the IFRB must follow in doing this. The procedures are basically simple. The assignments are registered if they are in complete conformity with the plan, or if the parameters of the assignments differ from those of the plan in ways that do not increase the prospect of interference to the planned or actual assignments of any other administration. If departures from the planned parameters would increase the potential for interference, the administration must either amend the parameters so as to eliminate the potential for increased interference or obtain agreement to a modification of the plan to bring it into line with the parameters which it wishes to assign.

Modification of the plan is covered by procedures in Article 4 of the same two RR Appendices. These procedures allow for planned assignments to be amended, or cancelled, or for new assignments to be added to the plan, subject to such changes having no unacceptable effect on the rights of other administrations, under the plan or under other provisions of the Radio Regulations. The procedures, other than for cancellations, are made rather complex by differences in the regulatory situation of the BSS in the various Regions and by the need to respect registered assignments to stations of other services with sharing allocations. However, the essence of the procedures for modifications and additions to the plan is as follows:

(*a*) The proposing administration notifies its intentions to the IFRB and supplies the Board with a list of the other administrations with which it is thought that co-ordination will be necessary, in accordance with technical criteria which are to be found in Annex 1 of both of the two RR Appendices.

(*b*) The Board reviews the proposed list of affected administrations, adds to the list if it considers necessary, and publishes details of the proposals, and the list of administrations which appear to be potentially affected, in the IFRB weekly circular. Any administration which, seeing the published proposals and the list, considers that its potential interest has been overlooked may ask the Board for its name to be added to the list.

(*c*) The proposing administration then co-ordinates the proposal with the other potentially affected administrations and all parties are expected to seek solutions in good faith to any real problems that are found. The Board has some oversight over this process, to ensure that consent to the proposals is not unreasonably withheld or delayed and to assist where necessary.

(*d*) If the co-ordination process is brought to a successful conclusion, the plan is amended as appropriate, and the proposing administration is free to notify to the Board the assignment, which should then be registrable, being in accordance with the plan.

9.3.4 Control of interference from sharing services

In all of the bands which have been planned for the BSS at 12 GHz there are also primary allocations for the BS and FS. In Regions 2 and 3 there are also primary allocations for the MS (except AeM) and RR 846[26] makes an allocation to FSS (space-to-Earth) in Region 2. In the 14·5 GHz band planned for feeder links in Regions 1 and 3 there are also primary allocations for FS and MS,. There are no primary allocations in the international Table of Frequency Allocations sharing with planned feeder link assignments at 17·3–17·7 GHz, but the remainder of the 18 GHz feeder link band has primary allocations to the FSS (space-to-Earth), FS and MS. In addition to all of these situations, in which sharing arises amongst the allocations which are applicable within a single Region, interference between stations of different services in different Regions may also need to be taken into account.

There is plenty of opportunity at present for these other allocations to be used. These opportunities will continue to exist far into the future, especially if the build-up of the implementation of satellite broadcasting continues to be slow, as at present. However, the administrations which have ratified the BSS and feeder link plans have undertaken to respect the frequency assignments, present and future, made in accordance with those plans, and frequency assignments to stations of other services sharing these bands do not have this high status. Thus, an administration which is considering making an assignment in one of these bands to a station of one of these sharing services needs to be able to ascertain whether such an assignment would be compatible with the planned satellite broadcasting assignments. If such an assignment would be incompatible with the plan, there will also be a need to know when the BSS or feeder links assignments in question will be implemented. Various provisions and criteria have been included in RR Appendices 30[26] and 30A[26 and 1] for these purposes and RR Articles 15[26] and 15A[26 and 1] require that these provisions be used when notifying to the IFRB an assignment to a sharing service.

Sharing between BSS systems and FSS networks

The only situation in which the BSS and the FSS (space-to-Earth) share a frequency band in the same Region and in the same direction of transmission arises in Region 2 and from RR 846[26]. However, RR 846 solves the interference control problem by limiting use of the band for the FSS to situations which can be treated, in respect of interference, as if they were part of the BSS plan.

Throughout the band which has been planned for the BSS in one Region or another (that is, 11·7–12·7 GHz), there are allocations to the FSS (space-to-Earth) in one or both other Regions. The methods which are used to co-ordinate the two kinds of system are reviewed on pp. 294–296.

In Regions 1 and 3, feeder links and FSS down-links share the band 17·7–18·1 GHz; in Region 2 the shared band is 17·7–17·8 GHz. The special methods used in co-ordinating systems of the two services are reviewed on pp. 296–297.

It will be noted that one of the possible problems arising from sharing part or all of the band 17·7–18·1 GHz between feeder links and FSS network down-links is interference between transmitters at feeder link earth stations in one country and receivers at FSS network earth stations in a neighbouring country. In the absence of special provisions to deal with this problem, an administration (Administration A) might be inhibited for many years from assigning frequencies in this band for reception at an earth station near its frontiers because it did not know where the feeder link earth station of a neighbouring country (Administration B) would be located. A procedure set out in Article 7 and Annex 4 of RR Appendix 30A[26 and 1] enables Administration A to require Administration B to declare within three months where its feeder link earth station will be. If this declaration is not made, Administration B will lose its right, under the plan, to choose a location for its feeder link earth station within co-ordination distance of the frontier without co-ordination with FSS network earth stations receiving in this band on the territory of Administration A.

The procedures of Article 4 of both RR Appendix 30[26] and Appendix 30A[26 and 1] provide for the negotiation of amendments to the satellite broadcasting plans (see pp. 337–338) and changes of satellite location are not excluded from this procedure. However, it could be greatly to the disadvantage of the FSS for the BSS plan to be amended by transferring a planned satellite to an arc of the GSO which would otherwise be generally available for satellites of the FSS, For this reason, Annex 7 of RR Appendix 30[26] provides as follows:

- Region 1 broadcasting satellites operating in the band 11·7–12·2 GHz are excluded from the arc extending from longitude 37°W westwards to longitude 146°E, to protect the interests of the FSS in Region 2. For the same reason, Region 1 broadcasting satellites relocated in the orbital arc from 37°W eastwards to 10°W are to be placed in or close to the clusters at 6° intervals planned at WARC-BS-77.
- Region 2 broadcasting satellites are excluded from certain arcs of the GSO to protect the interests of the FSS in Regions 1 and 3. These arcs are as follows: from longitude 44°W eastwards to 175·2°W if operating in the band 12·2–12·5 GHz and from longitude 54°W, eastwards to 175·2°W if operating in the band 12·5–12.7 GHz.

The powerful space station emissions from broadcasting satellites may be accompanied by correspondingly powerful out-of-band radiation. Some of this out-of-band energy may lie outside the band allocated to the BSS and it may interfere with FSS systems operating in adjacent bands; see CCIR Report 712[14]. However, despite the high status given by the agreed frequency assignment plans to the in-band emissions of the BSS and its feeder links, RR 343 contains a regulation of general applicability, as follows:

> The frequency assigned to a station of a given service shall be separated from the limits of the band allocated to this service in such a way that, taking account of the frequency band assigned to the station, no harmful interference is caused to services to which frequency bands immediately adjoining are allocated.

The frequency assignment plans reviewed on p. 000 have guard bands at the edges of the allocated bands to reduce out-of-band radiation to an acceptable level.

Sharing between BSS at 12 GHz and terrestrial services

The BSS frequency assignment plan agreement (RR Appendix 30[26]) protects the broadcasting signal from interference from terrestrial stations within the service area; see also RR 838 and RR 844[26]. Annex 5 to RR Appendix 30[26] limits the service area of a BSS assignment to the area 'in which the administration responsible for the service has a right to demand that the agreed protection conditions be provided'; this right does not normally extend beyond the territory under the jurisdiction of the assignee administration.

The procedure set out in RR Resolution 32[46] provides a mechanism, applicable to Regions 1 and 3 only, for ascertaining whether an established frequency assignment to a terrestrial station is likely to cause harmful interference when, in due course, the satellite broadcasting facility is set up. For new assignments to terrestrial stations that are planned for the BSS, RR Article 15[26] requires the co-ordination procedure set out in Article 6 of RR Appendix 30[26] to be followed where appropriate (see also pp. 133–134 and 221–222). The interference criteria agreed for use in co-ordination calculations are contained in Annex 6 of RR Appendix 30[26] and are discussed in more detail in CCIR Report 631[25].

The emissions from broadcasting satellites may cause interference to receiving stations of terrestrial services far outside the BSS service areas; see for example CCIR Report 789[15], which discusses interference to receiving stations of the FS, In general the terrestrial service has no remedy against this interference, provided that the BSS emissions conform to the terms of the BSS plan, although Article 9 of RR Appendix 30[26] places a constraint on the power flux density of the emissions from broadcasting satellites serving Region 2 in the direction of land in Regions 1 and 3. However, terrestrial services are protected against an increase in this interference due to amendments in the BSS plan, since Article 4 of RR Appendix 30[26] requires a co-ordination procedure to be carried out if an administration wishes to amend the plan in a way that would increase significantly the potential interference outside the service area.

Sharing between BSS feeder links and terrestrial services

Sharing constraints are applied to transmitting stations of the FS and the MS in BSS feeder link bands, to limit interference to feeder link space station receivers. Thus:

- RR 2510, RR 2511, RR 2505 and RR 2508 limit transmitting stations of the FS and MS operating in the feeder link bands to an EIRP of + 55 dBW and an antenna input power of +10 dBW, and
- RR 2510 and RR 2503 urge that systems of the FS and MS transmitting an EIRP exceeding 45 dBW in the 14·5 GHz band should be arranged so that the antenna beam is directed at least 1·5° away from the GSO

See also CCIR Report 1006[47].

A feeder link earth station could cause interference to a terrestrial receiving station in an adjacent country and an administration which is considering assigning a frequency for reception at an FS or MS station in the band 14·5–14·8 GHz or 17·7–18·1 GHz may wish to assess the prospective interference level before, perhaps long before, the earth station is built. A similar situation arises for the earth stations of FSS networks with down-links in the 17·7–18·1 GHz band; see pp. 296–297. Article 6 of RR Appendix 30A[26 and 1] applies a similar solution, requiring the administration holding the planned assignment for the feeder link earth station to declare where the station will be sited or lose its privileged status (see also p. 339).

9.3.5 Interim satellite broadcasting systems

RR Resolution 42[48]. which originated at RARC-SAT-83, provides for the possibility that some Region 2 administrations may wish, separately or jointly, to set up satellite broadcasting systems to operate at 12·2–12·7 GHz and 17·3–17·8 GHz which are not in complete conformity with the plan. Such systems would be called 'Interim Systems'. The Annex to the resolution provides a procedure for doing this, on a temporary basis, subject to the consent of all other administrations whose interests under the plan are affected.

Preliminary consideration was given at WARC-ORB-88 to extending the option to set up interim systems to Regions 1 and 3; see Resolution 519[49] of that conference. The matter is to be considered further at a later ARC. If administrations wish to set up interim systems sooner, the procedure for amending the plan, contained in Article 4 of RR Appendices 30[26] and 30A[26 and 1], might be used.

9.4 References

1 Final Acts adopted by the second session of the World Administrative Radio Conference on the use of the geostationary-satellite orbit and the planning of space services using it (ORB-88) (ITU, Geneva, 1988)

2 'Selection of a frequency band for use by the broadcasting-satellite service and intended for wide RF-band high definition television, and of an associated frequency band for HDTV feeder links, and the adoption of related provisions by a future competent conference'. Resolution 521; *ibid.*

3 'Satellite sound broadcasting with portable receivers and receivers in automobiles'. CCIR Report 955–1; Recommendations and Reports of the CCIR, 1986, Volume X/XI–2 (ITU, Geneva, 1986)

4 'Protection of terrestrial line-of-sight radio-relay systems against inteference from the broadcasting-satellite service (sound) in the band 1427–1530 MHz'. CCIR Report 941; *ibid.*, Volume IX–1

5 'Future change in Article 8 for the broadcasting-satellite service (sound) in the frequency range 500–3000 MHz'. Resolution 520; Final Acts of WARC-ORB-88; item[1] above

6 'Feeder links to space stations in the broadcasting-satellite service'. CCIR Report 561–3; Recommendations and Reports of the CCIR, 1986, Volume IV–1 (ITU, Geneva, 1986)

7 'Principles governing the use by States of artificial earth satellites for international direct television broadcasting'. Resolution 37/92 of the United Nations General Assembly; adopted 10 December 1982

8 'Relating to the establishment of agreements and associated plans for the broadcasting-satellite service'. Resolution 507; Radio Regulations, edition of 1982 (ITU, Geneva, 1982)

9 'Relating to the bringing into use of space stations in the broadcasting-satellite service, prior to the entry into force of agreements and associated plans for the broadcasting-satellite service'. Resolution 33; *ibid.*

10 'Relating to studies of the maximum permitted levels of spurious emissions'. Recommendation 66; *ibid.*

11 'Relating to spurious emissions in the broadcasting-satellite service'. Recommendation 507; *ibid.*

12 'Relating to the harmonics of the fundamental frequency of broadcasting-satellite stations'. Recommendation 506; *ibid.*

13 'Unwanted emissions from broadcasting-satellite space stations'. CCIR Report 807–2; Recommendations and Reports of the CCIR, 1986, Volume X/XI–2 (ITU, Geneva, 1986)

14 'Factors concerning the protection of fixed-satellite earth stations operating in adjacent frequency band allocations against unwanted emissions from broadcasting satellites operating in frequency bands around 12 GHz'. CCIR Report 712–1; *ibid.*, Volume IV–1

15 'Protection of terrestrial line-of-sight radio-relay systems against interference from the broadcasting-satellite service in the band 11·7 to 12·75 GHz'. CCIR Report 789–1; *ibid.*, Volume IX–1

16 'Factors relative to the establishment of spurious emission limits for space services'. CCIR Report 980; *ibid.*, Volume II

17 'Systems for the broadcasting-satellite service (sound and television)'. CCIR Report 215–6; *ibid.*, Volume X/XI–2

18 'Broadcasting-satellite service (sound and television) – technically suitable methods of modulation'. CCIR Report 632–3, *ibid.*, Volume X/XI–2

19 'Television standards for the broadcasting-satellite service'. CCIR Report 1073; *ibid.*, Volume X/XI–2

20 'Satellite transmission of multiplexed analogue component (MAC) vision signals'. CCIR Report 1074; *ibid.*, Volume X/XI–2

21 'The present state of high-definition television'. CCIR Report 801–2; *ibid.*, Volume XI–1

22 'High-definition television by satellite'. CCIR Report 1075; *ibid.*, Volume X/XI–2
23 'Multiplexing methods for the emission of several digital audio signals and also data signals in broadcasting'. CCIR Report 954–1; *ibid.*, Volume X/XI–2
24 'Relating to the criteria to be applied for frequency sharing between the broadcasting-satellite service and the terrestrial broadcasting service in the band 620–790 MHz'. Recommendation 705; Radio Regulations, edition of 1982 (ITU, Geneva, 1982)
25 'Frequency sharing between the broadcasting-satellite service (sound and television) and terrestrial services'. CCIR Report 631–3; Recommendations and Reports of the CCIR, 1986, Volume X/XI–2 (ITU, Geneva, 1986)
26 Final Acts of the World Administrative Radio Conference on the use of the geostationary-satellite orbit and the planning of space services using it, First Session, Geneva 1985 (ITU, Geneva, 1986)
27 'Relating to the establishment of the broadcasting-satellite service in Region 3 in the 12·5–12·75 GHz frequency band and to sharing with space and terrestrial services in Regions 1, 2 and 3'. Resolution 34; Radio Regulations, edition of 1982 (ITU, Geneva, 1982)
28 'Frequency sharing between the inter-satellite service when used by the fixed-satellite service and other space services'. CCIR Report 874; Recommendations and Reports of the CCIR, 1986, Volume IV–1 (ITU, Geneva, 1986)
29 'Sharing between the inter-satellite service and the broadcasting-satellite service in the vicinity of 23 GHz'. CCIR Report 951; *ibid.*, Volume X/XI–2
30 'Factors affecting the system design and the selection of frequencies for inter-satellite links of the fixed-satellite service'. CCIR Report 451–3; *ibid.*, Volume IV–1
31 Final Acts of the World Broadcasting-Satellite Administrative Radio Conference, Geneva 1977 (ITU, Geneva, 1977)
32 'Relating to the submission of requirements for the broadcasting-satellite service in Region 2'. Resolution Sat-9; *ibid.*
33 'Use by space stations operating in the 12 GHz frequency bands allocated to the broadcasting-satellite service, of the geostationary-satellite orbit and no other'. Resolution 506; Final Acts of WARC-ORB-88; item [1] above
34 'Relating to up-links for the broadcasting-satellite service'. Recommendation Sat-1; Final Acts of WARC-BS-77; item [31] above
35 'Relating to the convening of a Regional Administrative Radio Conference for the detailed planning of the broadcasting-satellite service in the 12 GHz band and associated feeder links in Region 2'. Resolution 701; Radio Regulations, edition of 1982 (ITU, Geneva, 1982)
36 'Concerning the drawing up of agreements and the associated plans for feeder links to space stations in the broadcasting-satellite service operating in the 12 GHz band under the plan adopted by the World Broadcasting-Satellite Administrative Radio Conference, Geneva, 1977, for Regions 1 and 3'. Resolution 101; *ibid.*
37 Final Acts of the Regional Administrative Conference for the planning of the broadcasting-satellite service in Region 2 (SAT-83) Geneva 1983 (ITU, Geneva, 1984)
38 World Administrative Radio Conference on the use of the geostationary-satellite orbit and the planning of space services using it, First Session, Geneva 1985; Report to the Second Session of the Conference (ITU, Geneva, 1986)
39 'Orbit and frequency planning in the broadcasting-satellite service'. CCIR Report 633–3; Recommendations and Reports of the CCIR, 1986, Volume X/XI–2 (ITU, Geneva, 1986)
40 'Considerations affecting the accommodation of spacecraft service functions (TTC) within the broadcasting-satellite and feeder link service bands'. CCIR Report 1076; *ibid.*, Volume X/XI–2
41 'Broadcasting-satellite service – Planning elements including those used in the establishment of plans of frequency assignments and orbital positions for the

broadcasting-satellites in the 12 GHz band'. CCIR Report 811–2; *ibid.*, Volume X/XI–2
42 'Computer programs for planning broadcasting-satellite services in the 12 GHz band'. CCIR Report 812–2; *ibid.*, Volume X/XI–2
43 'Factors to be considered in the choice of polarization for planning the broadcasting-satellite service'. CCIR Report 814–2; *ibid.*, Volume X/XI–2
44 'Technical characteristics of feeder links to broadcasting satellites – Elements required for the establishment of plans of frequency assignments and orbital positions for the broadcasting-satellite service and the associated feeder links – Sharing in the feeder-link bands'. CCIR Report 952–1; *ibid.*, Volume X/XI–2
45 'Energy dispersal in the fixed-satellite service'. CCIR Report 384–5; *ibid.*, Volume IV–1
46 'Relating to the use of frequency assignments to terrestrial and space radiocommunication stations in the band 11·7–12.2 GHz in Region 3 and in the band 11·7–12·5 GHz in Region 1, Resolution 32; Radio Regulations, edition of 1982 (ITU, Geneva, 1982)
47 'Fixed service EIRP limits for the protection of the broadcasting-satellite feeder links around 18 GHz, CCIR Report 1006; Recommendations and Reports of the CCIR, 1986, Volume IV/IX–2 (ITU, Geneva, 1986)
48 'Use of interim systems in Region 2 in the broadcasting-satellite and fixed-satellite (feeder link) services in Region 2 for the bands covered by Appendix 30 (Orb-85) and Appendix 30A (Orb-88)'. Resolution 42; Final Acts of WARC-ORB-88; item [1] above
49 'Possible extension to Regions 1 and 3 of provisions for interim systems'. Resolution 519; *ibid.*

Chapter 10

The mobile-satellite services

10.1 Survey of mobile-satellite service regulation

The first few paragraphs of Chapter 7 describe what the (terrestrial) mobile service is, explain the terms used for the various kinds of station that comprise it and show how the service is structured for regulatory purposes. The mobile-satellite service, made up of mobile earth stations, land earth stations and space stations, is closely analogous. Thus, a mobile earth station is an earth station which operates whilst in motion at sea, in the air or on land, or when halted at an unspecified location. A land station operates at a given location on land, communicating with mobile earth stations via a space station. Mobile earth stations also communicate with other mobile earth stations via a space station. The links between the satellites and land earth stations are called feeder links; see RR 109[1].

Three specialised mobile-satellite services have been defined for regulatory purposes; their names, their roles and the terms used for the earth stations which are involved in them will be apparent from Table 10.1.

The AeMSS is sub-divided into the AeMSS (R) and the AeMSS (OR). Some frequency bands are allocated specifically to the AeMSS (R). RR 35A[2] reserves AeMSS (R) bands for air traffic control and similar purposes, primarily along national or international civil air routes. RR 35B[2] indicates that the AeMSS (OR) provides for communications, including those relating to flight co-ordination, which are primarily outside national or international civil air routes.

RR 27 permits feeder links of the MSS to be transmitted in bands allocated to the MSS and this provides a network configuration which is particularly convenient in systems in which mobile earth stations often need to communicate with other mobile earth stations. Alternatively, RR 22 permits feeder links to be assigned frequencies in bands allocated to the FSS.

As with terrestrial mobile services, the involvement of the ITU with the mobile-satellite services goes beyond management of the use of the radio spectrum. In collaboration with the IMO and ICAO, many of the RR Articles referred to on p. 000 relating to operational matters and the safety of life involving ships and aircraft have been made relevant, not only to the MM and the AeM, but also to the MMSS and the AeMSS. These articles are not generally concerned with spectrum management, but several of them, and several CCIR texts which relate to safety of life and in particular to EPIRB systems, have regulatory aspects; note in particular CCIR Recommendations 632[3] and 633[4] and Reports 761[5] and 1046[6].

Table 10.1 Terms for earth stations in the mobile-satellite services, (formal definitions for terms are in the RR paragraphs indicated)

Service	Earth stations which operate while in motion	Earth stations which operate at permanent locations
Mobile-satellite (MSS) (RR 27)	Mobile earth station (RR 66)	Land earth station (RR 67A[1])
Maritime mobile-satellite (MMSS) (RR 31)	Ship earth station (RR 73) (see note)	Coast earth station (RR 71)
Aeronautical mobile-satellite (AeMSS) (RR 35)	Aircraft earth station (RR 79) (see note)	Aeronautical earth station (RR 77)
Land mobile-satellite (LMSS) (RR 29)	Land mobile earth station (RR 69A[2])	Base earth station (RR 68A[2])

Note: Survival craft and emergency position-indicating radio beacons may also participate in the MMSS and AeMSS.

Frequency allocations and sharing

Several mobile-satellite systems are in operation and many new systems are under development, some being intended for the use of the stations of more than one of the specialised mobile-satellite services. A substantial number of frequency bands have been allocated for the MSS or are shared between the various specialised mobile-satellite services. It is convenient therefore to review the allocations as a single series.

(*a*) RR 641 allocates the bands 235–322 MHz and 335·4–399·9 MHz for the MSS worldwide, frequency assignments being subject to agreements reached in accordance with RR Article 14 and subject to the MSS stations not causing harmful interference to existing or planned stations of the FS and MS, for which the bands are also allocated.

(*b*) The band 406–406·1 MHz is allocated exclusively worldwide for the MSS for Earth-to-space links. RR 649[2] limits the use of this band to low power EPIRBs intended for use with satellites. RR 649A[2] forbids assignments in this band for any emission which could cause harmful interference to reception of signals from these beacons. These measures for the protection of a safety service are reinforced by RR Resolution 205[7], which co-ordinates and strengthens international action to identify emissions which are made in disregard of this ban and to have them stopped.

(*c*) The band 806–890 MHz is allocated for the MSS in Region 2 by RR 700[2], for use on systems operating within national boundaries, subject to agreement of frequency assignments in accordance with RR Article 14 to protect stations of the other services to which the band is allocated, that is, the BS, FS and MS,. The same frequency band, with 942–960 MHz in addition, is allocated to the MSS (except the AeMSS) in Region 3 on similar terms by RR 701[2].

(*d*) Around 1600 MHz there is an important group of mobile-satellite allocations, with down-links at 1530–1559 MHz and up-links at 1626·5–1660·5 MHz. These bands are being used for the principal current commercial mobile-satellite system developments. International management of mobile-satellite spectrum is more complex and more advanced in these bands than elsewhere, and it is convenient to consider it in a separate section (see pp. 350–355).

(*e*) The bands 1610–1626·5 MHz. 5000–5250 MHz and 15·4–15·7 GHz are allocated for the AeRN. However, there are also various sharing allocations in these bands, mostly by footnote, including RR 733, which allocates all three bands for the AeMSS (R) worldwide, with primary status, but subject to agreement obtained under the RR Article 14 procedure. These bands are already in use, mainly for various AeRN systems, and the sharing situation is complex. Use of these bands for AeMSS (R) has not developed yet, nor are there any agreed means for facilitating sharing between the AeMSS (R) and other services with allocations in these bands.

(*f*) RR 812 allocates the bands 7250–7375 MHz (space-to-Earth) and 7900–8025 MHz (Earth-to-space) for the MSS, sharing with FSS (in the same directions of transmission), FS and MS and subject to agreements reached in accordance with RR Article 14. To facilitate sharing between MSS space stations and the terrestrial services, the sharing constraints that apply to the FSS in these bands (see p. 111) are applied also to the MSS.

(*g*) RR 859 allocates 14·0–14·5 GHz (Earth-to-space) for the LMSS worldwide, with secondary status. The primary allocations in this band are for FSS (Earth-to-space) and RN (14·0–14·3 GHz), both worldwide, plus the FS and MS (except AeMS) in various sub-bands and with various geographical distributions. There are also other secondary allocations.

This LMSS up-link allocation can perhaps most conveniently be seen as an adjunct to the FSS in the same band, usable for unidirectional links from transportable earth stations, which would be subject to the same sharing constraints as FSS earth stations.

(*h*) In the bands 19·7–20·2 GHz (space-to-Earth) and 29·5–30·0 GHz (Earth-to-space) there are worldwide secondary allocations for the MSS, the primary service in these bands being FSS worldwide with the same directions of transmission. RR 873 also allocates the 19·7–20·2 GHz band for the FS and MS on a primary basis in a large number of countries, but no sharing constraint is applied to the power flux density set up at the Earth's surface by space station transmitters. RR 883 also allocates the 29·5–30·0 GHz band for the FS and MS in a substantial number of countries, but terrestrial transmitting stations are required to observe the power limits given in RR 2505 and RR 2508 to protect space station receivers.

(*i*) The bands 20·2–21·2 GHz (space-to-Earth) and 30–31 GHz (Earth-to-space) are allocated worldwide for the MSS on a primary basis, there being worldwide primary FSS allocations sharing the same bands with the same directions of transmission. RR 873 and RR 883 provide the same terrestrial sharing allocations in these bands as in the adjacent bands (see (*h*) above) and the same sharing constraint situations also apply.

(*j*) Ten more frequency bands have been allocated for the MSS above 40 GHz, all shared with various other services, space and terrestrial. Substantial use of these allocations, and the establishment of methods for managing the sharing in them, seem unlikely to develop for some years.

Technical constraints and requirements

Some of the basic regulations that apply to networks of the FSS, reviewed on pp. 271–274, apply also to the mobile-satellite services. Thus RR 2612 demands telecommand facilities for switching off space station transmitters. RR 2613 requires mobile-satellite networks using non-geostationary satellites to not cause unacceptable interference to FSS networks using geostationary satellites which are operating in accordance with the RR. The carrier frequency tolerance of 50 p.p.m. which applies to FSS stations operating above 2450 MHz is also applicable to mobile-satellite networks; below 2450 MHz the tolerance is 20 p.p.m.

Below 960 MHz, RR 304 and RR Appendix 8 apply limits to the spurious emissions of space stations and earth stations of mobile-satellite networks (see Appendix A.5 below). However, footnote 11 to the table in RR Appendix 8 suggests that better standards of spurious emission suppression may have to be considered in some circumstances and mobile-satellite space station transmitters may be such a case. These constraints do not apply to EPIRBs. No limit is laid down in RR Appendix 8 for the spurious emissions of transmitters operating above 960 MHz, but it is urged that their level should be as low as is practicable.

The satellite station-keeping tolerances that the RR apply to geostationary satellites of the FSS are indicated on pp. 274–275. RR 2615 applies the same tolerances (that is ±0·1° of longitude relative to the nominal position) to geostationary satellites of the mobile-satellite services that use frequencies, typically for feeder links, that are allocated to the FSS. A less severe requirement, namely the capability of remaining within 0·5° of longitude of the nominal orbital position, is applicable to geostationary satellites of the mobile-satellite services which make no use of frequency bands allocated for the FSS; see RR 2620–RR 2623. RR 2628–RR 2630 require the pointing directions of the antenna beams of all geostationary satellites to be maintainable within 10% of the −3 dB beamwidth or within ±0·3° (whichever requirement is the less stringent) of the nominal pointing direction. Neither the station-keeping requirement nor the beam-pointing requirement need to be observed unless excessive interference to another network would otherwise occur.

A sharing constraint in the form of a limit on the PFD at the Earth's surface is applied to mobile-satellite space stations operating in the band 7250–7375 MHz to protect the terrestrial services which have sharing allocations in that band; see RR 2565–RR 2568.

International spectrum management

Networks of mobile-satellite services must be co-ordinated with other space networks using the same frequency bands, whether the other network is of the same service or another service with equal allocation status in the frequency band, before the frequency assignments can be registered in the MIFR. Similarly, land earth stations must be co-ordinated with terrestrial stations having equal allocation status in neighbouring countries. Mobile earth stations operating in such bands also have to be co-ordinated with nearby, foreign terrestrial stations, but not individually; it is usual to co-ordinate 'typical' mobile earth stations and the area in which they will be deployed; see in particular RR 1111 and RR 1493A[1]–RR 1494A[1]. The procedures for co-ordination and registration are set out in RR Articles 11 and 13. They are in principle the same as those used for the FSS (see pp. 118–131 and 281–294).

In some frequency bands the terms of the mobile-satellite allocation require assignments to be agreed between administrations in accordance with the terms of RR Article 14 in addition to those of RR Article 11. However, a footnote (RR 1060.1) added to RR Article 11 at WARC-ORB-88[1] indicates that agreement between two administrations reached in accordance with RR Article 14 will be deemed to constitute successful completion of the RR Article 11 procedure between those two administrations also.

However, co-ordination using the RR Article 11 procedure for networks of the mobile-satellite services is less straightforward than, for example, its use for the FSS. This is because the procedure itself, and many of the criteria involved in it, such as channel performance objectives and permissible interference levels, which are well developed for the FSS, are still at an early stage of development for the mobile-satellite services. Furthermore, where systems operating below 1 GHz are involved, the lack of complete, agreed, short-term propagation data may also cause difficulties. It may thus be necessary for the parties involved to devise their own procedures and agree their own criteria, guided by section 7 of RR Appendix 28, by appropriate CCIR texts (see in particular CCIR Recommendation 370[8] and CCIR Reports 569[9], 382[10] and 773[11]) and by the advice of the IFRB. The criteria developed for MMSS networks operating at around 1·6 GHz may also be helpful (see p. 355).

10.2 Mobile-satellite systems at 1600 MHz

Frequency allocations

The bands allocated to the mobile-satellite services around 1600 MHz are the ones that are being used for most current development of commercial systems. The first allocations for mobile-satellite services in this part of the spectrum were made at WARC-71 and a number of space stations designed to make use of these original allocations are still in operation. The allocations have been changed twice since then, at WARC-79 and WARC-MOB-87, and another revision seems likely to occur within a few years. However, the total bandwidth available for the various mobile-satellite services here is quite limited, and it is of interest to see how the allocations have been changed from time to time as perceived needs have changed.

The allocations made by WARC-71 were as follows, all worldwide and primary:

MMSS	1535–1542·5 MHz	
MMSS sharing with AeMSS (R)	1542·5–1543·5 MHz	(space-to-Earth)
AeMSS (R)	1543·5–1558·5 MHz	
MMSS	1636·5–1644 MHz	
MMSS sharing with AeMSS (R)	1644–1645 MHz	(Earth-to-space)
AeMSS (R)	1645–1660 MHz	

The use of the bands allocated to the AeMSS (R) was strictly reserved for traffic related to the regularity and safety of flight on civil air routes; their use for 'public correspondence' (such as telephone calls to ground made by passengers) was forbidden by the RR. Aircraft were however permitted to use the MMSS allocations for this purpose; see RR 963 and RR 3571.

Eight years later, by which time MARISAT, the first commercial maritime satellite system, was in operation, WARC-79 increased the bandwidth allocated to MMSS for both directions of transmission. This increase was made feasible in part by transferring to the MMSS bandwidth previously allocated to the AeMSS (R). An additional band 5 MHz wide was allocated to MMSS in the Earth-to-space direction, to provide for a foreseen need to transmit heavy flows of data from ships to shore, this traffic being not balanced by corresponding flows of data in the shore-to-ship direction. One possible source of such data would, for example, be ships making seismic surveys of the geological structure of the seabed. The 1 MHz bands separating the MMSS and AeMSS (R) allocations were re-allocated to the MSS and limited to distress and safety operations. Thus, the worldwide allocations became as follows;

MMSS	1530–1544 MHz	
MSS	1544–1545 MHz	(Earth-to-space)
AeMSS (R)	1545–1559 MHz	
MMSS	1626·5–1645·5 MHz	
MSS	1645·5–1646·5 MHz	(space-to-Earth)
AeMSS (R)	1646·5–1660·5 MHz	

The MMSS allocation in the band 1530–1535 MHz did not enter into effect until 1 January 1990.

After a further eight years, the MMSS was growing vigorously through the INMARSAT system, but WARC-MOB-87 was anxious to make allocations for the LMSS in this part of the spectrum and there was also a growing interest in providing better facilities for public correspondence from aircraft.

The agenda of WARC-MOB-87 did not permit the Conference to increase the total bandwidth available for the various mobile-satellite allocations in the vicinity of 1600 MHz. To provide a long term solution to these and other related problems, the Conference agreed a resolution which called for the convening of another WARC not later than 1992 which would have authority to revise the international Table of Frequency Allocations on a more general basis in this part of the spectrum; see RR Resolution 208[12]. However, to provide some limited short term solutions for the problems posed by insufficient bandwidth for LMSS, WARC-MOB-87 agreed the following changes to the frequency allocation table;

(*a*) A band 3 MHz wide for each direction of transmission, already allocated with primary status for MMSS, was also allocated with primary status for LMSS.
(*b* All other MMSS allocations in this part of the spectrum, which retain their primary status, are to be shared with LMSS, the latter having secondary status and being limited to non-speech, low-bit-rate data transmissions; see RR 726B[2].
(*c*) The bandwidth allocated for AeMSS (R) was reduced by 4 MHz in each direction of transmission, the spectrum so freed being allocated with primary status for the LMSS.
(*d*) Despite provisions elsewhere in the Radio Regulations which forbid the

use of aeronautical mobile (R) allocations for public correspondence, RR 729A[2] was agreed and RR 3571 was amended (see RR 3571[2]). permitting administrations to authorise the use of the remaining AeMSS (R) allocations in this part of the spectrum for that purpose, provided that priority was to be given to urgent safety and flight control traffic.

These changes entered into effect on 3 October 1989. Thus, when all of the impending changes had taken effect, that is by 1 January 1990, the mobile-satellite allocations near 1600 MHz became as shown in Fig. 10.1

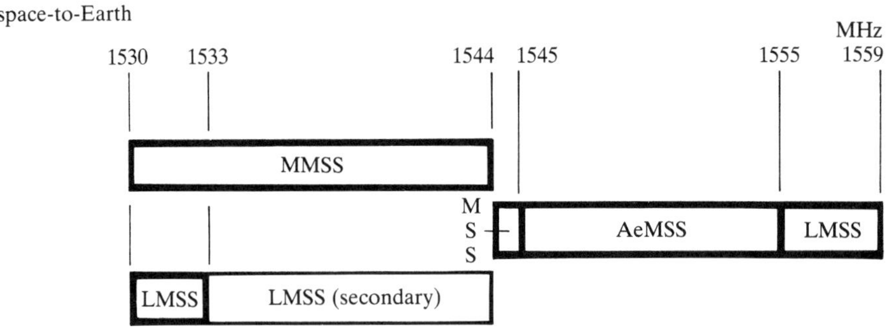

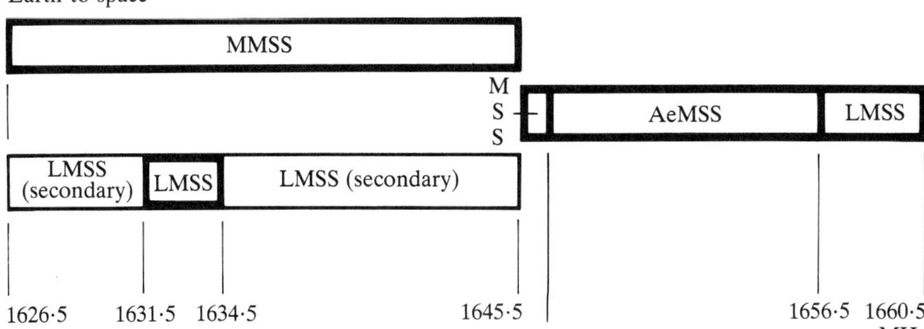

Fig. 10.1 Frequency allocations round 1600 MHz for the mobile-satellite services, effective from 1 January 1990. Allocations are primary except where a secondary allocation is indicated

In 1989 the ITU Plenipotentiary Conference resolved that a WARC should indeed be held in 1992 and a review of the allocations for mobile-satellite services in this part of the spectrum seems certain to be included in the agenda. If so, no doubt a case will be made for an increase, perhaps substantial, in the total bandwidth allocated for these services, in particular to meet a perceived need to provide for traffic, substantial in volume, with land vehicles and aircraft. There will, however, be a second important issue to consider; should these allocations be made specifically for maritime, aeronautical or land mobile use or should they be made for the MSS, to be used for whichever purpose administrations and system operators wish? No doubt the outcome will depend to some degree on the extent to which the total bandwidth available for satellite-mobile applications is increased. One arrangement that might meet most requirements would be to divide the available bandwidth into four parts, allocated as follows:

(*a*) A narrow band allocated for the MSS and reserved for distress and safety operations, as now
(*b*) A band allocated for the MMSS
(*c*) A band allocated for the AeMSS (R) and reserved for air traffic control purposes
(*d*) The remainder, perhaps the largest part, being allocated for the MSS and available for any satellite-mobile purpose except AeMSS (OR).

Frequency allocation sharing

Fig. 10.1 does not show the allocations to services other than the mobile-satellite services. These can be considered conveniently in two parts, namely sharing in the down-link bands around 1545 MHz and sharing in the up-link bands around 1645 MHz:

(*a*) The band 1530–1535 MHz is allocated worldwide for the SpO (space-to-Earth) and 1550–1559 MHz is allocated by RR 730 to the FS in a substantial number of countries, all of these allocations being primary. There are also various secondary sharing allocations. The primary sharing allocations raise some problems:

- No sharing constraints have been determined to protect SpO and MMSS down-links from one another. The usefulness for this purpose of network co-ordination for the SpO on the one hand and the MMSS and the LMSS on the other hand seems likely to be limited but no doubt the emission parameters used variously in these services and the antenna characteristics of SpO earth stations will ensure that interference is relatively low.
- There are no internationally agreed arrangements for protecting land mobile and aircraft earth station receivers from transmitting stations of the FS in the countries listed in RR 730, nor for protecting receiving stations of the FS from space station transmitters. The radiation patterns of satellite antennas are unlikely to be highly directional in this part of the spectrum and satellite-mobile systems need to provide a relatively powerful down-link signal to mobile stations. It seems likely that interference from the space station emissions will limit the performance of FS systems operating under RR 730. Similarly, the terrestrial transmissions seem likely to interfere with the space services in geographical areas that could be precisely defined. Through

planning of the use of this band for the FS, the administrations listed in RR 730 have a measure of control over where the space service signals will be usable.

(*b*) RR 730 also allocates the bands 1626·5–1645·5 MHz and 1646·5–1660 MHz to the FS with primary status in the same substantial group of countries, and the band 1660·0–1660·5 MHz is shared with the radio astronomy (RA) service, also with primary status. These sharing allocations also raise problems:

- To protect mobile-satellite space station receivers from interference, RR 2502, RR 2505, RR 2506, RR 2507 and RR 2509 apply the same power and pointing constraints to transmitting stations of the FS in this band as are applied to the FS in frequency bands shared with the FSS (see pp. 115–118).
- FS receivers will be protected from interference from the transmitters of the various kinds of mobile earth station partly by geographical and regulatory factors and partly by the directional properties of the FS receiving antennas. However, ship earth stations could not cause inteference to FS receivers located far from the sea and land mobile earth stations will not be a major problem to FS receivers at a distance from areas where the use of the LMSS is permitted. Interference from aircraft earth stations may be difficult to control, but it should, at worst, be intermittent.

RR 736 urges administrations to take all practicable steps to protect radio astronomy stations from inteference from aircraft earth stations in the band 1660·0–1660·5 MHz.

Feeder links

RR 27 permits frequency bands allocated to mobile-satellite services to be used for feeder links. However, the bands allocated to the MMSS, AeMSS and LMSS around 1545 MHz and 1645 MHz are narrow and the use made of them for links between space stations and mobile earth stations of various kinds is growing rapidly. In order that the further development of these facilities should not be avoidably impeded, RR 726A[2] forbids the use of these bands for feeder links under any but exceptional conditions.

The FSS bands at 4 and 6 GHz are used for most of the feeder links serving mobile-satellite systems operating around 1600 MHz. Concern was expressed at WARC-MOB-87 that difficulties will arise in the future in co-ordinating feeder links in these FSS bands, since the latter bands too are becoming congested with traffic. RR Recommendation 104[13] draws attention to this problem.

Network co-ordination

Networks of the mobile-satellite services operating near 1545 MHz and 1645 MHz are co-ordinated with other stations, space and terrestrial, in the frequency bands used for the satellite-to-mobile links and in the feeder link bands, in the ways described on pp. 349–390. However, a significant difference arises in this case, owing to the existance of CCIR texts which help to determine the criteria used in co-ordination, specifically for the MMSS and based upon the assumption that telephone circuits between ships and coast earth stations are to be extended into the public telephone network.

In keeping with the principles discussed on p. 260, CCIR Recommendation 546[14] defines an HRC for the MMSS. This MMSS HRC has much in common with the HRC for the FSS. Basically it consists of a single hop from the input of the coast earth station transmitter to the output of the ship earth station receiver, or from the input of the ship earth station transmitter to the output of the shore earth station receiver.

CCIR Recommendation 547[15] defines the noise objectives for the HRC under specified conditions. It is a relatively complex definition, necessarily so because, in order to minimise the pre-demodulator carrier-to-noise level requirement at the ship earth station and thereby to minimise the space station transmitter power requirement, it is assumed that a speech signal processing system will be used to improve the subjective signal-to-noise ratio, which will, in consequence, become a non-linear function of speech level; see also CCIR Recommendation 548[16].

Consistent with this HRC noise objective, CCIR Report 917[17] indicates the present state of work on the definition of interference criteria for the HRC. Provisional conclusions have been reached, and these may be finalised soon. Meanwhile, the report provides guidance for co-ordination in the short term.

10.3 References

1 Final Acts adopted by the second session of the World Administrative Radio Conference on the use of the geostationary-satellite orbit and the planning of space services using it (WARC-ORB-88) (ITU, Geneva, 1988)

2 Final Acts of the World Administrative Radio Conference for the Mobile Services, Geneva 1987 (WARC-MOB-87) (ITU, Geneva, 1987)

3 'Transmission characteristics of a satellite emergency position indicating radio beacon (satellite EPIRB) system operating through geostationary satellites in the 1·6 GHz band'. CCIR Recommendation 632; Recommendations and Reports of the CCIR, 1986, Volume VIII–3 (ITU, Geneva, 1986)

4 'Transmission characteristics of a satellite emergency position indicating radio beacon (satellite EPIRB) system operating through a low polar-orbiting satellite system in the 406 MHz band'. CCIR Recommendation 633; *ibid.*, Volume VIII–3

5 'Technical and operating characteristics of distress systems in the maritime mobile-satellite service'. CCIR Report 761–2; *ibid.*, Volume VIII–3

6 'Considerations on the feasibility of a single frequency band for satellite EPIRB operation in the FGMDSS'. CCIR Report 1046; *ibid.*, Volume VIII–3

7 'Relating to the protection of the band 406–406·1 MHz allocated to the mobile-satellite service'. RR Resolution 205, Final Acts of the World Administrative Radio Conference for the Mobile Services, Geneva 1983 (WARC-MOB-83; (ITU, Geneva, 1983)
8 'VHF and UHF propagation curves for the frequency range from 30 MHz to 1000 MHz'. CCIR Recommendation 370–5; Recommendations and Reports of the CCIR, 1986, Volume V (ITU, Geneva, 1986)
9 'The evaluation of propagation factors in interference problems between stations on the surface of the Earth at frequencies above about 0·5 GHz'. CCIR Report 569–3; *ibid.*, Volume V
10 'Determination of co-ordination area'. CCIR Report 382–5; *ibid.*, Volume IV/IX–2
11 'A concept of co-ordination and protection contours for use in the co-ordination of mobile earth stations'. CCIR Report 773; *ibid.*, Volume VIII–3
12 'Extension of the frequency bands allocated to the mobile-satellite and mobile services and their conditions of use'. RR Resolution 208, WARC-MOB-87; see [2] above
13 'Provision of frequency bands for feeder links in the fixed-satellite service for the mobile-satellite service or for the aeronautical, land or maritime mobile-satellite services in the bands 1530–1559 MHz and 1626·5–1660·5 MHz'. RR Recommendation 104, WARC-MOB-87; see [2] above
14 'Hypothetical telephone reference circuit in the maritime mobile-satellite service'. CCIR Recommendation 546–1; Recommendations and Reports of the CCIR, 1986, Volume VIII–3 (ITU, Geneva, 1986)
15 'Noise objectives in the hypothetical reference circuit for systems in the maritime mobile-satellite service'. CCIR Recommendation 547; *ibid.*, Volume VIII–3
16 'Overall transmission characteristics of telephone circuits in the maritime mobile-satellite service'. CCIR Recommendation 548; *ibid.*, Volume VIII–3
17 'Permissible levels of interference into telephone channels in the maritime mobile-satellite service'. CCIR Report 917–1; *ibid.*, Volume VIII–3

Chapter 11

The amateur services

11.1 Survey of regulation of the amateur services

The amateur service (AmS) is defined in RR 53 as:

> A radiocommunication service for the purpose of self-training, intercommunication and technical investigations carried out by amateurs, that is, by duly authorized persons interested in radio technique solely with a personal aim and without pecuniary interest.

Likewise, the amateur-satellite service (AmSS) is defined in RR 54 as:

> A radiocommunication service using space stations on earth satellites for the same purposes as those of the amateur service.

A substantial number of frequency bands have been allocated internationally for amateur use for terrestrial and space radio systems, and certain conditions, some agreed internationally and others determined nationally, are imposed on the use of these bands. Both the bands and the conditions are reviewed below.

In general, however, governments have perceived the wisdom of leaving the management of the amateur bands very largely to the amateurs themselves. Radio amateur associations have been set up in many countries to organise amateur activities and to plan and discipline amateur operation so that the AmS and AmSS allocations can be used effectively in ways that are of interest to amateurs. CCIR Report 906[1] provides a brief review of these activities. National radio amateur associations also provide a channel between amateurs and administrations for the discussion of matters of common interest. These national associations co-ordinate their interests worldwide through the International Amateur Radio Union (IARU), which attends with observer status those ITU conferences which are of concern to amateurs.

Frequency allocations and sharing

Many of the frequency bands allocated to the AmS or the AmSS in the international Table of Frequency Allocations (RR Article 8) are shared with other services. The administration of a country may decide not to license amateurs to operate in a part, or the whole, of one of these shared band. On the other hand, an administration may decide to make a national allocation to amateurs in a band which does not have an amateur allocation in the framed part of the international Table, and this national allocation may, or may not, be recognised internationally through a footnote to the international Table. Thus the list of bands available for the AmS and AmSS differs considerably from country to country and may differ, in any one country, from the list of allocations in the international Table. These variations should be borne in mind in considering the following list of the principal allocations in the international Table to the amateur services. Allocations are primary unless otherwise indicated.

(*a*) 1800–2000 kHz is allocated for the AmS in Region 3, sharing with the FS and various mobile and radiodetermination services. The same conditions apply in Region 2, except that 1800–1850 kHz is exclusive to the AmS. In Region 1, 1810–1850 kHz is allocated for amateurs, except in the countries listed in RR 490 and RR 493. In certain other Region 1 countries, listed in RR 491, amateurs share part of this band with the FS and mobile services but in most countries amateurs will have exclusive use of this band as soon as any remaining assignments to the FS, MM and LM have been transferred elsewhere; see RR 492 and RR Resolution 38[2]. RR 488 makes provision for additional allocations for the AmS in several Region 1 countries in the bands 1715–1800 kHz and 1850–2000 kHz at the discretion of the administrations concerned, subject to there being no interference to stations of the services to which these bands are allocated in other countries and subject to a limit of 10 W on the mean power of amateur transmitters.

(*b*) 3500–3900 kHz is allocated for the AmS in Region 3, sharing with the FS and MS. Similar conditions apply in most countries in Region 2 in the band 3500–4000 kHz, except that the AmS allocation is exclusive between 3500 and 3750 kHz. In Region 1, the AmS shares the band 3500–3800 kHz with the FS, MM and LM.

(*c*) 7000–7100 kHz is allocated for the AmS and AmSS, worldwide, except for a few countries listed in RR 527, and in general other services do not share this band. RR Resolution 641[3] specifically prohibits broadcasting in this band. There is an additional exclusive allocation for the AmS in Region 2 at 7100–7300 kHz. The band 7100–7300 kHz is allocated for the BS in Regions 1 and 3.

(*d*) 10·10–10·15 MHz is allocated worldwide for the AmS with secondary status, the primary allocation being for the FS.

(*e*) 14·00–14·25 MHz is allocated worldwide for the AmS and AmSS, sharing with no other service. There is a further worldwide allocation for the AmS at 14·25–14·35 MHz, and this is shared only with low-power stations of the FS in the countries listed in RR 535.

(*f*) 18·068–18·168 MHz and 24·89–24·99 MHz were allocated worldwide for the AmS and AmSS by WARC-79. The 18 MHz allocation is shared with the FS in one country; see RR 538. The 24 MHz allocation is shared with virtually no other service. However, RR Resolution 8[4] delayed the implementation of these new allocations until 1 July 1989, if necessary, to give time for stations of the services for which the bands were previously allocated, principally the FS and LM, to be assigned new operating frequencies.

(*g*) 21·00–21·45 MHz and 28·0–29·7 MHz are allocated worldwide for the AmS and AmSS, sharing with no other service.

(*h*) 50–54 MHz is allocated exclusively for the AmS in Region 2, in much of Region 3 and in some countries of Region 1; see RR 556–RR 560.

(*i*) 144–146 MHz is allocated worldwide for the AmS and AmSS, sharing with virtually no other service. In addition, 146–148 MHz is allocated exclusively for the AmS in Region 2; the same band is allocated for the AmS in Region 3, sharing with the FS and MS, with the exception of the few countries listed in RR 607.

(*j*) 220–225 MHz and 902–928 MHz are allocated for the AmS in Region 2 only, the 900 MHz allocation being secondary. Both of these allocations are shared, mainly with the FS and MS.

(*k*) 430–440 MHz is allocated for the AmS in Region 1, sharing with the RL and in many countries with the FS, MM and LM also. In the same band there is a primary AmS allocation in some countries in Region 2 (see RR 660) but in the rest of the Region and also in Region 3, the AmS allocation is secondary, the primary allocation being for the RL and, in many countries for the FS, MM and LM also. 435–438 MHz is allocated worldwide for the AmSS by footnote RR 664, subject to there being no harmful interference to stations of the other services for which the band is allocated and subject to there being effective arrangements for switching off the AmSS emissions if harmful interference should occur.

(*l*) There is a group of secondary allocations for the AmS between 1 and 6 GHz, as follows:

1240–1300 MHz (worldwide)
2300–2450 MHz (worldwide)
3300–3500 MHz (Regions 2 and 3)
5650–5850 MHz (worldwide)
5850–5925 MHz (Region 2)

The primary services with which these allocations are shared are mostly the RL, FS and MS. There is also sharing with the FSS at 3400–3500 MHz (space-to-Earth) and, in Region 1 only, at 5725–5850 MHz (Earth-to-space). Within four of these AmS allocations there are also allocations for the AmSS, made by RR 664, the conditions of the allocation being the same as for the band 435–438 MHz (see item (*k*) above). The frequency bands of the AmSS allocations are as follows:

1260–1270 MHz (worldwide) (Earth-to-space only)
2400–2450 MHz (worldwide)
3400–3410 MHz (Regions 2 and 3 only)
5650–5670 MHz (worldwide) (Earth-to-space only)

The feasibility of sharing between AmSS and various other services is discussed in CCIR Report 542[5].

(*m*) 10·0–10·5 GHz is allocated worldwide for the AmS with secondary status, the primary allocation being for the RL and also, in Regions 1 and 3, the FS and MS. 10·45–10·50 GHz is allocated for the AmSS, also on a secondary basis.

(*n*) Finally there are two groups of millimetre wave allocations for the amateur services. All of the allocations in one group are worldwide and primary; they are made for the AmS and the AmSS and for no other service; the frequencies are:

24·00–24·05 GHz 142–144 GHz
47·0–47·2 GHz 248–250 GHz
75·5–76·0 GHz

All of the allocations in the other group are worldwide and secondary, the primary allocation being to the RL. One of these allocations, at 24·05–24·25 GHz is made for the AmS only. The rest, at:

76–81 GHz
144–149 GHz and
241–248 GHz

are allocated for both the AmS and AmSS; see also CCIR Report 542[5].

Rules for amateur operation

In other services, the owner of a radio station is licensed to operate that specified station by the administration having jurisdiction over the station, but call signs and frequencies are assigned to stations, where appropriate. In the amateur services, the licence and the call sign are assigned to the person who operates the station.

RR 2731 recognises the right of an administration to forbid its amateurs to communicate with amateurs in other countries. RR 2732–RR 2734 require all amateur trans-national communication to be in plain language, and forbid the trans-national transmission of messages for third parties in the absence of special arrangements between the administrations concerned.

As a condition for licensing, RR 2735 and RR 2736 require administrations to ensure that amateur station operators possess the basic skills, technical and operational, that are necessary for the effective operation of their radio equipment. These skills include the ability to send Morse signals by hand and to receive Morse aurally, although RR 2735 does not insist on this capability for amateurs who are not permitted to use frequencies below 30 MHz.

In some countries amateurs play an important role in maintaining communication services in times of natural disaster. This role is recognised in RR Resolution 640[6]. This resolution, in association with RR 510, notes that, under such conditions, it may be desirable for stations which are not operated by amateurs to have the option of using AmS frequency allocations and that some communication links operating in this way may be trans-national. However, this practice is to be limited as much as possible and is only to be exercised in specified frequency bands, namely those at 3·5 MHz, 7·0 MHz, 10·1 MHz, 14·0 MHz, 18·068 MHz, 21·0 MHz, 24·89 MHz and 144 MHz.

Constraints are imposed by administrations on the types of emission which amateurs may use in the various bands and on the power of their transmitters, in order that the bandwidth allocated, which in some bands is quite small, shall be available to the extent possible for the large number of amateur operators who want to use it; see RR 2737, In shared frequency bands similar constraints may be imposed in order to protect the sharing services.

RR 2738 applies all the general rules of the ITU Convention and the Radio Regulations to amateur stations. Amongst the technical requirements laid down by the Radio Regulations, the following in particular may be noted:

Carrier frequency tolerance: No tolerances are given for the AmS in RR Appendix 7. RR 2738 and RR 2740 call for carrier frequencies to be 'as stable as the state of technical development for such stations permits'. However, regardless of the service to which they belong, RR Appendix 7 requires earth stations and space stations to maintain carriers below 2450 MHz within 20 parts per million (p.p.m.) of the nominal frequency, this tolerance being relaxed to 50 p.p.m. between 2450 MHz and 10·5 GHz and to 100 p.p.m. between 10·5 and 40 GHz.

Spurious emissions: Relatively complex constraints are applied by RR Appendix 8 to the spurious emissions of all terrestrial station transmitters operating below 17·7 GHz and space system transmitters operating below 960 MHz, not excluding those of the AmS and the AmSS (see Appendix A.5). For practical purposes, however, the direction in RR 2738, that amateur station emissions should be 'as free from spurious emissions as the state of technical development for such stations permits' seems appropriate, recognising that administrations may need to insist on exceptional measures to suppress spurious emissions in special circumstances.

Space station emissions: The requirement in RR 2612 that 'space stations be fitted with devices to ensure immediate cessation of their emissions by telecommand, whenever such cessation is required under the provisions of these Regulations' applies to AmSS satellites as well as to those of any other space service. RR 2741 goes further with a specific requirement that administrations authorising AmSS space stations 'shall ensure that sufficient earth command stations are established before launch to guarantee that any harmful interference which might be reported can be terminated by the authorising administration'.

Other constraints on space networks: RR Articles 28 and 29 list various constraints on space networks in addition to the telecommand requirement referred to in the previous item. However, for networks operating in frequency bands which are not shared with the FSS or the FS, these constraints should not affect the AmSS in the foreseeable future.

Frequencies for operation at AmS stations are not normally assigned by administrations, nor are the frequencies which such stations use notified for registration to the IFRB. The procedures of RR Articles 11 and 13 are however applicable for the co-ordination and registration of the frequencies used by earth stations and space stations of the AmSS; see also RR 2741 and RR Resolution 642[7]. These procedures are discussed on pp. 118–131 and 281–294 above in the context of the FSS. However, much of the complexity of these procedures will not arise for AmSS networks which do not use frequencies shared with the FSS, FS or MS, and especially if a non-geostationary orbit is used.

11.2 References

1 'Frequency usage in the amateur service'. CCIR Report 906; Recommendations and Reports of the CCIR, 1986, Volume VIII–1 (ITU, Geneva, 1986)
2 'Relating to the reassignment of frequencies of stations in the fixed and mobile services in the bands allocated to the radiolocation and amateur services in Region 1'. Resolution 38; Radio Regulations, edition of 1982 (ITU, Geneva, 1982)
3 'Relating to the use of the frequency band 7000–7100 kHz'. Resolution 641, *ibid.*
4 'Relating to implementation of the changes in allocations in the bands between 4000 and 27 500 kHz'. Resolution 8, *ibid.*
5 'Technical feasibility of frequency sharing by the amateur-satellite service'. CCIR Report 542–1; Recommendations and Reports of the CCIR, 1986, Volume VIII–1 (ITU, Geneva, 1986)
6 'Relating to the international use of radiocommunications in the event of natural disasters in frequency bands allocated to the amateur service'. Resolution 640; Radio Regulations , edition of 1982 (ITU, Geneva, 1982)
7 'Relating to the bringing into use of earth stations in the amateur-satellite service'. Resolution 642, *ibid.*

Chapter 12

Exploration and research by radio

12.1 The exploration and space operation services

12.1.1 Introduction

The exploration services

Systems of the exploration services obtain information about the Earth and its atmosphere, and about the universe at large, mainly in two ways:

(*a*] By passive observation, measuring at a distance the rate at which energy is emitted by matter at one or more frequencies within the radio spectrum. Instruments used for measuring the radiated power flux for such purposes are sometimes called passive sensors; for a definition of this term see RR 175.
(*b*) By active methods resembling radar, in which a radio signal is transmitted and the signal reflected by an object is received and studied in order to obtain information about that object. Instruments used for obtaining information in this way are sometimes called active sensors; for a definition see RR 174.

Thus radio astronomy is defined in RR 14 as 'astronomy based on the reception of radio waves of cosmic origin'; hence the radio astronomy service (RA), which is formally defined in RR 55. The RA is entirely passive. Radar is also used for exploring the solar system, but this activity would be categorised as RL, not RA.

The earth exploration-satellite service (EESS) is concerned with obtaining information about the Earth and its atmosphere, mainly by means of passive and active sensors, and with distributing such information. The formal definition of the service, to be found in RR 48, is as follows:

'A radiocommunications service between earth stations and one or more space stations, which may include links between space stations, in which:

- information relating to the characteristics of the Earth and its natural phenomena is obtained from active sensors or passive sensors on earth satellites,
- similar information is collected from airborne or Earth-based platforms,
- such information may be distributed to earth stations within the system concerned, and
- platform interrogation may be included.

This service may also include feeder links necessary for its operation.'

The meteorological-satellite service (MetS) is formally defined in RR 49 as a specialised application of the EESS; it is concerned with the collection and distribution of information relating to the weather.

The space research service (SRS) uses many techniques in common with the EESS but its field of activity is not confined to the Earth. It includes amongst its functions, for example, communication with and control of manned spacecraft and inter-planetary vehicles. The formal definition of the SRS, at RR 52, is in very broad terms.

The search for extra-terrestrial intelligence

Another activity that may increase man's knowledge of the universe lies in the search for radio signals of extra-terrestrial origin that might reveal intelligence that is not human. Technical ground rules that might guide the search for extra-terrestrial intelligence (SETI) are to be found in CCIR Report 700[1]. SETI is not a part of any ITU radio service, but RR 722 advises that passive SETI research is being conducted in the bands 1400–1727 MHz, 101–120 GHz and 197–220 GHz by some countries. RR Recommendation 702[2] proposes that consideration should be given at some future WARC to making some suitable provision for this use of these parts of the spectrum.

The space operation service

Spacecraft, once launched, are monitored and controlled through radio links, using tracking, telemetry and command (TTC) earth stations. Once a satellite has been established in a well-defined orbit, such links are usually operated in frequency bands allocated to the service to which the satellite belongs, by a TTC earth station which is dedicated to networks of that service. During the launch phase of a satellite, however, and throughout the operational lifetime of a spacecraft which does not attain a stable orbit, it is usual for this TTC function to be exercised by earth stations associated with the launching process. Such an earth station has to control spacecraft belonging to a variety of services, and the earth station equipment would be undesirably complicated if it were required to operate in all of the frequency bands which these various services have been allocated. The use for TTC, during the launch phase, of the relatively high microwave frequencies that are used for the normal operation of many satellite networks would also raise technical problems.

For these reasons many spacecraft have two sets of telecommand receivers and telemetry transmitters. One set is for use in normal operation with the dedicated TTC earth station, and it is assigned frequencies in the operational bands. The other set is used during the launch phase with frequencies which are standardised for the launch site TTC stations. To provide for the latter, frequency bands are allocated to the space operation service (SpO), which is defined in RR 25.

Thus the SpO is not an exploratory service. However, the management of the spectrum allocated to the SpO has much in common with the management of the spectrum used by the SRS for communication links and it is convenient to treat them together in this Chapter.

Frequency allocations

The frequency allocations which have been made to these five services fall into three groups. A review of the frequency bands that have been allocated to the EESS, MetS and SRS for passive sensors and to the RA follows in section 12.1.2. Allocations for active sensors and satellite-borne radars of the EESS, MetS and SRS are reviewed on pp. 372–373. The frequency bands allocated for the communication links which the EESS, MetS, SRS and SpO require are considered on pp. 373–378. The international registration of assignments in all of these bands is discussed on pp. 378–379.

12.1.2 Frequency allocations for passive sensing systems

Spectrum requirements for radio astronomy

Energy radiated naturally from extra-terrestrial sources extends over the whole of the radio spectrum that is observable from the Earth and indeed far beyond those limits. Radio astronomers want to observe, for each source of interest, the spectral intensity of this continuum of radiation at representative points across the radio spectrum. For such purposes there is a need for access to frequency allocations, free from man-made signals, at octave intervals or better, throughout the radio spectrum from about 3 MHz upwards. The precise frequencies defining the bands used for this kind of observation are not important, but wide bands are needed for sensitive measurements; see CCIR Reports 852[3] and 699[4].

Astronomers also observe enhanced radiation and absorption of energy at specific frequencies within the continuum spectrum. These emission and absorption 'lines' arise mainly from the presence of gas in inter-stellar space. There is a list in CCIR Recommendation 314[5] of the 42 most important lines that have been observed below 275 GHz, which is the effective upper limit of the international Table of Frequency Allocations, but there are many others; see also CCIR Report 852[3]. One line is near 327 MHz; the rest are above 1400 MHz. Astronomers would wish to have allocations made to the RA to facilitate the observation of these lines. The bandwidth of each line is very small, but the velocity of movement of sources, relative to the Earth, causes Doppler shift, broadening the frequency range within which phenomena are observable. For sources within our galaxy, a bandwidth of ±0·1% is sufficient to encompass the Doppler shift; for observation of distant galaxies a larger downward extension in frequency is required.

Spectrum requirements for Earth exploration

As with radio astronomy, passive Earth sensing involves continuum measurements, desirably at a large number of frequencies, mainly for information on conditions at and just below the Earth's surface, and also line measurements, mainly for information on the atmosphere. The frequency range of interest for continuum measurements is broadly from 1 to 100 GHz. The line frequencies of interest are mostly near 22 GHz and 55 GHz, plus about 20 lines between 100 and 275 GHz. However, whereas in the RA the bandwidth desirable for continuum measurements is large but a comparatively narrow band is sufficient for line observations, the reverse is true for Earth exploration; see CCIR Recommendation 515[6] and CCIR Reports 693[7] and 395[8]. In general the lines which are of particular interest for the RA are not of interest for Earth exploration, and vice versa, and the requirements of the MetS differ considerably from those of the EESS in general and from those of the SRS.

Frequency allocations

Perhaps inevitably, it has not been found feasible to withdraw from other use all of the bandwidth that it would be desirable to reserve for exploratory applications. Extensive allocations have indeed been made, specifically for passive systems, but there are sharing allocations in some of these bands, especially in the parts of the spectrum where use for other services is well established.

The spectrum provision made for passive systems of the RA, EESS (including the MetS) and SRS is complex, involving well over 100 allocations in all. The situation can, however, be summarised as follows:

(*a*) *For continuum measurements below 1 GHz:* For the RA there are worldwide allocations around 13·4 MHz, 25·6 MHz, 37·5 MHz and 408 MHz. Various allocations in Regions or individual countries between 80 and 220 MHz and around 608 MHz help to make good the gaps in the worldwide series. Most of these allocations are primary.

(*b*) *For continuum measurements above 1 GHz:* For the RA there are worldwide, mostly primary allocations around the following frequencies, all expressed in gigahertz:

1·41	10·65	23·8	89·0	224
2·68	15·38	31·65	110	270
4·9	22·35	43·0	166	

All of these allocations except those at 43 and 270 GHz are shared with passive sensors of the EESS and SRS. In some cases these latter services have extra bandwidth allocated, but some of these extension band allocations have secondary status. In addition, the EESS and SRS have been allocated bands, worldwide and primary, at:

18·7 21·3 36·5 55 and 64·5 GHz

No allocation has been made to bridge the gap between 5 and 10 GHz but RR 809 calls on administrations, in assigning frequencies between 6425 and 7250 MHz, to bear in mind the use of this band for passive sensors by the EESS and the SRS.

(*c*) *For line observations:* Some spectral lines observed in radio astronomy or earth exploration occur in the bands listed in (*b*) above for continuum observations. In addition there is a worldwide, primary allocation for RA line observations around 325 MHz. About 20 more narrow bands are allocated to the RA, mostly above 100 GHz and with primary status, for the observation of specific lines or groups of lines. In addition, several footnotes to the international Table of Frequency Allocations, whilst refraining from making formal allocations of additional bands for RA, draw the attention of administrations to the desirability of not making assignments to stations of other services which would cause harmful interference to astronomical observations on specified line frequencies. Perhaps the most important of these footnotes is RR 718, which relates to observations of lines due to neutral hydrogen in distant galaxies in the band 1330–1400 MHz; RR Recommendation 701[9] looks forward to some improvement in the status of RA use of this band at some future time. About ten primary allocations to the EESS and SRS, mostly above 100 GHz, provide for line observations relevant to Earth exploration.

Sharing, spurious emissions and radio astronomy

Most RA allocations are shared with allocations to the EESS and SRS, use being limited to passive sensors. For obvious reasons sharing between the various passive services raises no problems of interference and it has the positive advantage of broadening concern amongst administrations to maintain quiet radio conditions in those bands. However, many RA allocations are shared with services which use transmitters.

Radio astronomy stations use extremely sensitive receivers and for some purposes their methods press measurement technology to the limit of what is feasible. These high-sensitivity observations are very susceptible to interference from radio stations transmitting in the same frequency band and to spurious emissions radiated by stations operating in other bands. The opening words of CCIR Report 224[10] express the situation eloquently:

> 'The radiation measured in radioastronomy has a Gaussian probability distribution in amplitude, and qualitatively cannot be distinguished from the noise generated in the receivers or from thermal radiation from the Earth and its atmosphere. Furthermore, the level of cosmic radiation as received by an antenna is usually much lower than the system noise power, often by 30 dB or more. A full recognition of these facts is the key to understanding the interference problems encountered by the radio astronomy service. The radio astronomers' signal-to-noise ratio is −30 dB or worse; in extreme cases a signal-to-noise ratio as low as −60 dB may yield useful data.'

See also CCIR Report 697[11]. Sharing frequency bands with active services involves problems, only some of which are solvable by regulation.

The regulatory situation is as follows. Each administration is responsible for resolving interference problems arising between stations which are within its own jurisdiction. RR Article 22 provides a basis for resolving problems when harmful interference arises from foreign stations (see pp. 64–65). RR 163 defines 'harmful interference' as 'interference which ... seriously degrades, obstructs or repeatedly interrupts a radiocommunication service'. Uniquely among the 33 ITU radio services, the RA is not by definition a radiocommunication service, but RR 344 requires that it be treated as one for the purpose of resolving cases of harmful interference. Thus, in short:

- if an RA station suffers interference when operating in a band allocated to the RA with primary status, on a frequency assigned to the station and registered in the MIFR and
- if the interfering station can be identified and the frequency assignment which it is using does not have priority of registration in the MIFR relative to that of the RA assignment

then there is a good prospect of getting the interference suppressed. The problem becomes intractable, however, if interference arises, not from a single, identifiable station, but from many stations, the signals from individual stations being too weak for the sources to be identified. This situation will often arise for stations which are as sensitive to interference as those of the RA. Furthermore an observatory has no grounds for complaint if interference is traced to a foreign station of a service with primary allocation status in a frequency band in which the RA allocation is secondary.

Although RR 344 allows a radio astronomy station to apply its own very stringent criterion of 'harmful interference' to interference from an identifiable station operating in the same frequency band, the same paragraph of the regulations concludes with the words:

> '... protection from services in other bands shall be afforded the radio astronomy service only to the extent that such services are afforded protection from each other.'

Thus, a RA station cannot claim special protection, by reason of its own exceptional sensitivity, from spurious emissions from transmitters operating in different frequency bands. Spurious emissions from space station transmitters are a very serious hazard for radio observatories, and the ITU has determined no objective standards so far for the suppression of spurious emissions from space stations operating above 960 MHz.

The sharing situation for the RA allocations is briefly as follows:

(*a*) 13·36–13·41 MHz is shared with the FS on equal primary terms.
(*b*) 25·55–25·67 MHz was allocated for the exclusive use of the RA by WARC-79. However, RR 545 and RR Resolution 8[12], also agreed at that conference, allowed time for new frequency assignments to be found in other bands for stations of the services to which the band was previously allocated. The FS and MS (except AeM) were given until 1 July 1989 to vacate the 25·55–25·60 MHz band. The primary allocation to the BS in the 25·60–25.67 MHz band continues in force until frequency assignment planning has been implemented for HF broadcasting; as is noted on pp. 161–163, this has not happened yet.

Nevertheless, RR 545 also promises a long term solution in the 26 MHz band for the problem posed by large numbers of interfering signals which are so weak that they cannot be identified, by recording that:

> 'after completion of all the above-mentioned provisions, all emissions capable of causing harmful interference to the radio astronomy service in the band 25 550–25 670 kHz shall be avoided'.

In other words, in that band, administrations have surrendered the general right, admitted by RR 342, to assign frequencies contrary to the international Table of Frequency Allocations if no consequential harmful interference is demonstrable. Similar action has been taken with regard to a number of RA allocations higher in the spectrum; see examples in (*d*) and (*e*) below.
(*c*) Between 30 and 1000 MHz there are eight substantial allocations to the RA. Two of them, at 74 MHz and 611 MHz, both limited to Region 2, are virtually exclusive to the RA within that Region. Of the remainder, two are secondary and all are shared with primary allocations to the FS and some or all of the mobile services.

(*d*) Between 1 and 23 GHz the RA has six allocations which are used mainly for continuum measurements. In a part at least of each band the RA allocation is primary and thus the most difficult sharing situation does not arise; in some cases emissions of any kind are prohibited. Nevertheless, in all but one of these allocations (the exception being 1400–1427 MHz) either all or some substantial part of the band is shared with allocations to the FS and MS, including the AeM in some cases, and the band 2655–2690 MHz is shared with primary allocations to the FSS (space-to-Earth) and BSS. These sharing services are likely to raise substantial interference problems for the RA. In addition, several RA bands are liable to interference from harmonic radiation from space stations, and especially those transmitting in the bands 3·4–4·2 GHz (FSS), 7·25–7·75 GHz (FSS etc.) and 10·7–11·7 GHz (FSS).
(*e*) Above 23 GHz the RA has eight allocations which are required for continuum measurements. All are primary, and emissions of any kind are prohibited in most of these bands. However, the 23·8 GHz RA allocation is susceptible to interference from harmonic radiation from space stations transmitting in the 11·7–12·2 GHz band (BSS and FSS).
(*f*) Of the RA allocations made for line observations, mostly above 50 GHz, most have primary status but they share bands with services such as the FS and the RL with equal status.

Clearly the limitation of interference to RA observatories to meet the very stringent requirements of the service presents severe problems in some bands, even when the receiving system has been designed so as to minimise the impact of interference on the integrity of the output data; see in particular CCIR Report 696[13].

The difficulty is greatest in the HF bands, where useful observations will probably be limited to times when the ionosphere does not support the propagation of waves at the frequency on which observations are being made from distant, potentially interfering terrestrial transmitters to the observatory.

At higher frequencies, interference by line-of-sight or quasi-line-of-sight propagation can be largely avoided by locating a radio astronomy observatory at a site which is isolated from transmitting stations by distance and is screened by hills. Much can then be done by careful spectrum management. The number of observatories is not very great and most observatories operate in only a few of the bands allocated for the RA. The administration having jurisdiction over an observatory, if necessary in collaboration of the administrations of neighbouring countries, can then limit interference from in-band transmitters by refraining from assigning frequencies near the observatory's working frequencies to transmitting stations within interference range of the observatory. Above all, assignments for transmission should not be made to aircraft stations at frequencies close to RA assignments.

The determination of what is to be regarded as 'interference range' in such cases must take into account high-loss propagation modes such as diffraction round obstacles, ionospheric and tropospheric scatter, tropospheric ducting and other sporadic propagation modes arising from the vagaries of the ionosphere and troposphere. Propagation via these modes may cause severe interference to very sensitive receiving systems, even though the modes might be totally disregarded as a medium for supporting a communication link (see in particular Appendices B.3.3, B.4.5, B.4.6, and B.6.11).

It is incumbent on all administrations to make progress in limiting spurious emissions which can do harm to radio astronomy. One important field for current action is the determination of appropriate standards for the suppression of spurious emissions from space stations and aircraft stations and the implementation of those standards in the field. The necessary studies of standards are in progress; see in particular CCIR Reports 713[14], 807[15] and 697[11]. There is general guidance on these matters in RR Article 36 and CCIR Recommendations 314[5], 517[16] and 611[17].

Despite all that can be done to improve the conditions under which the RA operates, limitations will remain, imposed by nature or by human activity. In the HF band and below, observation would be limited by ionospheric effects and atmospheric noise even if there were no major problems of interference from transmitters. Between 30 MHz and 23 GHz (an arbitrarily chosen upper bound) interference from emissions may be reduced eventually through changes in frequency allocations and better control of spurious emissions, but it is unlikely to disappear and interference due to spurious emissions from space stations seems likely to remain a serious problem. Above 23 GHz interference seems likely to increase as millimetre-wave bands are taken into use for active services.

One possible way of escaping from these problems would be to establish observatories on the side of the Moon which is remote from the Earth. Consideration has been given to the spectrum management policies that should be followed to ensure that this remains a useful option for the future; see CCIR Report 539[18] and Recommendation 479[19]. The first steps have been taken to give regulatory force to the protection of the shielded zone of the Moon from emissions that could be harmful to future radio astronomy observatories there; see RR 2632–RR 2635.

Sharing with passive sensors for Earth exploration

Some of the bandwidth of every allocation made to the EESS and SRS for passive sensors below 23 GHz is shared with active services and in some bands the allocation for passive sensors has secondary status. Sharing is mostly with the FS and MS but one potentially important band is shared with RL and another is shared with the FSS (space-to-Earth). However, in the frequency bands above 23 GHz, the passive sensor allocations are primary and most are exclusive to passive services.

The thresholds beyond which interference from terrestrial services degrades the performance of satellite-borne passive sensors used for Earth exploration are considered in CCIR Report 694[20]. These thresholds are substantially higher than those given in CCIR Report 224[10] for the RA.

On the other hand the sharing conditions are in general such as make high levels of interference more likely in the bands used for passive sensors. Report 694 shows that in some circumstances the presence of a large number of FS transmitting stations spread over a large geographical area or just a few MS transmitters with non-directive antennas would make it impossible to obtain good passive sensor measurements for that same area; see also CCIR Report 942[21]. The area for which data is lost due to RL emissions would be greater. In general the passive sensor allocations have secondary status where the worst interference conditions occur. Nevertheless, even in those bands sharing does not make the securing of useful data totally impossible.

There is another significant interference mode for satellite-borne passive sensors. Radio frequency energy transmitted downwards by active satellites in high orbits is scattered upwards after having struck the ground. It can then cause considerable interference to passive sensors which are scanning the earth from low orbits. CCIR Report 850[22] shows that passive sensors would suffer severe interference via this mode from transmitters on space stations of the FSS in the band 18·6–18·8 GHz unless the PFD set up by the FSS at the Earth's surface was kept far below the constraint set by RR 2577–RR 2580 to protect the operation of receivers of the FS and MS (see p. 111). This poses regulatory problems, so far unsolved, since in Region 2 the allocations for passive sensors and the FSS have equal primary status; see also RR Recommendation 706[23].

In all bands passive sensors are liable to receive interference from spurious emissions from stations transmitting in other bands, not least the adjacent bands. The foreseeable effects of this interference are considered in CCIR Report 987[24]. In general it is found that serious interference can be avoided by designing the passive sensors to receive less than the full bandwidth of the allocation, leaving guard bands at the band edges to isolate the sensors from the most intense spectral density of spurious emissions.

12.1.3 Frequency allocations for active sensing systems

Active sensors and radars on satellite of the SRS, EESS and MetS are used in three basic modes (called the scatterometer, altimeter and imager modes) to acquire information on a wide range of the properties of the Earth's surface and atmosphere. Different kinds of radar response are produced by the use of different radio frequencies and some operational modes involve the simultaneous use of different frequencies. The physical basis for these techniques is reviewed briefly in CCIR Report 693[7] and the sensors used in a number of systems in current operation, or planned for use, in the EESS and the MetS are described in CCIR Reports 535[25] and 395[8]. A scheme of frequency allocations that would be desirable for radars and active sensors on satellites is set out in CCIR Recommendation 577[26].

A total of 11 frequency bands have been allocated internationally for active sensors and satellite-borne radars, around

1·26 GHz	9·65 GHz	24·15 GHz
3·2 GHz	10·0 GHz	35·55 GHz
5·3 GHz	13·7 GHz	78·5 GHz
8·6 GHz	17·25 GHz	

Most of these allocations have been made for both EESS and SRS. The EESS allocations are automatically available for MetS space stations. Exceptionally the 10 GHz band is limited to the MetS and the 24 GHz band is limited to EESS. Every one of these bands is shared with the terrestrial RL and in all but the two highest bands the active sensors have secondary allocations; however, CCIR Report 695[27] concludes that such sharing will not usually degrade the performance of the satellite systems unacceptably.

12.1.4 Frequency allocations for communication links

Although the SRS, EESS and MetS share many frequency bands which are allocated for sensing, sharing of allocations for communication links is more limited. In most situations these services, together with the SpO, have competing needs for spectrum for communication links and it has been found desirable to make separate allocations for them. These allocations are reviewed in convenient groups below.

Frequency allocations and sharing for the space operation service

Some functions of the SpO are carried out in the immediate vicinity of a spacecraft launching site and a number of allocations with limited geographical applicability have been made for these functions between 100 and 1000 MHz. The management of assignments in these bands involves sharing problems, although they can usually be resolved by the responsible administration.

In addition to these sub-Regional allocations, ten frequency bands have been allocated for the SpO with wide applicability. Five of these allocations are below 1 GHz, being around 137 MHz, 149 MHz, 270 MHz, 401 MHz and 450 MHz. Most of these allocations are worldwide but most of them have secondary status, all are narrow and all are shared with other services having equal or higher allocation status. Some involve RR Article 14 consultation.

Of the five SpO bands above 1 GHz:

- There is a pair of worldwide allocations at 1427–1429 MHz (Earth-to-space) and 1525–1535 MHz (space-to-Earth). Both are primary allocations, sharing with the FS and MS (except AeM in most situations) and also sharing with MMSS and LMSS (both space-to-Earth) in the sub-band 1530–1535 MHz.

- Two more worldwide allocations are made by RR 747 and RR 750, assignments in both being subject to consultation in accordance with RR Article 14. The bands are 2025–2110 MHz (Earth-to-space) and 2200–2290 MHz (space-to-Earth). Both bands may also be used for inter-satellite links. Sharing is with links of the same kinds belonging to the SRS and EESS and also with the FS and MS.
- The fifth allocation, at 7125–7155 MHz (Earth-to-space), made by RR 810, is limited to Region 2. Sharing is with FS and MS and assignments to the SpO are subject to consultation in accordance with RR Article 14.

Thus, the management of frequency assignments for the SpO involves sharing in all bands. All space-to-Earth assignments above 1 GHz are constrained by the PFD limits imposed by RR Article 28 and most bands involve RR Article 14 consultation. Sharing problems are discussed in CCIR Reports 845[28] and 396[29]. Substantial problems arise, and they have to be solved on a case by case basis by the responsible administration, often no doubt with the assistance of the administrations of neighbouring countries.

Frequency allocations and sharing for the meteorological-satellite service

Meteorological-satellite networks are complex, typically incorporating communication facilities for:

(*a*) collecting meteorological data from data collection platforms on land, at sea and in the air and relaying them to data acquisition earth stations,
(*b*) transmitting data collected by active and passive sensors on satellites to data-acquisition earth stations, and
(*c*) distributing information on meteorological conditions, through a large number of small-antenna earth stations, to users. This information is typically in the form of facsimile weather maps (WEFAX), based on data derived from (*a*), (*b*) and other sources after processing on the ground.

TTC facilities are also necessary; see CCIR Report 395[30]. Some of these satellites incorporate maritime and aeronautical search-and-rescue sub-systems. Many meteorological satellites are in low orbits but the use of geostationary satellites is increasing. The use of geostationary data relay satellites to collect data from satellites in low orbits for forwarding to a single data acquisition earth station is contemplated.

Eight main frequency allocations have been made for the MetS, all worldwide. Four of these are below 1 GHz and they are used mainly for data collection. The allocation around 400·5 MHz has primary status and those around 137 MHz, 402 MHz and 465 MHz are secondary. Sharing is mainly with MetA, FS and MS. Sharing between the MetS and the MetA is considered in CCIR Report 541[31]; the conclusion reached is that satisfactory sharing is feasible, given geostationary meteorological satellites and careful siting of stations. Successful sharing between the MetS and the FS and MS requires careful management of assignments on the part of the administration responsible, in co-ordination where necessary with the administrations of neighbouring countries.

The four allocations above 1 GHz are all primary and worldwide, as follows:

1670–1710 MHz (space-to-earth): The allocation is shared with a primary allocation for the MetA. In addition there are primary allocations for FS and MS (except AeM) at 1670–1690 MHz and for FS at 1700–1710 MHz. RR 740 and RR 741 make further allocations to the FS and MS (except AeM) at 1690–1700 MHz in a number of specified countries.

7450–7550 MHz (space-to-Earth) and 8175–8215 MHz (Earth-to-space): These allocations are shared with primary allocations to the FSS (in the same directions of transmission), the FS and the MS (excepting AeM at 7 GHz). There is also a primary allocation in Region 2 at 8 GHz to the EESS operating in the reverse direction of transmission, the EESS allocation being secondary in Regions 1 and 3.

18·1–18·3 Ghz (space-to-Earth): This primary allocation is made by footnote RR 870 and only geostationary meteorological satellites may use it. It is shared with the FSS (also space-to-Earth), the FS and the MS.

Thus, all space-to-Earth MetS allocations above 1 GHz are shared with the FS and MS, although this sharing is limited to a relatively small number of countries in the sub-band 1690–1700 MHz. The terrestrial services are protected from interference due to MetS down-links by sharing constraints in the form of limits on the MetS signal PFD at the Earth's surface; see RR 2556–RR 2560, RR 2565–RR 2568 and RR 2577–RR 2580 and also p. 111. A further sharing constraint, given in RR 2552–RR 2555, is applied to MetS down-links to protect the MetA in the band 1670–1700 MHz. This latter constraint is discussed in CCIR Report 541[31] and is found to be compatible with the operation of MetS systems but CCIR Report 851[32] shows that the more general constraint which protects the FS and the MS is too severe to permit the use of earth stations of least cost for receiving WEFAX data broadcasts.

The other problems raised by sharing are resolved by frequency co-ordination (see pp. 378–379).

Frequency allocations and sharing for research in deep space

Some important space research projects involve communication between Earth and distant spacecraft on inter-planetary paths. The reception of such links may pose particularly difficult interference problems, comparable with those of the RA. The degree of protection from interference which is required at earth stations engaged in deep space research is considered in CCIR Report 685[33]. Interference from the spurious emissions of space station transmitters on Earth satellites is considered in CCIR Report 844[34]. Conclusions are drawn on criteria for sharing and feasible sharing arrangements in CCIR Recommendation 578[35]. Because of these problems, a number of frequency bands allocated to the SRS have seen set aside for use for links to and from deep space, defined by RR 169[36] (following CCIR Recommendation 610[37] as 'space at distances from the Earth equal to or greater than 2×10^6 km.

The allocations of spectrum that are desirable for SRS (deep space) are discussed in CCIR Reports 536[38], 683[39] and 849[40] and in CCIR Recommendation 576[41]. The bands that have been allocated for this purpose are as follows:

2110–2120 MHz (Earth-to-space)
2290–2300 MHz (space-to-Earth)
5650–5725 MHz (both directions of transmission available)
7145–7190 MHz (Earth-to-space)
8400–8450 MHz (space-to-Earth)
12·75–13·25 GHz (space-to-Earth)
16·6–17·1 GHz (Earth-to-space)
31·8–32·3 GHz (space-to-Earth)
34·2–34·7 GHz (Earth-to-space)

None of these SRS allocations is exclusive, most of them are secondary in status and some require consultation on assignments under the terms of RR Article 14. Some are footnote allocations, limited to a few countries. Part of the 31·8–32·3 GHz band is shared with the ISS. But none of the other services which share these nine bands uses it for space-borne transmitters. Only a few earth stations operate in the SRS (deep space) bands and it is possible for the administrations with jurisdiction over these stations to manage the assignment of frequencies to stations of the sharing services so that interference is kept to an acceptably low level.

It should be noted that RR 2544–RR 2546 and RR 2551 impose special limits on the spectral EIRP density which earth stations of the SRS, transmitting in deep space bands, may radiate at low angles of elevation. A number of other constraints are applied by other paragraphs of RR Article 28 but they are unlikely to have any significance in the normal operation of the service.

Frequency allocations and sharing for other SRS applications and the EESS

The communications requirements of the SRS (other than deep space) and the EESS (other than MetS) have much in common. Their shared use of frequency bands is quite limited, but nevertheless it is convenient to consider the two groups of applications together.

Firstly the SRS has ten secondary allocations of very small bandwidth between 2 and 40 MHz. There are several more such allocations, mostly secondary but some primary, around 140 MHz and near 400 and 460 MHz.

Other secondary allocations around 400 and 460 MHz have been made to the EESS. These allocations are used mainly for spacecraft beacons and low-information-rate communication channels for low-orbiting satellites and spacecraft in sub-orbital trajectories.

The basic spectrum requirements of the SRS and the EESS for communication links are mostly above 1 GHz and they are discussed in CCIR Reports 548[42] and 535[43] respectively. The preferred frequencies for the two services are considered in CCIR Reports 984[44] and 692[45]. The special needs of data relay satellite systems are considered in CCIR Report 848[46].

The allocations that have been made for these two services above 1 GHz are, with few exceptions, secondary or subject to co-ordination under the conditions of RR Article 14. In most cases it is feasible for EESS and SRS systems to operate with such resources, given careful spectrum management to contain interference from nearby stations. The main allocations (worldwide but secondary or involving RR Article 14 consultation unless otherwise indicated) are as follows:

(*a*) 1525–1535 MHz, EESS

(*b*) 1690–1710 MHz, EESS

(*c*) 2025–2110 MHz, EESS and SRS (Earth-to-space) and 2200–2290 MHz, EESS and SRS (space-to-Earth): Both bands may also be used for inter-satellite links. These frequency bands are also allocated to the SpO for the same directions of transmission on the same terms (that is, subject to RR Article 14 consultation), as well as the FS and MS, which have primary status. The sharing constraints given in RR 2557 to RR 2560 apply to space station transmitters using these allocations. Studies summarised in CCIR Reports 846[47] and 981[48] indicate that interference problems, both among the space services and between space services and the terrestrial services which have primary allocations in these bands, should be solvable.

(*d*) 7190–7235 MHz, SRS (Earth-to-space)

(*e*) 8025–8400 MHz, EESS (space-to-Earth): Primary in Region 2 but secondary elsewhere. Sharing is with the FSS and MetS (both Earth-to-space) and the FS, all of these sharing allocations being primary worldwide. The MS also has a primary allocation worldwide in this band, but RR 814 eases some of the interference problems of the EESS by prohibiting transmission by aircraft stations in Region 2. The sharing problems for the EESS in the 8025–8400 MHz band are examined in CCIR Report 540[49]. The conclusion reached there is that satisfactory operation is feasible, even in Regions 1 and 3, provided that systems adhere to a number of constraints. RR 2631 imposes a constraint on the PFD that a non-geostationary Earth exploration satellite may set up at the GSO in this frequency band, to protect space station receivers of the FSS and MetS.

(*f*) 8450–8500 MHz, SRS (space-to-Earth), primary worldwide, shared with FS and MS (except AeM). CCIR Report 687[50] considers the sharing situation and finds no major problems; the situation is indeed typical of space service down-link bands.

(*g*) There are various worldwide secondary allocations for the SRS between 13·25 and 15·35 GHz, which in aggregate occupy most of that frequency range, sharing with a variety of primary services. CCIR Report 847[51] examines some of the sharing problems arising.
(*h*) 25·25–27·5 GHz, EESS (space-to-space). This worldwide secondary allocation is shared with allocations for the FS and MS, which are primary. The FSS has a primary allocation, (Earth-to-space) at 27·0–27·5 GHz.
(*i*) 29·95–30·0 GHz, EESS (space-to-space). RR 882 allocates this band for the EESS for TTC links on a secondary basis, sharing with the FSS (Earth-to-space) which has primary allocation status. RR 2614 requires geostationary data relay satellites of the EESS, communicating with non-geostationary satellites, to limit the power of their emissions in this band in the direction of the GSO so that interference to FSS space station receivers will not exceed acceptable limits.
(*j*) and (*k*) 31·0–31·3 GHz and 31·8–32·3 GHz. These are worldwide secondary allocations for the SRS, sharing with primary allocations for the FS and MS at 31 GHz and primary allocations for the RN and ISS at 32 GHz.
(*l*) 65–66 GHz. This worldwide primary allocation for the SRS and EESS is shared with no other primary service.

12.1.5 Basic regulations and international registration of assignments

Regulations and constraints

Several paragraphs of the RR applicable to networks of the FSS and referenced on pp. 271–275 and 279 apply also to networks of the SRS, the EESS and the MetS, although in some cases with minor variations:

- RR 2612 requires means for telecommanding the cessation of space station emissions to be provided.
- RR 2613 requires non-geostationary networks of any service to refrain from interfering with geostationary FSS networks.
- RR 303 and RR Appendix 7 apply a tolerance to the carrier frequencies of transmitters at earth stations and space stations:

  Below 2·45 GHz it is 20 p.p.m.
  Between 2·45 and 10·5 GHz it is 50 p.p.m.
  Between 10·5 and 40 GHz it is 100 p.p.m.

- RR 304 and RR Appendix 8 apply limits to spurious emissions below 960 MHz (see Appendix A.5 below) and urge the best feasible suppression of spurious emission at higher frequencies, see also CCIR Report 980[52].
- RR 2615–RR 2623 require geostationary satellites to be kept, if necessary, at their nominal longitudinal station within ±0·1° if operating in a frequency band which is shared with the FSS, or ±0·5° if the band is not shared with the FSS.

- RR 2628–RR 2630 require the antenna beams of geostationary satellites to be pointed, if necessary, within 0·3° or 10% of the half-power beamwidth (whichever is less stringent) of their nominal directions.
- Almost all frequency allocations made for the exploration services and the SpO for communication down-links are shared with the FS; accordingly the PFD constraints applicable to the FSS apply to these services also; see RR 2552–RR 2585 and p. 111 Similarly some parameters of earth station emissions may be constrained by RR 2540–RR 2548 (see p. 266).

Various specific sharing constraints also apply to these networks, depending on the sharing situation, but these are reviewed on pp. 000–000.

Frequency assignments

The intricate pattern of sharing in which most frequency allocations for the exploration services and the SpO are involved creates a difficult task of national spectrum management for administrations. In addition, frequency assignments will often require international consultation in accordance with RR Article 14, will usually be co-ordinated internationally and will often be notified to the IFRB for registration in the MIFR.

In addition to any necessary consultation under RR Article 14, frequency assignments to earth stations or space stations of the SRS, the EESS, the MetS or the SpO are co-ordinated with terrestrial stations and with other space networks by the procedures of RR Article 11. After successful co-ordination the assignments are notified to the IFRB for registration in the MIFR by the procedures of RR Article 13. Both of these Articles were significantly amended at WARC-ORB-88; see the Final Acts of that conference[36]. The procedures are, in principle, the same as those applied to the FSS (see pp. 118–131 and 281–294).

Frequencies assigned to radio astronomy stations are not co-ordinated. When an assignment is made, the information listed in Section IV of RR Appendix 3[36] may be notified to the IFRB (see RR 1492) and will be recorded in the MIFR (see RR 1546).

12.2 Standards of frequency and time

Some transmitting stations, used primarily for communication and typically of the FS, MM and BS, operating in the VLF, LF or MF bands, emit carrier waves of high frequency stability, the absolute value of the frequency being known by comparison with standard frequency sources. Some radionavigation transmitters have very stable carrier frequencies and some transmit modulating signals exhibiting precisely determined time intervals which may be correlated with standard time sources; this may also be true of the synchronising waveforms of VHF and UHF television broadcasting signals. Within the service areas of these stations the signals may be used for comparing and calibrating laboratory frequency sources and clock systems. There are lists of stations having emissions that can be used in these ways in CCIR Report 267[53]. In addition, of course, broadcasting stations transmit time signals for the use of the general public.

However, these facilities are not sufficiently precise for some purposes and they are not available worldwide. In particular, there is a need for means for very precise comparison of the various national standards of time and frequency and for the broadcasting of very precise time standards which are simple to interpret. The standard frequency and time signal service (SFS) has been defined (see RR 50) for these purposes. Stations of the SFS operating in these bands are listed in CCIR Report 267[53], together with information on signal parameters.

In the future there will be a need for even greater precision in the comparison of frequency and time sources and in the distribution of these standards. The principal factor limiting the precision of comparison and distribution through the SFS is the short-term variability of terrestrial radio propagation over long distances. It is foreseen that satellite communication will permit precision to be improved; see CCIR Recommendation 582[54] and CCIR Report 518[55]. Consequently, the standard frequency and time signal-satellite service (SFSS) has been defined; see RR 51.

The terrestrial service

The frequency bands allocated for the SFS are as follows:

19·95–20·05 kHz (worldwide)
2495–2505 kHz (Regions 2 and 3), 2498–2502 kHz (Region 1)
4995–5005 kHz (worldwide)
9995–10 005 kHz (worldwide)
14 990–15 010 kHz (worldwide)
19 990–20 010 kHz (worldwide)
24 990–25 010 kHz (worldwide)

The nominal standard frequencies radiated in most cases in these bands are 20 kHz and 2·5, 5, 10, 15, 20 and 25 MHz. These allocations are all primary and are not shared with any other service with primary status. However, in each except the 20 kHz band, a narrow sub-band which does not overlap with the nominal standard frequency is allocated with secondary status to the SRS.

In addition to these SFS allocations, footnotes to the international Table of Frequency Allocations (RR Article 8) recognise that standard frequency emissions and precision time signals may be transmitted in other specified bands by stations of the services to which those bands are allocated. RR 447 provides this option in a number of VLF and LF bands. RR 516, RR 529 and RR 536 provide the option in Region 3 for the bands 3995–4005 kHz, 7995–8005 kHz and 15 995–16 005 kHz.

A considerable number of countries provide standard frequency and time signal emissions in the SFS frequency allocations. There is considerable congestion. A few emissions have their carrier frequency offset by a few kilohertz from the nominal standard frequency, but most operate at the nominal standard frequency. Interference between stations considerably reduces the precision with which these signals can be used and this constitutes the chief spectrum management problem of the SFS. CCIR Report 732[56] discusses various ways of reducing interference within the service; see also CCIR Recommendations 374[57] and 537[58].

RR Article 33 requires new assignments to the SFS bands to be co-ordinated with existing assignments made previously by other administrations, so as to reduce interference to an acceptably low amount. With that done, assignments may be notified to the IRFB in accordance with RR 1214–RR 1217 and, if they are otherwise valid, they are registered in the MIFR without scrutiny as to interference prospects.

The space service

The frequency bands allocated for the SFSS are as follows:

(*a*) 400·05–400·15 MHz, worldwide, primary and exclusive, except for sharing on an equal primary basis with the FS and MS in countries listed in RR 647.

(*b*) 4200–4204 MHz (space-to-Earth) and 6425–6429 MHz (Earth-to-space) worldwide, subject to agreements obtained under the procedure set forth in RR Article 14; see RR 791.

(*c*) 13·4–14·0 GHz (Earth-to-space), worldwide but secondary. This allocation is shared with a primary allocation for the RL, a secondary allocation for the SRS and with the FS and MS in the countries listed in RR 854.

(*d*) 20·2–21·2 GHz (space-to-Earth), worldwide but secondary. This allocation is shared with primary allocations for the FSS and MSS (both in the same direction of transmission) and with the FS and MS in the countries listed in RR 873.

(*e*) 25·25–27·0 GHz (Earth-to-space), worldwide but secondary, shared with primary allocations for the FS and MS and a secondary allocation for the EESS (space-to-space).

(*f*) 30·0–31·3 GHz (space-to-Earth), worldwide but secondary. Below 31.0 GHz this band is shared with primary allocations for the FSS and MSS (both in the contrary direction of transmission) and with secondary allocations for the FS and MS in the countries listed in RR 883. Above 31·0 GHz the sharing services are the FS and MS, both primary, and the SRS, which is secondary in most countries but primary in the countries listed in RR 885.

There have been many studies already and extensive experimentation, exploring ways in which satellites might be used for improving time and frequency dissemination; see CCIR Report 518[55]. The problems of sharing in the allocations set out above have also been studied and appear to be conditionally solvable; see CCIR Report 736[59].

The deployment of systems in the SFSS has not begun yet. When it begins, the procedures that are used in the FSS for co-ordinating, notifying and registering frequency assignments to earth stations and space stations will be used in the SFSS also, modified as necessary to take into account the secondary status of most of the SFSS allocations and some of the sharing allocations; see RR Articles 11 and 13 and, where appropriate, RR Article 14 (see also pp. 000 and 000 above).

12.3 The meteorological aids service

The meteorological aids service (MetA) is defined in RR 47 as follows:

> A radiocommunication service used for meteorological, including hydrological, observations and exploration.

Typical stations of the MetA are of three kinds:

- Some, called radiosondes, are flown on free-floating balloons which are released every few hours from meteorological observatories to enable measurements to be made of meteorological conditions in the upper atmosphere.
- Others are carried by hydrological buoys at sea, where data on marine conditions are collected.
- The third kind of station is located on the ground, typically at a permanent site, being used to command the stations on balloons and buoys to transmit data on the environment, and to receive those data.

The principal frequency allocations that have been made for the MetA are as follows:

(*a*) 2025–2045 kHz: A secondary allocation, for Region 1 only, its use being limited to oceanographic buoy stations. The allocation is shared with the FS, LM, AeM (OR) and MM, all of which have primary status.
(*b*) 27·5–28 MHz: A worldwide primary allocation, shared with equal primary status with the FS and MS. However, RR Recommendation 620[60] asks administrations that are using this band for the MetA to transfer operations to appropriate higher bands as soon as possible.
(*c*) 153–154 MHz: A secondary allocation, for Region 1 only. The allocation is shared with the FS, LM, MM and AeM (OR), all of which have primary status.
(*d*) 400·15–406·0 MHz: A worldwide primary allocation. Part of this band is shared with SRS and MetS and another part of the band is shared with the SpO (all space-to-Earth), all also having primary allocation status. There are various sharing secondary allocations.

(*e*) 1668·4–1700·0 MHz: A worldwide primary allocation. The allocation is shared with various other services, the most significant being the primary allocations for the MetS, FS, LM and MM.
(*f*) 35·2–36 GHz: A worldwide primary allocation, shared with the RL and with the FS in the countries listed in RR 894.
(*g*) Other bands are allocated for the MetA in limited geographical areas in accordance with RR 540, RR 542 and RR 738.

The most important MetA allocations are those at 400 and 1680 MHz, and they are both shared with the MetS. The feasibility of sharing spectrum in this way has been considered by the CCIR and the conclusion reached is that sharing is conditionally feasible, provided that the MetS satellite is geostationary; see CCIR Report 541[31]. Sharing between MetA and terrestrial services, however, raises other problems. It will be necessary for an administration to take care in assigning frequencies in MetA bands to stations of the FS, LM and MM which are located close to MetA stations if interference is to be avoided.

Frequencies assigned to MetA stations at permanent sites are notified to the IFRB for scrutiny and registration in the MIFR in accordance with RR Article 12, in the same way as are those for the FS:

- For frequencies below 28 MHz see pp. 84–86.
- For frequencies above 28 MHz, not shared with space services, see pp. 103–106.
- For frequencies shared with space services, it may sometimes be necessary to implement the co-ordination procedures set out on pp. 118–131.

Frequencies assigned to balloon-borne and buoy stations are not notified to the IFRB, nor are they registered.

12.4 References

1 'Radiocommunication requirements for systems to search for extraterrestrial intelligence (SETI)'. CCIR Report 700–1; Recommendations and Reports of the CCIR, 1986, Volume II (ITU, Geneva, 1986)
2 'Relating to the use of the frequency bands 1400–1727 MHz, 101–120 GHz and 197–220 GHz for search for intentional emissions of extraterrestrial origin'. Recommendation 702; Radio Regulations, edition of 1982 (ITU, Geneva, 1982)
3 'Characteristics of the radioastronomy service and preferred frequency bands'. CCIR Report 852–1; Recommendations and Reports of the CCIR, 1986, Volume II (ITU, Geneva, 1986)
4 'Ionospheric limitations to ground-based radioastronomy below 20 MHz'. CCIR Report 699–2; *ibid.*, Volume II
5 'Protection for frequencies used for radioastronomical measurements'. CCIR Recommendation 314–6; *ibid.*, Volume II
6 'Preferred frequencies for passive sensing measurements'. CCIR Recommendation 515; *ibid.*, Volume II
7 'Preferred frequency bands for active and passive microwave sensors'. CCIR Report 693–2; *ibid.*, Volume II

8 'Radiocommunications for meteorological satellite systems'. CCIR Report 395–4; *ibid.,* Volume II
9 'Relating to the use of the frequency band 1330–1400 MHz by the radio astronomy service'. Recommendation 701; Radio Regulations, edition of 1982 (ITU, Geneva, 1982)
10 'Interference protection criteria for the radioastronomy service'. CCIR Report 224–6; Recommendations and Reports of the CCIR, 1986, Volume II (ITU, Geneva, 1986)
11 'Interference to the radioastronomy service from transmitters in other bands'. CCIR Report 697–2; *ibid.,* Volume II
12 'Relating to implementation of the changes in allocations in the bands between 4000 kHz and 27 500 kHz'. Resolution 8; Radio Regulations, edition of 1982 (ITU, Geneva, 1982)
13 'Feasibility of frequency sharing between radioastronomy and other services'. CCIR Report 696–1; Recommendations and Reports of the CCIR, 1986, Volume II (ITU, Geneva, 1986)
14 'Spurious emissions from earth stations and space stations of the fixed-satellite service'. CCIR Report 713–1; *ibid.,* Volume IV–1
15 'Unwanted emissions from broadcasting-satellite space stations'. CCIR Report 807–2; *ibid.,* Volume X/XI–2
16 'Protection of the radioastronomy service from transmitters in adjacent bands'. CCIR Recommendation 517–1; *ibid.,* Volume II
17 'Protection of the radioastronomy service from spurious emissions'. CCIR Recommendation 611; *ibid.,* Volume II
18 'The protection of radioastronomy observations in the shielded zone of the Moon'. CCIR Report 539–1; *ibid.,* Volume II
19 'Protection of frequencies for radioastronomical measurements in the shielded zone of the Moon'. CCIR Recommendation 479–2; *ibid.,* Volume II
20 'Sharing considerations and protection criteria relating to passive microwave sensors'. CCIR Report 694–2; *ibid.,* Volume II
21 'Frequency sharing between the fixed service and passive sensors in the band 18·6–18·8 GHz'. CCIR Report 942–1; *ibid.,* Volume IX–1
22 'Frequency sharing by passive sensors with the fixed, mobile except aeronautical mobile, and fixed-satellite services in the band 18·6–18·8 GHz'. CCIR Report 850–1, *ibid.,* Volume II
23 'Relating to frequency sharing by the Earth exploration-satellite service (passive sensors) and the space research service (passive sensors) with the fixed, mobile except aeronautical mobile and fixed-satellite services in the band 18·6–18·8 GHz'. Recommendation 706; Radio Regulations, edition of 1982 (ITU, Geneva, 1982)
24 'Interference to spaceborne remote passive microwave sensors from active services in adjacent and sub-harmonic bands'. CCIR Report 987; Recommendations and Reports of the CCIR, 1986, Volume II (ITU, Geneva, 1986)
25 'Technical and operational considerations for the Earth exploration-satellite service', CCIR Report 535–3; *ibid.,* Volume II
26 'Preferred frequency bands for active sensing measurements'. CCIR Recommendation 577–1; *ibid.,* Volume II
27 'Feasibility of frequency sharing between spaceborne radars and terrestrial radars in the radiolocation service'. CCIR Report 695–2; *ibid.,* Volume II
28 'Space operation systems – frequencies, bandwidths and protection criteria'. CCIR Report 845–1; *ibid.,* Volume II
29 'Maintenance telemetering, tracking and telecommand for developmental and operational satellites – possibilities of frequency sharing between Earth-satellite telemetering or telecommand links and terrestrial services'. CCIR Report 396–5; *ibid.,* Volume II
30 'Radiocommunications for meteorological satellite systems'. CCIR Report 395–4; *ibid.,* Volume II

31 'Feasibility of frequency sharing between a geostationary meteorological satellite system and the meteorological aids service in the region of 400 MHz and in the upper part of Band 9 (1 to 3 GHz)'. CCIR Report 541; *ibid.*, Volume II
32 'Power flux-density limitation in the band 1670–1710 MHz for dissemination of meteorological information to small earth terminals'. CCIR Report 851; *ibid.*, Volume II
33 'Protection criteria and sharing considerations relating to deep-space research'. CCIR Report 685–2; *ibid.*, Volume II
34 'Potential interference between deep-space telecommunications and fixed-satellite and broadcasting-satellite systems in harmonically related bands'. CCIR Report 844; *ibid.*, Volume II
35 'Protection criteria and sharing considerations relating to deep-space research'. CCIR Recommendation 578; *ibid.*, Volume II
36 Final Acts adopted by the second session of the World Administrative Radio Conference on the use of the geostationary-satellite orbit and the planning of space services using it (ORB-88) (ITU, Geneva, 1988)
37 'Classification of space distances for spacecraft utilization'. CCIR Recommendation 610; Recommendations and Reports of the CCIR, 1986, Volume II (ITU, Geneva, 1986)
38 'Telecommunication requirements for manned and unmanned deep-space research'. CCIR Report 536–3; *ibid.*, Volume II
39 'Frequency bands in the 1 to 20 GHz range that are preferred for deep-space research'. CCIR Report 683–2; *ibid.*, Volume II
40 'Frequency bands in the 20 to 120 GHz range that are preferred for deep-space research'. CCIR Report 849–1; *ibid.*, Volume II
41 'Preferred frequencies and bandwidths for deep-space research'. CCIR Recommendation 576; *ibid.*, Volume II
42 'Telecommunication requirements for manned and unmanned near-Earth space research'. CCIR Report 548–2; *ibid.*, Volume II
43 'Technical and operational considerations for the Earth exploration-satellite service'. CCIR Report 535–3; *ibid.*, Volume II
44 'Frequency bands preferred for transmission to and from manned and unmanned spacecraft for near-Earth space research'. CCIR Report 984; *ibid.*, Volume II
45 'Preferred frequency bands and power flux-density considerations for Earth exploration satellites'. CCIR Report 692–2; *ibid.*, Volume II
46 'Characteristics of data relay satellite systems in bands 9 and 10'. CCIR Report 848–1; *ibid.*, Volume II
47 'Data relay satellites – sharing with other space research systems near 2 GHz'. CCIR Report 846; *ibid.*, Volume II
48 'Sharing considerations near 2 GHz between satellites in the Earth exploration, space research and space operation services and terrestrial line-of-sight radio-relay systems in the fixed service'. CCIR Report 981; *ibid.*, Volume II
49 'Feasibility of frequency sharing between an Earth exploration-satellite (EES) system and fixed satellite, meteorological satellite and terrestrial fixed and mobile services'. CCIR Report 540–1; *ibid.*, Volume II
50 'Feasibility of frequency sharing between space research (near-Earth) and fixed and mobile services in the 7 to 8 GHz spectral region'. CCIR Report 687–1; *ibid.*, Volume II
51 'Data relay satellites – sharing with other services in bands 9 and 10'. CCIR Report 847–1; *ibid.*, Volume II
52 'Factors relative to establishment of spurious emission limits for space services'. CCIR Report 980; *ibid.*, Volume II
53 'Characteristics of standard-frequency and time-signal emissions in allocated bands and characteristics of stations emitting with regular schedules with stabilised frequencies, outside of allocated bands'. CCIR Report 267–6; *ibid.*, Volume VII

54 'Time and frequency reference signal dissemination and co-ordination using satellite methods'. CCIR Recommendation 582; *ibid.*, Volume VII
55 'Time/frequency dissemination and co-ordination via satellite'. CCIR Report 518–4; *ibid.*,Volume VII
56 'Proposed reduction of mutual interference between standard-frequency and time-signal emissions in bands 6 and 7'. CCIR Report 732–2; *ibid.*, Volume VII
57 'Standard-frequency and time-signal emissions'. CCIR Recommendation 374–3; *ibid.*, Volume VII
58 'Reduction of mutual interference between emissions of the standard-frequency and time-signal service on the allocated frequencies in bands 6 and 7'. CCIR Recommendation 537; *ibid.*,Volume VII
59 'Frequency sharing between the time-signal service and the radiolocation service, the fixed-satellite service and the fixed and mobile services near 14, 21, 26 and 31 GHz'. CCIR Report 736–1; *ibid.*, Volume VII
60 'Relating to the meteorological aids service in the band 27·5–28·0 MHz'. Recommendation 620; Radio Regulations, edition of 1982 (ITU, Geneva, 1982)

Chapter 13

The inter-satellite service

13.1 Survey of inter-satellite service regulation

Inter-satellite links (ISLs) are already in use in space research and their introduction into the systems of other services is often discussed. It is convenient to identify several kinds of ISLs according to the ray-path geometry.

- The data relay satellites already operating as part of the SRS use links between geostationary satellites and satellites in low earth orbit. The geometry of these 'data relay' links resembles that of MSS up-links and down-links, although of course the ISLs need not penetrate the Earth's atmosphere.
- The ISLs that are contemplated to extend the capabilities of, for example, networks of the FSS are likely to connect geostationary satellites. These 'geostationary ISLs' can conveniently be classified as long-span links connecting satellites having an orbital separation which will typically exceed 90°, short-span links connecting satellites which are separated in orbit by less than about 10° and links inter-connecting a cluster of satellites which are close together in orbit, perhaps having the same nominal orbital location.

The short-span geostationary ISLs may be expected to have high-gain antennas with beams which are approximately tangential relative to the GSO, with very low gain in the direction of the Earth. The long-span geostationary ISLs will also have high gain antennas; the antenna gain in the direction of the Earth will usually be low, but it may become significant if the orbital separation of the satellites exceeds about 140°; if the orbital separation were to reach 162·8°, the ray path between the satellites would be obstructed by the Earth. Cluster links are likely to have low gain antennas.

The definitions of some space services specifically include ISLs and in no case, except by implication the BSS, are ISLs specifically excluded. Accordingly it would often be proper to assign to ISLs frequencies in the bands allocated to the service to which the satellites that are to be linked belong. However, many frequency allocations are made to space services for a defined transmission mode, Earth-to-space or space-to-Earth; very few are for space-to-space transmission. For example, at present all allocations to the FSS are designated either for up-links or down-links or both; none is available for ISLs.

Thus there is a need for other spectrum provision for ISLs. The inter-satellite service (ISS) meets this need. The service is defined in RR 24. Frequencies in bands allocated to the ISS can be used for ISLs in any kind of satellite network.

The use of laser communication systems for ISLs is also under study; such links, being above the frequency limit (that is, 3000 GHz) adopted by the ITU in its definition of radio waves (see RR 6), would not fall within the present competence of the ITU.

Frequency allocations

The frequency allocations for the ISS are as follows; all are worldwide and primary:

22·55–23·55 GHz
32–33 GHz
54·25–58·2 GHz and 59–64 GHz
116–134 GHz
170–182 GHz and 185–190 GHz

Two further allocations for the ISS, made by RR 797 and relating to 5·0–5·25 GHz and 15·4–15·7 GHz, differ in principle from the allocations listed above, their use being limited to ISLs which are part of AeRN or AeM (R) systems. These two allocations are not considered further in this chapter.

Frequency allocation sharing

All of the ISS frequency allocations are shared with several other services, and these sharing services can be considered in three groups;

- The BSS shares part of the 23 GHz band with the ISS in Regions 2 and 3, subject to agreements reached under RR Article 14 (see RR 887).
- The EESS (passive sensors) and the SRS (passive sensors) share the 56 GHz band and much of the 125 GHz and 175 GHz bands with the ISS.
- Terrestrial services, which share all of the ISS bands.

Most of these sharing situations have already been studied in CCIR, with the following results:

(*a*) *The BSS at 22 GHz:* The technical feasibility of sharing spectrum between the ISS and the BSS at 22 GHz, the ISLs being assumed to be between geostationary satellites, has been studied in CCIR; see Reports 951[1] and 874[2]. The main conclusion reached in these studies is that interference from a BSS transmitter to an ISS receiver should not be unacceptably high if the satellites were several degrees apart in orbit, but sharing would not be compatible with heavy usage of the GSO and the frequency band by either service.

(*b*) *Passive sensors:* The technical feasibility of sharing spectrum between the ISS and passive sensors of the EESS and the SRS has been studied in CCIR; see Report 694[3]. The conclusion reached is that interference from ISS transmitters to the passive sensors would be negligible, whether the ISLs were geostationary or of the data relay type. Passive sensors do not transmit, and therefore they cannot cause interference at ISS receivers.

(*c*) *The FS, MS and RN below 50 GHz:* CCIR Report 791[4] shows that sharing criteria that would probably be acceptable could be developed to permit the 22 GHz band to be used independently by the ISS on the one hand and the FS and the MS on the other hand, assuming that the ISS were used for geostationary ISLs. Interference problems arising from data relay ISLs were not studied. CCIR Report 872[5] shows that no interference problems should arise between the ISS and the RN at 32 GHz if the ISLs using the ISS allocation are short-span geostationary links. With long-span geostationary links, significant interference might occur in both directions unless sharing constraints which may significantly limit system options are adopted. Interference problems arising between data relay ISLs and RN systems have not been fully studied. RR Recommendation 707[6] calls for urgent study of these problems.

(*d*) *Terrestrial services above 50 GHz:* Various combinations of the FS, MS and radiolocation services (RL) share the ISS bands above 50 GHz. It will be noted that all of these bands are at frequencies where molecular resonances in the atmospheric gases increase the path loss between stations on the Earth's surface and stations in space (see Appendix B.4.4). In some parts of these bands the additional path loss due to absorption is very large. As CCIR Report 791[4] indicates, this gaseous absorption adds to the confidence with which the conclusion on sharing between geostationary ISLs and stations of the FS, MM and LM, summarised in (*c*) above, can be extended to the bands above 50 GHz. It may also be assumed that the conclusions drawn in CCIR Report 872[5] in connection with the RN at 32 GHz (see (*c*) above) can be extended to the RL in bands above 50 GHz; an allowance made for gaseous absorption leads to optimistic expectations for sharing between ISS and RL stations on the Earth's surface. However, gaseous absorption between a space station and a high-flying aircraft is much less than the absorption that arises when the whole depth of the atmosphere has to be traversed by an interfering signal originating at ground level. Accordingly, RR 909 and RR 910 specifically require stations of the AeM and airborne radars of the RL to not cause harmful interference to ISLs operating in ISS bands above 50 GHz. RR Recommendations 709[7] and 710[8] call for study of these problems.

International spectrum management

The procedures of RR Articles 11 and 13 are available as an administrative basis for the co-ordination and registration of frequency assignments to ISLs and they have been used for data relay ISLs which have been assigned frequencies in SRS bands. Material in CCIR Report 451[9] can be seen as a beginning in the process of developing a technical basis for co-ordination. In time, no doubt, ground rules will be agreed for the systematic and efficient use of the ISS allocations. For the present, however, the ISS is a service that has not yet come into operation.

13.2 References

1 'Sharing between the inter-satellite service and the broadcasting-satellite service in the vicinity of 23 GHz'. CCIR Report 951; Recommendations and Reports of the CCIR, 1986, Volume X/XI–2 (ITU, Geneva, 1986)
2 'Frequency sharing between the inter-satellite service when used by the fixed-satellite service and other space services'. CCIR Report 874; *ibid.,* Volume IV–1
3 'Sharing considerations and protection criteria relating to passive microwave sensors'. CCIR Report 694–2; *ibid.,* Volume II
4 'Inter-satellite service sharing with the fixed and mobile services'. CCIR Report 791–1; *ibid.,* Volume IV/IX–2
5 'Sharing criteria between inter-satellite links connecting geostationary satellites in the fixed-satellite service and the radionavigation service at 33 GHz'. CCIR Report 872; *ibid.,* Volume IV–1
6 'Relating to the use of the frequency band 32–33 GHz shared between the inter-satellite service and the radionavigation service'. Recommendation 707; Radio Regulations, edition of 1982 (ITU, Geneva, 1982)
7 'Relating to sharing frequency bands between the aeronautical mobile service and the inter-satellite service'. Recommendation 709; *ibid.*
8 'Relating to the use of airborne radars in the frequency bands shared between the inter-satellite service and the radiolocation service'. Recommendation 710; *ibid.*
9 'Factors affecting the system design and the selection of frequencies for inter-satellite links of the fixed-satellite service'. CCIR Report 451–3; Recommendations and Reports of the CCIR, 1986, Volume IV–1 (ITU, Geneva, 1986)

Chapter 14

Location by radio

14.1 Introduction

Radiodetermination is defined in RR 10 as:

> The determination of the position, velocity and/or other characteristics of an object, or the obtaining of information relating to these parameters, by means of the propagation properties of radio waves.

The radiodetermination service (RD) and, where satellites are directly involved, the radiodetermination-satellite service (RDSS) comprise stations that have these functions; see RR 38 and RR 39.

Some, though not all, radiodetermination systems are used in the navigation of ships or aircraft. The safety of human life may depend on the satisfactory operation of navigation systems and special measures are included in the RR, designed to minimise the risk of interference to such systems and to ensure that interference that does arise is quickly stopped. To facilitate the application of these special measures where they are appropriate, and to ensure that they are not invoked where the circumstances do not justify them, the RR differentiate between radiodetermination stations which are used for navigation and those which are not. Stations which are not used for navigation are classified as radiolocation stations (see RR 12) and the RD has two sub-divisions, the radionavigation service (RN) and the radiolocation service (RL), defined in RR 40 and RR 46. Similarly, the radionavigation-satellite service (RNSS) has been defined (see RR 41), as the part of the RDSS which relates to navigation.

Thus, the distinction between RN and RL arises from the functions of the system, and not from the characteristics of its equipment or its emissions. Radiobeacon stations, radiodirection-finding stations, radars carried by ships or aircraft to assist navigation and systems which enable ships and aircraft to keep to a chosen course or to determine their own position are all proper to the radionavigation service. Ground radars used for airport surveillance, for air traffic control or for assisting ships to navigate narrow or hazardous sea lanes are also classified as radionavigation stations. However, a military radar which is used, not for these purposes but for defence, is a radiolocation station. Weather radars may be treated as RL or MetA stations.

It has not been found necessary to define a radiolocation-satellite service. Weather radars carried by satellites are part of the MetS. Devices which resemble radars, are carried by satellites, but are used for earth exploration, are called active senors and are treated by the ITU as part of the EESS.

Sub-divisions have been created within the RN and the RNSS to enable spectrum to be allocated specifically for radionavigation systems intended for the use of either ships or aircraft, as follows:

Maritime radionavigation service (MRN)
Maritime radionavigation-satellite service (MRNSS)
Aeronautical radionavigation service (AeRN)
Aeronautical radionavigation-satellite service (AeRNSS)

See RR 42–RR 45. It should, however, be noted that no frequency band has been allocated so far to either the MRNSS or the AeRNSS.

Different kinds of radiodetermination system make use of different radio wave propagation phenomena to achieve their purposes, and these phenomena are exhibited, in combinations which may be optimum for particular systems, in different parts of the spectrum. Also, over many years, designers of systems for the various radiodetermination services have found it desirable and technically feasible to use different frequency ranges at different times. Many long-established, but not yet obsolete, systems are still in use, as well as newly-invented systems. For these various reasons many frequency bands in all parts of the radio spectrum are currently allocated for the radiodetermination services.

Spectrum allocated for the RD and RDSS is managed through the ITU, in some respects, like that of the other radio services. Administrations assign frequencies to the stations within their jurisdiction and the ITU provides a medium for the registration of frequencies which are assigned for transmitting or receiving at stations at fixed locations, although not for mobile stations. Certain technical constraints are applied to the emissions of these stations through the RR. The CCIR provides reports bearing on these systems and makes recommendations within its competence. Furthermore, and very importantly, the RR provide an internationally agreed basis for the clearance of interference that may affect the operation of these systems. However, in other respects that bear upon spectrum management, the treatment of radiodetermination systems may differ from that of most other kinds of radio system. For example:

- Many radiodetermination stations are functionally a part of operational networks, the good functioning of which is of concern to specialised international agencies such as the International Civil Aviation Organisation (ICAO) and the International Maritime Organisation (IMO). Governments act mainly through these other agencies in deciding how the frequency bands allocated to the RD and the RDSS should be used.

- Frequency bands allocated to the RD, the RDSS and related services are used in many instances for a single kind of radiodetermination facility, and the efficient use of spectrum or the good operation of systems may depend on the maintenance of this degree of segregation. More diversity in the systems using a frequency band is usually found in other services.
- The bandwidth occupied by some RD systems is comparable with the width of the whole allocated band. The usual concepts of spectrum management through frequency assignment cannot be effectively applied in such bands.
- Some radionavigation systems with wide geographical coverage may be set up by a small number of government agencies or perhaps by only one. The operators of these systems may, in fact, have the exclusive use of a frequency band. The operating agencies may be the sole source of expertise on these systems and the only bodies capable of planning spectrum use or system development. The advisory role of the CCIR, and the usefulness of constraints imposed through the Radio Regulations, may be rather limited in such cases.
- Some RD systems, such as pulse radars, are capable of causing interference to radiocommunication systems at great distances but they may be relatively immune to interference from other, similar, RD systems. Other RD systems are rather susceptible to interference. The consequences of interference may be particularly serious when the system is used for navigation. For these reasons, there is relatively little sharing of RD frequency allocations with other services.
- Many radiodetermination systems are military, and this leads to some lack of openness in international reporting of the technology involved.

For reasons such as these, the approach of the ITU to the management of some of the frequency bands allocated to the RD and RDSS is much less intensive than is usual in bands which are allocated to other services. However, where principles or constraints have been determined in ICAO or IMO for the spectrum management of specific frequency bands, these decisions are sometimes incorporated into the ITU Radio Regulations.

The principal frequency allocations for the various RD and RDSS applications, and the ITU management of the assignments that are made in them, are reviewed briefly on pp. 394–407. There are, of course, local variations and special provisions, made by footnotes to the international Table of Frequency Allocations (RR Article 8) but they are, in general, too numerous to review here. General ITU provisions for spectrum management for the RD and RDSS are reviewed on pp. 407–409.

14.2 Frequency allocations, sharing and systems

14.2.1 Position-fixing systems operating at VLF and LF

The band 9–14 kHz is allocated exclusively for RN worldwide. It is used mainly for hyperbolic position-fixing systems depending on carrier phase comparison, such as OMEGA.

The whole of the band 70–130 kHz is allocated for RN or MRN. Parts of this band are exclusive for RN or MRN but other parts are shared, principally with FS and MM, although use for these latter services is declining. Like the VLF band, these LF RN allocations are used mainly for carrier phase comparison systems, such as Loran C and Decca Navigator.

In Region 1, RR 451 limits the RN and MRN operating in the bands 70–86 kHz and 112–130 kHz to continuous wave systems; see also RR 453. The same applies in Regions 2 and 3 in the bands 70–90 kHz and 110–130 kHz. In these same bands, in Region 2, an administration making a new MRN assignment must seek agreement with the other administrations affected, in accordance with the procedures of RR Article 14, before notifying the assignment to the IFRB for registration; see RR 452. To facilitate sharing in these bands, RR 448 and RR 454 limit the MM and FS to classes of emission that should cause least interference. CCIR Report 915[1] and CCIR Recommendation 589[2] discuss the carrier-to-interference ratios required to protect Loran C and Decca Navigator systems from interference from the sharing services.

Despite these regulatory requirements and sharing constraints, interference to RN and MRN systems from stations of other services continues to cause concern. WARC-MOB-87 deleted a secondary allocation to the MM in the band 90–130 kHz and sought in its Final Acts[3] the deletion by a competent future conference of the secondary FS allocation in the same band; see RR Resolution 706[4]. WARC-MOB-87 also resolved that administrations should in future take more care to avoid causing interference when making new assignments between 70 and 130 kHz; see RR Resolution 705[5].

14.2.2 Radiobeacons and direction finding at LF and MF

Radiobeacon systems

Some recently introduced LF maritime radiobeacon systems use phase comparison at the mobile receiver of carrier waves from different transmitting locations to provide hyperbolic position-fixing facilities; see CCIR Report 913[6] and CCIR Recommendation 631[7]. However, most radiobeacons provide an easily identified signal, radiated from a known location, which a mobile station receives with direction-finding equipment, deducing the bearing of the mobile station relative to the beacon site from the bearing of the beacon as measured at the mobile station. Recently introduced scanning receivers provide automatic measurement and display of the bearings of all beacons within range of the mobile station and some automatically compute the location of the mobile station from the intersection of the best bearing indications; see CCIR Report 913[6]. The use of scanning receivers in Region 1 has been facilitated by CCIR Recommendation 588[8] which calls for the carrier frequencies of beacons to be multiples of 100 Hz.

The navigation of a mobile station with direction-finding equipment may be assisted by observations made on signals emitted from any station of known location. Stations, typically those of the FS and land stations of the MM and AeM and operating in frequency bands allocated to those services, may be designated by administrations for such use; see RR 2847, RR 2849 and RR 2851. However, stations are also set up specifically for use as radiobeacons, maritime or aeronautical, and these radiobeacon stations, properly so-called, are confined to particular frequency bands and subject to specific constraints; see RR 2847, RR 2848, RR 2850 and RR 2852.

The basic signal of a radiobeacon consists of an unmodulated carrier emitted for 25 seconds, preceded by a morse identification signal and followed by a 'long dash' to complete a cycle of 60 seconds, the whole cycle being repeated indefinitely. The identification signal may be an A1A emission but A2A and H2A are also used. The 'long dash' phase of the cycle is sometimes keyed to carry a message, but CCIR Recommendation 487[9] insists that aeronautical radiobeacon signals should not be operated in this way unless airborne automatic direction-finder performance is shown not to be unacceptably degraded; the practice is not deprecated for maritime radiobeacons unless it causes interference to other beacons; see CCIR Report 1037[10].

The frequency range that is optimum for beacons of this kind is quite narrow and it is also in demand for various other purposes, especially broadcasting and maritime radiotelegraphy. The total bandwidth allocated for maritime radiobeacons is only about 30 kHz, plus a further 10 kHz shared with the AeRN and also used for radiobeacons; this spectrum must be used efficiently to ensure that essential requirements can be met. Rather more bandwidth has been allocated for aeronautical radiobeacons, since some of the means by which the maritime service economises in the use of spectrum are not feasible in aeronautical operations.

To the extent made necessary by local circumstances, various means are used to enable limited allocated bandwidth to meet the operational requirements, including:

(*a*) The bandwidth of individual beacons may be minimised by the use of A1A emission only, and the frequency spacing between the nominal carrier frequencies is made small.

(*b*) A group of stations may time-share a nominal carrier frequency, each transmitting for one minute in turn.

(*c*) The carrier power level is constrained, no station radiating more power than is needed to cover its service area. The day-time field strength to be set up by an aeronautical or maritime radiobeacon at the limit of its service area is stated, as a function of Region and latitude, in RR 2852–RR 2864.

(*d*) The maximum use is made of spectrum shared between the AeRN and other services, especially the MM. This involves significant interference problems; however, see CCIR Report 910[11].
(*e*) The assignment of carrier frequencies to stations may be planned and co-ordinated nationally, internationally or Regionally, to promote efficient spectrum use. For aeronautical radiobeacons co-ordination is usually done under the aegis of ICAO.

The problems of congestion are greatest in Europe, and this has led to significant differences in spectrum management practice between Region 1 on the one hand and Regions 2 and 3 on the other hand.

Radiobeacons in Region 1

The MF allocations for radionavigation in Region 1 which are used for radiobeacons are as follows:

255–283·5 kHz; AeRN (permitted) sharing with BS (primary)
283·5–315 kHz; MRN (primary) and AeRN (permitted)
315–325 kHz; AeRN (primary) and MRN (secondary)
325–405 kHz; AeRN (primary)
415–435 kHz; AeRN (primary) sharing with MM (permitted)
505–526·5 kHz; AeRN (permitted) sharing with MM (primary)

For the second band in this list, the lower limit was 285 kHz until 1 February 1990, providing an opportunity for a re-arrangement of Region 1 LF broadcasting frequencies; see RR 458 and p. 142 above.

The band 148·5–283·5 kHz is allocated for broadcasting in Region 1 but that allocation is little used in much of Africa. Accordingly, in the countries listed in RR 462, the AeRN has an exclusive primary allocation between 200 and 283·5 kHz.

The four 'permitted' allocations listed above were all newly created by WARC-79. Firm assignments to stations of the permitted services depend on frequency plans prepared for the service with the primary allocation. This raised no substantial problem for RN assignments below 283·5 kHz, since the broadcasting plan prepared in 1975 (see pp. 149–153) had only to be adjusted to take into account the 1·5 kHz reduction in carrier frequencies referred to on p. 142. In the other cases, however, new planning was required. Thus:

- Before WARC-79 the band 285–315 kHz had been allocated for MRN (primary) and AeRN (secondary). The basic frequency assignment plan for MF maritime radiobeacons for the European Maritime Area, still at that time operative, was contained in the Paris Arrangement, 1951. (The ITU European Maritime Area is defined formally in RR 405; approximately, it is the part of Region 1 which lies north of 30° North latitude and west of 55° East longitude.) WARC-79 recommended that the Paris Arrangement be brought up to date, technically and operationally;

see RR Recommendation 602[12]. The Conference also recommended formal planning for the African area; see RR Recommendation 603[13].

- Before WARC-79 there was an exclusive allocation for MM in the band 415–490 kHz. WARC-79 made a primary allocation for AeRN at 415–435 kHz, the MM allocation being reduced to permitted status in that sub-band. It was subsequently realised that the MM had a need to optimise its use of this band within the limits imposed by the new AeRN allocation, and so the preparation of a frequency assignment plan for the AeRN was urgent.
- Before WARC-79 there was a primary MM allocation in the band 510–525 kHz, AeRN being secondary. WARC-79 up-graded the AeRN allocation to permitted status, the MM allocation remaining primary and both allocations were extended upwards to 526·5 kHz. Frequency assignment plans for the MM were thus required, in order that the AeRN might make the best use of this band, within the terms of its allocation.

In its Final Acts[14], WARC-MOB-83 proposed that a RARC be called to plan MRN beacons in the band 283·5–315 kHz in the European Maritime Area; see RR Recommendation 602[15]. RR Resolution 704[16], also of WARC-MOB-83, called for another RARC to prepare frequency assignment plans for the whole of Region 1 for the AeRN and the MM in the bands 415–435 kHz and 505–526·5 kHz, respectively. Both of the RARCs were held in 1985. The recommendation of WARC-79 on planning MRN beacons for the African area has not been implemented yet; WARC-MOB-87 renewed the recommendation for action in its revision of RR Recommendation 603[17].

One of the 1985 RARCs prepared a maritime radio beacon plan for the 283·5–315 kHz band in the European Maritime Area. The frequency assignment plan and the Agreement of which it is the heart are to be found in the Final Acts of the Conference[18]. The plan divides the band into 62 channels, 500 Hz wide and identifies a number of beacons, typically about ten, for each channel, stating the power required for the stated service radius. All beacons will use an A1A emission. When an administration assigns a frequency to a beacon in accordance with the plan, the assignment is notified to the IFRB in accordance with Section IIA of RR Article 12. If such a notification is in strict conformity with the terms of the plan it will be registered automatically by the Board unless it is incompatible with a notification, already registered, from an administration which is not a party to the Agreement. Articles 4 and 6 of the Agreement contain procedures for amending the plan and for accommodating new assignments for aeronautical beacons respectively. The Agreement is to enter into force on 1 April 1992. WARC-MOB-87, however, recognised that technical factors may already have created a need to revise the Plan; see RR Resolution 602[19].

The Final Acts of the other RARC-85[20] contain the Agreement relating to Region 1 AeRN radiobeacons at 415–435 kHz and the MM in Region 1 at 505–526·5 kHz. The Conference also included in the Agreement a plan for aeronautical radiobeacons at 505–526.5 kHz. The Agreement is considered from the MM standpoint on pp. 209–211. For the AeRN the Agreement is very similar to the maritime beacon plan for the band 283·5–315 kHz. The principal differences were the use of channels 1 kHz wide and of A2A emissions. This Agreement also enters into force on 1 April 1992.

Radiobeacons in Regions 2 and 3

In Regions 2 and 3 the band 285–325 kHz is allocated with primary status for the MRN and with permitted status for the AeRN. The rest of the spectrum between 190 and 405 kHz is allocated to the AeRN with primary status; there are no other primary allocations in these bands.

There are additional MF allocations for RN in these Regions, used in part for radiobeacons, namely:

130–160 kHz, allocated for RN in Region 3
505–526·5 kHz, allocated for AeRN in Region 3
510–535 kHz, allocated for AeRN in Region 2

However, these latter bands are shared with equal status with various other services.

The RD allocations around 2 MHz

There are extensive allocations for RN and RL in Regions 2 and 3 between 1·6 and 2·0 MHz, shared with various other services on equal terms. There are narrower RL allocations in the same frequency range in Region 1, including some just above 2 MHz. These allocations are used mainly for aeronautical radiobeacons and various position-fixing systems.

MF direction finding facilities

There is a world-wide primary RN allocation at 405–415 kHz. The centre frequency, 410 kHz, is designated for maritime radio direction-finding (DF). Coast stations offering the facility use direction-finding receivers to measure, on demand, the bearing of a mobile station emitting a signal at 410 kHz; see RR 2841 and RR Appendix 41. RR 468 requires that other RN systems using this band shall cause no harmful interference to direction-finding between 406·5 and 413·5 kHz. MF DF facilities are also available to ships within the MM bands, at 500 kHz and 2182 kHz; see RR 2841 and RR 2842.

14.2.3 Aircraft navigation systems

Various radio systems are used directly in the navigation of aircraft. Some of these are radar systems, including Doppler radars carried by the aircraft, and these are assigned frequencies in bands reviewed on pp. 401–403. Others are LF and MF radiobeacons (see pp. 394–398). A third group of aircraft navigation facilities, including for example route-finding and position-fixing systems, radio and radar altimeters, landing aids and secondary surveillance radar (SSR), are assigned frequencies in various other frequency bands allocated to the AeRN, the more important of which are reviewed in this Section.

These latter bands are, in most cases, allocated exclusively to the AeRN and they are managed, to a large extent, by governments co-operating through ICAO. However, there is a growing hunger for bandwidth for other kinds of radio system in the same part of the spectrum and this is leading to increased ITU concern for an involvement in the management of the AeRN bands.

Instrument landing aids

The established standard Instrument Landing System (ILS) uses three VHF/UHF frequencies:

(*a*) The spot frequency 75·0 MHz is used for marker beacons. The band 74·8–75·2 MHz is allocated for the AeRN, providing guard bands of 200 kHz to protect the aircraft marker signal receivers, some of which are old and have poor selectivity. RR 572 provides some further protection from interference by requiring administrations to refrain from assigning frequencies in the vicinity of the guard bands to stations of other services which have powerful transmitters or are close to airports.

(*b*) The band 108–117·975 MHz is allocated for the AeRN. The lower portion of this band, typically 108–111·975 MHz, is used for ILS VHF localiser signals. Increasing use for FM sound broadcasting of the BS band just below 108 MHz has given rise to concern that interference from broadcasting transmitters may enter aircraft localiser receivers via various spurious modes; see RR Recommendation 704[21], CCIR Recommendation 591[22], CCIR Report 929[23] and pp. 177–178 above. Studies continue on the extent of this risk and means for reducing it.

(*c*) The band 328·6–335·4 MHz is allocated for the AeRN and RR 645 limits its use to ILS glide path signals.

Each of these ILS bands is allocated worldwide with primary status to AeRN and there are no other primary allocations for these bands. Nevertheless, the introduction of a new standard instrument landing system operating in the microwave frequency range is in sight and presumably ILS will eventually be phased out. Foreseeing that the ILS bands will therefore become available one day for other uses, RR 572A[3],

RR 590A[3] and RR 645A[3] allocate them also for MS with secondary status in a number of countries, the assignment of frequencies to mobile stations being subject the application of the procedure of RR Article 14 to ensure that any remaining ILS systems will not suffer interference.

The band 5000–5250 MHz, allocated to AeRN, has been little used so far. RR 796 indicates that this band is reserved for an international standard system for precision approach and landing and that the requirements for this, the 'Microwave Landing System' (MLS), shall take precedence over other uses of this band. The MLS is currently being installed in the band 5030–5090 MHz; it will be used, typically, in association with Distance Measuring Equipment (DME), which is operated in the band 960–1215 MHz (see below).

WARC-MOB-87 asked ICAO to consider whether some of the 5000–5250 MHz band might be surplus to aeronautical radionavigation requirements, in which case it might be put to other use, and whether sharing with other services would be feasible in the spectrum remaining allocated to AeRN; see RR Recommendation 607[24]. RR 797A[3] and RR 797B[3] in some degree anticipate the outcome of this approach to ICAO, the first by an allocation in the band 5150–5216 MHz for RDSS (see pp. 405–407) and the second by an allocation of the band 5150–5250 MHz for the MS, with primary status but subject to the procedures of RR Article 14.

Route-finding and position-fixing systems

VHF Omni-directional Radio Range (VOR) beacons operate in the 108–117·975 MHz AeRN allocation and in particular in the part above 111·975 MHz. VOR guides the pilot of an aircraft along a predetermined route towards his destination but it does not indicate where the aircraft is along that route. Thus the system does not, in itself, enable the position of the aircraft to be fixed. The comments above about possible interference to aircraft ILS localiser receivers from broadcasting transmissions below 108 MHz apply to the VOR receivers also.

Other systems, such as Distance Measuring Equipment (DME) which is operated in the band 960–1215 MHz band, enable the pilot to determine how far the aircraft is from the DME station. Many DME stations are co-located with VOR beacons, the two systems complementing each other to provide a geographical fix. There are also other systems serving these purposes, some operating in other AeRN bands, including various footnote allocations of limited geographical cover below 1 GHz.

Secondary surveillance radar (SSR), which is used to provide air traffic control stations with information on the identity of aircraft which are under observation by radar, is operated in the 960–1215 MHz band.

Already the spectrum below 1 GHz is very heavily loaded with operating systems of all kinds, and new aeronautical navigation systems, no doubt providing more precise information to users and probably needing wide bandwidth, will have to operate in higher bands. The bands 1610–1626·5 MHz, 5000–5250 MHz and 15·4–15·7 GHz, all allocated worldwide for AeRN, have been identified for 'the development of airborne electronic aids to air navigation' since 1959. One possible perceived need, for example, is a collision avoidance system; see RR Recommendation 601[25]. Provision has also been made in principle for such future systems to use satellite communication techniques in the same bands; see RR 732, RR 733 and RR 797. However, so far developments of this kind have been very limited and erosion of these reservations is beginning to occur as spectrum has to be found for new systems of other kinds that are ready for operation; see for example the comments on Microwave Landing Systems above and on the recent RDSS allocations on pp. 405–407.

In addition to those three bands identified for future development, many bands above 20 GHz are allocated to RN. Most of these bands will, no doubt, be used ultimately for radars but there will no doubt be opportunities to set one or more millimetre wave bands aside for position-finding systems, at some time in the future, if the need is found to arise.

Radio altimeters

The band 4200–4400 MHz is allocated with primary status for AeRN and RR 789 limits the use of the band by that service to radio altimeters. The band is used typically for CW radar altimeters using sawtooth frequency modulation of the carrier and it is the principal band used for that purpose. Pulse radar altimeters are operated in various other RD bands.

There is no other primary allocation in the 4200–4400 MHz band. However, this allocation has remained unchanged since 1947 and the view has been expressed in the ITU that too little attention may have been given to efficient use of spectrum in the design of altimeters. RR Recommendation 606[26] calls for study by the CCIR of the possibility of operating altimeters with sufficient accuracy in a bandwidth of less than 200 MHz; the recommendation invites ICAO to study the matter also and looks forward to the possibility of re-allocating part of this spectrum to other services at some time in the future.

14.2.4 Primary radars

Frequency allocations and sharing

A substantial number of frequency bands allocated for RL, RN or one or other of the specialised radionavigation services are used mainly for radars. Some allocations have secondary status with Regional or worldwide applicability. Others have limited geographical application, mostly made by footnotes to the international Table of Frequency Allocations. However, below 36 GHz, there are also eight main blocks of primary, worldwide allocations used for radars, sub-divided between various specialised applications. These eight blocks are reviewed below. There are, in addition, a number of allocations, not yet in substantial use, in the millimetre-wave spectrum above 36 GHz.

(*a*) *1215–1400 MHz:* All of this block of spectrum except 1300–1350 MHz is allocated for RL. The whole of the block except 1350–1400 MHz is also allocated for RN or AeRN, worldwide or over large geographical areas, through RR 712 and RR 714. There are sharing primary allocations for FS and MS below 1300 MHz (geographically limited, see RR 711) and above 1350 MHz (Region 1 only). However, the RL and RN stations operating in this block are typically ground radars and interference can be limited by geographical separation. There is also a sharing allocation for RNSS (space-to-Earth) at 1215–1260 MHz. This allocation has a nominal primary status, but RR 710 requires that its stations cause no harmful interference to RN stations operating under RR 712. Interference prospects are discussed in CCIR Report 766[27]; see also p. 404 below.

(*b*) *2700–3400 MHz:* With limited variations made by footnotes to the international Table of Frequency Allocations, the allocations in this block are for AeRN in the band 2700–2900 MHz, RN in the band 2900–3100 MHz and RL in the band 3100–3400 MHz. Below 3300 MHz these bands are not shared with other services and the allocations to other services above 3300 MHz are secondary or geographically limited.

(*c*) *5250–5850 MHz:* The basic frequency allocations in this block are for RL at 5250–5350 MHz and 5650–5850 MHz and for one or other of the radionavigation services at 5350–5650 MHz, but various footnote allocations complicate the situation. The sub-band 5725–5850 MHz is shared with FSS (Earth-to-space) in Region 1; this is the only significant sharing situation for the RD in this block. It is evident that radar emissions of the RL could cause interference to space station receivers of the FSS and this interference might be severe. FSS earth station emissions might also interfere with RL receivers. CCIR studies of these sharing situations have not been completed yet; however, see Part II, Annex 5.3.3 of the Report of the CCIR Conference Preparatory Meeting[28] in 1984, addressed to WARC-ORB-85, and also CCIR Reports 827[29] and 828[30].

(*d*) *8500–10 500 MHz:* The whole of this block, except 9000–9200 MHz and 9300–9599 MHz, is allocated for RL. Between 8750 and 9800 MHz there are allocations for one or other of the radionavigation services. It may be noted that the 8750–8850 MHz allocation for AeRN is reserved for airborne Doppler navigation aids (see RR 821) and that preference is given to weather radars in the band 9300–9500 MHz (see RR 825). At the upper and lower ends of this block of spectrum there is sharing with the FS and MS. There are no established criteria for sharing with these services.

(*e*) *13·25–14·3 GHz:* The AeRN allocation at 13·25–13·4 GHz is limited to Doppler navigation aids; see RR 851. The remainder of this block is allocated for RL (13·4–14·0 GHz) or RN (14·0–14·3 GHz). The RN allocation at 14·0–14·3 GHz is shared with the FSS (Earth-to-space) and RR 856 requires that stations of the RN be operated so that space station receivers of the FSS are sufficiently protected from interference. CCIR Report 560[31] and Recommendation 496[32] provide the necessary sharing criteria. No criteria have been developed for the protection of RN receivers from FSS earth station transmitters; substantial geographical separation may be needed for this.

(*f*) *15·7–17·3 GHz:* The entire band 15·7–17·3 GHz is allocated for RL. There are primary footnote allocations of the whole band for the FS and MS in a substantial number of countries; see RR 866 and RR 867.

(*g*) *24·05–25·25 GHz:* There is an allocation for RL at 24·05–24·25 GHz and one for RN at 24·25–25·25 GHz. The RN allocation is exclusive worldwide and the RL allocation is not shared with any service with primary status.

(*h*) *31·8–36·0 GHz:* The band 31·8–33·4 GHz is allocated for RN and 33·4–36·0 GHz is allocated for RL. At 32·0–33·0 GHz RN shares the band with ISS. RR 897 permits radars located on spacecraft to be operated with primary status at 35·5–35·6 GHz. At 35·2–36·0 GHz there is a sharing allocation for MetA; it seems likely that this allocation would be used mainly for radars also. In addition, RR 894 allocates 33·4–36·0 GHz for FS and MS in a number of countries with primary allocation status. The CCIR has begun studies of the sharing criteria that would permit RN and ISS to use the band that they share without co-ordination. The results obtained to date indicate that no constraints would be required on either service if the inter-satellite links were between geostationary satellites separated by no more than 90° of orbital longitude; see CCIR Report 872[33]. A limit might have to be placed on the power of the RN transmission to avoid interference to the ISS receiver if the latter were used for longer links between geostationary satellites. The situation that would arise if the inter-satellite links involved one or more non-geostationary satellites has not been studied yet.

Efficient spectrum utilisation by primary radars

Six of the eight groups of frequency bands reviewed above comprise 33% of the whole spectrum between 1 and 20 GHz. There are other RD allocations in this frequency range and major allocations at higher frequencies also. A radar capable of providing good resolution of targets occupies a wide bandwidth. In busy shipping lanes and on air routes, in harbours and at airports, there are concentrations of operating radars and interference tends to be considerable. Even for fixed radars it is not always feasible to assign frequencies so that interference does not arise. However there are various methods for reducing interference and for reducing the extent to which any interference which remains denies reliable information to the operator. These methods are discussed in CCIR Report 914[34].

14.2.5 Satellite radiodetermination

Radionavigation using satellites

The principal frequency allocations for satellite radiodetermination are all for the RNSS. They are as follows, all primary and worldwide:

149·9–150·05 MHz and 399·9–400·05 MHz; exclusive
1215–1260 MHz (space-to-Earth); shared with RL, also primary
1559–1610 MHz (space-to-Earth); shared with AeRN, also primary

The 1200 MHz band, wholly or in part, is also allocated with primary status for various terrestrial services, including RN, FS, MS and AeRN, in a number of countries; see RR 711, RR 712 and RR 714. The 1600 MHz band is also allocated with primary status for the FS in a number of countries and a narrow slot within the band is allocated for the AeM; see RR 730 and RR 731B[3]. There is also a secondary allocation for the RNSS at 14·3–14·4 GHz, the principal sharing primary service being FSS (Earth-to-space).

RR 710 requires that stations of the RNSS operating in the 1200 MHz band shall not cause harmful interference to stations of the RN in the countries listed in RR 712. CCIR Report 766[27] examines the feasibility of sharing between RDSS down-links and RN receivers on the terms of RR 710 and concludes that interference might arise; much depends on the characteristics and requirements of the terrestrial systems. Stations of other terrestrial services sharing these bands with the RNSS are likely to be less sensitive to interference. In the reverse direction, CCIR Report 766 finds that receivers of the RNSS are not likely to suffer interference from pulse radars in the 1200 MHz band but continuous wave terrestrial service emissions may cause interference over wide areas. This interference must be regarded as a serious unresolved problem for a safety service like the RNSS.

Several millimetre-wave frequency bands have been allocated for RNSS, the lowest being 43·5–47 GHz. All of these bands are shared with equal primary status with the MS, MSS and RN; none is in substantial use at present.

The 150 and 400 MHz RNSS allocations are used by the Transit system. The 1200 and 1600 MHz RNSS allocations are being taken into use for such systems as NAVSTAR GPS and GLONASS. All of these systems enable a mobile station to determine its location by observing signals emitted by transmitters on several satellites. No up-link signal is required from the mobile station. Up-links and down-links used for control purposes between the satellites and fixed earth stations are feeder links (see RR 109) and frequencies may be assigned for them in bands allocated to the FSS (see RR 22).

Other satellite radiodetermination systems

Other satellite radiodetermination systems have been announced recently such as GEOSTAR and LOCSTAR. These systems determine the location of a mobile station by measurements made at a central fixed earth station of the transmission delay of the two-way transmission paths to and from the mobile station via two or more satellites. The information on location becomes available at the central station, offering a surveillance facility. The information can also be transmitted through the system for information at the mobile station. The design of systems may also allow the transmission of short messages in a store-and-forward mode over the links which make up the system. Such systems need up-links and down-links between mobile stations and the satellites, plus feeder links, up and down, between the satellites and the central station. Frequencies for the feeder links of these latter systems could be assigned in bands allocated to the FSS.

These systems have not been classified as part of the satellite radionavigation service, and until recently there have been no allocations of spectrum for, for example, the RDSS, from which frequency assignments for the links between mobile stations and satellites could be made. With systems of this kind in view, the administration of the USA made national allocations in 1985 as follows:

1610–1626·5 MHz for RDSS (Earth-to-space)
2483·5–2500 MHz for RDSS (space-to-Earth)

for links between the satellites and the mobile stations, and

5117–5183 MHz for FSS (space-to-Earth)

for feeder down-links, it being intended that frequencies around 6433 MHz, already allocated for the FSS (Earth-to-space), should be assigned for feeder up-links. At WARC-MOB-87 the USA proposed that these allocations should be made worldwide, sharing with the services for which the bands were already allocated. Studies had already been made of some of the sharing problems and the results, with a description of the GEOSTAR system, are to be found in CCIR Report 1050[35].

Some delegations at WARC-MOB-87 welcomed the proposal, seeing benefit in the facilities which such systems offered. Other delegations opposed the proposal, principally because of foreseen sharing problems. In particular these latter delegations foresaw difficulty in protecting the AeRN from interference from the RDSS in the 1600 MHz and 5100 MHz bands, and saw no acceptable way of protecting the RDSS from interference from the stations of services to which the bands are already allocated.

A complex package of compromises and constraints was agreed in an attempt to reconcile these views. The more important elements of this package are as follows:

(i) The band 1610–1626·5 MHz is allocated for RDSS (space-to-Earth) in Region 2 with primary status, except for Venezuela (where the allocation is secondary) and Cuba. The same allocation has secondary status in Regions 1 and 2, except for the countries listed in RR 733B[3], where the allocation has primary status, subject to agreements reached with other countries under RR Article 14 and except in Sweden where there is no RDSS allocation (see RR 733F[3] and RR 731[3]). The RDSS shares the band with the AeRN worldwide with primary status and there are various other footnote allocations, mostly of limited geographical applicability.

(ii) The band 2483·5–2500 MHz is allocated for RDSS (Earth-to-space) with primary status in Region 2, except for Cuba. The same allocation has secondary status in Regions 1 and 3, except for the countries listed in RR 753C[3], where the allocation is primary; see RR 733F[3]. In certain Region 3 countries that were listed in RR 733B[3] but not in RR 753C[3], a primary allocation for the RDSS has been made in the band 2500–2516·5 MHz instead; see RR 754A[3]. The implementation of assignments under primary allocations, where they apply in Regions 1 and 3, is subject to the procedures of RR Article 14. The RDSS shares these bands at 2·5 GHz with worldwide primary allocations for the FS and mobile services. There is also a primary allocation for RL below 2500 MHz in Regions 2 and 3. In Region 1 the allocation for RL below 2500 MHz has secondary status, but RR 753B[3] has the effect of up-grading the status of RL, relative to RDSS, in Region 1 countries which are not listed in RR 753C[3]. The band 2400–2500 MHz is also designated for ISM applications; see RR 752. In Region 3 above 2500 MHz, in addition to terrestrial services, the band is shared with primary allocations for down-links from satellites of the FSS and BSS.

(iii) The band 5152–5216 MHz is allocated for RDSS (space-to-Earth) with the same allocation status, Region by Region and country by country, as has been given to the RDSS allocations at 1600 and 2500 MHz; see RR 797B[3]. The band is shared with a primary worldwide allocation for the AeRN and there are various footnote allocations for other services.

(iv) In the bands 1610–1626·5 MHz and 2483·5–2500 MHz, RR 733A[3] and RR 753A[3] explicitly deny to stations of the RDSS the special protection from interference accorded by RR 953 to safety services.

(v) Sharing constraints are applied to RDSS signal levels in various bands to protect receiving station of sharing services. The EIRP of the RDSS mobile earth stations in the 1600 MHz band is not to exceed −3 dBW in any 4 kHz sampling band in any direction; see RR 2548A[3]. The PFD at the Earth's surface from RDSS satellite emissions in the 2483·5–2500 MHz band is not to exceed a value between −154 and −144 dBW per square metre in any 4 kHz sampling band, the actual limit depending on the angle of elevation of the signal reaching the ground; in the 2500–2516·5 MHz band the corresponding range of PFD limits is from −152 to −137 dBW; see RR 2557 to RR 2564 as modified in the Final Acts of WARC-MOB-87[3]. The corresponding PFD figure in the 5 GHz band is −159 dBW, regardless of the angle of elevation; see RR 797A[3]. These various constraints are listed in RR Resolution 708[36], which calls for further consideration of the matter at the next competent WARC after further study by the CCIR.

(vi) RR 734 makes a secondary allocation for RA in the band 1610–1613·8 MHz. It is obvious that mobile earth stations of the RDSS, transmitting in this band, have a potential for causing severe interference to radio astronomy observatories, despite the EIRP limit of −3 dBW. RR 733E requires that harmful interference be prevented from arising in this way in Regions 1 and 3 but does not stipulate how this should be done. One possible solution is explained in section 2.1.2 of CCIR Report 1050[35].

(vii) Guidelines for the co-ordination of mobile earth stations with foreign terrestrial stations at 1600 and 2500 MHz are set out in RR 1107.2[3].

If systems such as GEOSTAR and LOCSTAR are found to provide desirable facilities, these sharing arrangements and constraints seem likely to impede system deployment. It is to be hoped that a future WARC will find ways of making more satisfactory allocations.

14.3 Regulatory provisions for radiodetermination stations

Some general regulatory provisions for the RD and RDSS are to be found at RR 2831–RR 2840. These deal in large part, not with spectrum management, but with operational responsibilities. Information on radio stations operating in the RD and RDSS is published by the ITU as one of its 'service documents'; see RR 2212–RR 2214.

Notification and registration of frequency assignments

The special provisions for registering frequency assignments made to radiobeacon stations in accordance with formal frequency assignment plans are mentioned on p. 397. More generally, and in the absence of special requirements such as agreement in accordance with RR Article 14, frequencies assigned to RD stations operating at a fixed location for transmitting or receiving are notified to the IFRB for registration in the MIFR in accordance with Section IIA of RR Article 12 if the responsible administration considers that the emissions are capable of causing interference to a receiving station abroad or if the administration wishes to obtain international recognition for their use (RR 1215, RR 1217 and RR 1219). The corresponding procedures of RR Articles 11 and 13 apply for assignments to stations of the RDSS. These various procedures are reviewed, in the context of the FS and the FSS, on pp. 84–86, 103–104, 118–131 and 281–294 above.

Interference to safety services

In general, systems of the RN and RNSS provide 'safety services', that is their use may involve the safeguarding of human life and property (RR 56). RR 953 emphasises the need for special measures to ensure freedom from interference for frequency assignments to safety services. RR 1952 and RR 1957 require immediate action to be taken by administrations to eliminate harmful interference (that is, in this context, interference which endangers the functioning of safety services; see RR 163) if it should occur.

Carrier frequency tolerance

The frequency tolerances applied by RR 303 to all radiodetermination stations that are assigned specific frequencies are shown in Table 14.1. The tolerances above 470 MHz are rather lax. Indeed, the operators of radars which are not assigned specific frequencies are merely required to ensure that the bandwidth occupied by the emission remains wholly within the allocated band. However, licensing authorities may require tighter tolerances to be achieved in some instances, in order to limit interference, and many systems of the RN have an operational need to maintain extremely precise carrier frequencies.

Table 14.1 Frequency tolerances for radiodetermination stations (RR Appendix 7)

Frequency range	Tolerance applicable from 1990 (parts per million)
9–535 kHz	100
1·6–4 MHz	
power < 200 W	20
power > 200 W	10
29·7–100 MHz	50
100–470 MHz	50
470–2450 MHz	500
2·45–10·5 GHz	1250
10·5–40 GHz	5000

Spurious emissions

The provisions of RR 304 and RR Appendix 8 limiting spurious emissions from any radio transmitter are summarised in Appendix A.5. However these provisions are not well suited to RD transmitters, in particular because the maximum levels of spurious emissions are related in RR Appendix 8 to the mean power level of the wanted emission, and this form of specification is not suitable for application to the many RD pulse emissions. Accordingly, paragraph 5 of RR Appendix 8 calls for the lowest practicable power level of spurious emissions to be achieved at terrestrial radiodetermination stations 'until acceptable methods of measurement exist'. The same objective, within limits set by technical and economic constraints, is laid down in Note 13 to the Table in RR Appendix 8 for RDSS transmitters operating above 960 MHz.

14.4 References

1 'Interference between fixed, maritime mobile and radionavigation services in the bands between 70 kHz and 130 kHz'. CCIR Report 915–1; Recommendations and Reports of the CCIR, 1986, Volume VIII–2 (ITU, Geneva, 1986)
2 'Interference to radionavigation services from other services in the bands between 70 kHz and 130 kHz'. CCIR Recommendation 589–1; *ibid.*, Volume VIII–2
3 Final Acts of the World Administrative Radio Conference for the Mobile Services (MOB-87) Geneva, 1987 (ITU, Geneva, 1988)
4 'Operation of the fixed and maritime mobile services in the band 90–110 kHz'. Resolution 706; Final Acts of WARC-MOB-87; see [3] above
5 'Mutual protection of radio services operating in the band 70–130 kHz'. RR Resolution 705; *ibid.*
6 'Technical characteristics of maritime radiobeacons'. CCIR Report 913–1; Recommendations and Reports of the CCIR, 1986, Volume VIII–2 (ITU, Geneva, 1986)
7 'Use of hyperbolic maritime radionavigation systems in the band 283·5–315 kHz'. CCIR Recommendation 631; *ibid.*, Volume VIII–2
8 'Characteristics of maritime radio beacons (Region 1)'. CCIR Recommendation 588; *ibid.*, Volume VIII–2
9 'Use of radio-beacon stations for communications'. CCIR Recommendation 487; *ibid.*, Volume VIII–2
10 'Choice between the FSK and MSK techniques for data transmission from maritime radiobeacons'. CCIR Report 1037; *ibid.*, Volume VIII–2
11 'Sharing between the maritime mobile service and the aeronautical radionavigation service in the band 415–526·5 kHz. CCIR Report 910–1; *ibid.*, Volume VIII–2
12 'Relating to maritime radiobeacons'. Recommendation 602, Radio Regulations, edition of 1982 (ITU, Geneva, 1982)
13 'Relating to technical provisions for maritime radiobeacons in the African Area'. Recommendation 603; *ibid.*
14 Final Acts of the World Administrative Radio Conference for the Mobile Services (MOB-83) Geneva, 1983 (ITU, Geneva, 1983)

15 'Relating to the planning of frequencies in the band 283·5–315 kHz used for maritime radiobeacons in the European Maritime Area'. RR Recommendation 602; Final Acts of WARC-MOB-83; see item [14] above
16 'Relating to the holding of a Regional Administrative Radio Conference to prepare frequency assignment plans for the maritime mobile service in the bands between 435 kHz and 526·5 kHz and in parts of the band between 1606·5 kHz and 3400 kHz in Region 1 and to plan for the aeronautical radionavigation service in the band 415–435 kHz in Region 1'. Resolution 704; Final Acts of WARC-MOB-83; see item [14] above
17 'Technical provisions for maritime radiobeacons in the African Area'. Recommendation 603; Final Acts of WARC-MOB-87; see item [3] above
18 Final Acts of the Regional Administrative Conference for the planning of the maritime radionavigation service (radiobeacons) in the European Maritime Area, Geneva, 1985 (ITU, Geneva, 1986)
19 'Data transmission from maritime radiobeacons for differential radionavigation systems'. Resolution 602; Final Acts of WARC-MOB-87; see item [3] above
20 Final Acts of the Regional Administrative Conference for the planning of the MF maritime mobile and aeronautical radionavigation services (Region 1) Geneva, 1985 (ITU Geneva, 1986)
21 'Relating to the compatibility between the broadcasting service in the band 100–108 MHz and the aeronautical radionavigation service in the band 108–117·975 MHz'. Recommendation 704; Radio Regulations, edition of 1982 (ITU, Geneva, 1982)
22 'Compatibility between the broadcasting service in the band of about 87–108 MHz and the aeronautical services in the band 108–136 MHz'. CCIR Recommendation 591–1; Recommendations and Reports of the CCIR, 1986, Volume VIII–3 (ITU, Geneva, 1986)
23 'Compatibility between the broadcasting service in the band of about 87–108 MHz and the aeronautical services in the band 108–136 MHz'. CCIR Report 929–1; *ibid.*, Volume VIII–3
24 'Future requirements of the band 5000–5250 MHz for the aeronautical radionavigation service'. Recommendation 607; Final Acts of WARC-MOB-87; see item [3] above
25 'Concerning the matter of providing a suitable frequency allocation for a collision avoidance system in the aeronautical radionavigation service'. Recommendation 601; Radio Regulations, edition of 1982 (ITU, Geneva, 1982)
26 'The possibility of reducing the band 4200–4400 MHz used by radio altimeters in the aeronautical radionavigation service'. Recommendation 606; Final Acts of WARC-MOB-87; see item [3] above
27 'Feasibility of frequency sharing in the 1215–1240 MHz band between the NAVSTAR GPS and terrestrial radiolocation and radionavigation systems', CCIR Report 766–1; Recommendations and Reports of the CCIR, 1986, Volume VIII–3 (ITU, Geneva, 1986)
28 'Technical bases for the World Administrative Radio Conference on the use of the geostationary-satellite orbit and the planning of the space services utilizing it'. Report of the CCIR Conference Preparatory Meeting (CPM) 1984 (ITU, Geneva, 1984)
29 'Radar spectrum utilization'. Report 827; Recommendations and Reports of the CCIR, 1986, Volume I (ITU, Geneva, 1986)
30 'Theoretical and experimental results for spectrum sharing between FDM-FM and radar systems using pulse blanking with frequency diversity reception'. Report 828–1; *ibid.*, Volume I
31 'Sharing criteria for the protection of space stations in the fixed-satellite service receiving in the band 14·0–14·4 GHz'. Report 560–2; *ibid.*, Volume IV–1
32 'Limits of power flux-density of radionavigation transmitters to protect space station receivers in the fixed-satellite service in the 14 GHz band'. Recommendation 496–2; *ibid.*, Volume VIII–3

33 'Sharing criteria between inter-satellite links connecting geostationary satellites in the fixed-satellite service and the radionavigation service at 33 GHz'. Report 872; *ibid.*, Volume IV–1
34 'Efficient use of the radio spectrum by radar stations in the radiodetermination service'. CCIR Report 914–1; *ibid.*, Volume VIII–2
35 'Technical and operational considerations for a radiodetermination satellite service in bands 9 and 10'. Report 1050; *ibid.*, Volume VIII–3
36 'Criteria for sharing between the radiodetermination-satellite service and terrestrial services in the bands 1610–1626·5 MHz, 2483·5–2500 MHz' and 2500–2516·5 MHz'. Resolution 708; Final Acts of WARC-MOB-87; item [3] above

Appendix A

Terminology in radio spectrum management

A.1 Introduction

Terminology presents a major challenge to the ITU. The work of the Union involves complex regulatory and technical agreements, binding upon the governments that ratify them and these agreements much be unambiguous and clearly understood. The CCIR is a forum where experts from many different telecommunications fields come together to discuss problems which are usually complex and often subtle. Some of these discussions are at the forefront of technical development, where the use of technical terms may be at a formative stage. Moreover many delegates at ITU conferences do not have one of the working languages of the ITU as their mother tongue, and some have linguistic problems.

For these reasons the ITU attaches great importance to the precise use of language and the standardisation of terms and symbols. Thus:

- RR Articles 1 to 4 define a large number of terms, conventions and symbols, regulatory and technical, used in the Regulations. Other formal documents, such as frequency assignment plans, contain definitions of important terms, used in the document, which are not defined elsewhere.
- Many of the CCIR Study Groups have found it necessary to draw up a report or recommendation standardising the special technical vocabulary used in their various fields of activity.
- CCIR Recommendation 573[1] gathers together a large number of the more widely used technical terms used in radio, with their definitions. CCIR Recommendation 662[2] lists a number of more general telecommunications terms defined in common by the CCIR and CCITT. CCIR Recommendation 574[3] establishes a preferred usage for logarithmic quantities used in telecommunications. CCIR Recommendation 666[4] lists some of the abbreviations in common use in telecommunications.

The concern of the ITU to achieve worldwide standardisation of the use of terms in telecommunications does not end with the regulatory process. The International Electrotechnical Commission (IEC) is drawing up a multilingual vocabulary, the International Electrotechnical Vocabulary (IEV)[5], defining terms for general technical use in all fields of electrical technology, in consultation with various other international organisations and national standards organisations. The ITU is participating with the IEC in drafting the chapters concerned with telecommunication, to support this important work and to ensure that the technical terminology of the ITU remains well aligned with general practice; see CCIR Recommendation 662[2]. The IEV Chapters that will relate to radio are as follows:

Chapter 705: Radio wave propagation
Chapter 712: Antennas
Chapter 713: Radio transmitters, receivers, networks and operation
Chapter 723: Broadcasting
Chapter 725: Space radiocommunication
Chapter 726: Transmission lines and waveguides

It would not be appropriate to reproduce this extensive material on regulatory and technical terminology in this book. The author has sought to use terms defined by the ITU in the defined way, and the ordinary dictionary meaning of words is a sufficient guide to their ITU usage in most circumstances. Formal definitions of key terms are quoted where appropriate where this seems to be desirable. However it may be helpful to discuss in a general context the ITU usage of a few terms and to introduce some widely used symbols. That is the purpose of Appendices A.2–A.7.

A.2 The terms telecommunication and radiocommunication

The ITU definitions of 'telecommunication' and 'radiocommunication' are broad. Thus RR 4 defines *telecommunication* as:

> Any transmission, emission or reception of signs, signals, writing, images and sounds or intelligence of any nature by wire, radio, optical or other electromagnetic systems.

and *radiocommunication* is defined in RR 7 as:

> Telecommunication by means of radio waves

An upper limit is placed on the frequency range which is of concern to the ITU by RR 6, which defines *radio waves* as:

> ... electromagnetic waves of frequencies arbitrarily lower than 3000 GHz propagated in space without artificial guide.

although the terms of reference of the CCIR (see p. 21) are not restrictive in this way. However, some governments have already taken powers to regulate nationally the use for communication of frequencies around the visible range and it may be only a matter of time before the introduction of long distance laser communication systems, used for example for inter-satellite links, creates a need for ITU involvement.

Radiocommunication, according to these definitions, embraces the reception of intelligence of any nature by radio. Thus the regulation, for example, of the use of radio for navigation systems, radar and for studying earth resources by satellite falls within the province of the ITU.

A.3 Radiocommunication services

In layman's usage, a radio service is one specific product of the telecommunication or broadcasting industry, involving the use of radio, made available to users outside the organisation which provided the service. Thus, for example, a programme of medium wave sound broadcasting or one particular kind of radio paging facility would be called a radio service. The word is sometimes used in that way in texts of the ITU but the usual and much more important usage of the term in the Union is a specialised one.

Section III of RR Article 1 comprises RR 20 to RR 57. The first of those paragraphs, RR 20, defines a *radiocommunication service* as:

> a service as defined in this Section involving the transmission, emission and/or reception of radio waves for specific telecommunication purposes.

Each of the other 37 paragraphs in Section III defines a radio service. Of these definitions, 33 are administrative categories into which radio stations fall; they are listed in Table A.1, together with the number of the chapter of this book in which the management of the spectrum used for each service is discussed. Reasons for categorising stations in these ways are discussed in Chapters 1 and 3.

Table A.1 also indicates the abbreviations which are used in this book for these services. These abbreviations have not been standardised in the ITU but most of them are widely used in ITU texts and in the technical literature.

Four other definitions in Section III of RR Article 1 remain. All arguably use the word 'service' in the layman's sense discussed above. RR 32 and RR 33 define the *port operations service* and the *ship movement service*; these are special products within the MM and for basic spectrum management purposes they can be considered as part of the MM. RR 56 provides for the name *safety service* to be applied to any radio system which is used, permanently or temporarily, for the safeguarding of human life and property and which is therefore entitled, for example, to specially favourable treatment if interference arises. RR 57 provides for the name *special service* to be applied to a system which does not fit into any of the 33 services listed in Table A.1.

Of the 33 services, those which involve the use of satellites or spacecraft are classified as *space radiocommunication* and all the others (except the radio astronomy service) are classified as *terrestrial radiocommunication*; see RR 8

Table A.1 The ITU radio services

Service name and abbreviation		Chapter in which discussed
Aeronautical fixed service	(AeFS)	5
Aeronautical mobile service	(AeM)	7
Aeronautical mobile-satellite service	(AeMSS)	10
Aeronautical radionavigation service	(AeRN)	14
Aeronautical radionavigation-satellite service	(AeRNSS)	14
Amateur service	(AmS)	11
Amateur-satellite service	(AmSS)	11
Broadcasting service	(BS)	6
Broadcasting-satellite service	(BSS)	9
Earth exploration-satellite service	(EES)	12
Fixed service	(FS)	5
Fixed-satellite service	(FSS)	8
Inter-satellite service	(ISS)	13
Land mobile service	(LM)	7
Land mobile-satellite service	(LMSS)	10
Maritime mobile service	(MM)	7
Maritime mobile-satellite service	(MMSS)	10
Maritime radionavigation service	(MRN)	14
Maritime radionavigation-satellite service	(MRNSS)	14
Meteorological aids service	(MetA)	12
Meteorological-satellite service	(MetS)	12
Mobile service	(MS)	7
Mobile-satellite service	(MSS)	10
Radio astronomy service	(RA)	12
Radiodetermination service	(RD)	14
Radiodetermination-satellite service	(RDSS)	14
Radiolocation service	(RL)	14
Radionavigation service	(RN)	14
Radionavigation-satellite service	(RNSS)	14
Space operation service	(SpO)	12
Space research service	(SRS)	12
Standard frequency and time signal service	(SFS)	12
Standard frequency and time signal-satellite service	(SFSS)	12

and RR 9. The *radio astronomy service* is alone in a third category.

In this book, to the extent feasible, the word 'service' is used as a noun only in the sense of the 38 paragraphs of RR Article 1, Part III.

A.4 The terms allocation, assignment and allotment

Two terms which are very basic to the radio regulatory process are defined by the ITU as follows:

> *Allocation* (of a frequency band): Entry in the Table of Frequency Allocations of a given frequency band for the purpose of its use by one or more terrestrial or space radiocommunication services or the radio astronomy service under specified conditions. This term shall also be applied to the frequency band itself (RR 17).
> *Assignment* (of a radio frequency or a radio frequency channel): Authorisation given by an administration for a radio station to use a radio frequency or radio frequency channel under specified conditions (RR 19).

A number of frequency assignment plans have been produced within the ITU for specific frequency bands and specific services, it having been calculated that an acceptably low level of interference would arise if administrations made the planned frequency assignments to the specified stations within their jurisdiction. However, it is not always convenient to implement such a restrictive plan; for example, it may not always be feasible to foresee the location of a new station or the parameters of its emissions at the time when the plan is drawn up. Accordingly, plans have been produced for other applications which allow administrations greater, though not unlimited, freedom to determine the conditions of an assignment at the time when it is to be taken into use, without endangering interference margins. Such plans may be called frequency allotment plans, reflecting the following definition:

> *Allotment* (of a radio frequency or radio frequency channel): Entry of a designated frequency channel in an agreed plan, adopted by a competent conference, for use by one or more administrations for a terrestrial or space radiocommunication service in one or more identified countries or geographical areas and under specified conditions (RR 18).

A.5 Radiation, emission and associated terms

Two terms, basic in spectrum management, are defined as follows:

Radiation: The outward flow of energy from any source in the form of radio waves (RR 131)
Emission: (i) Radiation in the case where the source is a radio transmitter. (ii) Radio waves or signals produced by a radio transmitting station (CCIR Recommendation 573[1]).

The use of the word 'transmission' to mean the radiation from a transmitter is deprecated. In the ITU the term *transmission* is used for the conveyance of intelligence all of the way from a source to its destination, as in the following definition:

> The action of conveying from one point to one or more other points, signs, symbols, documents, pictures, sounds, or information of any nature, by means of signals (CCIR Recommendation 662[2])

Thus, all of the radiation flowing from a transmitter is called its 'emission' and this comprises the wanted signal (sometimes more than one wanted signal) plus any radiation, such as harmonics of the signal and parasitic oscillations, that was not wanted.

The bandwidth of emissions

For most types of emission, and in the absence of distortion within the transmitter, the energy in the sidebands remote from the carrier could be reduced, although at some cost in loss of information or of performance margin at the receiver. The *necessary bandwidth* is defined as follows:

> for a given class of emission, the width of the frequency band which is just sufficient to ensure the transmission of information at the rate and the quality required under specified conditions (RR 146)

RR Appendix 6 contains largely empirical rules for calculating the necessary bandwidth of a large number of different classes of emission, with many worked examples; see also CCIR Reports 836[6] and 837[7]. Thus, for practical purposes the necessary bandwidth can be regarded as the smallest theoretically attainable value for the bandwidth of an emission. The RR direct that this is the bandwidth that should be assigned to an emission, after allowances have been made for other sources of spectrum spreading that cannot readily be avoided. Thus, the *assigned frequency band* is defined in RR 141 as follows:

> The frequency band within which the emission of a station is authorised; the width of the band equals the necessary bandwidth plus twice the absolute value of the frequency tolerance. Where space stations are concerned, the assigned frequency band includes twice the maximum Doppler shift that may occur in relation to any point on the Earth's surface.

RR 303 and RR Appendix 7 place limits on the frequency tolerance of emissions but the tolerances that have been agreed vary from service to service and with other factors; the limits are discussed in general in CCIR

Report 181[8] and they are reviewed, service by service, in Chapters 5 to 14. The Doppler shift due to the movement of a space station will depend greatly on the orbit used.

The power spectra of emissions are rarely ideal. There will usually be some spectral energy outside the limits of the necessary bandwidth, although in the vicinity of the wanted signal, caused by the use of sub-optimum band-limiting filters for the modulating signal, amplitude non-linearity in the transmitter power amplifiers and so on; this is called *out-of-band emission*, the term being defined as follows:

> emission on a frequency or frequencies immediately outside the necessary bandwidth which results from the modulation process but excluding spurious emissions ((RR 138)

In order to permit the out-of-band emission to be quantified in a practical way, the term *occupied bandwidth* has been defined, as follows:

> the width of a frequency band such that, below the lower and above the upper frequency limits, the mean powers emitted are each equal to a specified percentage of the total mean power of a given emission (RR 147).

The specified percentage is usually 0·5%. The relationship between necessary bandwidth and occupied bandwidth has been studied by the CCIR in depth; see CCIR Report 977[9]. CCIR Recommendation 328[10] sets targets for out-of-band emission that are considered to be attainable and urges administrations to seek to have them achieved at radio stations within their jurisdiction.

The *assigned frequency* is, by definition, at the centre of the assigned frequency band; see RR 142. This is a convenient convention and particularly so for emissions which radiate a strong spectral line due to the carrier wave at the centre of the emitted spectrum, as many do. In such cases it is easy to relate the measured frequency of the carrier to the assigned frequency. However, some emissions do not radiate a carrier wave nor any other consistently identifiable spectral component at the centre of the assigned frequency band; a single sideband amplitude modulated telephony emission is a common example of this. If there is a permanent spectral component elsewhere in the emission, for example a reduced-level carrier in a single sideband emission, it may be identified as the *characteristic frequency* (RR 143) and its frequency may be declared as a *reference frequency* (RR 144).

Fig. A.1 illustrates some of these terms, using as models a carrier keyed on-off by a telegraph signal and a single-sideband telephony emission with a reduced-level carrier.

f_C = carrier frequency
δf = frequency tolerance (an allowance for Doppler shift would be included for a space station emission)
F_A = assigned frequency
F_C = characteristic frequency
F_R = reference frequency

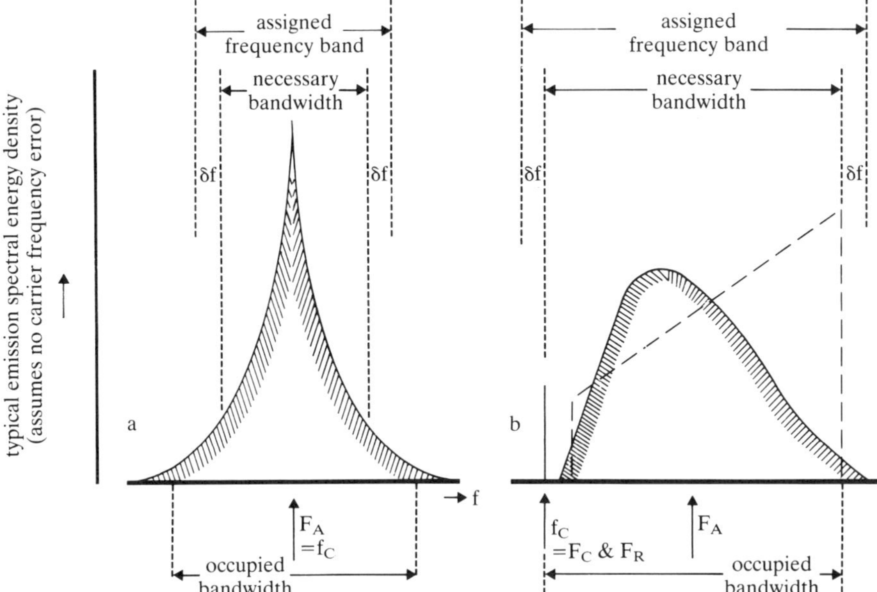

Fig. A.1 Sketches to illustrate terms used in describing features of the spectrum of emissions
a A carrier, keyed on–off by a telegraph signal
b An SSB telephony emission with a reduced-level carrier

Channels and circuits

When a plan is drawn up for a number of emissions to be operated in a frequency band, the band is divided into *radio channels*, each defined by an upper frequency limit f_U and a lower frequency limit f_L. Each radio channel is intended to be used for one emission, and the bandwidth of the radio channels $(f_U - f_L)$ is usually equal to the occupied bandwidth of the emissions, or approximately so (see Fig. A.2).

In most plans the frequency separation between the centres of adjacent radio channels is equal to the radio channel bandwidth. Sometimes, however, the frequency separation between radio channel centres is made less than the radio channel bandwidth; the radio channels are said to be *interleaved* (see Fig. A.2). This enables more stations to be assigned frequencies in a given frequency band and it does not necessarily lead to unacceptable interference between the emissions assigned to adjacent radio channels. Thus:

- For most kinds of modulation the spectral power density falls away to a low value near the limits of the occupied bandwidth, as shown in Fig. A.1. In such cases, relatively little interference power from one emission falls into the adjacent radio channel assigned to another emission.
- In selecting the stations to which each radio channel is to be assigned it is often possible to reduce interference further by assigning adjacent radio channels to stations which are distant from one another, which have directional antennas which are optimised to serve different locations or which are technically different in some other way, for example in the polarisation mode of the radiated signal.

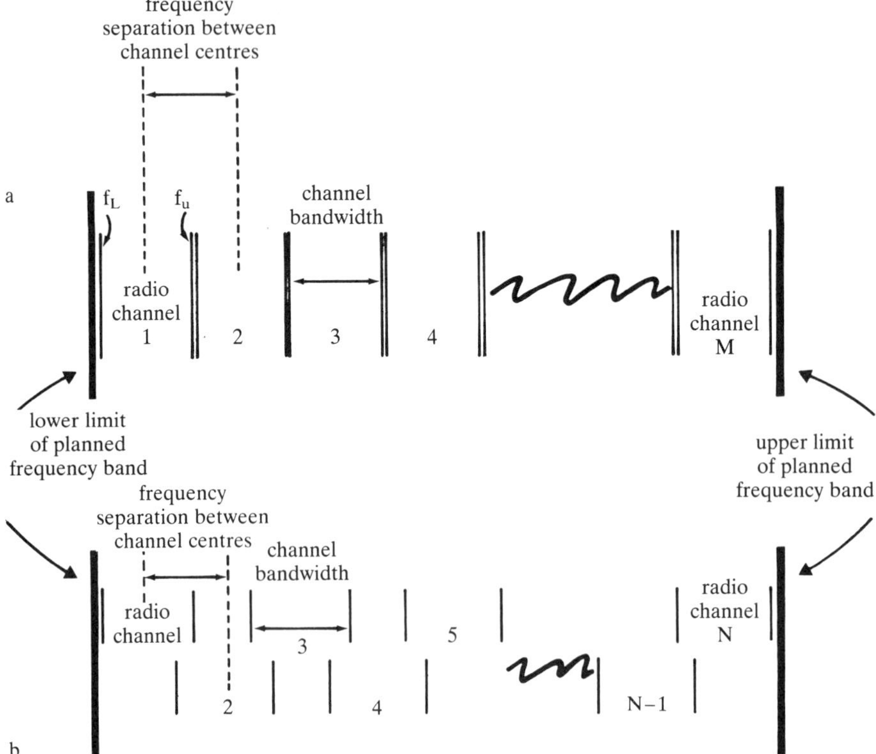

Fig. A.2 Radio channels. At (*b*) the channels are interleaved

A transmitter at point A and a receiver at point B may provide a *radio link*. Such a radio link may be capable of carrying unidirectionally from A to B one speech signal, one teleprinter signal, one video signal or whatever else the design of the link and the available bandwidth allow. Any of these facilities constitutes a transmission channel, usually called just a *channel*. Some radio links are designed so that they can carry more than one channel; a link is said to be *diplex* if it can carry two independent channels simultaneously and *multiplex* if it can carry more than two independent channels.

If, in addition to a channel provided by the transmitter at A and the receiver at B, there is another channel of the same kind in the opposite direction, provided typically by a transmitter at B and a receiver at A, the two channels can be coupled together to provide communication in both directions; the two channels then form a *circuit*. Some circuits can transmit in both directions simultaneously; this is called *duplex* operation. Other circuits cannot be operated simultaneously in both directions and instead they transmit, first in one direction and then in the other; this is called *simplex* operation.

Spurious emissions

The unwanted components in the emission which are not in the immediate vicinity of the wanted emission are called *spurious emissions*, the term being defined as follows:

> emissions on a frequency or frequencies which are outside the necessary bandwidth and the level of which may be reduced without affecting the corresponding transmission of information. Spurious emissions include harmonic emissions, parasitic emissions, intermodulation products and frequency conversion products but exclude out-of-band emissions (RR 139)

Out-of-band emissions and spurious emissions, taken together, are called *unwanted emissions* (RR 140).

The level of spurious emissions which arises with well-designed transmitters is discussed in CCIR Report 838[11]. CCIR Recommendation 329[12] proposes limits which should be achieved and advises on methods for measuring the power of the spurious components which a transmitter radiates. With minor amendments, WARC-79 adopted these recommendations (to the extent that they are applicable) for new transmitters installed after 1984 and for all transmitters after 1993; see RR 304 and RR Appendix 8. With the omission of a considerable amount of detail, these limits can be conveniently represented in diagrammatic form (see Fig. A.3).

The limits on spurious emissions in RR Appendix 8 are not applied to certain categories of transmitter. The more important exemptions are transmitters with digital modulation operating above 960 MHz, transmitters in space services operating above 960 MHz and all transmitters operating above 17·7 GHz. Furthermore, the limits are, in general, specified in relation to the mean power of the wanted emission; for pulse transmitters such as radars, this method is unsatisfactory and so RR Appendix 8 acknowledges that the limits which it sets cannot in general be applied to radiodetermination transmitters. In each of these cases administrations are urged to ensure that the lowest feasible levels of spurious emission are achieved, pending the setting of formal limits in the light of the studies now in progress in CCIR; see also RR Recommendation 66[13].

Reference is made where appropriate in Chapters 5 to 14 above to the progress being made in CCIR in determining appropriate limits for spurious emissions in cases where they do not exist at present.

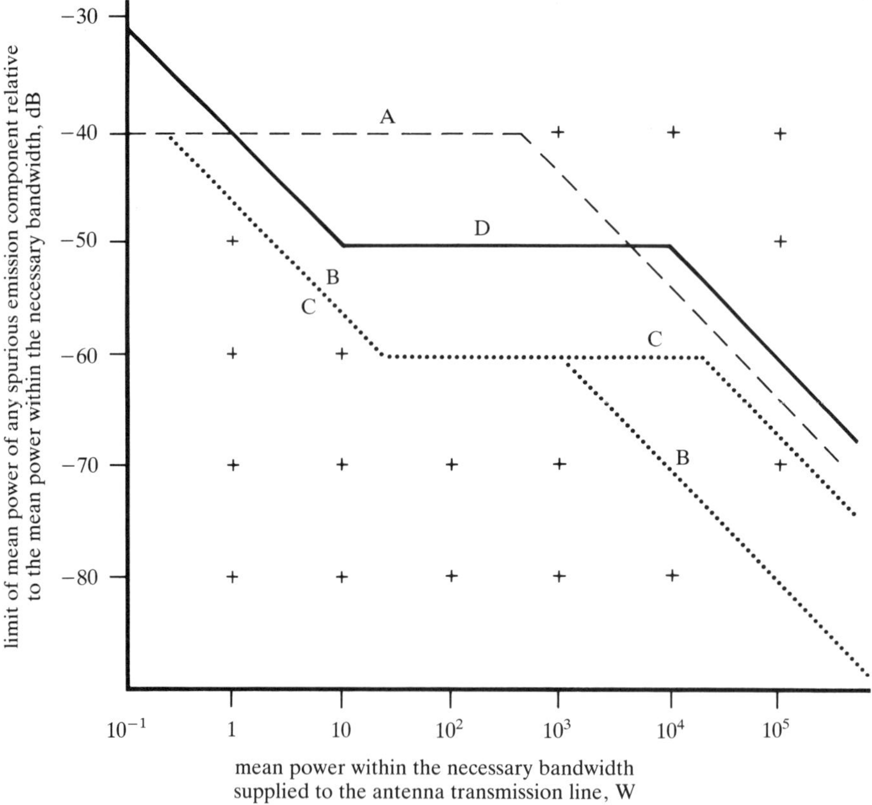

Fig. A.3 Simplified account of the limits imposed by RR Appendix 8 on the power of spurious emissions radiated by terrestrial service transmitters operating below 17·7 GHz and space service transmitters operating below 960 MHz (after CCIR Recommendation 329[12])
Curve A applies between 9 kHz and 30 MHz
Curve B applies between 30 MHz and 235 MHz
Curve C applies between 235 MHz and 960 MHz
Curve D applies (terrestrial services only) between 960 MHz and 17·7 GHz

A.6. The designation of emissions

The ITU method for indicating the nature of an emission is set out in RR Article 4. A designation usually consists of seven characters, letters or figures, expressed as a single group. Eighth and ninth characters may be included in some cases. Thus a complete emission designation takes the form:

wxyzABCmn

where

- the first four characters, wxyz, give the necessary bandwidth,
- the next three characters, ABC, give the classification of the emission according to the type of modulation and the nature of the modulating signal, and
- the final two characters, mn, if added to the class of emission designator, indicate additional characteristics of the modulation.

The 'necessary bandwidth' designator comprises three numerical characters and one letter. The three figures, treated as a three-digit number, give the bandwidth to three significant figures:

in hertz for bandwidths between 0·001 Hz nd 999 Hz
in kilohertz for bandwidths between 1·00 kHz and 999 kHz,
in megahertz for bandwidths between 1·00 MHz and 999 MHz and
in gigahertz for bandwiths between 1·00 GHz and 999 GHz.

These four bandwidth ranges are denoted by upper case letters H, K, M and G, respectively. The order of magnitude of the bandwidth is indicated by inserting this letter into the group of digits in place of the decimal point. Thus, H020 represents a necessary bandwidth of 0·020 Hz, 54H5 represents 54·5 Hz, 6K00 represents 6·00 kHz, 105M represents 105 MHz and 1G15 represents 1·15 GHz.

The three mandatory characters of the 'class of emission' designator have the following significance:

- The first character, an upper case letter, indicates the type of modulation of the main carrier; the code used is in Table A.2.
- The second character, in most cases a numerical digit, indicates the nature of the signal modulating the main carrier (see Table A.3).
- The third character, another upper case letter, indicates the type of information to be transmitted (see table A.4).

If the two optional characters are used, the fourth character of the classification of the emission gives more information on the nature of the modulating signal and the fifth character states what kind of multiplexing is employed. The codes for the optional characters are given in RR Appendix 6, Part A.

RR Appendix 6, Part B lists as examples the designations of a wide variety of type of emission.

Table A.2 Classification of emissions. 1: Type of modulation

Type of modulation, main carrier	Code
An unmodulated carrier	N
The main carrier is amplitude-modulated	
Double-sideband	A
Single-sideband, full carrier	H
Single-sideband, reduced or variable level carrier	R
Single sideband, suppressed carrier	J
Independent sidebands	B
Vestigial sideband	C
The main carrier is angle modulated	
Frequency modulation	F
Phase modulation	G
The main carrier is amplitude-modulated and angle-modulated, simultaneously or in a pre-established sequence	D
Pulse emissions	
A sequence of unmodulated pulses	P
Modulated in amplitude	K
Modulated in width/duration	L
Modulated in position/phase	M
With angle-modulation of carrier during pulses	Q
Other pulse modulation methods or combinations of methods	V
Hybrid modulation systems not covered above, involving two or more basic modulation techniques, amplitude, angle or pulse	W
Other cases	X

Table A.3 Classification of emissions. 2: Nature of modulating signal

Modulating signal	Code
No modulating signal	0
A single channel of quantized or digital information	
without the use of a sub-carrier	1
with a sub-carrier	2
A single channel containing analogue information	3
Two or more channels containing quantized or digital information	7
Two or more channels containing analogue information	8
A composite system containing both analogue and digital channels	9
Other cases	X

Table A.4 Classification of emissions. 3: Type of information

Type of information transmitted	Code
No information transmitted	N
Telegraphy – for aural reception	A
Telegraphy – for automatic reception	B
Facsimile	C
Data transmission, telemetry, telecommand	D
Telephony (including sound broadcasting)	E
Television	F
Combinations of the above	W
Other cases	X

It will be noted from Tables A.2. to A.4 and from RR Appendix 6 that the designation of types of emission is not sufficiently discriminating to enable complex types of modulation to be precisely identified. This is particularly true with regard to the various sophisticated methods which have been developed in recent years for the bandwidth-efficient transmission of digital signals. It should be appreciated that these ITU codes are intended to meet regulatory needs with a minimum of characters; they cannot take the place of a technical description of the modulation technique for all purposes.

Work on the further refinement of these codes for classifying emissions is continuing; see RR Recommendtion 62[14] and CCIR Question 1/1[15].

A.7 The designation of ranges of the radio spectrum

A point in the radio spectrum could be identified equally precisely in terms of frequency or wavelength, the permittivity of the medium being known, but frequency is always used in spectrum management. When it is necessary to identify with precision a specific slot in the spectrum, perhaps the band to which a frequency assignment or a frequency allocation applies, this too is done by stating the frequencies of the band limits.

However, it is often necessary to identify a part of the radio spectrum in broad terms. The ITU has adopted various conventions for doing this. Schemes covering 3 kHz to 3000 GHz are set out in RR Article 2 and these schemes have been extended down to 300 Hz and up to 3000 terahertz by the CCIR. The International Union of Radio Science (URSI) has extended these schemes down to 0·03 Hz; see CCIR Recommendation 431[16].

For all of these schemes the spectrum is divided into contiguous frequency ranges, the frequency of the upper limit of each range being ten times the frequency of the lower limit. Examples are 3–30 kHz, 30–300 kHz and 300kHz–3 MHz. Various sets of names are applied to these ranges (see CCIR Recommendation 431), and the following are the more commonly used ones:

(*a*) The set of symbols in general use are abbreviations of phrases describing the ranges in terms of the order of frequency, such as MF (from

'Medium Frequency') and VHF (from 'Very High Frequency'). Some of these phrases now seem anachronistic, but the symbols are well-established. (*b*) The 'band number' is also used, particularly in technical texts. The band number of a range can easily be remembered because it is the power to which 10 must be raised to give the frequency in hertz at the middle of the range.

The symbols and band numbers covering the radio spectrum which currently concerns the ITU are shown in Table A.5.

Table A.5 Designation of ranges of frequency

Band number	Symbol	Frequency range
4	VLF	3–30 kHz
5	LF	30–300 kHz
6	MF	300 kHz–3 MHz
7	HF	3–30 MHz
8	VHF	30–300 MHz
9	UHF	300 MHz–3 GHz
10	SHF	3–30 GHz
11	EHF	30–300 GHz
12	–	300–3000 GHz

In ITU texts and in regulatory affairs generally some limited use is made of other conventions for identifying frequency bands in broad terms, but these usually, perhaps always, relate to the frequency allocations for particular services. Thus, the bands allocated to the broadcasting service at VHF and UHF are often identified by the Roman numbers I to V; (see p. 141).

Other terms are sometimes used to indicate an order of frequency:

- Such terms as long wave band and short wave band have been used for many years in broadcasting.
- Terms more recently coined for higher frequencies, such as 'microwave' and 'millimetre-wave' are widely used; their significance is not standardised but many engineers would understand the first to cover about 1 GHz to 20 GHz and the second to cover about 20 GHz to 60 GHz.
- Various schemes for identifying parts of the spectrum above about 200 MHz by means of a letter code are often used in the technical literature.

Ambiguity may, however, arise from these terms and their use in a regulatory context is deprecated.

A.8 References

1 'Radiocommunication vocabulary'. Recommendation 573–2; Recommendations and Reports of the CCIR, 1986, Volume XIII (ITU, Geneva, 1986)
2 'Terms and definitions'. Recommendation 662; *ibid.*, Volume XIII
3 'Use of the decibel and the neper in telecommunications'. Recommendation 574–2; *ibid.*, Volume XIII
4 'Abbreviations and initials used in telecommunications'. Recommendation 666; *ibid.*, Volume XIII
5 'International Electrotechnical Vocabulary'. Publication 50 of the International Electrotechnical Commission, Geneva, chapters being published separately as available.
6 'Necessary bandwidth calculations'. Report 836–1; Recommendations and Reports of the CCIR, 1986, Volume I (ITU, Geneva, 1986)
7 'Methods for calculating pulsed radar emission spectrum bandwidth'. Report 837; *ibid.*, Volume I
8 'Frequency tolerance of transmitters'. Report 181–4; *ibid.*, Volume I
9 'Spectra and bandwidth of emissions'. Report 977; *ibid.*, Volume I
10 'Spectra and bandwidth of emissions'. Recommendation 328–6; *ibid.*, Volume I
11 'Spurious emissions from transmitters'. Report 838–1; *ibid.*, Volume I
12 'Spurious emissions'. Recommendation 329–5; *ibid.*, Volume I
13 'Relating to studies of the maximum permitted levels of spurious emissions'. Recommendation 66; Radio Regulations, edition of 1982 (ITU, Geneva, 1982)
14 'Supplementing the additional characteristics for classifying emissions and providing additional examples for the full designation of emissions, both as given in Appendix 6'. Recommendation 62; *ibid.*
15 'Designation of emissions'. CCIR Question 1–2/1; Recommendations and Reports of the CCIR, 1986, Volume I (ITU, Geneva, 1986)
16 'Nomenclature of the frequency and wavelength bands used in telecommunications'. CCIR Recommendation 431–5; Recommendations and Reports of the CCIR, 1986; *ibid.*, Volume XIII

Appendix B

Radio propagation and noise

B.1 Survey of factors affecting radio transmission

In free space a radio signal is propagated in a straight line from a transmitting antenna to a receiving antenna. The power flux density (PFD) declines as distance from the source increases, due to the spreading of the signal power over an expanding wave-front. On and near the Earth's surface, the presence of the atmosphere and of solid objects, above all the solid bulk of the Earth itself, may obstruct the propagation of the signal or degrade the signal in various other ways. There are, however, many mechanisms by which a signal in a terrestrial environment, whether it is a wanted signal or interference, may make its way round or through obstacles. Radio noise from various external sources will tend to degrade the signal received.

Fig. B.1 shows a signal passing as a direct ray through the troposphere from a transmitting antenna to a receiving antenna which is within a clear line-of-sight of the transmitting antenna. Both antennas are shown in the figure as if they were at fixed locations but either or both might be mobile, on land, at sea or on an aircraft. It usually happens that the ray path is refracted downwards a little due to the normal variation with height in the electrical properties of the atmospheric gases. This may permit communication by the direct ray between antennas which are not quite in line-of-sight of one another. In the course of propagation through the troposphere the signal may be attenuated, refracted, scattered and distorted by the atmospheric gases, raindrops, clouds, solid particles etc. which it encounters, in addition to the decline in PFD due to the expansion of the wavefront.

There will often be a second ray from the transmitting antenna which reaches the receiving antenna after reflection from the ground, combining algebraically with the direct ray signal. Depending on the geometry of the ray paths and the height of the antennas as a function of wavelength, it may be convenient to treat this ground-reflected wave either as one of the parameters which determine the radiating characteristics of the antennas

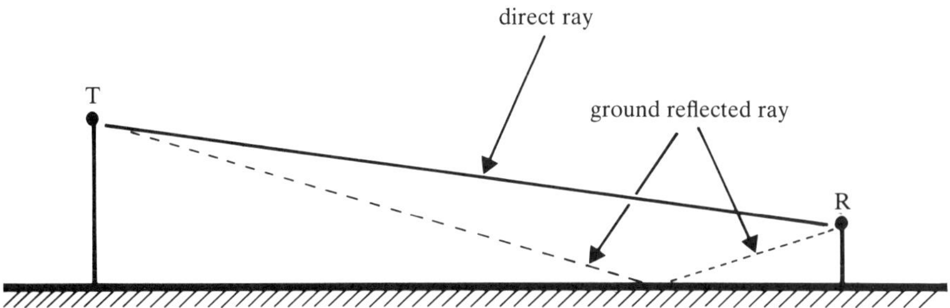

Fig. B.1 Line-of-sight propagation, typical of short distance links above about 50 MHz

or as a cause of signal level variation due to wave interference with the direct ray. Other rays reflected from buildings or other man-made or natural surfaces may also be received, causing further wave interference.

Other propagation situations are illustrated in Fig. B.2. Here the transmitting and receiving antennas may be further apart, perhaps out of sight of one another because of the curvature of the Earth's surface.

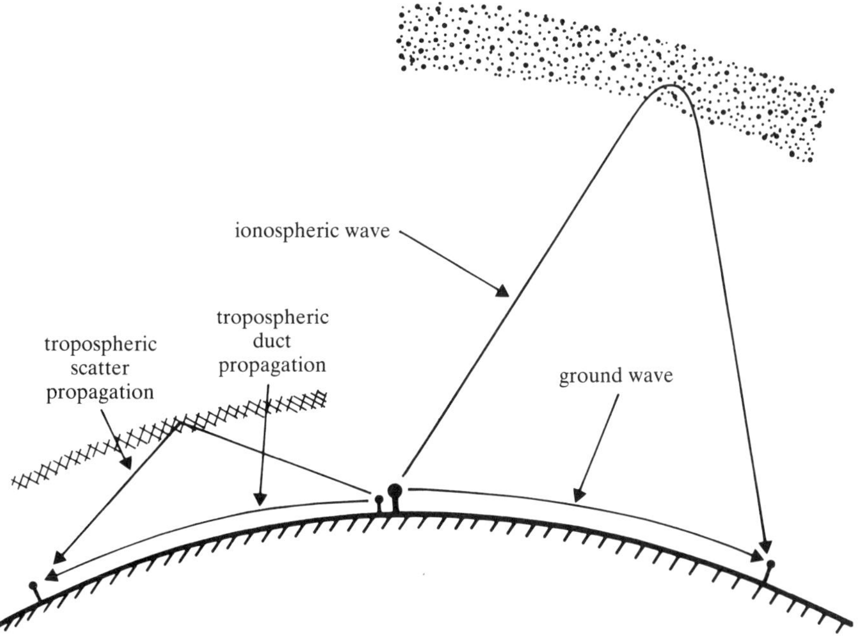

Fig. B.2 Trans-horizon propagation by ground wave, tropospheric duct, ionospheric wave and tropospheric scatter

However, by various mechanisms a ray leaving the transmitting antenna in an approximately horizontal direction may follow the curvature of the Earth and reach the receiving station with enough strength to allow communication or to cause interference. Such transmission may be by the 'ground wave', by effects akin to refraction below the Earth's surface, by diffraction or by transmission through a tropospheric duct. Alternatively, waves leaving the transmitting antennas at angles of elevation above the horizontal may be reflected or scattered back to Earth and so to the receiving station by layers of ionised air in the upper atmosphere, the so-called ionosphere or by meteor trails. Trans-horizon propagation may also occur owing to scattering lower down, in the troposphere, the scattering agent being irregularities in the physical properties of the atmospheric gases, or by raindrops. Such signals are, in most cases, more degraded than line-of-sight signals; depolarisation, changes of velocity and, in some circumstances, discontinuities, may occur.

A third set of ray path geometries is illustrated in Fig. B.3. In this case one station T_1 is assumed to be on the Earth's surface and the other R_1 is high above the surface, in space, typically on board a satellite. In cases of interest, the frequency is usually high enough to pass through the whole of the atmosphere, troposphere and ionosphere alike, but various kinds of signal degradation may be observed. Fig. B.3 also shows a link T_2–R_2, between two stations in space; this link will enjoy what is virtually free space propagation.

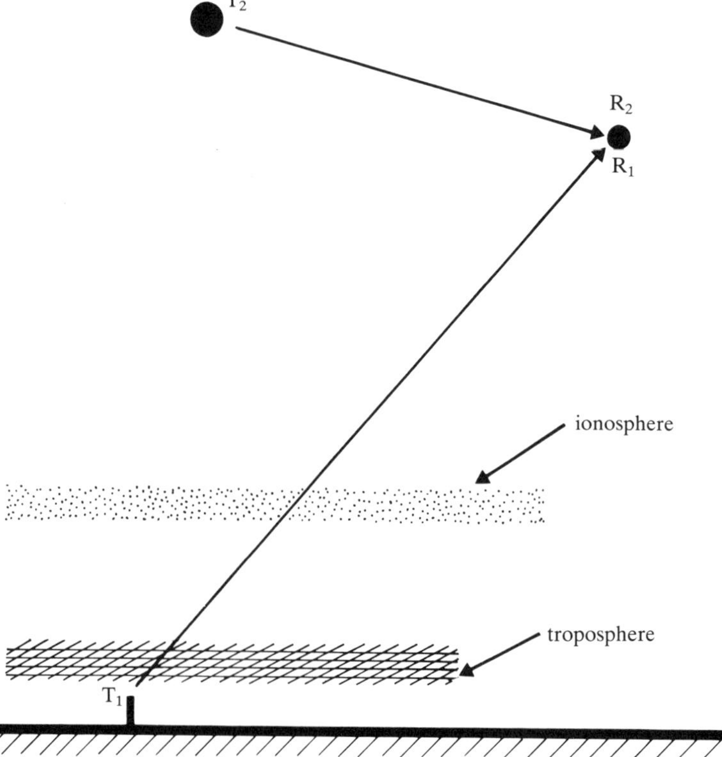

Fig. B.3 Earth–space and space–space propagation

A radio spectrum management engineer must be aware of the various modes of propagation that may be operative on a link, and to know how their valuable properties may be utilised and their disadvantages minimised. Spectrum management also requires the ability to forecast the strength of wanted signals propagated by the direct path from a transmitting antenna to a receiving antenna or by the various indirect paths. The strength of foreseen interfering signals and external noise at the receiver may also have to be determined.

One way of obtaining such information is to measure signal and noise levels at the working stations. However, these signal and noise levels must often be estimated, with whatever precision is feasible and necessary, before the radio stations have even been built, since one reason for acquiring this information is to determine whether a proposed station site is likely to be satisfactory. Prediction of radio propagation conditions may also be important for another reason; many propagation modes are variable with time, and predictions, based on long observation of basic phenomena, may give statistical information on the signal and noise levels in prospect which would take an impracticably long time to gather for each real system. Without reliable estimates of signal and noise levels, there will be a tendency for radio systems to be designed with insufficient or excessive performance margins and for interference levels between systems assigned the same frequency to be needlessly low or objectionably high. Where the spectrum is crowded, this constitutes a shortfall in the efficiency of radio spectrum management.

The geophysical processes which are at the heart of these radio propagation phenomena are complex and some are erratic. Study Groups 5 and 6 of the CCIR play a major role in stimulating their systematic study. The CCIR has also done much to develop convenient ways of applying basic propagation data to practical problems. In some situations, calculations of received signal strength can be based on a firm foundation of theory. In other situations so many factors would have to be taken into consideration, some not fully understood and some unquantified at present, that an exact theoretical treatment is not yet feasible. In these latter situations, empirical predictions have to be used, often in a statistical form, related to a data base built up from large numbers of measurements and buffered by any essential margins. The results of this work are presented in reports and recommendations of the CCIR in ways that can be used by engineers who do not necessarily have a profound understanding of the physical processes involved.

Just as wanted signals are usually weakened and distorted by propagation effects on their way to the receiver, and unwanted signals may arrive at the receiver by various routes, the wanted signal may be further degraded by the addition of electrical noise. This electrical noise comes from various sources, near to the receiver or remote, and it is propagated to the receiver via the same propagation mechanisms as are operative for signals.

It is convenient to review these propagation phenomena in this Appendix in three groups, ground wave propagation (pp. 436–440), ionospheric propagation (pp. 440–462) and tropospheric propagation (pp. 463–486). The various forms of radio noise are reviewed on pp. 486–494. For particular operational situations, however, it is often convenient to bring together data on all of the relevant propagation phenomena and noise sources that may have to be taken into account in predicting the received state of a particular kind of received signal. The CCIR texts which present material in this way are reviewed on pp. 494–502.

These CCIR methods and data are up-to-date, relatively easy to use and as accurate as any available and generally applicable easy-to-use techniques are likely to be. They have a special significance in radio regulation because, when mature and presented in the form of formal recommendations or of reports backed by recommendations, they are internationally approved and therefore provide an agreed basis for resolving international frequency assignment and planning problems.

Nevertheless, many radio propagation parameters depend on natural phenomena, the statistics of which vary greatly from place to place. This is true, for example, of tropospheric ducting, of ground wave propagation at frequencies where the nature and condition of the sub-soil beneath the ray path have a large effect on the field strength, of the absorption of SHF signals by rain and on sky-wave propagation via the F2 layer. The statistics of these phenomena are well researched for some parts of the world but much less research has been done elsewhere. The use of generalised data, intended for use anywhere, gives predictions which are less precise than those that could be made for a well-researched area using data which is specific to that area. In some situations, predictions of propagation conditions, more accurate than the CCIR methods can provide, could be made using methods of analysis and computation which may be too complex for general use. In parts of the world where propagation conditions have been particularly well researched, and in operational situations where the use of more precise predictions methods is desirable and feasible, countries which agree to base predictions which concern them both on these other methods and data are, of course, free to do so.

The power of emissions and transmission losses

In the management of spectrum it is convenient for some purposes to treat a transmitting station as a signal source of specified power. In other situations it is more appropriate to consider the field strength which the transmitter sets up at a known distance. In yet other situations it is necessary to estimate the signal strength set up by a specified transmission in a specified receiving antenna; this is most conveniently treated as the attenuation that would be measured between the output terminals of the transmitter and the input terminals of the receiver or in some similar way. In each of these three approaches the ideal formulation will depend to some degree on the frequency range and the situation; some of these formulations are outlined below.

The relationship between the power supplied to, or available from, an antenna and the field strength in the immediate vicinity of the antenna is made complex by the simultaneous presence of both a radiation field and an induction field. However, the induction field weakens rapidly as the distance from the antenna increases, becoming negligible when the distance equals a few tens of wavelengths, and it is not taken into consideration in this book.

The transmitting antenna as a signal source

An isotropic antenna is the fundamental antenna standard, but other reference antennas are also found to be convenient for use as standards in some circumstances, in particular:

A Hertzian dipole in free space
A half-wave dipole in free space
A Hertzian dipole near a perfectly conducting ground plane
A short vertical monopole on a perfectly conducting ground plane
A quarter-wave monopole on a perfectly conducting ground plane

The standard symbols for the gain of an antenna in a specified direction, expressed in decibels relative to the gain of an isotropic antenna or the maximum gain of a half-wave dipole, are 'dBi' and 'dBd' respectively. The maximum gains of these standard antennas, in arithmetical units and logarithmic units, are compared in Table B.1.

Table B.1 The directivity of typical reference antennas (from CCIR Recommendation 341)

Reference antenna	g_t	$G = 10 \log g_t$ (dB)
Isotropic in free space	1·0	0
Hertzian dipole in free space	1·5	1·75
Half-wave dipole in free space	1·65	2·15
Hertzian dipole near a perfectly conducting ground plane, or a short vertical monopole on the ground plane	3·0	4·8
Quarter-wave monopole on perfectly conducting ground plane	3·3	5·2

The strength of the signal produced by some of these standard antennas in the direction of their maximum gain when the antenna is radiating a reference power level, say 1 W, is also used as a standard of comparison for defining the strength of the signal radiated in a specified direction by a real transmitting antenna radiating a known power level. Thus:

- The signal strength from a real antenna radiating the real power, relative to that of an isotropic antenna radiating 1 W, is called the equivalent isotropically radiated power (EIRP) of the source, typically expressed in dBW.

- Signal strength related to that of a half-wave dipole is called the effective radiated power (ERP).
- Signal strength related to that of a short vertical monopole is called the effective monopole radiated power (EMRP).

The field strength at a distance from a transmitting antenna in free space

The field strength E of a linearly polarised wave at a specified point is expressed by:

$$E = \sqrt{30\, w_t}/d \quad \text{volts (rms) per metre} \qquad \text{(B.1)}$$

where w_t = EIRP of the transmitter in the direction of the point, W
d = distance from the transmitter to the point, m

For a circularly polarised wave, E is replaced by $\sqrt{2}\ E$ in expression (B.1). For an elliptically polarised wave, E is the root sum of the squares of the major and minor axes of the voltage ellipse.

The power flux-density (PFD) of a linearly polarised plane wave at a specified distance from a transmitting antenna in free space is expressed by

$$S = w_t/4\pi\, d^2 \quad \text{watts per square metre} \qquad \text{(B.2)}$$

where S = PFD
and other symbols are as in expression (B.1)
The power available to a matched load from an isotropic receiving antenna located in such a field is given by

$$w_r = \lambda^2 w_t/(4\pi\, d)^2 \quad \text{watts} \qquad \text{(B.3)}$$

where w_r = received power
λ = the wavelength, metres
and other symbols are as in expression (B.1).

Transmission loss and the real world

The signal that reaches a radio receiver is always weaker than the signal that had been delivered to the transmitting antenna. This reduction of the power of the signal is due to a combination of factors; the spreading of the energy of the signal over an expanding wavefront, absorption in the propagation medium, a mismatch between the directional properties of antennas and the optimum propagation path, losses in the antennas and feeders, electrical mismatches within the transmitting and receiving systems and so on. The CCIR has defined terms (see CCIR Recommendations 525[1] and 341[2] to facilitate separate consideration of these factors.

The most fundamental concept of loss in a radio link is the ratio between the RF signal power supplied by the transmitter to the transmitting antenna and the resulting signal power made available to the corresponding receiver. This is called the 'total loss' (L_L) and it is usually expressed as a positive quantity in decibels. Thus

$$L_L = 10\log_{10} w_t/w_r \quad \text{dB} \tag{B.4}$$
$$= P_t - P_r \quad \text{dB} \tag{B.5}$$

where w_t = signal power at RF supplied by the transmitter, W
w_r = signal power made available to the receiver, W
P_t = $10\log_{10} w_t$, dBW
P_r = $10\log_{10} w_r$, dBW

The total loss can be resolved into components; thus:

$$L_L = L_{ft} + L_{at} + L_{bf} + L_m - G_T - G_R + L_{ar} + L_{fr} \quad \text{dB} \tag{B.6}$$

where L_{ft} = loss in the transmitting antenna feeders and associated filters etc, dB
L_{at} = loss in the transmitting antenna, dB
L_{bf} = the transmission loss (see below) that would be observed if the antennas were replaced by isotropic antennas with the same polarisation characteristics as the real antennas and if these isotropic antennas were in free space but separated by the same distance as the real antennas, dB
L_m = loss, in additional to L_{bf}, that would be observed if the isotropic antennas were located in the same propagation environment as the real antennas, dB
G_T = gain of the real transmitting antenna in the operative direction, dBi
G_R = gain of the real receiving antenna in the operative direction, dBi
L_{ar} = loss in the receiving antenna (dB)
L_{fr} = loss in the receiving antenna feeders and associated filters etc., dB

The 'system loss' L_s is defined as the ratio between the RF signal power input at the terminals of the transmitting antenna and the resulting power available at the terminals of the receiving antenna:

$$L_s = L_L - L_{ft} - L_{fr} \quad \text{dB} \tag{B.7}$$

The 'transmission loss' L is defined as the ratio between the power radiated by the transmitting antenna and the power that would be available at the receiving antenna output terminals if there were no losses in the receiving antenna:

$$L = L_s - L_{at} - L_{ar} \quad \text{dB} \tag{B.8}$$

L_{bf} is called the 'free space basic transmission loss' and it is given by

$$L_{bf} = 20\log_{10}(4\pi\ d/\lambda) \quad \text{dB} \tag{B.9}$$

L_m is called the 'loss relative to free space'. It arises in various ways, for example through

Absorption in the ionosphere, the atmospheric gases or rain
Diffraction losses
Imperfect reflection, scattering or focusing due to curvature of reflecting layers
Polarisation coupling loss
Wave interference

It is sometimes convenient to combine L_{bf} and L_m into a single quantity called the 'basic transmission loss' L_b; thus:

$$L_b = L_{bf} + L_m \quad \text{dB} \tag{B.10}$$

Finally, it is sometimes necessary to define a particular form of the basic transmission loss, applicable to a specific ray path, instead of the more generalised quantity arising from the definition of G_T and R_R for eq. (B.6) above. Thus the 'ray path transmission loss' L_t is given by

$$L_t = L_b - G_t - G_r \quad \text{dB} \tag{B.11}$$

where G_t = gain of the transmitting antenna for the direction of the specified ray, dBi
G_r = gain of the receiving antenna for the direction of the specified ray, dBi

B.2 Ground wave propagation phenomena

B.2.1 Introduction

A radio wave produced by a transmitting antenna at or just above the Earth's surface may be propagated in all directions which do not intersect the Earth and there may be some limited penetration of the wave into the ground. However, the energy which is radiated approximately parallel to the surface of the Earth is affected by the nearby presence of the ground, causing its path to curve downwards. Some further downwards curvature of the path of the wave may be caused by a regular variation with height of the physical properties of the gases forming the atmosphere. Consequently, the radio wave can be received beyond the horizon as seen from the transmitting antenna. This mode of propagation is called the ground wave.

The transmission loss by ground wave propagation relative to free space for distances of a few hundreds of kilometres may be relatively small at low frequencies (that is, below about 300 kHz) and this propagation mode may also be operative for systems using frequencies up to several megahertz. The ground wave may indeed be quite strong at distances up to say 100 km from the transmitter at much higher frequencies. However, as frequency increases the ray-bending effect of the ground on the wave decreases whilst the various effects of the atmosphere increase. The behaviour of trans-horizon signals at these higher frequencies is therefore best interpreted in terms of tropospheric propagation phenomena. In this section consideration of ground wave propagation is limited to frequencies below 30 MHz. It should however be noted that ionospheric propagation modes are also operative in this frequency range.

Two phenomena contribute significantly to the formation of the ground wave, namely:

- refraction due to non-uniformity of the index of refraction of the atmosphere, and
- coupling between the wave propagated in the atmosphere and the wave propagated below the Earth's surface.

These phenomena are considered below. A third mechanism, diffraction, is also operative, owing to the presence of the Earth, but its effect can be disregarded at frequencies below 30 MHz in comparison with these other phenomena.

B.2.2 Atmospheric refraction

The index of refraction of air is more than, but very close to, unity. The magnitude of that small difference is affected by variations in the atmospheric pressure and temperature and the partial pressure due to water vapour. Because of these factors the mean refractive index falls approximtely exponentially with increasing height within the troposphere, becoming relatively constant above about 7 km. Given data on these factors, means are available for computing their effects on radio wave propagation; see CCIR Report 714[3]. However, for many purposes it is sufficient to establish baseline conditions and the CCIR has adopted the following expression as a 'reference atmosphere for refraction', to be used as a model representing typical conditions:

$$n(h) = 1 + 0{\cdot}000315 \times \exp(-0{\cdot}136\,h) \qquad \text{(B.12)}$$

where $n(h)$ is the refractive index n at height h km above sea level, see Recommendation 369[4]. This baseline is sufficiently accurate for most ground wave field strength studies.

At frequencies above about 10 MHz the effect on ground wave propagation between two antennas, both near to the ground, of an atmosphere conforming to this model is well represented by assuming that the ray path is straight but the radius of the Earth is equal to 4/3 times its actual value. Below 10 MHz the increasing thickness of the layer of the atmosphere which is involved in the propagation of the wave results in a reduction in the effective mean rate of variation of the refractive index with height, as a result of which the equivalent radius of the Earth becomes less than 4/3 times the actual radius. This trend becomes marked below 3 MHz. At 10 kHz the atmosphere up to a great height is involved in the propagation of the ground wave and the effect of atmospheric refraction is small.

B.2.3 The effect of the ground

Any horizontal component of the electrical field of the wave close to the ground is rapidly attenuated by losses in the ground; thus the ground wave is predominantly vertically polarised. However, the wave also penetrates into the ground. The velocity of propagation of the wave below ground is less than its velocity above ground and it may be considered that this difference in velocity is the source of a secondary wave above the surface, the electrical vector of which is approximately in line with the direction of propagation of the ground wave. The combination of this secondary wave with the original wave produces a downward tilt in the wavefront close to the ground, so sustaining the ground wave round the Earth's curved surface.

The velocity of propagation of the wave below the Earth's surface is determined by the permeability, permittivity and conductivity of the surface layers of the Earth. The permeability of common sub-soil minerals is approximately equal to the permeability of free space, so this factor can be ignored in most circumstances. The relative permittivity (that is, the quotient of the permittivity of a substance by the permittivity of vacuum) and the conductivity of some typical soil and sub-soil conditions are shown in Table B.2. It will be noted that the moisture content of soil has a major effect on its conductivity and a significant effect on its permittivity, as a result of which the ground wave typically suffers seasonal variations in strength. Note also that permittivity tends to fall and conductivity to rise at frequencies above 30 MHz. For more information see CCIR Recommendation 527[5] and Report 229[6].

Table B.2 The relative permittivity and the conductivity of typical subsoil materials

	Relative permittivity	Conductivity (Siemens/m)
Sea water (average salinity), 20°C	75	5·0
Fresh water, 20°C	80	0·03
Wet ground	30	0·01
Dry ground	15	0·001
Desert and dry rock	3	0·0001
Ice (fresh water), −1°C	3	0·00006
−10°C	3	0·00002
Sea ice	see text	see text

It will be noted that the conductivity of sea water is exceptionally high in comparison with any other ground condition. The conductivity varies considerably with salinity, a range from 1 S/m in parts of the Black Sea to 6 S/m in the Red Sea having been observed, but the effect of differences of water temperature is not great.

The conductivity and the relative permittivity of ice are very low but variable in complex ways. Below 100 kHz the relative permittivity of fresh water ice rises considerably with decreasing frequency, reaching values between 10 and 30 at 10 kHz, depending on the temperature. The relative permittivity and conductivity of sea ice varies considerably with the temperature and the age of the ice, and with frequency. Values for both parameters are somewhat higher than for fresh water ice at 100 kHz, the relative permittivity tending to fall towards the value for fresh water ice and the conductivity tending to rise further with increasing frequency; see CCIR Report 229[6].

In the absence of ice, an oversea path provides markedly good ground wave propagation conditions above about 1 MHz. The beneficial effect of fresh water is much less marked.

CCIR Recommendtion 368[7] provides ground wave propagation curves for various frequencies between 10 kHz and 30 MHz, assuming ground with various electrical properties.

The ground wave signal strength reaching a receiving antenna remote from a transmitting antenna is influenced not merely by the permittivity and conductivity of the sub-soil beneath the great circle path between the two antennas; it is also affected to some degree by sub-soil conditions to either side of the great circle path. The first Fresnel half-wave zone might be taken as a guideline to the area which should be considered, that is, the ellipse having its foci at the two antennas and having major and minor axes equal to $(D + \lambda/2)$ and $\sqrt{D\lambda}$ respectively, where D is the distance between the antennas and λ is the wavelength.

Similarly it may be necessary to take into account the electrical characteristics of deep-lying strata. The signal becomes weaker as it penetrates more deeply into the ground but it may be assumed that the effect of the subterranean wave on the ground wave is significant, and therefore the nature of the geological strata is significant, down to a depth where the strength of the wave has fallen to $1/e$ (that is to 37%) of its strength at the surface. The depth at which this condition is reached, called the penetration depth, depends mainly on the conductivity of the ground and the frequency. At 10 kHz the penetration depth may be as great as 1000 m if the conductivity of the ground is very low. In the MF band, typical penetration depths are from 5 to 100 m. At 30 MHz the penetration depth may be as little as one metre, and perhaps much less if the surface layer is moist.

Given the necessary data it is feasible to calculate composite values for the relative permittivity and the conductivity of the ground at a given location and for a given frequency which take into account to an appropriate degree the characteristics of all of the known geological formations in the immediate vicinity which will affect the ground wave.

These composite values are called the effective relative permittivity and the effective conductivity. Through the ITU, a world atlas of effective ground conductivity, suitable for use in the VLF, LF and MF parts of the spectrum, is being compiled. CCIR Resolution 73[8] and Report 717[9] refer. A considerable amount of information, complete for the VLF band and substantial for the LF and MF bands, is already available in Report 717.

B.2.4 The phase of the ground wave

The phase of the ground wave at the lower end of the radio spectrum is coherent and relatively predictable. Thus, for example, the phase difference Φ (radians) between the signal at the transmitting antenna and a receiving antenna d distant from it, at a given instant, can be represented by

$$\Phi = 2\pi\, nd/\lambda + \Phi_s \qquad \text{(B.13)}$$

where n = refractive index of the atmosphere at the Earth's surface
d = distance between the antennas
λ = wavelength, in the same units as d
Φ_s = an adjustment factor called the 'secondary phase', which allows for departures from completely regular propagation, radians

At frequencies below about 3 MHz the value of Φ_s is small for transmission over sea. Over land, Φ_s is affected by the rocks lying below the surface; there are means for calculating the scale of the effect, outlined in CCIR Report 716[10] and set out in more detail in various other CCIR Reports and various papers referenced in Report 716.

The phase of the signal reaching the receiving antenna may be substantially affected if a significant part of the received signal has been propagated by paths other than the ground wave.

B.3 Ionospheric propagation phenomena

B.3.1 Introduction

The ionosphere is the outer part of the Earth's atmosphere where the air is partially ionised, mainly by radiation from the sun. Its lower limit is about 50 km above the ground. Its upper limit can be taken arbitrarily to be at a height of 2000 km. Such atmosphere as there is above this level, merging with plasma from the solar wind, is also ionised but it has relatively little direct effect on radio propagation.

Ionisation modifies the behaviour of the atmosphere towards radio waves. In particular the permittivity of the air becomes a function of the free electron density, allowing a horizontal layer of ionised air to deviate by refraction the path of a radio wave passing into it (as with the Ray ABC in Fig. B.4) or to reflect it back to Earth (as ray ADE). Furthermore, whereas non-ionised air absorbs virtually no energy from a radio wave passing through it (except at frequencies above 20 GHz), there are circumstances in

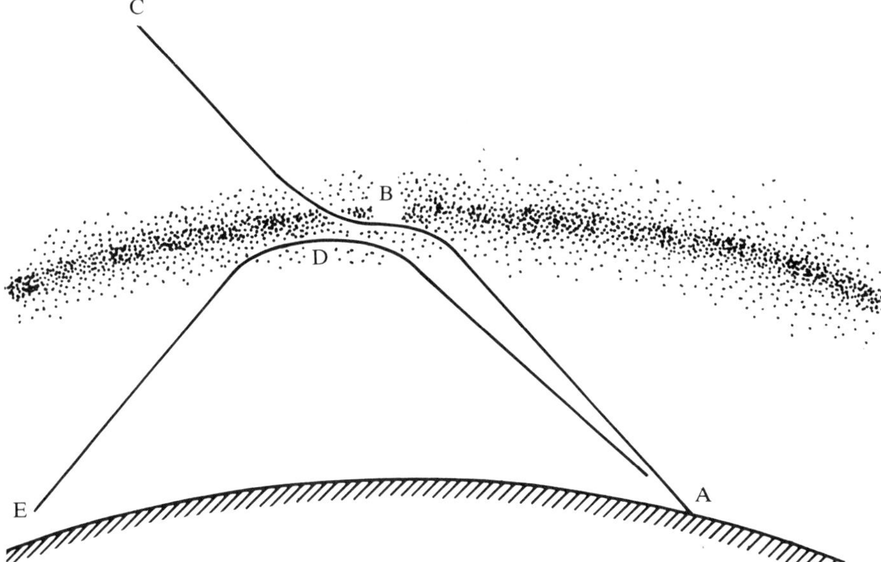

Fig. B.4 Ray paths through an idealised, ionised layer

which ionised air absorbs a considerable amount of energy from the wave. A combination of atmospheric ionisation and the Earth's magnetic field have further effects on a wave propagated in the ionosphere.

A much-simplified review of ionospheric refraction follows and the other phenomena are touched on later in this section.

Reflection from an idealised ionised layer

Fig. B.5 shows a layer of ionised air above the Earth's surface. Let it be assumed that the layer and the Earth's surface are concentric spheres and let the Earth's magnetic field be disregarded for the moment. The vertical distribution of the density of free electrons through the layer can be assumed to be parabolic, the density being at a maximum at a height h_m above the surface.

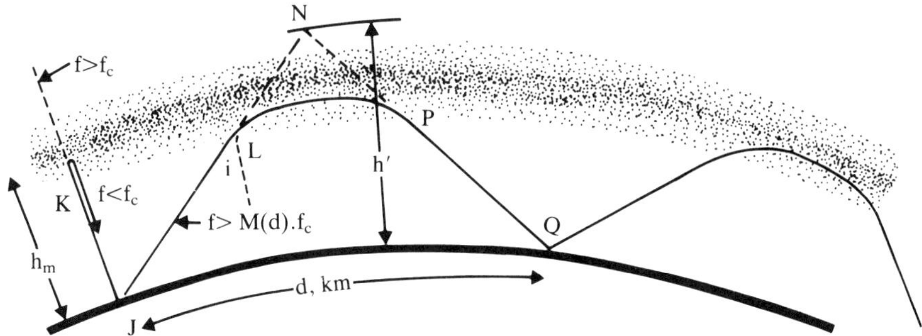

Fig. B.5 Reflection of HF waves by any idealised, ionised layer

A wave emitted from a transmitting antenna at J on the ground, incident perpendicularly on the underside of the layer at K, will be reflected downwards at a level below h_m returning to the ground at J, provided that the frequency of the wave is not too high. If the frequency of the wave is raised, the wave will penetrate more deeply into the layer before being reflected, reaching levels where the free electron density is greater. When the frequency is so high that the wave penetrates to h_m, the wave passes through the layer and emerges from the upper surface. The frequency at which penetration occurs is called the vertical incidence critical frequency, f_c.

Consider now a ray of the wave transmitted from J which enters the ionised layer at L obliquely with an angle of incidence i. If the frequency of the wave is equal to f_c, the ray will not penetrate to h_m and then pass through the layer as it did at vertical incidence; instead it will be reflected downwards lower in the layer, emerging from the under-side at P with an angle of emergence equal to the angle of incidence, to strike the Earth's surface at Q, a distance d km from J. The ray path JLPQ is said to be a 'single hop'. The frequency of the incident ray would have to be raised to f_c multiplied by $M(d)$ before it would penetrate the layer at that angle of incidence i, where

$$M(d) = \sec i \tag{B.14}$$

$M(d)f_c$ is called the Maximum Usable Frequency (MUF) for distance d and $M(d)$ is called the MUF factor for that distance.

Reaching point Q on the Earth's surface, the ray might be reflected again upwards towards the ionised layer. The total transmission path of a ray might thus consist of one hop to a receiver at Q or several hops to more distant receivers, produced by successive reflections from the ionised layer and from the ground.

While the wave is within the ionised layer, its phase velocity is increased, relative to the velocity of the wave in free space, but the group velocity falls below the free space velocity. Thus, the transmission time of a signal along the curved ray path is greater than would be expected if it were assumed that the path was in free space. It is as if the signal had been reflected, not at the maximum height to which the ray had penetrated, but at N, higher in the layer and perhaps even above its top. The height of N is called the effective height h' for the frequency of the wave and its angle of incidence at the bottom of the layer. To a close approximation, the geometry of the ray paths outside the ionised layer, JL and PQ, is also consistent with straight-line propagation and reflection at a surface a distance h' above the ground.

For any frequency f which exceeds f_c there is an area of ground in the vicinity of the transmitting station which a signal cannot reach after regular reflection by the ionised layer, since a ray which is incident on the lower surface of the layer at the angle which is geometrically required for reflection into this area will, in fact, pass right through the layer. This area is called the skip zone. The radius of the skip zone, called the skip distance d_s is the value of d for which $M(d) = f/f_c$. The signal may, however, enter the skip zone by ground wave propagation or by various scattering mechanisms.

So far the Earth's magnetic field has been disregarded. In fact, the magnetic field affects the behaviour of a wave in the ionised layer significantly. In particular, the wave on entering the layer is split into two components, the ordinary wave and the extraordinary wave, which then follow different paths. The critical frequency for the ordinary wave is lower than that of the extraordinary wave. The vertical incidence critical frequencies of the ordinary and extraordinary components have the symbols f_c and f_x respectively. Also, the plane of polarisation of each component is rotated.

The regions and layers of the ionosphere and the sources of ionisation

It is convenient to divide the ionosphere into three regions, called the D, E and F regions respectively. The D region covers the height range from 50 to 90 km above the Earth's surface, the E region is from 90 to 130 km and the F region is from 130 km upwards. Ionised layers form in these regions and they are discussed below as if the various layers were discrete, separated by non-ionised layers. In fact they are not separate; during the daytime there is ionisation present throughout the ionosphere and the named layers are more like ledges in a graph of free electron concentration as it climbs from a low density around 50 km height to a maximum at a height between 300 and 500 km. At night much of the ionisation below about 200 km disappears, neutral atoms of the atmospheric gases being formed by recombination of positive and negative ions. However, the diurnal behaviour of ions at various levels differs, and it is convenient to consider the layers as if they were discrete.

An ionised layer forms regularly in the D region, called the D layer. The ionisation is caused mainly by solar ultra-violet radiation, although cosmic rays contribute to the ionisation in the lower part of the region. The atmospheric pressure at this height is high (relative to the pressure in the E and F regions). Consequently the mean free path of free electrons is relatively short and recombination of the ions due to solar radiation takes place quickly after sunset, ions being reformed at sunrise. The ionisation due to cosmic rays tends to be sustained during the night.

A regular layer forms in the E region, called the E layer. It is often called the normal E layer, to distinguish it from sporadic E ionisation. Its ionisation is caused mainly by solar ultra-violet and X-ray radiation but energetic particles emitted by the sun are another source of E region ionisation. As in the D layer, recombination is rapid after sunset and relatively little ionisation remains at night, except in polar regions where a significant residue of ionisation due to solar particles usually persists through the night.

In addition to the regular E layer, sporadic ionisation occurs in the E region. Three forms of sporadic E region ionisation (E_s) are recognised, as follows:

- Equatorial E_s forms frequently around the geomagnetic dip equator during daytime.

- Auroral E_s tends to form a ring round each geomagnetic pole, principally at night, peak ionisation occurring at about 70° geomagnetic latitude. E_s also forms a polar cap within the ring of auroral E_s.
- Temperate zone E_s is chiefly a summer daytime phenomenon.

The causes of E_s are complex and are not completely understood at present; geomagnetism is clearly involved, but metallic ions, originating in meteors, which accumulate in the E region, and large-scale movements of the atmosphere probably play substantial roles.

Finally, meteor trails are strongly ionised and develop mainly in the E region.

A thick regular layer forms in the F region. It is often convenient to distinguish two layers here, the F1 layer lying between 130 and 210 km above the Earth's surface and the F2 layer, the maximum free electron density of which is found between 300 and 500 km altitude. The main source of ionisation in both layers is solar radiation in the ultra-violet and X-ray bands. As in the regular E layer, the F1 layer ionisation level follows closely the incoming solar energy flux. In the F2 layer the ion recombination rate is much lower than in the other layers and consequently the daily pattern in the distribution of free electron density is influenced, not only by the solar radiation level, but also by large scale movements in the gases of the atmosphere itself. This produces a pattern of daily and seasonal variation in the F2 layer which is much more complex than the variations which the other layers exhibit. A much greater day to day variability is also observed.

For more information on the structure of the ionosphere, see CCIR Reports 725[11] and 886[12]. The characteristics of the regular layers are reviewed on pp. 446–452. The irregular phenomena and disturbances of the regular layers are reviewed on pp. 452–456.

Solar activity indices

The Sun is a variable star and the level of solar ultra-violet and X-ray flux reaching the Earth varies with a period of about 11 years, rising from a base level, passing through a maximum and returning to the base level again before the commencement of a new cycle. The period is erratic, and the range of solar activity from maximum to minimum is variable from cycle to cycle.

Most of the ionisation in the Earth's atmosphere is caused by solar radiation. If the geographical, diurnal and seasonal variations in the critical frequencies of the ionised layers are discounted, there remains a long term variation in the critical frequencies which follows the same 11-year cycle as the Sun. It is important to be able to predict critical frequencies for use in the operation of radio systems using the HF part of the spectrum. Therefore, it is necessary to be able to forecast the level of activity of the Sun, the causative agent. Effective forecasting of solar activity requires knowledge of the past history of solar activity variation, preferably over many 11-year cycles.

Before there were artificial earth satellites, it was not possible to measure the level of solar radiation in the ultra-violet and X-ray parts of the spectrum directly because virtually all such radiation reaching the Earth is absorbed in the atmosphere. Thus, there is a need to identify a solar phenomenon which can be observed on Earth, which varies in sympathy with the variation of that part of the solar radiation which causes ionisation in the atmosphere and which, desirably, has been observed for long enough to provide a basis for predicting future trends in that variation. Two possibilities have been studied:

(*a*) It has been found that the long term variation in the level of sunspot activity is one such phenomenon. There has been systematic observation of sunspots since 1749 and records of the years when maxima and minima of sunspot activity occurred go back to 1610. An index of sunspot activity, developed by Rudolf Wolf (1816–93), calculated from the number, size and disposition of the sunspots which are visible each day and averaged monthly, is called the 'sunspot number' and has the symbol R. There is a wide scatter in the month by month values of R, which is not closely representative of the general trend of solar activity, and to eliminate this scatter from the statistics, a 12-month running mean of the values of R is used as the index of solar activity for a month at the middle of the 12-month period. This smoothed sunspot number has the symbol R_{12}. Thus the most recent monthly index that is available at any time is always at least six months out of date. R_{12} approaches zero at the minimum phase of the sunspot cycle, and values as high as 150 have been exceeded at a few maxima. After a long period during which sunspot observations made at Zurich Observatory were used for assessing solar activity, data are now available from the Sunspot Index Data Centre at Brussels.
(*b*) The solar radio noise flux at about 3 GHz is observed regularly at several observatories. Information has been available on this factor since 1947. There is relatively little scatter of the day to day measurements, so it is acceptable to use monthly mean values for the prediction of future trends. The variations of noise flux also correlate well with variations in critical frequencies, especially E layer critical frequencies. Monthly means of measured daily values of flux issued by the National Research Council, Ottawa, are given the symbol Φ. Twelve-month running means of Φ, having the symbol Φ_{12}, are also calculated.

CCIR Report 340[13] contains predictions of the worldwide distribution of F2 layer critical frequencies for notional solar activity states of $R_{12} = 0$ and $R_{12} = 100$. Critical frequencies for any other sunspot number can be predicted from these data by linear interpolation or extrapolation. However, in addition to criticisms of the accuracy of the data for $R_{12} = 0$ and $R_{12} = 100$ (see CCIR Report 430[14]), it is now thought that significant inaccuracies may arise from linear interpolation and extrapolation for the predicted value of R_{12}. An alternative measure of solar activity, based on observations of F2 layer critical frequencies, might give a better basis for prediction. One formulation, discussed by Minnis[15], based on observations of noon values of F2 layer critical frequencies recorded at a number of long-established observatories, has been shown to provide a good correlation with subsequent observations. Monthly means of the daily values of this index have the symbol I_{F2} and are published by the Rutherford Appleton Laboratory, Chilton, United Kingdom. Other formulations of an ionospheric index are also being studied.

There is growing confidence that the use of forecasts of an ionospheric index could improve forecasts of F2 layer critical frequencies from the basic data in Report 340. Means have been developed for translating I_{F2} into a form in which it can be directly substituted for R when using the Report 340 data; in Liu *et al.* (1983)[16] this new index was called the Global Effective Sunspot Number (GESSN) and it has since been given the symbol I_G; see CCIR Opinion 82[17]. The symbol I_{G12} is used for the twelve-month running mean of I_G.

Various procedures have been developed for forecasting these indices up to the time of the next minimum phase of the solar activity cycle, based on the history of past cycles and the latest trends of the current cycle, plus a substantial judgmental element. Accuracy tends to decline as the forecasting period lengthens. Recent observations and short term forecasts of R_{12}, I_{G12}, Φ and Φ_{12} are pubished monthly in the ITU *Telecommunication Journal*[18]. No accurate method has yet been found for forecasting any of these indices beyond the end of the current cycle.

At the present time the CCIR recommends the use of forecasts of R_{12} for all long term predictions of critical frequencies. Forecasts of Φ and I_{G12} are also recommended as an alternative to R_{12} for medium term F2 layer predictions, that is for periods up to six months from the date of the last observed values, and with less confidence for a further six months. However, for medium term predictions of normal E and F1 critical frequencies, forecasts of Φ provide the best index now available; see CCIR Recommendation 371[19].

B.3.2 The regular ionospheric layers at HF

The D layer

The free electron density in the D layer is very low compared with that of the higher layers and reflection of waves at frequencies above 1 MHz does not take place. However, energy imparted to free electrons by HF waves passing through the D layer tends to be lost in collisions between the electrons and the molecules of the atmospheric gases. Since the density of the atmosphere in the D region is high, relative to that of the E and F regions, the absorption of energy from the wave by this mechanism tends to be substantial. This loss of energy is called non-deviative absorption, to distinguish it from the deviative absorption which occurs in the layers where waves are reflected.

The degree of non-deviative absorption suffered by a wave is proportional to the product of the collision frequency and the free electron density and is approximately inversely proportional to the square of the wave frequency. The free electron density depends closely on the angle between the Sun and the zenith and on the level of solar activity. Fig. B.6 charts the zenith angle of the Sun against latitude and longitude at one particular instant in the annual cycle and corresponding data for other seasons are in CCIR Report 340[13]. Non-deviative absorption can be assumed to be negligibly small when the zenith angle of the Sun is about 110° (that is, when the Sun is substantially below the horizon).

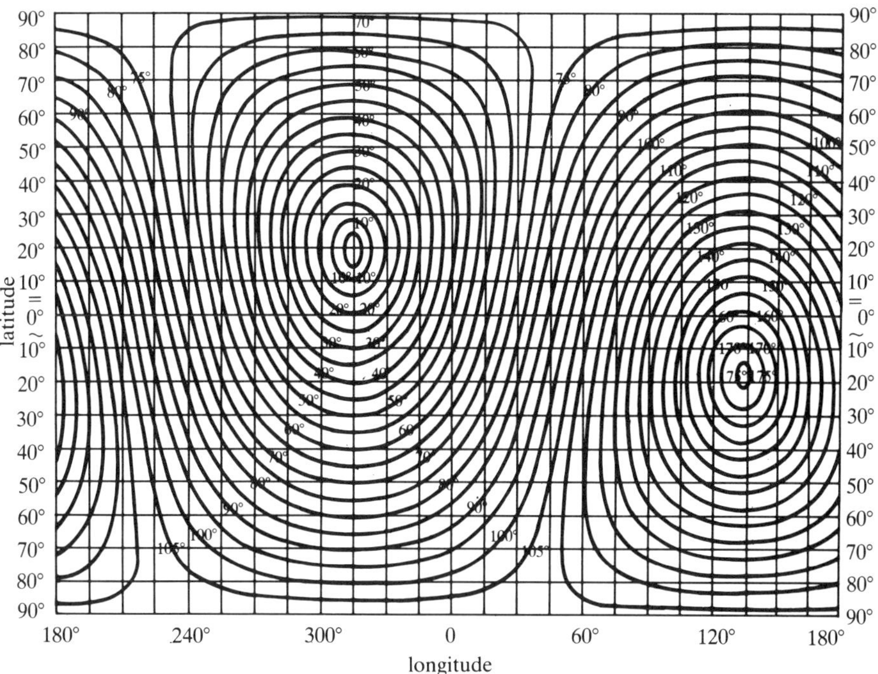

Fig. B.6 Zenith angle of the Sun at 15 hours UT in May for an location on Earth (after CCIR Report 340[13])

Methods for estimating the value of the absorption arising in the D layer and elsewhere in the ionosphere are discussed in CCIR Report 252[20] and are reviewed briefly on p. 497.

The E layer

Muggleton[21] shows that the E layer critical frequency (f_{oE}) at a given place and time can be represented precisely by functions of four factors:

A solar activity factor, preferably related to Φ

A seasonal factor related to the solar declination and the geographical latitude of the point of reflection
A latitude factor
A time-of-day factor, related to the solar zenith angle, but with minimum values below which the factor does not decline at night.

Data for calculating f_{0E} are given in CCIR Report 340[13]. The results obtained can be used with high confidence; errors from all sources including day-to-day variations are not likely to exceed 5%. The E layer MUF factor, used to convert vertical incidence critical frequencies to the basic MUF for a single hop transmission path having length d km, is based on the assumption that the level of maximum ionisation of the layer occurs at a height of 110 km. Fig. B.7 shows some examples of single hop E layer critical frequencies.

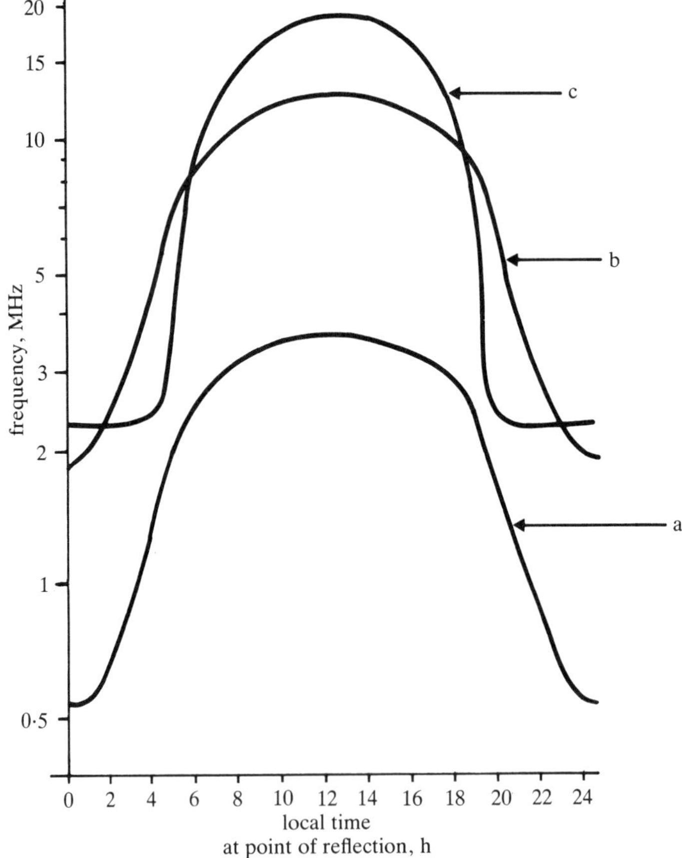

Fig. B.7 Predicted E layer critical frequencies for May when $\Phi = 120$
a f_{0E} for reflection point at 50° North latitude
b MUF for a transmission distance of 1000 km, the reflection point being at 50° North latitude
c MUF for a transmission distance of 2000 km, the reflection point being at 10° North latitude

When the distance between the transmitter and the receiver exceeds 2300 km, propagation by means of a single hop via the E layer is not possible. Multiple hop propagation may be possible for longer distances. To find the critical frequency for multi-hop propagation, the great circle distance between the transmitter and the receiver is divided into two or more equal parts, each representing a single hop, and the basic MUF for the complete path is the lowest of the basic MUFs for the separate hops. Non-deviative absorption arising in repeated journeys through the D layer may limit the number of E layer hops that will support communication but in such circumstances propagation may be possible with fewer hops by reflection from a higher layer, possibly at a higher frequency, instead.

There may be significant deviative absorption as the wave is reflected in the E layer and this absorption tends to be higher when the frequency of the wave is close to the basic MUF.

The F1 layer

As with the E layer, f_{0F1} varies primarily with the solar zenith angle. The F1 layer also resembles the E layer in that it may be assumed, without too much inaccuracy, that the height of maximum ionisation does not vary with time of day, so the F1 MUF factor varies in a relatively simple way with distance d. The F1 layer may provide higher maximum usable frequencies than the E layer, particularly when d is greater than 2000 km, and single hop transmission may take place via the F1 layer at values of d up to about 3400 km. Multiple hop transmission may also take place via the F1 layer.

There are some differences in the behaviour of the E and F1 layers. The variation of f_{0F1} with solar activity is represented rather better by I_{F2} than by Φ, but R_{12} also gives acceptable accuracy. The variation of f_{0F1} with solar zenith angle is also dependent on the level of solar activity and the geomagnetic latitude. Furthermore, the height of maximum ionisation rises as solar activity increases, causing the MUF factor for a given distance to fall. For fuller data see CCIR Report 340[13]. The predictability of F1 layer characteristics is good and the scatter of day-by-day values is small, but in neither respect is the F1 layer quite as regular as the E layer.

The F2 layer

The vertical incidence critical frequency for the F2 layer is higher than that for the lower layers by day and ionisation is sustained more completely through the night. However, the pattern of variation, diurnal and seasonal, is irregular, although it is clearly related to the incidence of solar radiation. Some examples of the diurnal variation of MUF for particular routes (taken from predictions) are shown in Fig. B.8. The longest single hop transmission path which is normally feasible via the F2 layer is about 4000 km.

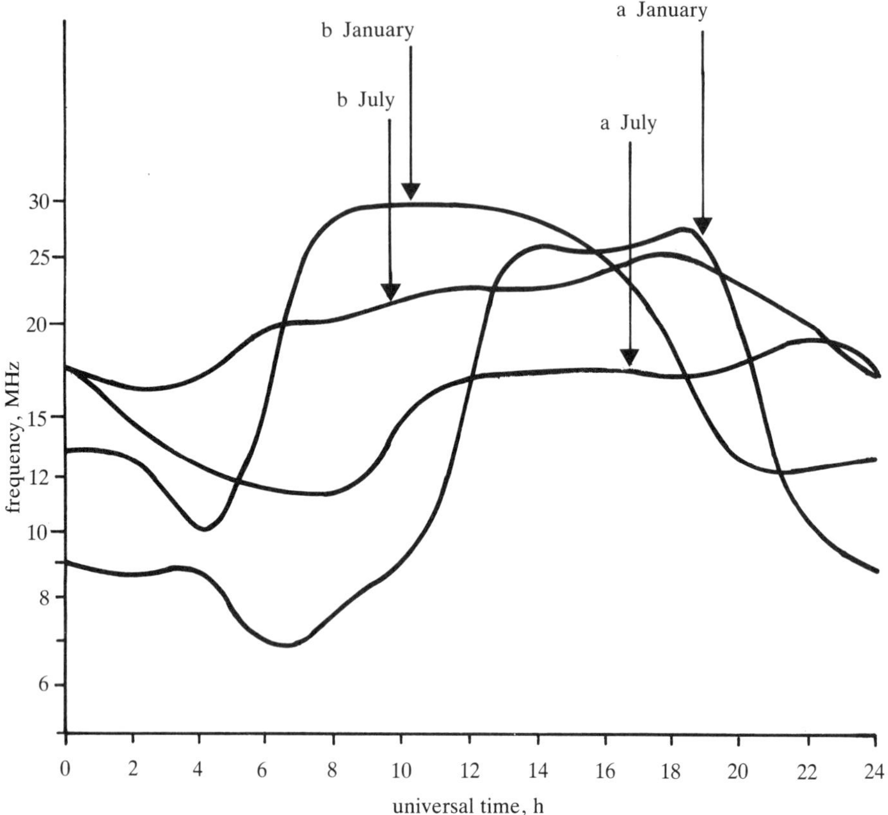

Fig. B.8 Predicted F2 layer basic MUFs for January and July for radio links London–New York (*a*) and Madrid–Cairo (*b*) when R_{12} = 60, based on CCIR Report 340[13]

The vertical incidence critical frequency (f_{0F2}) and the height of maximum ionisation of the F2 layer vary with geographical location, time of day, season and level of solar activity and the patterns of these variations are much more complex than those of the lower layers. The variations of these parameters have not been reduced to an analytical system of mathematical relationships. Predictions of future conditions are mainly empirical, being based, at present, on extrapolation from measurements of f_{0F2} and h'_{F2}, made over many years at a large number of observatories on the ground. Predictions are available for all parts of the world, although there are large areas, particularly in the southern hemisphere, where the data base is not as comprehensive as would be desirable.

Data from ionosphere sounders at these ionospheric observatories have been used in CCIR Report 340[13] for the synthesis of worldwide charts of monthly mean f_{xF2} (designated F2(0)MUF for this purpose) for notional values of R_{12} of 0 and 100; 12 charts have been prepared for each calendar month, representing conditions on every even hour of universal time (UT). Further sets of charts have been prepared showing the corresponding values of the MUF for a single hop 4000 km long, designated F2(4000)MUF. Fig. B.9 shows an example of these charts. To predict the MUF at any point in time and place and for a forecast sunspot number for a hop shorter than 4000 km, procedures have been devised for interpolating and/or extrapolating from these charts. Procedures are also available for predicting MUFs for transmission paths longer than 4000 km. Studies of means for improving the accuracy of these predictions continue; see, for example, Rush *et al.*[22] and CCIR Report 430[14].

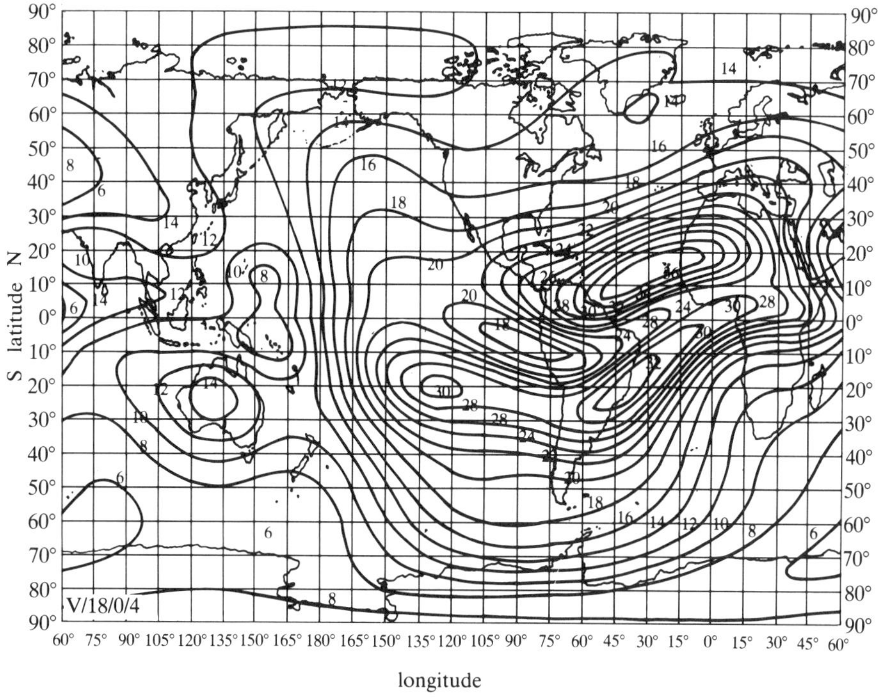

Fig. B.9 Predicted world distribution of F2(4000)MUF (in MHz) in May at 18.00 hours UT at sunspot minimum (R_{12} = 0); CCIR Report 340[13]

There is a considerable scatter of day-to-day values of f_{xF2} about the monthly mean value at any given place, partly due to short-term variations in the flux of energy reaching the Earth from the Sun and partly due to irregularities in the behaviour of the upper atmosphere. The ratio between the monthly lower decile value of the F2 layer MUF and the monthly mean value is usually taken to be 0·85 although observed values show considerably greater scatter in some circumstances; see the Supplement to CCIR Report 252[23].

For a significant fraction of the time when solar activity is high, F2(4000)MUF may be more than 50 MHz over large areas of the Earth, and it reaches 70 MHz in some places, especially in tropical regions. On such occasions powerful interference may occur between radio systems operating thousands of kilometres apart in the lower part of the VHF band; see CCIR Report 259[24]. Similar problems arise in temperate latitudes due to sporadic E ionisation (see pp. 453–454).

Scattering in ionospheric propagation

In general, the reflection of radio waves by ionospheric layers and, for multi-hop propagation, at the ground can be regarded as specular and the reflecting surfaces can be assumed to be horizontal, Consequently propagation on terrestrial links is along great circle routes. Nevertheless, some scattering of energy does take place, mainly at the ground but also in the ionosphere, and some of this scattering happens off the great circle route. For most of the time the weak scattered signals are not evident on working radio links because of the presence of a much stronger specularly reflected component reaching the receiver. However, when the transmitted frequency is above the MUF for the great circle route, a weak signal that has been scattered on the ground or in the ionosphere off the great circle route may be received if propagation conditions between the transmitter and the scattering location and also between the scattering location and the receiver both permit propagation at the frequency transmitted. Such signals may be strong enough to support communication. CCIR Report 726[25] discusses scattering of this kind and contains an extensive list of references.

Ionospheric scattering by another mode is also observed. Irregularities in the density of ionisation, mostly at a height of about 85 km, scatter back to Earth some of the energy of signals, even though the carrier frequency may be too high for reflection by the normal mode. This propagation mode provides a weak signal along the great circle route, and the transmission loss rises rapidly as the radio frequency rises. Persistent interference may arise in this way; see CCIR Report 260[26].

For frequencies between 30 and 60 MHz a reliable radio link of limited information capacity could be obtained in most parts of the world using this latter mode of propagation, with hop lengths of about 1000 to 2000 km, if very high power were radiated at the transmitter. However, it has been agreed that this propagation mode will not be used for radio links, because of the widespread interference it would cause.

B.3.3. The irregular ionospheric phenomena

Auroral ionisation

Under normal conditions the Earth's magnetic field protects most of the ionosphere from the effects of the stream of ionised gas which is constantly poured out by the Sun, the so-called solar wind, although this protection is overwhelmed to some degree when the Sun is disturbed. However, ions from the Sun can penetrate relatively freely down to the E and D regions near Earth's poles, forming very variable concentrations of ionised gas which may reflect signals but may also absorb them strongly. These concentrations are particularly heavy in the auroral zones, at about 60° to 70° of geomagnetic latitude. Concentrations also develop as polar caps within the rings formed by the auroral zones and there is some tendency for them to spread to a small extent below 60° of geomagnetic latitude when the Sun is disturbed.

Polar E region conditions are discussed briefly in the next section on sporadic E ionisation. In the D region, extremely high absorption occurs intermittently, owing to large discrete clouds of relatively intense ionisation which are often associated with visible displays of aurora. In the F region, the effect of electrons from the solar wind is to raise, intermittently but often considerably, the level of both F1 and F2 layer ionisation with corresponding effects on critical frequencies.

These various high-latitude effects are reviewed in CCIR Report 886[12]. However a complete understanding of the behaviour of the ionosphere at high latitudes is lacking at present. This subject is under intensive study and CCIR Report 1012[27] outlines some of the research programmes which are in progress.

Sporadic E ionisation

As indicated on pp. 443–444, sporadic E region ionisation (E_s) occurs in three main forms, equatorial E_s, temperate zone E_s and high latitude or auroral E_s.

Equatorial E_s ionisation is present for most of the time during daylight hours in a well-marked narrow zone to either side of the geomagnetic equator. Vertical incidence critical frequencies of 10 MHz are commonly observed, more than twice as great as the corresponding value for the regular E layer.

Temperate zone E_s occurs most frequently in summer daytime. The diurnal variation exhibits maxima in the mid-morning hours and near sunset. Ionisation is distributed in patches varying in horizontal extent from several kilometres to 1000 km and the thickness is typically less than 2 km. The probability of occurrence and the intensity of E_s are reviewed in CCIR Recommendation 534[28]; see also CCIR Report 340[13]. To generalise from Recommendation 534; it may be said that f_{0E_s} is likely to exceed f_{0E} for from 10 to 50% of the time, depending on the location, the season and the time of day. For small percentages of the time f_{0E_s} may be as high as 20 MHz. Oblique incidence propagation due to the sporadic E layer is often supported by scattering rather than specular reflection.

At geomagnetic latitudes higher than about 60°, temperate zone E_s characteristics give way abruptly to auroral E_s conditions. At night sporadic

E layer ionisation is present for a considerable percentage of the time, but f_{0Es} varies rapidly and widely, being sometimes as high as 15 MHz. It should be noted that within the polar caps the night-time ionisation of the E layer itself is regularly higher than would be expected from the Sun's zenith angle; see CCIR Report 886[12]. However the intensity of both of these forms of ionisation is much affected by ionospheric disturbances.

Propagation by reflection from sporadic E ionisation is important for radio links operating in parts of the world where the regular layers provide poor communication conditions, but obviously the sporadic nature of this propagation mode raises system design and operating problems if high circuit availability is required. Propagation of interference by sporadic E reflection is a significant problem at frequencies up to about 150 MHz; see CCIR Report 259[24].

F2 layer anomalies: 'spread-F ionisation'

As indicated on p. 449, the configuration of the F2 layer is much more variable than that of the other regular layers. This is mainly because F2 layer ionisation persists many hours after local sunset has removed the main cause of ionisation, providing an opportunity for winds and large scale vertical movements of air to affect the height, concentration and geographical location of ions. The effects of these variations on radio propagation are most conveniently expressed empirically in the forecasts of the regular critical frequencies of the layer.

However, in addition to the normal irregularities of the F2 layer and the polar effects touched on under auroral ionisation above, there is another important phenomenon. Just after local sunset and before local sunrise, mainly in tropical regions, the ionisation of the F2 layer sometimes breaks up into discrete clouds, producing a characteristic pattern of echoes on ionospheric sounders which has led to the phenomenon being given the name 'spread-F' ionisation. This causes scattering of signals on long distance HF radio links.

Ionospheric disturbances

The radiation of energy from the Sun has short term variations as well as the cyclical variations associated with the 11-year solar activity cycle. Some of the variations of f_{0F2} have been correlated with variations of the solar noise flux at 3 GHz. However, other events on the Sun can also effect ionospheric conditions.

The enhanced flux of X-ray radiation which accompanies a large solar flare may cause a Sudden Ionospheric Disturbance (SID) involving an immediate increase in D layer ionisation over the Earth's sunlit hemisphere. The consequential large increase in the absorption of ionospherically propagated HF signals usually persists for several tens of minutes and may sometimes continue for several hours. SIDs are sometimes followed, after an interval of two or three days, by an ionospheric storm due to interaction between the Earth's magnetic field and a temporarily enhanced solar wind. The effect of an ionospheric storm on radio propagation conditions is typically to depress for a few days the critical frequencies of the regular ionospheric layers and to increase absorption, particularly at high latitudes and within the auroral ring (Polar Cap Absorption).

Ionospheric storms are also associated with conditions of the solar corona which are called 'coronal holes'. These coronal holes are relatively persistent, being sometimes present for several months and they are a feature of the declining phase of the sunspot cycle. Coronal holes are the source of high speed solar wind streams (HSSWS). The plasma contained in a HSSWS travels at several times the velocity of the background solar wind. As a stream sweeps across the Earth with the rotation of the Sun, an ionospheric storm may occur, lasting two or three days and possibly recurring after 27 days, the mean period of the Sun's rotation.

Considerable success has been achieved in forecasting short term variations of critical frequencies, the incidence of SIDs and the course of ionospheric storms, using observations of ionospheric and geomagnetic parameters and visual observations of the Sun's disc; see CCIR Reports 727[29] and 888[30] and CCIR Recommendation 313[31].

Meteor trails

The trails of meteors, which are observed at a height of about 100 km, are strongly ionised and they reflect a signal transmitted from one point to a distant receiving station if the orientation of the trail is favourable.

Very large numbers of meteors enter the Earth's atmosphere during meteor showers. However, there is also a considerable flux of meteors at other times, and meteor burst communication systems depend on specular reflection from the line of ionisation produced by these latter, so-called sporadic, meteors. For astronomical reasons, there is a diurnal cycle in the rate at which sporadic meteors enter the atmosphere at any particular location, with a maximum at about 0600 local time.

The duration of each trail is of the order of one second. There are not enough sporadic meteor trails to provide continuous communication, but, given suitable equipment, the intermittent communication which successive trails provide can be used in a burst mode at a relatively high instantaneous information rate, permitting a significant average information rate to be achieved. Radio links up to about 2200 km long are feasible via this mode. Frequencies in the range 30–90 MHz are optimum; see CCIR Report 251[32].

Artificial modification of ionospheric conditions

The propagation of a powerful amplitude modulated radio wave through an ionised medium may cause temporary changes in the behaviour of the ions in the medium within the period of the modulation. As a result, the propagation of another wave through the same space may be affected, modulation from the first wave being transferred to the second wave. This process of ionospheric cross-modulation is particularly likely to occur when the ionospheric reflection of the second wave takes place in the lower levels of the ionosphere and significant deviative absorption arises. Thus, the wave suffering cross-modulation is usually relatively low in frequency; see CCIR Report 574[33] and CCIR Recommendation 498[34].

A related process is observed sometimes when a very powerful wave at a frequency slightly less than f_{0F2} transfers so much energy to the ions in the F2 layer that the structure of the layer breaks up temporarily into discrete ionised clouds. These clouds will scatter waves at frequencies much higher than f_{0F2}, permitting the propagation of signals over distances of several thousands of kilometres by an artificial ionospheric scattering mechanism; see CCIR Report 728[35]. While there is clearly a possibility of taking advantage of this mechanism to permit VHF and UHF carriers to transmit wide basebands over long distances, it can also create great problems by causing interference where there would have been none at a significant level if the ionosphere had remained in its natural state. CCIR Recommendation 532[36] draws attention to the risks that arise from the use of high transmitted power levels. Related phenomena are foreseen from the transmission of very powerful fluxes of power at much higher frequencies through the ionosphere, such as would arise if solar power satellites fed electrical power to Earth by means of intense microwave beams; see CCIR Report 893[37].

Finally, CCIR Reports 893 and 1011[38] discuss effects that are produced in the ionosphere by the release of products of combustion by chemical rockets. The effects are shown to be complex, and to involve substantial depletion of free electrons over a wide area surrounding the rocket launch site, in particular in the F region; this depletion may persist for several hours.

B.3.4 Some operational aspects of ionospheric propagation at HF

HF radio signals are sometimes propagated over great distances by abnormal modes. There are several mechanisms by which signals may be propagated by ionospheric reflection, without intermediate reflection at the ground, between radio stations which are separated by more than the conventional maximum distance for a single hop. These mechanisms are reviewed in CCIR Report 250[39]. Also, some signal energy is scattered at the ground and in the ionosphere; it may subsequently reach the receiving antenna having traversed a path which is not the great circle between the transmitting and receiving antennas.

However, in general signals may be assumed to follow identifiable ray paths along the great circle route, being reflected specularly at the effective height for the signal frequency in the various operative ionospheric layers, with intermediate reflection at the ground if propagation is multi-hop. Thus for example there might be single hop F2 layer propagation, conventionally symbolised as the '1F2' mode, on a link between radio stations 3000 km apart; the 2F2, 1F1 and 2E propagation modes might, for example, be present also. A link 10 000 km long might be served simultaneously by the 3F2, 4F2 and 5F2 propagation modes. For some propagation modes both the ordinary and the extraordinary rays might be receivable. Each ray path is associated with a specific angle of elevation at the transmitting and receiving antennas, although these angles are subject to change within limits as ionospheric conditions change.

If necessary, a basic MUF can be predicted for each of these propagation modes, subject to uncertainties due to errors of prediction and the short-term (day-by-day and hour-by-hour) variability of the layer concerned (see p. 497). The transmission loss for each propagation mode can also be estimated. The situation changes minute by minute, hour by hour, seasonally and in accordance with the 11-year solar activity cycle as the transmission medium follows its regular pattern of change. Propagation conditions may also be affected by the irregular ionospheric phenomena.

A number of consequences of operational importance flow from these various circumstances.

Choice of working frequency

The MUF at an instant in time for each propagation mode between two stations is the highest frequency that mode could support, given optimised radio equipment at each station. The 'basic MUF' for that instant is defined as the highest of these MUFs. However, the working conditions at the two stations may not be ideal; for example, the gain of one or both antennas may be low at an angle of elevation which a low-loss propagation mode requires. The loss of signal power on some propagation modes may be so high, owing to propagation conditions or equipment shortcomings, that these modes are not capable of supporting the transmission of information. Accordingly, the MUF obtainable via some propagation modes may have to be disregarded. The 'operational MUF' is defined as the highest frequency that would permit acceptable operation under specified working conditions.

Clearly optimum radio circuit peformance demands transmitting and receiving antennas with performance matched to the propagation conditions in which they will be used. Conversely, a reliable prediction of usable frequencies depends on knowledge of certain characteristics of the radio equipment, and in particular the antennas, which is to be used. This matter is discussed in CCIR Report 891[40].

It will usually be necessary for a complement of two or more frequencies, selected from appropriate parts of the HF frequency range, to be assigned to any but quite short-distance radio links, each frequency to be used when required. The choice of the frequency to be used at any time from amongst the frequencies of the complement would be based on predictions of propagation conditions (see pp. 496–498), often supported in real time by observation of the state of received signals.

Fading and inter-symbol distortion

For much of the time HF radio links are supported by more than one propagation mode. Thus, the received signal consists of two or more components which have followed different routes from the transmitting antenna. The separate components may be varying in amplitude owing to variations in the propagation medium. The plane of polarisation of each component is likely to be randomly related to the plane of polarisation to which the receiving antenna is most sensitive, causing further amplitude variations of the level of the separate components at the input terminal of the receiver. The various components will arrive at the receiver with random relative carrier phase relationships and will have suffered different time delays. They will also exhibit small Doppler shifts of frequency due to movements of the effective planes of reflection.

The aggregate of components which is presented to the receiver input terminals will be a distorted reproduction of the signal that had been transmitted, showing amplitude variations due to the algebraic combination of the components and a spread in the time dimension of the modulation which the signal is carrying. The amplitude variations are called fading. If the differences between the transmission times of the major components is large, the fading cycles of different spectral components of the received signal will be out of step, producing selective fading. For digital signals, the spread in the time dimension produces inter-symbol distortion. The effects of these phenomena tend to be least severe for propagation paths of moderate length, say 2000 km; for shorter paths and longer paths it often happens that two or three ray paths arrive at the receiving antenna with signal strengths which are approximately equal but with major differences in propagation time, causing severe selective fading and inter-symbol distortion.

The fading arising within each separate component tends to be relatively slow, but the amplitude range may be large, particularly when the carrier frequency is close to the MUF for the propagation mode followed by that component. Multi-path fading typically has a period ranging from a fraction of a second to several seconds and, depending on the nature of the wave interference situation, it may be deep or shallow and frequency-selective or unselective (flat fading). The statistical nature of fading is examined in CCIR Report 266[41]; see also CCIR Report 304[42].

Fading and inter-symbol distortion are major limitations of HF ionospheric communication but there are various ways by which the effects can be mitigated. These ways can be classified as

- diversity reception techniques which make use of the fact that troughs of wave-interference fading do not occur simultaneously at all locations in the neighbourhood of the receiving station, at all frequencies within the frequency band occupied by the wanted signal and in all planes of polarisation,
- ray-selection techniques, which eliminate some of the harmful multi-path combination modes,

- system designs which reduce the susceptibility of the signal to fading and inter-symbol distortion, typically by converting high-speed digital signal streams into several or many parallel streams of lower speed and by the use of bit-error correction techniques, and
- operational procedures which reduce exposure to multi-path propagation.

The most appropriate technique for use in any specific system depends greatly on the circumstances of the application.

Dynamic HF circuit management

Long distance HF radio links have several radio frequencies assigned to them, forming a 'complement of frequencies', the frequencies of the complement being chosen with the intention that one frequency or another will be suitable for use at any time that the link is required to operate. Where operating conditions permit, observation of the state of received signals and long term forecasts of propagation conditions reinforce the experience of the staff operating the link, helping them to decide which of the frequencies in the complement to use in real time. There are two further aids to the maintenance of communication at HF, and they are especially valuable for optimising the performance of circuits for which high availability is required in parts of the world where ionospheric parameters are particularly unstable.

Firstly, as discussed briefly on p. 455, it is feasible to make useful short-term forecasts of ionospheric disturbances, based on observations of geomagnetic and ionospheric parameters, visual observations of the Sun's disc and the recent history of ionospheric disturbances. Such forecasts give valuable guidance as to the best action to take when propagation conditions are sub-normal.

Secondly, various techniques have been devised for making an objective assessment in real time of the propagation conditions which are being obtained either at the working frequency, or at all of the frequencies in the complement or in general across the whole HF spectrum. These techniques are called Real Time Channel Evaluation (RTCE). Some RTCE techniques involve transmitting test signals over the working circuit. Others produce an oblique incidence sounding of ionospheric conditions between the two stations concerned. A third group makes real time use of observations made at a network of ionospheric observatories in the geographical area in which the radio link operates; see CCIR Reports 889[43], 249[44] and 890[45].

B.3.5 Ionospheric propagation below about 2 MHz

There is a mechanism by which very low frequency waves can penetrate the ionosphere and return, penetrating the ionosphere again, to Earth, having travelled in a curved path through the outer regions of the Earth's magnetic field; see 'Whistler mode' propagation below. However, in general, waves below about 2 MHz are trapped between the Earth's surface and the ionosphere.

In the VLF band and in much of the LF band, the wave is reflected from the D layer, by night and day, with little absorption. Worldwide propagation is quite possible at such frequencies. The effective height is typically about 70 km by day and around 90 km by night, the daytime height of reflection being somewhat lower during a Sudden Ionospheric Disturbance. At 10 kHz the wavelength in non-ionised air is 30 km and it is realistic to view long distance ionospheric propagation at VLF as if it were the excitation of a wave in a waveguide cavity made up of two concentric spheres, consisting of the Earth's surface and the bottom of the D layer.

At 1 MHz, however, the wavelength in non-ionised air is only 300 m and absorption on reflection from the ionosphere is sometimes large. For signals in the MF band it is more realistic to consider ionospheric propagation is taking place by discrete hops, as for higher frequencies.

In the upper part of the LF band and in the MF band, reflection takes place at a higher level, mostly in the E region but sometimes in the F region at the upper end of this frequency range. The absorption in the ionosphere is relatively small at night. Around 300 kHz there is a significant amount of absorption by day and for much of the MF band the daytime absorption is quite severe, such that the ionospheric wave can usually be disregarded for practical purposes. Propagation conditions are affected by the Earth's magnetic field and perhaps by the level of solar activity. There is also a considerable scatter of day by day conditions. See CCIR Reports 895[46] and 265[47].

'Whistler mode' propagation

The Whistler mode is so called because it was first observed propagating broad-spectrum noise impulses radiated from lightning flashes to distant untuned receivers. The higher frequency components of the noise, travelling faster, reached the receiver before the slower low frequency components, producing in the receiver output a sound of descending pitch.

Irregularities in the plasma in and beyond the ionosphere are aligned by the Earth's magnetic field, and this alignment provides a low velocity dispersive propagation medium which could link transmitters on the Earth's surface to receivers at approximately the same geomagnetic latitude in the opposite hemisphere. The effect is particularly strong at night; by day the D layer tends to cut off access to this guided-wave medium. It operates under favourable circumstances for radio frequencies up to about 35 kHz. This mode of transmission has received considerable study, but at present it is used little, if at all, in operational radio systems; see CCIR Report 262[48].

B.3.6 Ionospheric phenomena on Earth–space paths

Most spacecraft in sub-orbital trajectories and most space stations on Earth satellites are moving within the ionosphere or beyond it. Thus radio signals passing between Earth and any satellite pass through some or all of the ionosphere. The ionosphere has various effects on these signals, including absorption, scintillation and a group of other effects, the severity of which is related to the total electron content (TEC) of the whole signal path through the ionosphere. All of these effects are proportional to an inverse function of the radio frequency of the signal.

Absorption

Powerful VLF signals from transmitters on Earth have been received on satellites in and above the ionosphere. In general, however, if the ionospheric critical frequency along the straight line joining an earth station and a satellite is higher than the frequency of the signal, the signal will be reflected, giving the appearance of very high absorption between Earth and space.

If the signal frequency is above the critical frequency, the signal will penetrate the layer. There will be some non-deviative absorption if the signal frequency is close to the critical frequency but this absorption, owing to the regular layers in their normal state, will usually be negligible above 70 MHz except when the angle of elevation of the ray path, seen from the earth station, is very low. Figures quoted in CCIR Report 263[49] show significant auroral absorption at 127 MHz for a significant proportion of the time between earth stations at high latitudes and geostationary satellites. High space–Earth absorption is also associated with Polar Cap Absorption (PCA) events.

TEC-related effects

It is straightforward in principle to calculate from measurements of vertical incidence ionospheric characteristics the total number of free electrons in a vertical column, 1 m^2 in cross-section, up to the height of maximum ionisation density of the F2 layer. Other methods can be used to derive a similar measure of the TEC for a column of 1 m^2 cross-section all the way from an earth station to a satellite. Some specimen maps of TEC are to be found in CCIR Report 725[11]. TEC is found to vary fairly closely with f_{0F2}, but irregular ionospheric effects, especially those occurring near the poles, can also have a significant sporadic impact on TEC.

The intensity of three groups of effects vary with the TEC.

(*a*) The group velocity of the radio wave within ionised regions, including ionisation encountered above the ionosphere, is less than the velocity of the wave in free space, and it varies inversely with the frequency squared. The dependence of group delay on frequency makes an ionised medium dispersive, and this distorts the waveform of the modulating signal. These effects, if not compensated for in the design of the system, reduce the efficiency of communication systems and the accuracy of navigation systems at VHF and potentially in the lower part of the UHF band. These effects are not significant above 1 GHz.

(*b*) The presence of the Earth's magnetic field in the ionosphere causes the plane of polarisation of a wave to be rotated to an extent which is a function of the TEC and an inverse function of the square of the frequency. This phenomenon is called Faraday rotation. At VHF the extent of the rotation over the complete Earth–space path is many complete turns and variations of the TEC from minute to minute ensure that the plane of polarisation of a linearly polarised wave is always rapidly changing. At 5 GHz the extent of the variation of the plane of polarisation is much less but not negligible. It is negligible for most purposes above 10 GHz. Circular polarisation may be used to avoid the consequences of Faraday rotation.

(*c*) Deviation of the ray path from a straight line also occurs, owing to refraction in the ionospheric layers. The effect is perceptible at VHF but not at SHF, and it causes little practical difficulty at any frequency.

Scintillation

In addition to the effects, mentioned above, which may be strong at VHF but which become less important or completely negligible above 1 GHz, where most space radio services operate, there is a group of effects resembling fading which, though stronger at VHF and below, nevertheless remain significant at much higher frequencies. These effects are the various forms of scintillation, in which irregularities of the free electron density in the medium through which a wave passes cause variations in the amplitude, phase, polarisation and angle of arrival of the signal.

The incidence of scintillation depends considerably on the location of the earth station observing it. At equatorial earth stations scintillation is a night-time phenomenon, rather frequent in occurrence and rather severe. It is most prevalent at the equinoxes and in years of high solar activity. In mid-latitudes scintillation also occurs, mainly at night, apparently associated with the spread-F formation, and tends to be mild in degree. At high latitudes rather strong scintillation occurs in winter, during times when the ionosphere is disturbed and in years of high solar activity.

The effects of fluctuations of signal amplitude of up to several decibels which occur at SHF due to scintillation can be overcome by the provision of an enhanced margin of carrier level relative to the noise level. Various forms of diversity at the earth station are used to combat the more severe fluctuations which occur at lower frequencies; see CCIR Report 263[49] for a detailed study of these phenomena.

B.4 Tropospheric propagation phenomena

B.4.1 Introduction

In its most basic form, tropospheric propagation involves propagation between a transmitting antenna and a receiving antenna which are in clear line-of-sight of one another. In such a case the transmission loss is usually close to the value given by expression (B.9), that is, the free space basic transmission loss, after allowance has been made for the gains of the two antennas in the relevant directions and for the effect of reflection at the ground. Nevertheless, the transmission loss may be affected by such phenomena as:

- refraction, multipath transmission, scintillation, ducting and defocusing due to non-uniformities in the temperature and humidity of the air, and consequently in its permittivity (see pp. 464–470),
- absorption or scattering in rain and other hydrometeors and in sand or dust suspended in the air (see pp. 470–477), and
- absorption in the atmospheric gases (see pp. 477–479).

Other characteristics of the signal, such as its state of polarisation, may also be affected during propagation in the troposphere.

However, not all communication is between antennas which are in line-of-sight, and interference often arises by indirect modes. In such cases signals pass from a transmitting antenna to a receiving antenna by various mechanisms, including:

- tropospheric scatter and scattering by precipitation (see pp. 480–483);
- diffraction round obstacles, penetration into buildings or through the trees of a forest, scattering by buildings, reflection from aircraft and so on (see pp. 483–486).

Statistics in tropospheric propagation

Many of the phenomena of tropospheric propagation degrade the performance of telecommunication systems, and their incidence is in most cases sporadic, often associated with weather conditions and usually subject to seasonal variations. In order that the performance criteria of systems may take such degradations efficiently into account, these criteria are often specified by requiring that a stated degree of degradation be not exceeded for more than a stated percentage of a given period of time.

The simplest 'given period' to adopt would be one year, since this would take into account a full cycle of seasonal variations. There is, however, a disadvantage to using so long a sampling period for this purpose. It happens in some climates that virtually all of the weather conditions which cause the performance of a system to fall below the specified threshold of acceptable performance arise in one or two months of the year. It is judged that a larger total period of degradation below the specified threshold can be accepted if it is spread more evenly in time. Thus, by specifying that stated thresholds of degradation should not be exceeded for a given proportion of a shorter period, typically one month, the user of systems located in places where adverse conditions are concentrated into a small fraction of the year may not be disadvantaged and the providers of systems in locations where the climate is more equable can enjoy some relaxation of the performance requirements.

Where the degradation of radio system performance arises from seasonal factors, the CCIR and CCITT usually recommend performance standards which should be attained for at least a stated percentage of 'any month', which is taken to mean the worst calendar month in any period of twelve consecutive calendar months. Since climatic conditions often vary considerably from year to year, the meaning of the term is sometimes further refined, making it signify the long term average of annual worst-month conditions; see CCIR Recommendation 581[50] and Report 723[51].

Reference should be made to CCIR Report 1007[52] for a more general treatment of the use of statistics in radiowave propagation.

B.4.2 Atmospheric refraction, multipath transmission and ducting

Ray-bending in the troposphere

A ray path in the troposphere would be a straight line if the electrical properties of the air in the vicinity of the path were uniform. This is not the usual condition. Ray paths are typically bent owing to tropospheric non-uniformities, the curvature being most marked in the vertical plane. This bending is slight, but it may have an important effect on radio propagation conditions, especially if the line-of-sight between a receiving antenna and the wanted transmitting antenna passes close to an obstruction or if a solid body which screens a receiving antenna from a source of interference only just intercepts the line-of-sight. There are techniques of analysis for investigating tropospheric ray bending and the methods which have been devised to simplify graphical analysis are outlined below.

The index of refraction of air, n, is given by $n = c/v$, where c and v are the velocities of propagation of electromagnetic waves in free space and non-ionised air respectively. v is somewhat smaller than c. However, n is very close to unity, and it is convenient to define a term 'Radio Refractivity' N by the relationship

$$N = (n - 1) \times 10^6 \quad \text{(in '}N\text{ units')} \qquad \text{(B.15)}$$

Thus, if $c/v = 1{\cdot}00033$, the Radio Refractivity is said to be 330 N units.

The Radio Refractivity (see Bean and Dutton[53]) is given approximately by

$$N = 77{\cdot}6\ P/T + 3{\cdot}73 \times 10^5\ e/T^2 \qquad \text{(B.16)}$$

where P = atmospheric pressure, mb
T = absolute temperature, K
e = water vapour pressure, mb

The value of N at sea level is usually in the range 300 to 350, rising to 400 in some ocean areas when the weather is hot. In CCIR Report 563[54] there are sample maps which show, in very general terms, the worldwide distribution of the monthly mean value of N at sea level.

The troposphere tends to be horizontally stratified with respect to temperature and humidity. As h, the height above a reference level, increases, N usually falls, typically exponentially, due mainly to falling atmospheric pressure. N approaches zero at the upper limit of the troposphere. The rate of change of N with increasing height (dn/dh) is called the 'lapse rate for N'; thus, if N falls with increasing height, dn/dh is negative.

CCIR Report 563[54] contains world maps of the observed monthly mean lapse rate for N, averaged over a layer 1 km thick based at the Earth's surface. Figures of from -30 to $-70\,N$ units per km are typical. However, if the lapse rate is averaged over a smaller vertical interval, much steeper gradients are found, sometimes with wide variations. For a small percentage of the time, a large positive value (say $+100$ or $+200\,N$ units per km) is sometimes observed in a shallow layer close to the ground and for another small percentage of the time a large negative value (say -200 or $-300\,N$ units per km) may occur. The occurrence of rapid changes of N with height depends on local meteorological conditions.

The path of a ray in the lower portion of the atmosphere may be bent because of variations of N with height. The curvature of the ray path, $1/\rho$, at any point is given by

$$\frac{1}{\rho} = \frac{(\cos \phi)}{n} \frac{dn}{dh} \tag{B.17}$$

where ϕ is the angle of elevation of the ray at the point. For a horizontal ray, eq. (B.17) is reduced to

$$1/\rho = dn/dh \tag{B.18}$$

If it is necessary to study the geometry of the ray-paths between two antennas close to the ground, it is feasible to do so using a true vertical profile of the path between the antennas and using ray paths having a curvature calculated from a knowledge of dn/dh or an assumption of the lapse rate for N. However, the geometry of the problem would be made more convenient for graphical analysis if the co-ordinates of the profile could be adjusted so that the ray paths became straight. This can be done, subject to the simplifying assumption of a single value for the lapse rate; the value chosen might be the average over the lowest 1 km of height. Thus, by a transformation of expression (B.18), it can be shown that propagation can be considered to be rectilinear in the troposphere above a hypothetical earth of radius ka, where

$$1/ka = 1/a + dn/dh \tag{B.19}$$

a = radius of the real Earth

The assumption of a single value for dn/dh, and therefore a single value for k, gives results of acceptable accuracy when a generalised assessment of ray paths between two antennas is being made.

CCIR Recommendation 369[4] establishes expression (B.12) as the reference atmosphere for refraction, for use where information specific to the location is not available. For this reference atmosphere, $dn/dh = -40\,N$ units per km for the lowest layer 1 km thick, and $k = 4/3$. The conditions which exist in this reference atmosphere are called 'standard refraction'. The terms 'sub-refraction' and 'super-refraction' are applied to conditions in which dn/dh is respectively greater (that is, less negative) or less (that is, more negative) then the standard conditions (see Figs. B.10 and B.11).

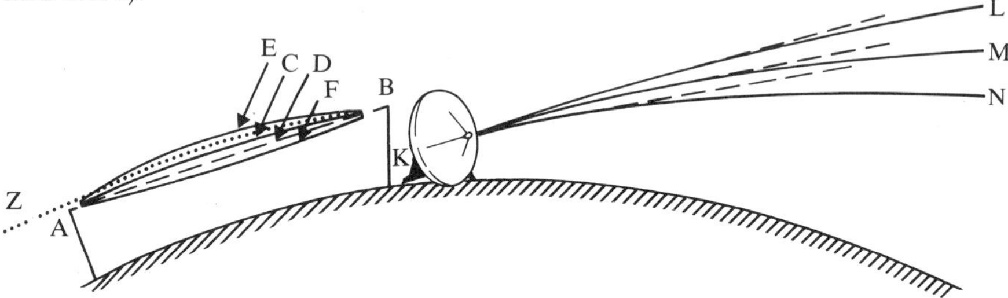

Fig. B.10 Tropospheric ray bending. The Earth may be supposed to be represented here at its true radius

Ray ACB links stations A and B under standard refraction conditions; note the curvature of the ray, emphasised here by the straight line ADB

Rays AEB and AFB show the direct ray path under super-refractive and sub-refractive conditions respectively

BZ shows a ray path when $dn/dh = -157$ N units per km; the ray is curved, remaining parallel to the Earth's surface

Three rays, KL, KM and KN, from earth station K exhibit ray bending and consequent beam defocusing

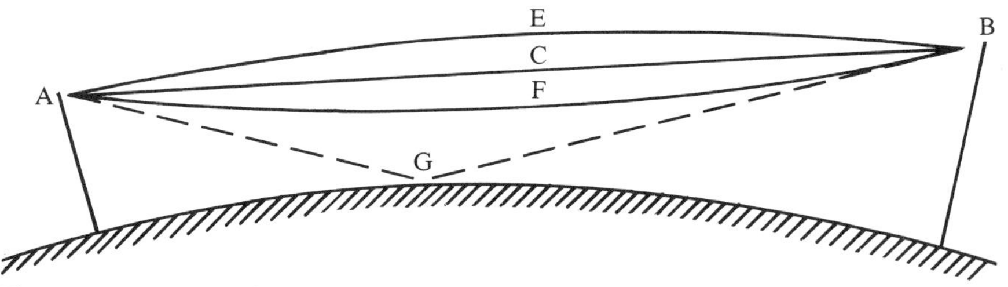

Fig. B.11 Tropospheric ray bending. The Earth may be supposed to be represented here at 4/3 times its true radius. Rays ACB, AEB and AFB relate to tropospheric conditions which are the same as for the corresponding rays in Fig. B.10

Note that ray ACB is now shown as straight. The ground-reflected ray AGB is also shown.

Another hypothesis can be used to facilitate the study of ray paths when account is to be taken of the effect of variation of dn/dh with height. When $dn/dh = 10^6/a = -157\,N$ units per km, the curvature of the path of a ray which is initially horizontal equals the curvature of the Earth's surface. Under such circumstances, the ray path would remain parallel to the Earth's surface for as far as these conditions were obtained. If the geometry and the parameters of this situation were hypothetically adjusted so as to make this ray path straight, the hypothetical Earth's surface would be flat. To implement these assumptions, it is convenient to define terms analogous to n and N which take the real curvature of the Earth's surface into account. Thus the Modified Refractive Index m for a given height above the ground is given by

$$m = n + h/a \tag{B.20}$$

where a = Earth's radius km
h = height, km

The Refractive Modulus M, which for this flat-Earth model is equivalent to N, is given by

$$M = (m - 1) \times 10^6 \quad \text{(in } M \text{ units)} \tag{B.21}$$

$$= N + 157\,h \quad \text{(in } M \text{ units)} \tag{B.22}$$

There are formal definitions of these and other terms used in the study of tropospheric propagation in CCIR Recommendation 310[55]. For a general discussion of tropospheric ray bending, see CCIR Report 718[56].

Ray bending and beam de-focusing on slant paths

When the ray path is not horizontal, typically for an Earth–space path, the change of ray direction due to passage through the atmosphere down to the level of the earth station makes the angle of elevation of the satellite as seen from the earth station appear to be larger than the true value. When the true angle of elevation is very small, say 1° or 2°, the ray will be bent through an angle of about 0·5°. Ray bending is less for higher angles of elevation. The extent of ray bending depends on the value of N at the earth station site, so bending will usually be less if the earth station site is high above sea level; see CCIR Report 718[56] and Fig. B.10. It should be noted that bending of Earth–space rays also arises through a different mechanism in the ionosphere (see p. 462); the ionospheric effect is significant at VHF and lower frequencies but negligible at SHF and above.

Since the extent of ray bending is reduced as the angle of elevation increases, the divergence in the lower atmosphere of a bundle of rays from an earth station antenna becomes greater after it has emerged into space (see Fig. B.10). This defocusing effect is significant for an angle of elevation of less than 3°. It causes a loss of the order of 0·8 dB at 1° above the horizontal, rising to a maximum somewhat over 2 dB for a horizontal ray (Crane[57]).

Ducting

As indicated above, if $dn/dh = -157\,N$ units per km, a horizontal ray has the same curvature as the Earth's surface. If the ray path is not obstructed, say by hills, the ray will be propagated parallel to the ground for as far as the same atmospheric conditions persist. If the lapse rate for N is more negative than $-157\,N$ units per km, the ray may, in favourable circumstances, be propagated round the curved surface of the Earth by successive reflections at the upper and lower surfaces of the layer in which these conditions are obtained. This is called tropospheric ducting.

A horizontal layer of ducting air may have significant effects on radio wave propagation, both on unobstructed paths and on paths which would not provide line-of-sight propagation under standard conditions of tropospheric refraction. For unobstructed paths, ducting layers can focus the wavefront radiated by a transmitting antenna, causing individual rays to intersect, as a result of which there are locations, possibly occupied by a receiving antenna, at which wave interference occurs, causing multi-path propagation. Rays may be deflected away from other locations altogether, leaving no signal where there is a clear line-of-sight from the transmitting antenna.

When ducting occurs on paths which would not be line-of-sight under conditions of standard refraction, a very potent mechanism may be set up for trans-horizon interference. In free space propagation the energy of a wave spreads out in two dimensions orthogonal to the direction of propagation as the wave progresses, and the transmission loss is related to the square of the distance. In a duct there is usually little spread of energy in the vertical plane, and the basic transmission loss in the duct L_b over distance d' is related to the basic free space transmission loss L_{bf} as follows:

$$L_b = L_{bf} - 10\log d' + A \tag{B.23}$$

where A represents various losses, due for example to leakage from the duct through irregularities in the duct boundaries and losses on reflection at the ground. Sometimes A is small, in which case the basic transmission loss in a duct may be much less than the free space value for the same distance.

A duct has some of the properties of a lossy waveguide, including a critical frequency, related to the duct thickness, below which transmission in the ducted mode is rapidly attenuated.

Under some meteorological conditions there is a tendency for stratification of the lower troposphere to occur. A layer with a strong ducting gradient a few tens or hundreds of metres in vertical thickness may lie on top of a layer of non-ducting air, the air above the ducting layer being non-ducting also. The interfaces between such layers are often turbulent. Surface ducts may also arise with no layer of non-ducting air between the ducting layer and the ground.

Over land, meteorological conditions are not usually favourable to the establishment of extensive surface ducts. Turbulence due to wind and irregularities of the ground tends to prevent uniform stratification of the atmosphere. However, a shallow surface duct often forms over the sea or a large lake, due to the high humidity of the air layer which is in immediate contact with the water. A thick surface duct often forms over a warm sea in the vicinity of land in the evening, when a dry breeze blows off the land. A surface duct, exhibiting an inversion of the fall in air temperature with increasing height which is normally present, may develop over land at night, when the air in contact with the ground cools more rapidly than the air at higher levels.

Elevated ducts sometimes develop in the morning from a surface duct formed by a night-time temperature inversion when the Sun's heat warms the ground, the duct being raised bodily by convection. Elevated ducts are more commonly formed, however, in anticyclonic weather conditions, by a temperature inversion caused by the descent, compression and warming of an air mass over cold low-lying air.

For an analysis of ducted propagation in terms of the modified refractive index m and the refractive modulus M, see Dougherty and Hart[58] and CCIR Report 718[56].

Tropospheric scintillation

As indicated above, tropospheric stratification can cause multipath propagation between transmitting and receiving antennas. However, small scale non-uniformities in the refractivity of the air normally associated with atmospheric turbulence can also disturb propagation, particularly at SHF. As a result, the wavefront reaching the receiving antenna may not closely approximate to a plane wave produced by the regular expansion with distance of the spherical wave which left the transmitting antenna. Also, multipath transmission is usually present and small, rapid variations may occur in the angle of arrival of signals at a receiving antenna. These latter phenomena are known collectively as tropospheric scintillation.

The variations in the arrival angle of signals due to tropospheric scintillation are at worst only a small fraction of 1°. They are mainly in the vertical plane, the change in horizontal arrival angle being very small. They are likely to affect radio link performance significantly only in terrestrial systems involving very long hops or in space–Earth systems operating with a very low angle of elevation at the earth stations, the antennas in use in either case being very large relative to the wavelength.

The loss of coherence of the wavefront can cause some loss in the effective gain of antennas. There is some consideration of the extent of this loss of gain in CCIR Report 718[56], but it can be assumed to be small except for antennas which are very large compared with the wavelength, operating on terrestrial paths or on space–Earth paths at low angles of elevation.

However, multipath transmission due to scintillation causes selective fading and, on systems using digital transmission, inter-symbol distortion. This may be of major importance on any radio relay system using long hops. Sometimes a large number of active ray paths are present, but the deepest fading is likely to be associated with conditions where there are two dominant rays present and it is usually sufficient to assume that not more than three rays are significant. The severity of the effects of multipath transmission on radio relay systems increases with frequency in the SHF band, and increases rapidly with the length of the hop. Studies aiming at modelling these multipath conditions and the signal degradations which they produce are reported in CCIR Report 718[56]; see also CCIR Report 784[59].

Clear-air cross-polarisation

On occasion a serious deterioration of the cross-polarisation discrimination has been observed on an SHF radio link when there is no evident cause, such as precipitation, in the ray path. Possible causes of such events are listed in CCIR Report 722[60], but the matter is not, at present, properly understood. It seems probable that cross-polarisation on line-of-sight paths is almost always due to precipitation of some kind (see pp. 476–477).

Ground reflection

In most circumstances signal energy will pass from the transmitting antenna to the receiving antenna, not only by the direct ray but also by a ray which is reflected at the ground; see, for example, the ray AGB in Fig. B.11. Typically the reflected signal can be considered to consist of a specular component and a diffusely reflected component. The magnitude of the specularly reflected component and the relative proportions of the two components depend on the character of the reflecting surface, the polarisation of the wave relative to the reflecting plane, the angle of incidence of the wave on the reflecting plane and the frequency; see CCIR Report 1008[61].

The magnitude of the specularly reflected component may vary considerably with the angle of incidence if the plane of polarisation is vertical. The magnitude is then least at a certain angle of incidence, called the Brewster angle, which itself is a function of the frequency and the relative permittivity and conductivity of the reflecting surface. For angles of elevation greater than the Brewster angle, there is a phase reversal of the wave on reflection.

The ground reflected signal, and in particular its specularly reflected component, combines algebraically with the direct ray at the input terminals of the receiver. At VHF, for example in the land mobile service, it may be an important aspect of system design to ensure that the rays combine constructively for links with stations in important operational locations. At SHF, typically for radio relay systems, the relative phase of the two rays depends to some extent on propagation conditions; there may be no fixed arrangement of antennas that avoids destructive combination for some of the time, causing multi-path fading. Therefore, at SHF it is often a major consideration of system design to ensure that the ground reflected signal is weak; if this cannot be done, diversity operation, usually employing vertically spaced antennas, may be necessary.

B.4.3 Absorption in precipitation and other atmospheric particles

Raindrops in the ray-path of a line-of-sight radio system absorb some of the energy of the wave, scatter some more and cause some degree of depolarisation. Mist and cloud have similar effects, the extent of which is quite limited below 30 GHz. Dry snow, hail and other frozen particles in the ray path cause little absorption, because the permittivity and the conductivity of ice are so much lower than those of water, although accumulations of snow on antenna and random surfaces may degrade radio link performance in other ways. The water associated with wet snow, however, can cause serious absorption. Ice particles in the atmosphere can, in some circumstances, cause significant cross-polarisation. Scattering from rain is considered on pp. 481–482 but these other phenomena are discussed in this section.

Extensive and dense sand and dust storms occur in many countries and it may be that the presence of these solid particles in the ray path can affect radio propagation. At present, however, the evidence from calculations and laboratory experiments indicates that absorption and scattering due to such particles is less than that from light rain and no reported field measurements indicate the contrary; see CCIR Report 721[62].

Attenuation by rain

The acceptability of an SHF or EHF radio link is often determined by its performance when the rainfall in the propagation path is the highest that is commonly experienced. For some kinds of system the criterion might, for example, be the performance when attenuation due to rain is the worst that is to be expected for 99% of the time in the worst month of an average year. For other kinds of system, and especially where large numbers of links are to be connected in tandem and the acceptability of the system as a whole is determined by aggregating the degradations of all of the individual links, it would be necessary to set a much more stringent criterion for the individual links. Thus, for planning purposes it is necessary to forecast the incidence of rainfall, and the attenuation that it will cause, at rainfall rates which are so bad that they occur for no more than a very small percentage of the time.

The specific attenuation γ_R of a wave by rain of given characteristics (that is, the attenuation per kilometre of ray path exposed to that rain) depends on the radio frequency, the rain intensity and the size distribution, temperature, shape and terminal velocity of the raindrops. Studies have been made of the ways in which these characteristics affect γ_R; see for example CCIR Report 563[54]. However, rain characteristics differ greatly from moment to moment and from place to place. In considering the impact of rain on radio systems it is necessary to make use of broad and approximate generalisations. It is found that the following relationship is approximately valid (Olsen *et al.*[63]):

$$\gamma_R = kR^{\alpha} \quad \text{dB/km} \tag{B.24}$$

where R = rainfall rate, mm/hour

k and α = co-efficients, varying with the frequency, the path elevation angle and the polarisation tilt angle (that is, the angle between the plane of polarisation and the horizontal plane).

Values for k and α are given in CCIR Report 721[62]. Fig. B.12 reproduced from Report 721, relates γ_R to frequency for a range of rainfall rates.

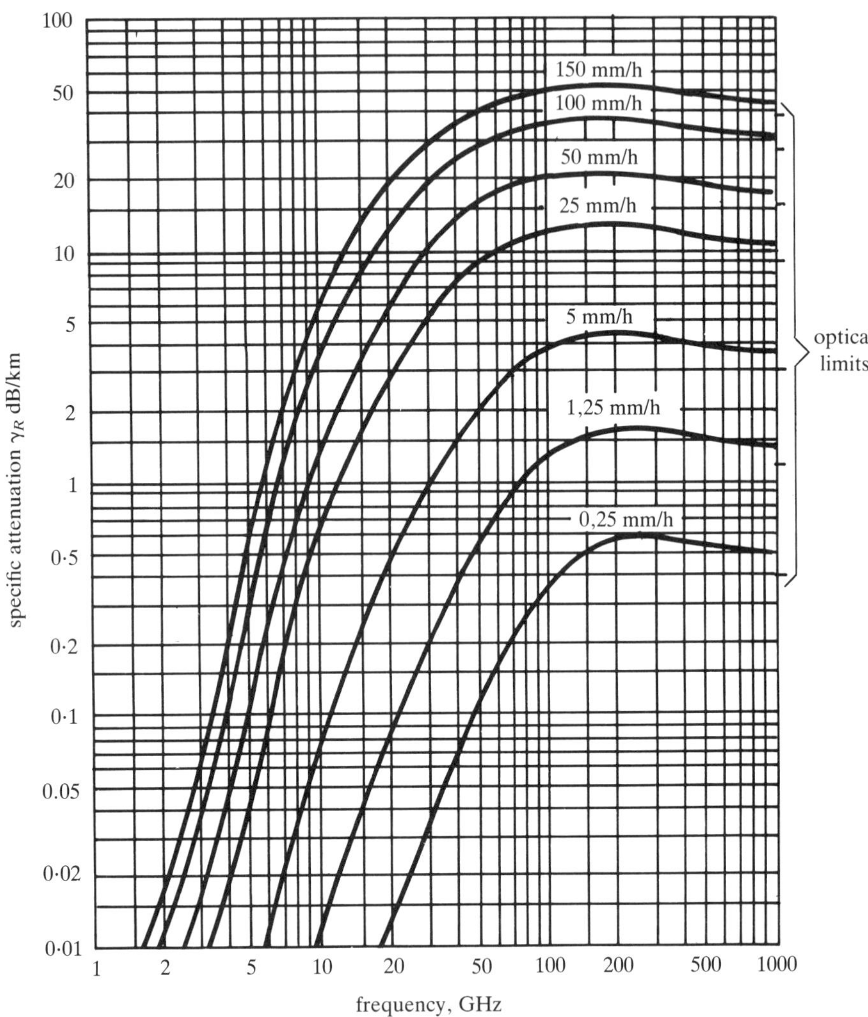

Fig. B.12 Specific attenuation due to rain; CCIR Report 721[62]
Raindrop size distribution (Laws and Parsons, 1943)
Terminal velocity of raindrops (Gunn and Kinzer, 1949)
Index of refraction of water at 20°C (Ray, 1972)
Spherical drops

It is feasible to divide the world into rain-climatic zones and to forecast for each zone the rainfall rate that will be exceeded for any desired percentage of the year or the worst month. Such forecasts can form the basis of forecasts of radio system performance. CCIR Report 563[54] contains world maps showing 14 rain-climatic zones, ranging from Zone A (the polar regions and sub-tropical deserts), where rain is very rare, to Zone P (tropical rain-forest regions), where very heavy rain is prolonged. An example of these maps is reproduced in Fig. B. 13. The Report contains estimates of the rainfall rate exceeded for percentages of a year ranging from 1·0% to 0·0001%; examples are to be found in Table B.3. Other maps in the Report show contours within which various rainfall rates are exceeded for 0·01% of the time.

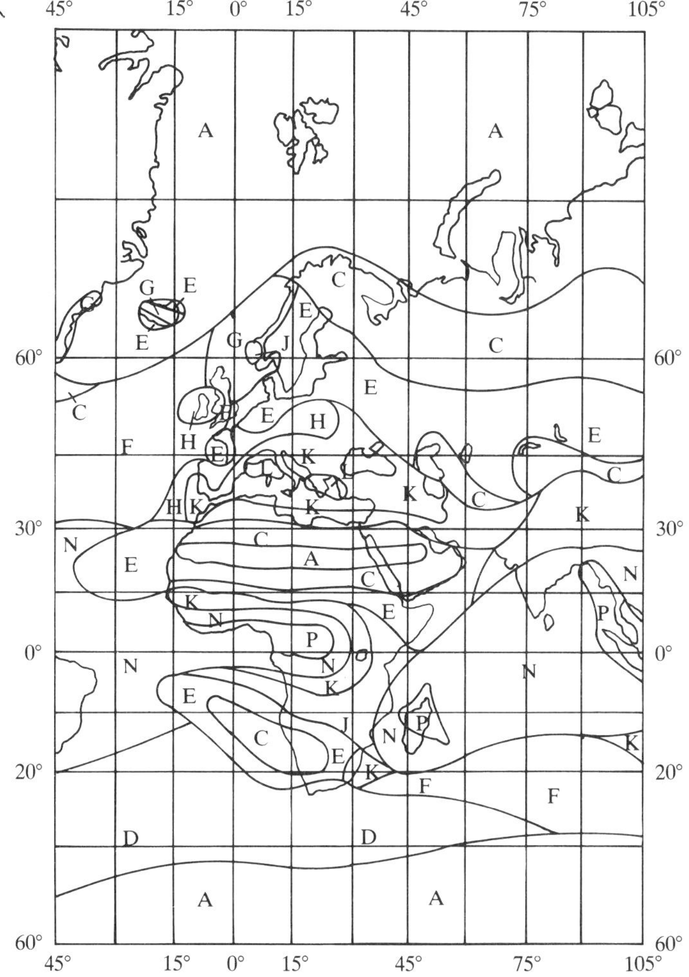

Fig. B.13 Rain climatic zones between 45° West longitude and 105° East longitude; CCIR Report 563[54]

Table B.3 Rainfall intensity in various rain climatic zones (CCIR Report 563[54])

Percentage of time	Rainfall intensity (in mm per hour) exceeded for the stated percentage of the year in the stated rain climatic zones					
	Zone A	Zone C	Zone E	Zone H	Zone K	Zone P
1·0	< 0.5	2	1	2	2	12
0·3	1	3	3	4	6	34
0·1	2	5	6	10	12	65
0·03	5	9	12	18	23	105
0·01	8	15	22	32	42	145
0·003	14	26	41	55	70	200
0·001	22	42	70	83	100	250

Total rainfall varies considerably from year to year, and forecasts of radio system performance must be treated as forecasts of average conditions, taking one year with another. Correspondingly, the rainfall data used for system performance forecasts should, wherever possible, be based on observations spread over several or many years. The rainfall estimates in Report 563 are based on meteorological records which are supported, at many locations, by a long history of measurements and so for such locations they are well-founded. Nevertheless, they have serious limitations as a means of forecasting the attenuation statistics for specific radio links:

- For the broad geographical areas where published meteorological records are lacking or have short time series only; it has been necessary to make intelligent estimates of the most likely climatic zone in drawing up the maps of Report 563.
- The rain-climate within a large area, such as one of the zones defined in Fig. B.13, is not uniform. Depending on the direction of the prevailing winds and the local topography, the climate of two places which are not far apart may be significantly different.
- Measurements carried out with fast-response rain gauges indicate that rainfall of high intensity is often concentrated into short periods of time, typically lasting a few minutes only. It is short periods such as these which determine the acceptability of the performance of many kinds of radio system. Conventional meteorological records, such as those from which the estimates in Report 563 are derived, integrate rainfall over much longer periods. Studies have suggested ways of converting hourly rainfall rates to peak, short-period values but it cannot be said at present that such relationships are well established. There is a considerable amount of information available on rainfall rates measured with fast-response rain gauges at a number of locations and this data base is growing, both in the length of the time series of

measurements and the geographical spread of observations. Nevertheless, the available basis for forecasting short-term rainfall rates for a single point anywhere in the world is far from satisfactory at present.

If specific information on local climate is available it is often profitable to study it when designing radio systems. It may even be desirable for a programme of measurements to be carried out at a specific location thought to be suitable for an important radio station, for example, a major earth station, before a final decision to build at that location is taken. These measurements may be of rainfall, in which case a fast-response rain-gauge should be used. Alternatively and preferably, a device for measuring directly the specific attenuation due to rain or some parameter closely related to it might be used. For a terrestrial radio system this might take the form of a line-of-sight test link measuring the transmission loss. For a radio system using space–Earth paths, a test earth terminal measuring the signal strength of a satellite beacon might be used or alternatively a sky-noise radiometer.

Unfortunately, it is desirable for a programme of such measurements to be maintained over several years before important conclusions are drawn from the results. The measurements should preferably be made in the frequency range intended for the operation of the planned radio station, but if this is not possible the methods discussed in CCIR Report 721[62] for scaling measurements made at one frequency for application in another part of the spectrum should be considered.

Rainfall events are also geographically inhomogeneous. Severe rain storms usually consist of small areas of intense rain, called volume cells, typically only a few kilometres in diameter, imbedded in larger areas of relatively moderate rainfall. Two consequences flow from this:

- It would usually be pessimistic to assume that the whole of a radio link is exposed simultaneously to the most severe rainfall rate; see CCIR Report 338[64].
- In radio systems for which peak rain attentuation is an important economic factor it may be desirable to use space diversity to ensure that, if one ray path is suffering severe attenuation in a volume cell, an alternative ray path is available which is very unlikely to be encountering severe attenuation due to the same, or a different, volume cell at the same instant in time. CCIR Reports 564[65] and 552[66] discuss the application of earth station diversity on space–Earth paths and Report 338[64] contains a brief discussion of route diversity in radio relay systems.

Studies using radars show that the absorption of radio waves varies little with height in a rain storm up to the level of the 'melting layer' (also called the 'bright band' because of the strong radar response it provides). Above this level, the temperature of the cloud is below freezing point and the water present is typically in the form of hail. Attenuation due to hail is relatively small. At tropical latitudes the melting layer is at a height of around 5 km. At higher latitudes it is lower and its height depends to a considerable extent on seasonal and meteorological conditions. In calculating the length of an Earth–space path which is exposed to attenuation due to rain, it is necessary to allow for the fact that the part of the path which is above the melting layer does not suffer significant loss from this cause.

Attenuation in snow and hail

Dry snow causes little attenuation below 50 GHz but studies by Misme[67] indicate that there may be significant effects at higher frequencies. There have been reports of attenuation by hail at much lower frequencies, but such events are probably statistically unimportant. However, wet snow can cause worse attenuation than rain of equivalent water content.

Attenuation in cloud, mist and fog

The specific attenuation γ_C due to cloud, mist or fog is a function of the liquid water content of the medium and it can be represented by the expression

$$\gamma_C = K_l M \quad \text{dB/km} \tag{B.25}$$

where K_l = co-efficient which varies with frequency and temperature
M = mass of liquid water, grammes/m^3

The water content of typical clouds and mist is in the range 0·05–0·5 grammes/m^3. The values of γ_C given in CCIR Report 721[62] indicate that such conditions will cause negligible attentuation at frequencies below 50 GHz, but losses of the order of 1 to 5 dB may arise on space–Earth paths at and above 100 GHz.

De-polarisation due to rain

When rain is present in the ray path of a plane-polarised wave, some of the energy of the wave may be transferred to other polarisation states. A similar effect happens with circularly polarised waves. The severity of the de-polarisation depends on the size and shape of the raindrops, the rainfall rate, the polarisation characteristics of the wave and the frequency. De-polarisation by rain can have serious effects on systems designed to 're-use' the frequency spectrum by means of dual polarisation, but studies of its incidence are at present incomplete. There is an account of the present state of knowledge on the matter in CCIR Report 722[60].

Cross-polarisation due to ice crystals in the atmosphere

Sporadic cross-polarisation of space–Earth signals has been observed in the absence of rain. This effect is attributed to ice crystals within thunder clouds, located above the melting layer and physically aligned by electrostatic potentials within the cloud. This effect is currently under study; see CCIR Report 722[60].

B.4.4 Absorption in the atmospheric gases

In specific frequency bands, all above 20 GHz, there is substantial absorption of energy from radio signals in the atmosphere, due to molecular resonances in the atmospheric gases. Unlike the absorption due to precipitation, discussed in the previous section, this absorption is present all of the time and in some frequency bands it is severe.

There is a weak absorption line due to a molecular resonance in water vapour centred on 22·3 GHz. Two very strong water vapour absorption lines are centred on 183·3 and 323·8 GHz and the absorption due to water vapour in the gaps between these two lines is substantial and rises with frequency. Owing to oxygen there is a broad and very strong absorption line centred on 60 GHz and a relatively weak line centred on 118·7 GHz. The specific absorption at sea level due to each of these lines, under moderately humid conditions, is shown in Fig. B.14. It will be seen from the Figure that the absorption is much less in the windows between the lines but it is considerable at all frequencies between 50 and 350 GHz.

There are many absorption lines above 350 GHz, due to oxygen, water vapour and other atmospheric gases, so that the specific absorption at sea level throughout most of the electromagnetic spectrum between the top of the ITU table of frequency allocations and the infra-red band is very high; see CCIR Report 719[68].

As would be expected, the specific attenuation due to water vapour depends on the absolute humidity of the atmosphere. Some data on the geographical distribution of water vapour are to be found in CCIR Report 563[54] and Report 719 provides means for relating humidity to specific attenuation.

The density of the atmosphere, and consequently the specific attenuation due to molecular resonances, decrease with height above ground. Furthermore, at heights at least 20 km above the Earth's surface, the broad band of absorption at 60 GHz due to oxygen breaks up into a series of peaks of absorption at frequencies at intervals of about 300 MHz, extending between about 55 and 65 GHz, the absorption in the troughs between the peaks being relatively small.

As a first order approximation to the total attenuation for a vertical path from sea level upwards through the atmosphere to space, it may be assumed that the oxygen present in the whole atmosphere is represented by a layer 6 km thick at sea level pressure. Similarly the water vapour for the whole atmosphere may be represented by a layer 2·2 km thick at typical sea level pressure. However, this approximation is inaccurate close to the absorption lines (see also Fig. B.15). CCIR Report 719[68] provides data for calculating the attenuation for slant paths through the atmosphere.

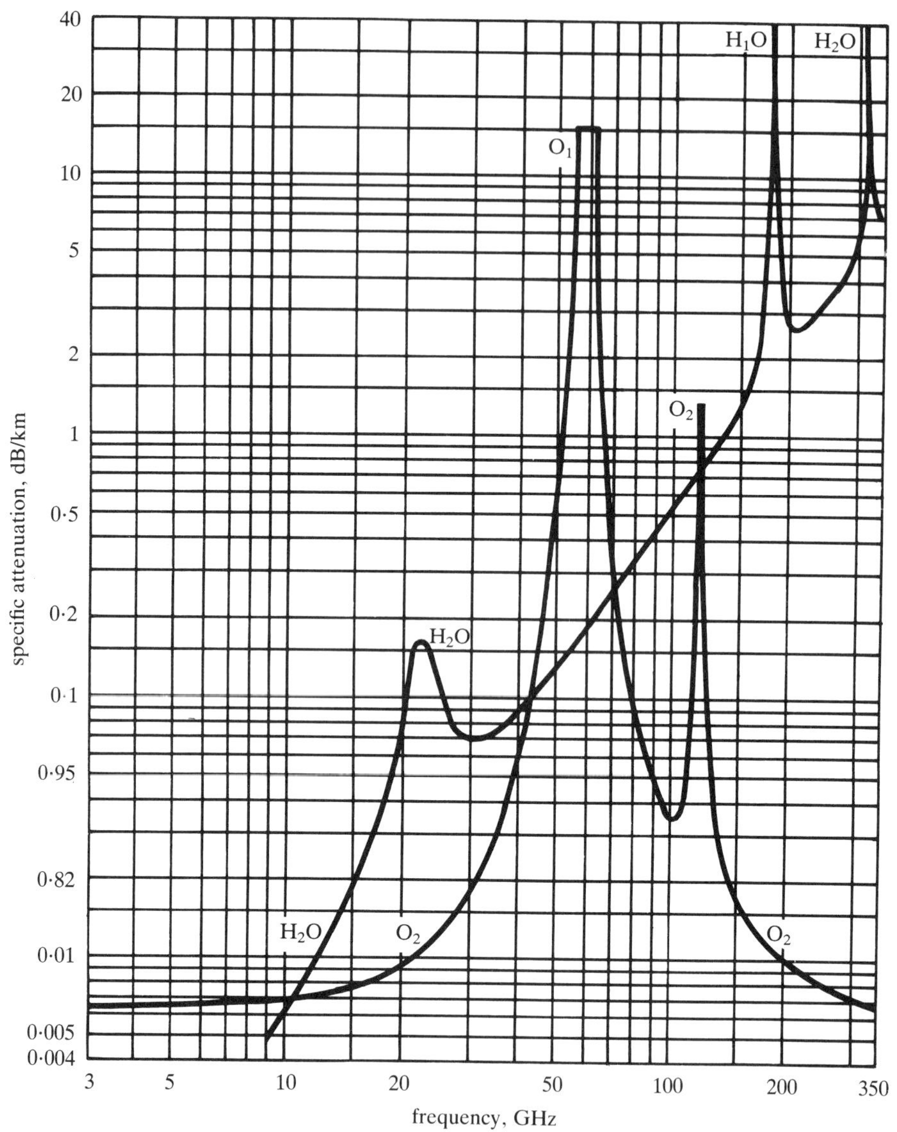

Fig. B.14 Specific attenuation due to atmospheric gases; CCIR Report 719[68]
Pressure = 1 atmosphere
Temperature = 20°C
Water vapour = 7·5 g/m^3

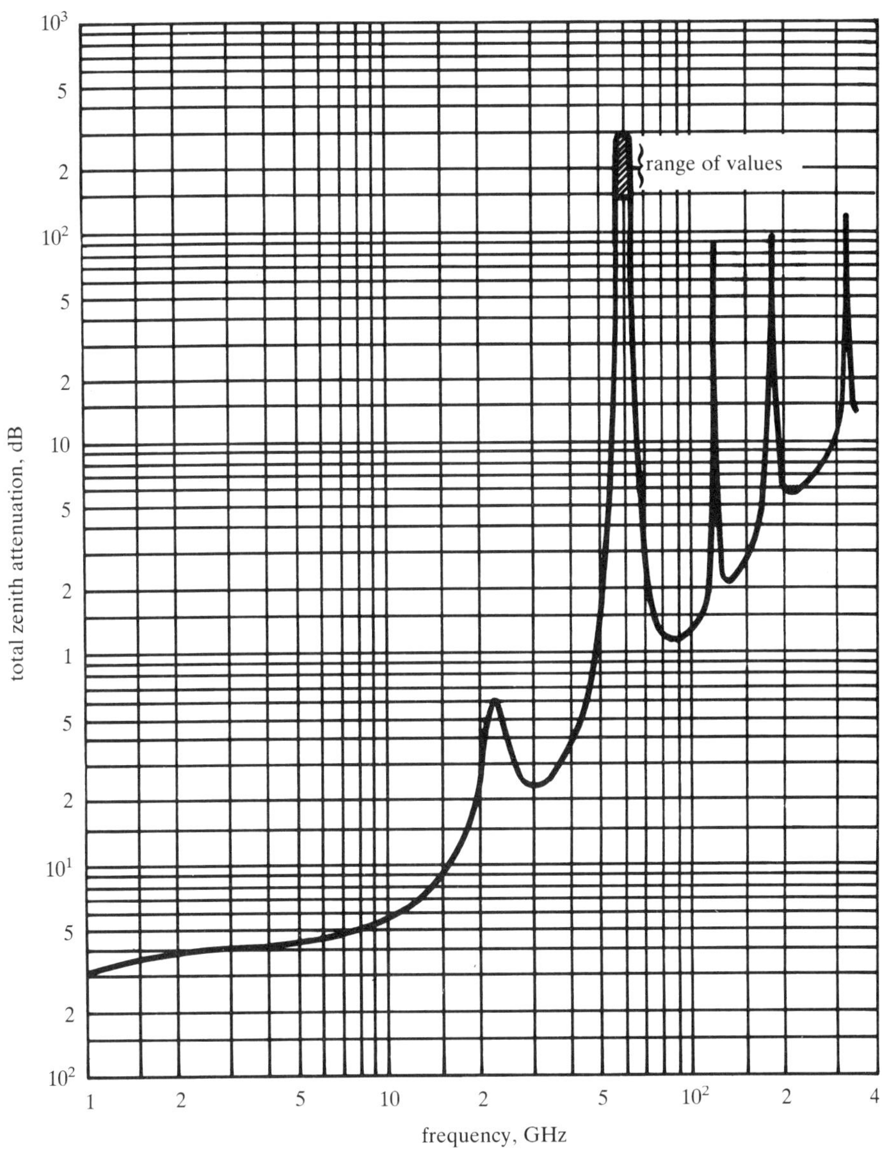

Fig. B.15 Total attentuation due to atmospheric gases in a zenithal path from sea level out to space; CCIR Report 719[68]

Pressure = 1 atmosphere

Temperature = 20°C

Water vapour = $7{\cdot}5\,g/m^3$

B.4.5 Scatter-mode propagation in the troposphere

Scatter-mode propagation by two different mechanisms takes place in the troposphere, Both modes allow signals to pass from transmitting antennas to receiving antennas which are not within line-of-sight, even after allowance is made for ray bending due to atmospheric refraction. Thus:

- Energy is scattered at small-scale irregularities in the refractive index of the air. This scattering is occurring all the time, and it may provide a reliable communication medium, although the transmission loss is high. Persistent interference may also be propagated in this way. This propagation mode is called tropospheric scatter.
- Interference may arise from scattering by raindrops, hail, snow and indeed by any form of precipitation. This interference is, of course, sporadic since it occurs only when the appropriate weather conditions are present. This mechanism is called hydrometeor scatter, precipitation scatter, or most commonly, rain scatter.

In addition, interference may be caused by signals reflected from aircraft. These mechanisms are considered in this section.

Tropospheric scatter

The signal reaching a receiving antenna by the tropospheric scatter mode is made up of a very large number of very weak elements, each of which has been scattered at a location within the troposphere where there is a discontinuity in the refractive index of the air and which is visible from both the transmitting antenna and the receiving antenna. The volume of air in which this scattering can occur, called the common scattering volume, bounded by the upper limit of the troposphere (called the tropopause) and the angular range within which the gain of the transmitting and receiving antennas is sufficiently high, is illustrated in Fig. B.16.

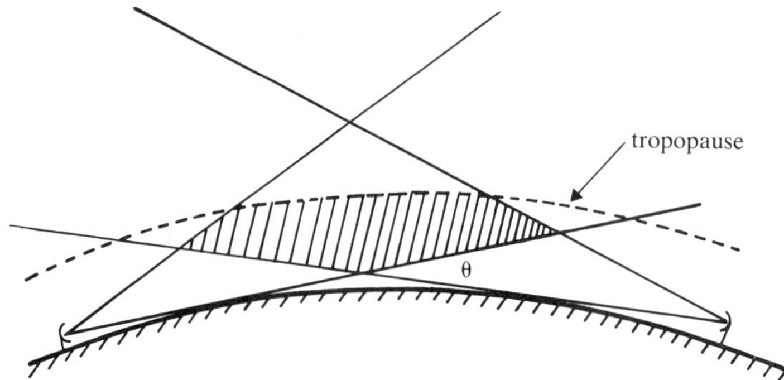

Fig. B.16 Common volume of the main lobes of two antennas coupled by tropospheric scatter
θ = minimum scattering angle

Most of the energy which is scattered in this way is deviated from its original path by no more than a few degrees. However, geometrical considerations set a minimum angle, shown as θ in Fig. B.16, through which energy must be deviated if it is to reach the receiver. Given favourable path geometry and sufficient EIRP, tropospheric scatter is capable of supporting reliable communication over distances up to a few hundreds of kilometres.

The increase of the transmission loss with increasing distance between the antennas is faster than the free-space inverse-distance-squared law, as would be expected, since, as the distance increases, the common scattering volume becomes smaller and the minimum scatter angle increases. The transmission loss also increases with frequency; it is generally agreed that the increase in the loss is approximately proportional to the third power of the frequency up to about 3 GHz and that the increase is less rapid at higher frequencies. The fragmented nature of the received signal leads to deep, selecting fading, to inter-symbol distortion in digital systems, and also to some loss of effective antenna gain (sometimes called the aperture-to-medium coupling loss), especially for large antennas.

The prediction of the transmission loss for any specific radio path depends to a large extent at the present time on empirical projections from measurements of loss made on other paths. A good deal of information, providing a basis for forecasting, is to be found in CCIR Reports 238[69] and 569[70]; see also CCIR Report 285[71]. Several empirical expressions, differing from one another in detail, have been developed for determining the median value of the hourly mean transmission loss, as a function of the characteristics and location of a radio link; see Report 238. The differences between the results which the various expressions give are not very great and studies are in progress in CCIR to ascertain which expression is the most accurate. Other data, also to be found in Report 238, permit estimates to be made of the departures from this mean transmission loss, due to fading, that will be experienced for some of the time, for use in system studies and interference studies.

Rain scatter

Energy is scattered from a radio wave by hydrometeors in the space through which the wave is passing. Some of the scattered energy may reach a receiving antenna which is not directly illuminated by the wave, causing sporadic interference. Scattering by precipitation differs from tropospheric scatter in that precipitation scatters energy more or less equally in all directions. Consequently it is not necessary for the location in which scattering is taking place to be on or near the great circle path between the antenna which is the source of the scattered signal and the antenna which is suffering the interference.

A body of rain sufficiently intense to cause significant scatter can also be expected to absorb energy from the signal which it is scattering. The scattering cross-section of an elemental volume containing precipitation and the attenuation due to absorption in precipitation are both related directly to the intensity of the precipitation but they produce opposing effects. The relative importance of the two factors depends on the frequency. One study by Awaka[72] found a minimum in the transmission loss, giving maximum coupling of interference, at a rain rate near 10 mm per hour for typical earth-station/terrestrial-station propagation geometry at frequencies in the 12–14 GHz range, a lower rain rate having a smaller scattering cross-section and a higher rain rate causing greater attenuation within the rain cell before and after scattering.

Expressions for the size of the scattering volume and the amount of energy scattered in unit volume and means for taking into account the attenuation of the interference by rain within the rain cell are given in CCIR Report 882[73]. Broad conclusions that can be drawn from these expressions include:

- precipitation scatter may be a significant source of interference if the main lobes of the two antennas concerned have a common volume at or below the melting layer when a volume cell of rain is present in that common volume, and
- significant interference is not likely to arise because a volume cell is simultaneously in the sidelobes of one of the antennas and the main lobe of the other unless the cell is close to one of the stations and the gain of the mainlobe involved is high.

Precipitation scatter is most likely to be significant when one of the stations concerned is an earth station operating at a low angle of elevation and the other station is either a radio relay station or another earth station which is using frequency bands in the reverse mode. It may be possible to avoid interference, or greatly reduce its incidence, by choosing sites for stations so that their main antenna lobes do not have a common volume within the troposphere.

Reflection from aircraft

Passive reflection of signals by aircraft shows itself in several ways, for example:

- It can cause sporadic interference between radio stations which are located so that no significant signal passes between them when no aircraft is present. There are some statistics on the incidence of interference of this kind on a specific tropospheric scatter link in CCIR Report 569[70]. A model which could be used for calculating the basic transmission loss for this mode of transmission is given in a document submitted to CCIR in the study period 1978–82[74].
- In a similar way signals reflected from an aircraft can cause multi-path fading when the direct signal reaching a receiving station is weak, due

for example to screening by hills or buildings, if the aircraft is illuminated by a strong signal from the transmitter.

- It might be feasible for regular communication to be maintained between well-separated radio stations by reflection from a tethered balloon.

B.4.6 Propagation near solid objects

In the absence of large solid obstacles it is convenient to use a simple ray concept in considering the propagation of radio waves, the rays being assumed to be straight if the medium has uniform electrical characteristics. This concept does not adequately represent wave propagation in the presence of obstacles which are opaque to the wave.

A concept, originating with Huygens, postulates that every point in a wavefront acts as a centre for a new spherical wavelet, the flux density at any other point being the sum of the contributions from all of these elementary wavelets. This approach accounts for rectilinear propagation in a uniform medium. In addition, however, it explains effects produced by an obstacle on the flux density at a point which is inside, or just outside, the geometric shadow of that obstacle. These phenomena, called diffraction, ensure that the shadow cast by an obstruction is not as sharp as a simple ray concept indicates; see CCIR Report 715[75]. The angular range, measured from the obstruction, over which diffraction has a significant influence over the field strength depends on the frequency, becoming less as the frequency rises.

It commonly happens that a radio wave is propagated past or round several or many obstacles before being received. This situation arises in particular in the broadcasting and land mobile services and when radio is used within forests, buildings and mines. It is usually necessary to employ empirical methods for predicting the additional transmission loss due to these obstacles; see later in this section and CCIR Report 236[76]. The field strength of diffracted signals can, however, be quantified if the geometry of the propagation path can be exactly determined and the configuration is mathematically tractable. Some particular cases are considered below.

Diffraction over a smooth, spherical Earth

If the surface of the Earth around a transmitting station is smooth (more specifically, if the vertical dimensions of the roughnesses of its surface are small compared with the wavelength) and if a straight line from a transmitting antenna to a receiving antenna passes close to the ground, then the signal strength reaching the receiving antenna from the transmitting antenna will be affected by diffraction. The signal strength may be made somewhat weaker than the free space value in some circumstances, but it may also be made somewhat more, depending upon which of the Fresnel zones are obstructed by the Earth's surface and to what degree. For antennas which are not in sight of one another, the transmission loss will, nevertheless, be greater than the free space value.

Quantitively, the effect of diffraction over the Earth's surface, in the absence of major topograhical features, depends on the distance between the antennas, the height–gain factors of the two antennas, the frequency, the polarisation of the wave and the electrical characteristics of the Earth's surface at the relevant point. Formulae and nomograms for calculating the signal level relative to free space propagation are to be found in CCIR Report 715[75].

Diffraction over hills or around single obstacles

As indicated above, if the extent of diffraction past obstacles is to be estimated with some precision, it is usually necessary for the geometry of the propagation path to be expressible in a form which is mathematically tractable. For example, where the obstacle is a hill, it may be possible to represent it sufficiently well as an obstacle topped by a knife edge or by a rounded top of known radius of curvature. Means of calculating the transmission loss in such cases, relative to the free space value, are set out in CCIR Report 715[75].

It may be noted that the presence of a knife edge obstacle in a long propagation path may cause the transmission loss to be considerably less than the loss that would arise if the terrain were smooth, propagation depending solely on diffraction over the spherical surface of the Earth. This so-called 'obstacle gain' is greatest when both antennas are in sight of the knife-edge ridge.

Report 715 also contains some discussion of approaches that could be used to estimate the effects of diffraction in the much more difficult situations that arise where these simple models do not represent sufficiently well the geometry of the propagation path.

The effects of irregular terrain and buildings on propagation

When the number of obstacles is large and they are not small compared with the wavelength, it will seldom be feasible to determine by which combination of propagation modes a signal from a transmitting antenna reaches a receiving antenna. Furthermore the stations and some of the obstacles may be moving and the dominant propagation mode may change very frequently. However, it may not be important to calculate the transmission loss between any two points at any instant in time in such cases. What may be important and is becoming feasible is to estimate in statistical terms the distribution of field strength from a transmitting antenna in such an environment. Techniques are being developed for dealing with such problems at VHF and UHF, specifically for the broadcasting and land mobile services, although a general treatment of the problem is not yet available.

To take account of irregularities of the terrain, it has been found convenient to define as a parameter 'terrain roughness'. CCIR Recommendation 310[55] defines terrain roughness as 'the difference in heights exceeded by 10% and 90% of the terrain along the total path or a specified section of the path'. Terrain roughness has the symbol Δh. It may also be desirable to take into account the mean slope of the ground relative to the ray path from the transmitting antenna, since clearly the roughness of the terrain will have a greater effect on the transmission loss if the angle of elevation, relative to the mean plane of the ground, is very small. There is discussion of these issues in CCIR Report 239[77] and the factor Δh is included in the propagation data for the frequency range 30–1000 MHz to be found in Recommendation 370[78] (for broadcasting) and Report 567[79] (for land mobile). These texts contain estimates of the additional transmission loss exceeded at various given percentages of locations due to roughness of terrain, Δh being a parameter.

Regarding the effects of buildings, Reports 239 and 567 contain various data on the additional transmission loss found in the shadow of individual large buildings and there are also reports of measurements which compare field strengths in towns with field strengths in open country, the other circumstances being similar. For the broadcasting service these latter measurements show small differences only, it being assumed that fixed domestic broadcasting antennas will be located at a height above ground (10 m) where the most severe screening effects of surburban buildings will not occur. It is recognised that the effect of buildings will be more severe in typical land mobile situations and for car radios used to receive VHF broadcasting. Report 567 shows the results of some measurement of field strength received at a mobile station with an antenna at 3 m above the ground.

CCIR Report 1067[80] reports that the sudden reductions in field strength experienced when receiving VHF (100 MHz) FM broadcasting signals in a car being driven or urban streets can be mitigated by the use of spaced-antenna diversity. However, the studies of these effects have not yet reached maturity.

Radio propagation in forests

The effect of a forested area on propagation to and from radio stations located within the forest depends greatly on the frequency; see CCIR Report 236[76]. In brief,

- Below about 2 MHz forest has little direct effect on propagation, although there may be an indirect effect due to the influence of vegetation on the amount of moisture in the sub-soil, and therefore on the ground conductivity.
- Between 2 and 30 MHz a forested area may be regarded as a slab of lossy dielectric material, the wave being propagated through the slab or via a so-called lateral wave above the tree tops. The situation is

made more complex by the possibility of ionospheric propagation, largely bypassing the effect of the forest.

- Above 30 MHz the additional transmission loss due to the presence of vegetation in the direct transmission path rises with frequency. For horizontal polarisation, measurements reported in Report 236 indicate an average loss in forest of a few decibels per kilometre at 30 MHz, rising to 100 dB per km at 500 MHz and continuing to rise as frequency is further increased. For vertical polarisation the loss, for a given frequency, is considerably greater.

Clearly, in any specific case much will depend on the characteristics of the forest and the radio stations, and the information in these CCIR texts can be regarded as no more than an indication of the order of magnitude of the additional transmission loss.

Radio propagation in buildings, tunnels and mines etc.

A radio wave may penetrate into a building and may permeate through the building, typically passing through windows, doorways and partitions and suffering multiple reflections within the building. The extent of the transmission loss in these modes is of interest, for example for paging and cordless telephone systems, and the results of various measurement programmes are to be found in CCIR Reports 567[79] and 880[81]. Most of these measurements relate to frequencies around 900 MHz.

In tunnels, mines and similar places, the walls can to some degree support propagation by a waveguide mode, provided that the frequency used is above the equivalent waveguide cut-off frequency. However, losses are severe, particularly at corners in tunnels and when there is a vehicle in the tunnel which seriously disturbs the properties of the tunnel as a waveguide. Less lossy propagation can be achieved by the use of a radiating transmission line or its equivalent along the tunnel; see Delogne[82 and 83] and CCIR Reports 880[81] and 902[84].

B.5 Radio noise

B.5.1 Introduction

Together with the wanted signal and any interference that may be received, the pre-demodulator stages of a receiver deliver to the demodulator some unwanted electrical fluctuations called radio noise. Some of this noise may have been present in the signal as transmitted and some will have been generated inadvertently in the receiver itself, but the third component, which is often the largest, arises outside the radio link and reaches the receiving antenna as electromagnetic waves, like the signal.

This externally generated radio noise arises from various sources but for practical purposes it is sufficient to identify the following:

Atmospheric noise: Large amounts of energy from lightning discharges are converted into electromagnetic waves having the form of radio noise.
Thermal noise of terrestrial origin: This consists largely of noise generated in and radiated by the ground and the atmosphere.
Extra-terrestrial noise: The main elements are galactic noise and solar noise. Noise radiated by other planets, the Moon and the most powerful of the 'radio stars' is also detectable on Earth, in communication systems using receivers with particularly low internally generated noise, as well as by radio astronomy observatories.
Man-made noise: Radiated from electrical apparatus of many kinds.

These four categories of radio noise are reviewed briefly in this section. See Fig. B.17 for a broad indication of the significance of these different kinds of noise in the various parts of the radio spectrum.

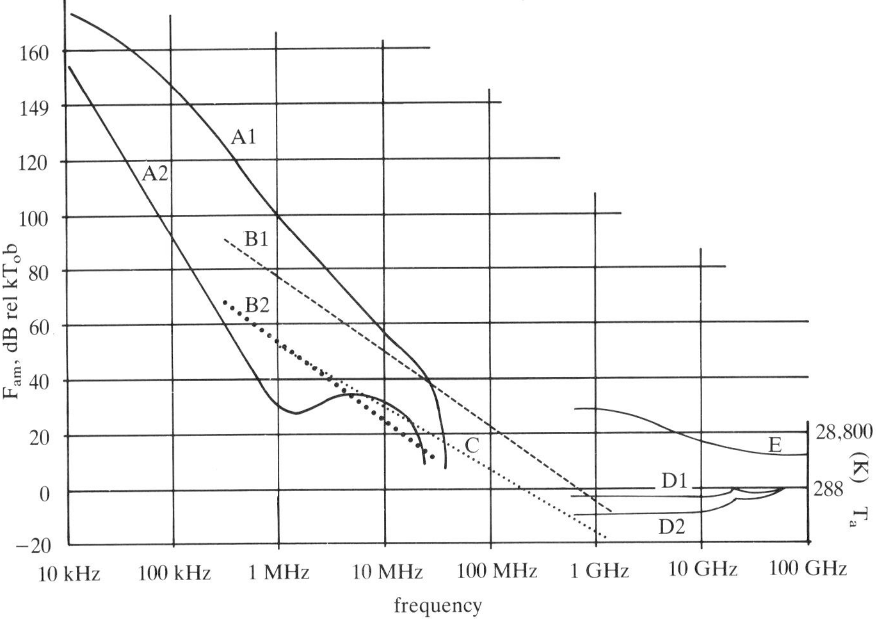

Fig. B.17 Typical radio noise levels, expressed as a median effective antenna noise factor F_{am} or an an effective antenna noise temperature T_a
A1 and A2: Predicted range of atmospheric noise levels; CCIR Report 322[86]
B1 and B2: Reported man-made noise levels for different environments; CCIR Reports 258[90] and 670[91]
C: Galactic noise level predicted for low gain antennas; CCIR Report 322[86]
D1 and D2: Thermal noise of terrestrial origin to be expected from typical terrestrial and earth station antennas under clear sky conditions
E: Order of magnitude of noise power received in a high-gain antenna directed at the Sun (CCIR Report 670[91])

It is found to be convenient to relate p_a, the mean radio noise power delivered by an antenna to a radio receiver, to p_0, the noise power that would be generated by thermal agitation (Johnson noise) in a hypothetical resistor matched to the input terminals of the receiver, if the resistor were connected in place of the antenna. The resistor is assumed to have an arbitrarily chosen reference temperature of 288 K.

The available noise power generated in such a resistor would be

$$p_0 = kT_0b \quad \text{Watts} \qquad \text{(B.26)}$$

where k = Boltzmann's constant = $1{\cdot}38 \times 10^{-23}$ J/K
T_0 = reference temperature, namely 288 K
b = effective bandwidth of the noise measured, Hz

Thus, with the antenna connected to the receiver and disregarding losses in the antenna, the effective antenna noise figure f_a is defined by

$$f_a = p_a/p_0 \qquad \text{(B.27)}$$
$$= p_a/kT_0\text{b} \qquad \text{(B.28)}$$

Alternatively the effective antenna temperature in the presence of external noise, T_a, can be defined by

$$T_a = p_a/kb \quad \text{K} \qquad \text{(B.29)}$$

therefore

$$f_a = T_a/T_0 \qquad \text{(B.30)}$$

Finally the same information can be expressed in logarithmic form as F_a, the antenna noise factor, by

$$F_a = 10\log_{10}f_a \qquad \text{(B.31)}$$
$$= 10\log_{10}p_a - 10\log_{10}b - 204 \quad \textit{dB relative to } kT_0b \qquad \text{(B.32)}$$

Just as the external noise can be conveniently related to the Johnson noise in a hypothetical resistor which replaces the antenna, a similar treatment is convenient for noise added to the signal within the receiver. As before, it is assumed that the antenna is replaced by a matched resistive load with a temperature of T_0 (= 288 K). Then the receiver noise figure n, and the receiver noise factor N, are given by

$$n = n_1/n_2 \qquad \text{(B.33)}$$
$$\text{and } N = 10\log_{10}n \qquad \text{(B.34)}$$

where n_1 = total noise power, measured at some appropriate point in the receiver in bandwidth b (Hz), when the antenna is replaced by a matched load at temperature T_0
n_2 = that part of the total noise power, measured at the same point, which is due solely to Johnson noise in the load which had replaced the antenna (= kT_0b)

Thus the noise added within the receiver is given by $(n - 1)\ kT_0b$, which leads to a definition of the noise temperature of the receiver, T_r, as follows:

$$T_r = (n - 1)\ T_0 \quad \text{K} \qquad \text{(B.35)}$$

To illustrate these relationships, if the internal noise of a receiver is equal to twice the Johnson noise that would be generated in a matched resistive load replacing the antenna, then $n = 3$, $N = 5$ dB and the receiver noise temperature $T_r = 576$ K.

In a practical situation, the signal reaching the demodulator of a receiver will be accompanied by the internal noise of the receiver and noise from one or more external sources. If the noise contributions from these various sources are all expressed as effective noise temperatures in a common hypothetical resistor, they can be summed arithmetically to obtain a measure of the total noise.

B.5.2 Atmospheric radio noise

Atmospheric noise has its origin in lightning discharges, mostly occurring in thunderstorms, many of which are in progress all of the time somewhere on Earth. The electromagnetic wave energy generated by lightning covers a broad spectrum. The maximum spectral power density occurs below 10 kHz but the amount of power that is generated at much higher frequencies, at least up into the VHF part of the spectrum, is quite significant.

Most thunderstorms take place in certain well-defined geographical areas which are shown in CCIR Report 254[85], and atmospheric noise is distributed to locations remote from a storm by the propagation mechanisms, mostly ionospheric, that also operate for radio signals in this part of the spectrum. These propagation mechanisms modify the spectral distribution of noise power, attenuating some parts and failing to propagate the higher frequency components over long distances.

Atmospheric radio noise is characterised by large, rapid fluctuations of level, but if the noise power is averaged over a period of several minutes, the average values are found to be nearly constant during a given hour, variations rarely exceeding ± 2 dB except near sunrise or sunset or when there is a local thunderstorm.

Radio noise level measurements made in research programmes co-ordinated by URSI and CCIR, described briefly in Report 254, have been used to produce the predictions of atmospheric noise level which are to be found in CCIR Report 322[86]. The principal parameter predicted is F_{am}, the median hourly value of the long-term average of F_a for a short vertical antenna over a perfectly conducting ground plane, for any terrestrial location and any frequency above 10 kHz, local thunderstorms being disregarded. Fig. B.17 shows the envelope within which these predictions lie. The predictions of F_{am} are complemented in Report 322 by predictions of various statistical parameters of atmospheric noise.

No doubt the levels of atmospheric noise change from year to year in sympathy with the solar activity cycle, like other ionospheric phenomena, but these changes have not been quantified yet.

B.5.3 Thermal noise of terrestrial origin

The Earth's surface, the permanent gases of the atmosphere and water in the various forms in which it is to be found in the atmosphere all radiate noise and this noise is perceptible in sensitive radio receiving systems operating in the upper part of the UHF band and at higher frequencies.

This is often called 'sky-noise', but it should be distinguished from the noise of extra-terrestrial origin which is discussed on pp. 491–493.

The level of the noise emitted by the Earth's surface depends on the physical characteristics of the surface in addition to its temperature. The surface may also reflect radiation incident from elsewhere. To take account of these phenomena it is convenient to define the brightness temperature T_B of a part of the surroundings of an antenna such that

$$T_B = p_r/kB\eta \quad \text{Watts} \tag{B.36}$$

where p_r = noise power received by the antenna (in watts) from relevant parts of the antenna's surroundings
k = Boltzmann's constant = $1{\cdot}38 \times 10^{-23}$, joule/K
B = bandwidth in which p_r is measured, Hz
η = antenna efficiency with regard to the relevant parts of the antenna's surroundings

The value of η lies between 0 and 1. It approaches 1 if the solid angle formed by the main lobe of the antenna is fully occupied by the specified part of the surroundings having the stated value of T_B and it approaches 0 if the whole of the specified part of the surroundings of the antenna is in the sidelobes.

The way in which the brightness temperature of the Earth's surface varies with the frequency and the temperature, roughness and permittivity of the ground is complex. It is discussed in CCIR Report 720[87] but in brief:

- The brightness temperature of the sea is considerably lower than its physical temperature; it is about 100 K in the microwave band and somewhat higher at millimetre wave frequencies.
- The brightness temperature of a land surface is usually about 90% of the physical temperature of the ground for high angles of elevation, but it may be substantially less for low angles of elevation and for very wet soil.

Just as the gases of the atmosphere, particularly oxygen and water vapour, absorb energy from a wanted signal, they also radiate thermal noise. The brightness temperature of the atmosphere under clear-sky conditions can be calculated for a given frequency f with sufficient accuracy from the expression

$$T_b = T_m\,(1 - 10^{-A/10}) \quad \text{K} \tag{B.37}$$

where T_m = mean temperature (K) of the atmospheric gas, typically about 270 K
A = attenuation (dB) through the atmosphere at frequency f along the principal axis of the antenna

In a similar way, the attenuation due to rain and other hydrometeors in the propagation path also has its counterpart in radiated noise. Below 10 GHz expression (B.37) can be used with sufficient accuracy for this purpose also. At higher frequencies the use of (B.37) may lead to an overestimate of T_b, since a significant part of the attenuation above 10 GHz may be due, not to absorption, but to scattering, which does not contribute to the brightness temperature; see CCIR Report 720[87].

Taking these various phenomena together, the antenna noise temperature due to terrestrial thermal noise can be estimated in particular cases. Typical cases are as follows:

- *T* due to terrestrial thermal noise is nowhere likely to exceed 300 K and so it is not likely to have a major effect on radio links using radio receivers which do not have low-noise first-stage amplifiers.
- For a radio receiver operating a terrestrial service, η for thermal radiation from the ground will probably exceed 0·5. T_a for noise from that source is likely to be of the order of 200 to 250 K. This is likely to exceed the man-made noise even at a relatively noisy location at frequencies above about 1 GHz, and it is likely to exceed the galactic noise at a site which is relatively free from man-made noise at frequencies above about 0·6 GHz. Thus, given a low-noise first-stage amplifier, terrestrial thermal noise may be dominant. T_a due to these causes may approach 300 K above 15 GHz in heavy rain and in the millimetre wave absorption bands.
- The earth station antenna case is discussed in CCIR Reports 720[87], 208[88] and 390[89]. The value of η for radiation from the ground will usually be small with a well-designed antenna unless the angle of elevation of the main lobe is very low, and the contribution to T_a from this source will often be as low as 20 K. When heavy rain is present in the main lobe of the antenna the increase in T_a will be small (perhaps 20 to 50 K) below 10 GHz but it may well be quite large (50 to 150 K) between 10 and 15 GHz. Above 15 GHz, rain and even clouds may have a major effect on T_a when they are present, particularly when the angle of elevation of the main lobe is low. The molecular absorption line at 22·3 GHz due to water vapour, which causes relatively little absorption of space-to-Earth signals, nevertheless causes substantial noise emission all of the time in most climates. The noise emission from oxygen, which is always present, and water vapour, which is almost always present to some degree, produces values of T_a ranging from 130 to 290 K between 45 and 350 GHz for low angles of elevation even under clear sky conditions, depending on frequency and climate, and the corresponding range for high angles of elevation is from about 50 to 290 K.
- Spacecraft and satellite antennas directed towards the Earth will usually have a value of η for radiation from the Earth which approaches 1·0. Because of the large fraction of sea surface that will often be present in the main lobe footprint of the antenna, T_a will often be somewhat less than 290 K, except in the molecular absorption bands of the gases in the Earth's atmosphere.

B.5.4 Extra-terrestial noise

A vast amount of radio noise is generated by various mechanisms in stars and in the interstellar gas in the Milky Way, our galaxy. Most of this galactic noise is generated in the centre of the galaxy and reaches the Earth from the direction of the constellation Sagittarius; see CCIR Report 720[87] for information on the variation with celestial direction and frequency of the strength of the noise flux, treated as a brightness temperature as on p. 000.

The Sun is also a strong radio noise source. The brightness temperature is about 1 000 000 K at VHF when the Sun is not disturbed. At 4 GHz the noise temperature is at least 23 000 K under quiet Sun conditions. At 12 GHz and 18 GHz the quiet Sun temperatures are found to be 12 000 and 9200 K respectively. For 1% of the time at sunspot maximum, radio noise bursts increase the noise temperature at 4 GHz by about 50% relative to that of the quiet Sun, and for smaller percentages of the time the increase is considerably greater.

The Moon has a brightness temperature, seen from the Earth, between 150 and 370 K, depending on radio frequency and lunar phase. The brightness temperatures of planets other than the Earth range up to 600 K. A number of discrete 'radio stars' within the galaxy have brightness temperatures of the same order; the brightest and best known of these stars are Casseopeia A, Taurus A and Cygnus A.

The effect of these celestial noise sources on receiver noise levels depends on the frequency, the gain of the receiving antenna, the angular magnitude and brightness temperature of the noise source, and the level of noise from other sources.

Fig. B.17 provides a broad indication of the galactic noise level averaged over a large area of sky; i.e. the extra-terrestrial noise level likely to be received by low-gain antennas. The other celestial noise sources, even the Sun, make a comparatively insignificant contribution if the receiving antenna gain is low. Galactic noise is the dominant external noise between 25 MHz and 1 GHz, in locations where the man-made noise is very low; in some circumstances, galactic noise may exceed atmospheric noise even at frequencies low in the HF part of the spectrum.

A high gain antenna directed towards the Sun would receive a strong radio noise emission. The same is true to a lesser extent for a high gain antenna directed at the galactic centre. Noise from the Moon will be perceptible if the antenna gain is high and the receiver noise factor is low. However, unless such an antenna tracks the Sun and Moon as they move across the sky, their noise emission will be perceived for no more than a few minutes or tens of minutes (depending on the gain of the antenna) on those days of the year, if any, on which these sources pass through the antenna beam. The angular size of the galactic noise source is bigger that that of the other sources and so its noise will be the most persistent, but it too will be perceived through a high-gain antenna for no more than a rather short daily period. All of these noise sources are strong in the UHF band but decline with increasing frequency, becoming less significant in comparison with noise radiated by the atmosphere in the upper part of the SHF band.

It may be noted that the brightest part of the galactic noise source in Sagittarius never passes behind geostationary satellites as seen from the Earth. Noise interference from the Sun affects links from a geostationary satellite to an earth station typically for no more than some tens of minutes per year (see CCIR Report 390[89]). Because of their orientation, many terrestrial station antennas never experience severe interference from major extra-terrestrial sources.

A high-gain antenna with low sidelobe response, directed upwards to the zenith, would receive a very low level of noise, perhaps no more than the brightness temperature 2·7 K which is the cosmic backround radiation level, if one of the specific radio noise sources were not passing through the beam at the time.

The planetary and stellar noise sources all subtend solid angles to antennas on Earth that are small compared with the beamwidth of the largest antennas used for conventional radio services. Also, as the Earth rotates, the planets and stars are present for brief periods at most in the field of view of antennas which do not track them. Consequently, these sources make no significant contribution to antenna noise temperatures. Nevertheless they provide signals of precisely known flux density which can be used in measuring the figure of merit (G/T) of earth station antennas of high gain which are equipped with low-noise first-stage amplifiers; see CCIR Report 390[89].

B.5.5 Man-made noise

Some industrial, scientific or medical apparatus generates radiation which is not unlike a radio signal, being mainly concentrated into a relatively narrow band and often coherent in waveform. Interference from sources of this kind is considered on pp. 39–40.

However, broad-band, noise-like electromagnetic radiation is generated by many kinds of industrial and domestic apparatus, by electric traction systems, overhead electric power transmission systems, and so on. One of the most important sources of man-made noise is the ignition systems of petrol-driven motor vehicles.

Studies done in USA and summarised in CCIR Report 258[90] have shown that F_{am}, the median value of the noise power from this source available from a simple antenna, can be represented by

$$F = c - d \log f \quad \text{dB relative to } kT_0b \tag{B.38}$$

where c and d = constants which are related to the environment
f = frequency, MHz
k = Boltzmann's constant = $1{\cdot}38 \times 10^{-23}$, J/K
T_0 = reference environmental temperature = 288 K
b = bandwidth in which the measurement is made, Hz

For 'business' areas, a category covering, for example, industrial and commercial districts, values of 76·8 and 27·7 were found for c and d respectively, applicable for the frequency range from 300 kHz to at least 250 MHz. The correspnding values found for 'quiet rural' area were 53·6 and 28·6, and these were applicable from 300 kHz to 30 MHz. Various intermediate area categories are also defined in Report 258.

The curves of F_{am} against frequency found in this study for 'business' and 'quiet rural' areas are indicated on Fig. B.17. It will be seen that man-made noise is likely to exceed galactic noise, and all other commonly occurring kinds of noise from natural sources, in business areas for all frequencies from around 10 MHz to over 1 GHz.

However, studies of man-made noise done in other countries have produced results differing in some degree from the USA study. References to, and brief summaries of, some of these other studies are also to be found in Report 258.

B.6. Prediction of radio link conditions

B.6.1 Introduction

There is extensive documentation in ITU texts of methods for predicting the conditions that will be obtained on a given radio link or a given interference path. Some of this material will have been prepared by a CCIR Conference Preparatory Meeting (CPM) and is to be found in the report of that CPM to the relevant Administrative Radio Conference (ARC). ARCs having the function of preparing a frequency assignment plan may include the essence of such radio propagation material in their plan document for later reference. If it is necessary to incorporate propagation data into a mandatory ITU procedure, such as the procedure for co-ordinating the use of shared frequency bands by space and terrestrial radio services, the data may be included in the Radio Regulations; see in particular RR Appendix 28. In general, however, this material is to be found in recommendations and reports of the CCIR. This material, and in particular the material in the most readily available sources, namely the Radio Regulations and the CCIR texts, is reviewed below.

B.6.2 VLF propagation, 10 to 30 kHz

CCIR Recommendation 368[7] contains propagation curves relating ground wave field strength to distance for 1 kW radiated from a short vertical monopole antenna over various ground conditions for various frequencies at VLF. Annex II to Recommendation 368 and CCIR Report 717[9] provide an approach to the calculation of field strengths over ground of non-uniform characteristics. Report 717 also contains an atlas of effective ground conductivity data. CCIR Report 716[10] considers the variation of ground wave phase with distance and the characteristics of the ground.

When the distance is no more than few hundred kilometres, the ground wave field strength variation with distance differs little from a simple inverse distance relationship and the ionospherically propagated wave can be disregarded. At greater distances it is essential to take account of ionospheric propagation. This raises problems of great complexity, which are susceptible to relatively precise solutions in specific cases but which have not been reduced to a simple presentation of generally applicable solutions; see CCIR Report 895[46].

Atmospheric noise is very strong at VLF; see CCIR Report 322[86].

B.6.3 LF and MF propagation, 30 kHz to 2 MHz

The first paragraph of Section B.6.2 applies with equal force to ground wave transmission at frequencies between 30 kHz and 2 MHz.

Ionospheric propagation is very significant in this frequency band:

- At the lower end of the band it extends the operating range of transmitters.
- At the upper end of the band the D layer absorbs signals severely by day. At night D layer absorption is much less and so the sky wave is propagated much more strongly; this causes severe fading at distances where the ground wave and sky wave field strengths are approximately equal or where two strong sky wave propagation modes are active simultaneously. This fading may make the signal useless for some purposes, such as broadcasting, at some times.
- The strong night-time sky wave can cause significant interference at great distances.

A lot of work has been done, seeking to establish a theoretical basis for predicting the field strength of ionospheric signals in this part of the spectrum, and CCIR Report 265[47] summarises the results of these studies, having particular reference to frequencies below 500 kHz. Nevertheless, at the present time it is necessary to depend largely on empirical predictions of sky-wave field strength in this part of the spectrum; see CCIR Report 432[92]. CCIR Report 575[93] considers methods, partly theoretical and partly empirical, which are used for the prediction of sky wave field strength in the 150–1600 kHz frequency range and CCIR Report 431[94] tests these methods against the results of a number of field strength measurement programmes. CCIR Recommendation 435[95] contains recommended curves of sky-wave field strength against distance and frequency, with additional data to enable the user to take account of other relevant factors; see pp. 455–456 for references to ionospheric cross-modulation effects.

The collection and reduction of the data referenced in the previous paragraph has been carried out largely by organisations involved in broadcasting. Clearly, however, these texts provide material that can be applied, within appropriate limits, to other services using this part of the spectrum.

The dominant form of external noise between 30 kHz and 2 MHz is atmospheric noise, although man-made noise may limit performance in some locations, especially above 1 MHz; see CCIR Reports 322[86] and 258[90].

B.6.4 HF propagation, 2 to 30 MHz

Ground wave propagation may be operative on short-distance HF links. CCIR Recommendtion 368[7] contains propagation curves relating ground wave field strength to distance from 1 kW radiated from a short vertical monopole antenna over various ground conditions for various frequencies between 2 and 30 MHz. Annex II to the Recommendation provides an approach to the calculation of field strengths over ground of non-uniform characteristics. Factors for applying data referenced to a short vertical monopole to other basic types of antenna are to be found in CCIR Recommendation 341[2]. In calculating ground wave field strengths, particularly at the higher end of this frequency range, it may be necessary to take account of large obstacles, such as hills, obstructing the propagation path close to either end of the radio link; guidance in this matter may be found in CCIR Report 715[75].

Long distance radio links using frequencies in the HF band employ ionospherically propagated waves. At any given time the frequency range from within which the operating frequency must be chosen is defined by E or F layer MUFs at the upper end and by absorption (mostly non-deviative absorption in the D layer) and noise external to the receiver at the lower end. Both limits are affected, to some degree, by the nature of the communication system and the performance of the radio equipment. On radio links which are very long or are inadequately equipped, there may be times when the signal-to-noise ratio is too low for satisfactory operation at any frequency which the reflecting layers are able to propagate. Such conditions may also arise relatively often within the auroral regions.

In some operational circumstances the choice of operating frequency may be made in real time in the light of the performance of the link and other factors. More generally, for operating sysems and for frequency assignment planning purposes, it is necessary to identify optimum working frequencies by predicting propagation conditions. Ideally, prediction of working frequencies for long HF links would involve four elements:

- prediction of the upper limit of usable frequencies, due allowance being made for the statistical scatter of propagation conditions,
- prediction of the signal strength at the receiver input for each available frequency assignment, assuming that the frequency is not above the upper limit of propagation,
- prediction of the external noise level at the receiver input for each available frequency assignment, and
- consideration, for each available frequency assignment, of whether circuit conditions, such as signal-to-noise ratio and multi-path propagation, will permit the radio link to provide the facilities required in a satisfactory manner.

It is usually convenient to make such predictions for each calendar month and for each two-hour interval throughout the daily schedule of operation. However, methods for making completely satisfactory predictions of some of these elements are not available yet, and experience with specific equipment on specific routes may have to be added to the guidance provided by predictions to enable the best use to be made of the medium.

Approximate operational MUFs and Optimum Working Frequencies (OWFs) can be predicted using procedures and data which are set out in CCIR Report 340[13]. The methods used are simple and they can be applied manually or by computer. For radio links which are more than 4000 km long, the two-control-point method is used for predicting F2-layer MUFs. More precise methods are available for predicting the Operational MUF, and in particular the F2-layer MUF, for long links (see CCIR Reports 255[96], 888[30] and 889[43]) but, having regard to the short-term variability of the medium, the methods given in Report 340 are sufficiently accurate for most purposes.

Many attempts have been made to devise a method for calculating the field strength that a given transmitting station would set up at a given receiving station at a given frequency. It is a rather intractable problem, and agreement has yet to be reached on an acceptable method. A number of previously published methods were reviewed in CCIR Report 252[20], published in 1970. This Report also contained details of an interim CCIR method, published for study, drawing on much of what had been published previously. This interim method sought to provide means to compute the 'system loss' of a radio link, that is the loss between the input terminals to a transmitting antenna and the output terminals from a receiving antenna at a given frequency, both antennas being of known characteristics, assuming that the frequency was below the operational MUF. Some reports on the accuracy of this method are in CCIR Report 571[97].

A Supplement to CCIR Report 252[23] was published in 1980. This contains an extensively revised method for computing field strengths, which takes account of a number of factors which were not covered by the method of 1970. This new method does not replace the 1970 method; it is another provisional method, published for study and comment. Information on the experience of administrations in using this 1980 method is also to be found in CCIR Report 571.

Work in this area continues in CCIR. No doubt a definitive method will emerge in time. CCIR Report 729[98] contains a record of recent progress. Meanwhile the two CCIR interim methods are available for tentative use, along with various other published methods reviewed in the CCIR reports referenced above.

External noise entering the receiver will usually be atmospheric noise, but man-made noise often predominates at noisy receiving locations and galactic noise may be detectable where both atmospheric and man-made noise are low; see CCIR Reports 322[86] and 258[90].

CCIR Report 892[99] discusses methods for calculating the reliability of HF radio systems. Any successful calculation of this kind must depend ultimately on knowledge of the extent of the propagation degradations etc. that will be tolerated by the communication system in question without unacceptable deterioration of the standard of performance provided and on knowledge of the signal-to-noise ratio that will be obtained. In many cases data relevant to such problems are to be found among the reports of the CCIR study group having concern with the radio service in question. However, as indicated above, the day-by-day average signal-to-noise ratio is not precisely predictable at present, and the variation of conditions from day to day is considerable; both of these factors militate against accurate forecasts of reliability.

There is a review of these factors in CCIR Report 894[100]. The review was prepared with the specific needs of the broadcasting service at HF in mind, but it will also be found helpful in other contexts.

B.6.5 VHF ionospheric propagation

Below 30 MHz the ionosphere provides a regular, if somewhat variable, transmission medium. Above 30 Mhz there are also ways of using ionisation in the atmosphere to support regular communication services. Meteor trail communication is one such way; see CCIR Report 251[32]. Radio amateurs make extensive use of various modes of ionospheric propagation at VHF. However, in general, VHF frequencies are assigned to stations for short-distance facilities by tropospheric modes of propagation, and the ionosphere acts as a medium for sporadic long-distance interference between these short-range services. Consequently, the principal objective of the prediction methods developed for ionospheric propagation in this part of the spectrum has been to forecast the probability of interference at a receiving station, given the location of an unwanted transmitting station.

Long distance interference by normal reflection from the regular ionised layers it not uncommon. Frequencies above 30 MHz are not likely to be propagated by the regular E layer at any time, but regular F1-layer reflection can occasionally take place at extreme range (the skip distance being around 3000 km) and regular F2 layer MUFs may be considerably higher for long distance propagation. The highest probability of regular F2 layer propagation arises near sunspot maximum, near local noon, near the equator. Under these conditions, regular daily propagation may take place up to about 40 MHz and propagation may occur for 1% of the time up to 60 MHz; see CCIR Report 259[24].

Long distance north–south propagation via the F2 layer has been observed on many occasions at frequencies well above the regular F2 MUF. The mechanism of propagation appears to be related to the 'equatorial F2 anomaly'; see CCIR Report 259[24].

There is extensive information on the probability of incidence of temperate zone sporadic E ionisation, and on the transmission loss to be expected, in CCIR Report 259[24] and CCIR Recommendation 534[28].

On relatively rare occasions the density of meteor trails is much higher than usual, owing to the incidence of a meteor shower. At these times interference via this mechanism would be considerably increased, but data on the probability of these events are lacking.

The references listed in CCIR Report 260[26] contain data which could provide a basis for calculating the level of interference to be expected via ionospheric forward scatter.

B.6.6 Terrestrial broadcasting and mobile services, 30 to 1000 MHz

CCIR Recommendation 370[78] provides propagation curves relating field strength to distance from the transmitter. These curves explicitly refer to two frequency bands, 30–250 MHz and 450–1000 MHz, since they have been prepared specifically for use in the context of terrestrial broadcasting, but they could also be used, with interpolation where necessary, for any service operating between 30 and 1000 MHz for which the assumptions on which the curves are based would be appropriate; see for example the next paragraph. There is provision for applying a parameter to allow for terrain roughness between the transmitting station and the area in which reception is required. The field strength for 1 kW effective radiated power from a half-wave dipole is expressed in statistical terms for various antenna heights. The bases on which these propagation curves were prepared, and data to enable the user to extend their application without unacceptable loss of accuracy, are set out in CCIR Report 239[77].

CCIR Recommendation 616[101] recommends the application of the curves and data in CCIR Recommendation 370[78] and CCIR Report 239[77] for use in the context of the maritime mobile service. In the land mobile service context, however, the screening effect of buildings on car-borne antennas in towns must be taken more fully into account; accordingly, CCIR Report 567[79] contains propagation curves prepared to serve the particular needs of the land mobile service. A third set of situations arises in the aeronautical mobile service, which requires propagation data on ground-to-air and air-to-air links; this is provided for spot frequencies of 125 and 300 MHz in Recommendation 528[102].

Throughout the band 30–1000 MHz, man-made noise is the dominant form of external noise in the noisier locations, and galactic noise predominates elsewhere; see CCIR Report 258[90].

B.6.7 Terrestrial broadcasting and mobile services above 1000 MHz

Studies have been made in a few countries of the propagation problems arising in terrestrial broadcasting at 12 GHz. It has become clear that obstruction by and reflection from buildings are major problems in towns in this band. The studies are in an early stage at present; see CCIR Report 562[103].

Reference was made above to the propagation curves in CCIR Recommendation 528[102] for aeronautical mobile services at VHF. The same text also contains propagation curves for 1·2, 5·1, 9·4 and 15·5 GHz, appropriate to the aeronautical mobile service.

Above 1 GHz the effective antenna temperature of a receiver used for terrestrial services is likely to be close to the temperature of the environment.

B.6.8 Terrestrial line-of-sight fixed links above 30 MHz

CCIR Report 338[64] provides detailed information on the propagation issues which arise in the design of line-of-sight fixed service links. Some aspects of line-of-sight propagation are examined, from a different standpoint, in CCIR Report 784[59]. For consideration of the propagation aspects of interference to and from these stations, see pp. 501 and 502.

Below 1 GHz, external noise at the receiver is likely to be dominated by galactic noise or man-made noise; see CCIR Report 322[86]. Above 1 GHz the effective noise temperature of the antenna is likely to be close to the temperature of the environment.

B.6.9 Terrestrial trans-horizon fixed links above 200 MHz

CCIR Report 238[69] provides detailed information on the propagation issues which arise in the design of trans-horizon radio links using the tropospheric scatter propagation mode. It should be noted that the data in this Report have not been validated above 4 GHz. Some of these issues are examined, from a different standpoint, in CCIR Report 285[71]. For consideration of the propagation aspects of interference to and from these stations, see pp. 501 and 502.

For the external noise at the receiver, see the previous section.

B.6.10 Space–Earth links

CCIR Report 564[65] is the basic text on space–Earth propagation, offering data on all the significant phenomena which attenuate or degrade a signal passing directly along a clear line-of-sight path from a satellite antenna to an earth station antenna or vice versa and covering virtually the whole range of frequencies currently of interest for satellite services.

The satellite broadcasting service has a particular interest in narrow sections of the field covered by Report 564. Accordingly, CCIR Report 565[104] has been prepared; it places special emphasis on propagation phenomena at 12 GHz and the particular problem of receiving signals from a satellite within a building and amongst buildings.

Two other space radio services also have specialised concerns. Both the maritime mobile-satellite and the land mobile-satellite services use earth stations, some of which use antennas of low directivity, and accordingly the signal reflected from the ground in the vicinity of the earth station may play a large part in determining the state of the signal received. Also, there are system advantages in the use of frequencies for mobile-satellite services at the lower end of the frequency range used for space services in general, which gives emphasis to the propagation problems, mainly those due to transmission through the ionosphere, which are most evident below 2 GHz. CCIR Reports 884[105] and 1009[106] contain material, additional to that in Report 564, required for the maritime mobile-satellite and land mobile-satellite services respectively.

There is a discussion of external noise sources of concern for earth stations with high gain antennas in Report 564. The effective noise temperature of the low gain antennas typical of mobile-satellite earth station receivers is likely to be close to the temperature of the environment.

B.6.11 Interference between stations on the Earth's surface

Below 30 MHz, interference may be by the ground wave or the sky wave. The methods reviewed on pp. 494–498 are usable, to some degree, for calculating the probability of interference and its level. Between 4 and 30 MHz, where frequency assignments are primarily for long distance links, the probability of interference on any specific frequency may be less than might be expected, because of a lack of coincidence of times when the same frequency is wanted for use on different radio links. These differences of frequency usage arise mainly from differences between links as to length and geographical situation and in the way in which propagation conditions on the links are affected by the daily, annual and solar activity cycles of ionospheric conditions.

Between 30 and about 100 MHz various ionospheric propagation modes are sometimes active and this may lead to interference at long distances (see pp. 498–499). It is characteristic of several of these modes that they provide powerful interfering signals at long distance for some percentage of the time, but interference may be virtually absent for the rest of the time. In such cases the essential criterion of acceptability of interference may be the probability of its occurrence, not the level of interference when it is present.

Between about 100 and about 500 MHz, CCIR Recommendation 452[107] advises the use of the propagation curves in CCIR Recommendation 370[78] for the calculation of the interference levels. For mobile service applications, the data in CCIR Recommendation 528[102] or CCIR Report 567[79] may be more appropriate and are also recommended. Unfortunately, the present state of knowledge of the field strength exceeded for a small percentage of the time at long distances in this range of frequencies is limited; estimates for percentages of the time smaller than 1% are not available.

Above 500 MHz, up to at least 12 GHz, information on the propagation modes which are active for very small proportions of the time is available for some geographical regions and this information can be extrapolated and applied with caution to the rest of the world. CCIR Report 569[70] reviews these data in detail. The particular case of interference between earth stations and terrestrial radio relay stations, including some propagation aspects, is discussed in CCIR Report 448[108].

The administrative process of co-ordinating the use of shared frequency bands by space and terrestrial services requires propagation data prepared in a way that facilates the determination of the distance (the 'co-ordination distance') from an earth station beyond which interference at an unacceptable level to a terrestrial station or from a terrestrial station is very unlikely to arise, no matter how unfavourable the characteristics, emission parameters and antenna orientation of the two stations may be. RR Appendix 28 contains data and technical procedures for this purpose, approved at the WARC in 1979. CCIR Report 382[109] is, in effect, a new version of the material in Appendix 28, revised to take into account more recent CCIR data and improvements in the technical procedures; no doubt Appendix 28 will be amended by a competent future WARC to take these improvements, or later further improvements, into account. CCIR Report 724[110] contains the latest relevant propagation information, no doubt destined to form the basis of further revision of Report 382.

For information on the interference situation which arises between two earth stations, one of which is transmitting in a frequency band which the other receives, see CCIR Reports 1010[111] and 999[112].

B.6.12 Interference between space stations and stations on Earth

If the atmosphere was perfectly transparent for all radio frequencies used in space radio systems, the calculation of interference levels as between space stations and stations on Earth would require merely the application of the inverse distance squared equation to calculate the transmission loss; see in particular expression (B.9). CCIR Report 885[113] provides means for allowing for the imperfections of the atmosphere as a transmission medium in making such calculations.

B.7 References

1 'Calculation of free-space attenuation'. CCIR Recommendation 525–1; Recommendations and Reports of the CCIR, 1986, Volume V (ITU, Geneva, 1986)

2 'The concept of transmission loss for radio links'. CCIR Recommendation 341–2; *ibid.*, Volume V

3 'Ground-wave propagation in an exponential atmosphere'. CCIR Report 714–1; *ibid.*, Volume V

4 'Reference atmosphere for refraction', CCIR Recommendation 369–3; *ibid.*, Volume V

5 'Electrical characteristics of the surface of the Earth'. CCIR Recommendation 527–1; *ibid.*, Volume V

6 'Electrical characteristics of the surface of the Earth'. CCIR Report 229–5; *ibid.*, Volume V

7 'Ground-wave propagation curves for frequencies between 10 kHz and 30 MHz'. CCIR Recommendation 368–5; *ibid.*, Volume V

8 'World atlas of ground conductivities'. CCIR Resolution 73; *ibid.*, Volume V

9 'World atlas of ground conductivities'. CCIR Report 717–2 (ITU, Geneva, 1988)

10 'The phase of the ground wave'. CCIR Report 716–2; Recommendations and Reports of the CCIR, 1986, Volume V (ITU, Geneva, 1986)

11 'Ionospheric properties'. CCIR Report 725–2; Recommendations and Reports of the CCIR, 1986, *ibid.*, Volume VI

12 'Special properties of the high latitude ionosphere affecting radiocommunications'. Report 886–1; *ibid.*, Volume VI

13 'CCIR Atlas of ionospheric characteristics'. CCIR Report 340–5 (ITU, Geneva 1987)

14 'Improvement in the worldwide ionospheric observing programme for numerical mapping purposes'. CCIR Report 430–4; Recommendations and Reports of the CCIR, 1986, Volume VI (ITU, Geneva, 1986)

15 'MINNIS, C.M.: 'Ionospheric indices', in SAXTON, J.A: 'Advances in radio research. Vol II' (Academic Press, London and New York, 1964)

16 LIU, R.Y., SMITH, P.A. and KING, J.W.: 'A new solar index which leads to improved f_{oF2} predictions using the CCIR Atlas', *Telecom. J., 1983,* **50**, pp. 408–414

17 'Use of an ionospherically derived solar activity index (IG) for the prediction of f_{oF2}'. Opinion 82; Recommendations and Reports of the CCIR, Volume VI, (ITU, Geneva, 1986)

18 *Telecommunications Journal.* Published monthly by ITU, Geneva

19 'Choice of indices for long-term ionospheric predictions'. CCIR Recommendation 371–5; Recommendations and Reports of the CCIR, 1986, Vol VI, (ITU, Geneva, 1986)

20 'CCIR interim method for estimating sky-wave field strength and transmission loss at frequencies between the approximate limits of 2 and 30 MHz'. CCIR Report 252–2 (ITU, Geneva, 1970)

21 MUGGLETON, L.M.: 'A method of predicting f_{oE} at any time and place', *Telecom J., 1975,* **42**, pp. 413–418

22 RUSH, C.M., POKEMPNER, M., ANDERSON, D., STEWART, F and PERRY, J.: 'A revised set of coefficients to represent the global variation of F_{oF2}'. Third International Conference on Antennas and Propagation, April 1983 (IEE, London, 1983)

23 'Second CCIR computer-based interim method for estimating sky-wave field strength and transmission loss at frequencies between 2 and 30 MHz'. Supplement to CCIR Report 252–2; see item 20 above (ITU, Geneva, 1980)

24 'VHF propagation by regular layers, sporadic E or other anomalous ionisation'. CCIR Report 259–6; Recommendations and Reports of the CCIR, 1986, Volume VI (ITU, Geneva, 1986)

25 'Ground and ionospheric side- and back-scatter'. CCIR Report 726–1; *ibid.*, Volume VI
26 'Ionospheric scatter propagation'. CCIR Report 260–3; *ibid.*, Volume VI
27 'Operational modelling of HF radio propagation conditions at high latitudes'. CCIR Report 1012; *ibid.*, Volume VI
28 'Method for calculating sporadic-E field strengths'. CCIR Recommendation 534–2; *ibid.*, Volume VI
29 'Short-term prediction of solar-induced variations of operational parameters for ionospheric propagation'. CCIR Report 727–2; *ibid.*, Volume VI
30 'Short-term forecasting of critical frequencies, operational maximum usable frequencies and total electron content'. CCIR Report 888–1; *ibid.*, Volume VI
31 'Exchange of information for short-term forecasts and transmission of ionospheric disturbance warnings'. CCIR Recommendation 313–5; *ibid.*, Volume VI
32 'Communication by meteor-burst propagation'. CCIR Report 251–4; *ibid.*, Volume VI
33 'Ionospheric cross-modulation'. CCIR Report 574–2; *ibid.*, Volume VI
34 'Ionospheric cross-modulation in the LF and MF broadcasting bands'. CCIR Recommendation 498–1; *ibid.*, Volume X–1
35 'Ionospheric modification by high power HF transmissions'. CCIR Report 728–2; *ibid.*, Volume VI
36 'Ionospheric modification by high power transmissions'. CCIR Recommendation 532; *ibid.*, Volume VI
37 'Solar power satellites and the ionosphere'. CCIR Report 893; *ibid.*, Volume VI
38 'Artificial modification of the ionosphere by chemical injections'. CCIR Report 1011; *ibid.*, Volume VI
39 'Long-distance ionospheric propagation without intermediate ground reflection'. CCIR Report 250–6; *ibid.*, Volume VI
40 'Antenna characteristics important to the analysis and prediction of sky-wave propagation paths'. CCIR Report 891–1; *ibid.*, Volume VI
41 'Ionospheric propagation characteristics pertinent to terrestrial radiocommunication systems design (Fading)'. CCIR Report 266–6; *ibid.*, Volume VI
42 'Fading characteristics for sound broadcasting in the tropical zone'. CCIR Report 304–2; *ibid.*, Volume X–1
43 'Real-time channel evaluation of ionospheric radio circuits'. CCIR Report 889–1; *ibid.*, Volume VI
44 'Ionospheric sounding at oblique incidence'. CCIR Report 249–6; *ibid.*, Volume VI
45 'The operational use of side-scatter and back-scatter'. CCIR Report 890–1; *ibid.*, Volume VI
46 'Sky-wave propagation and circuit performance at frequencies below about 30 kHz'. CCIR Report 895–1; *ibid.*, Volume VI
47 'Sky-wave propagation and circuit performance at frequencies between about 30 kHz and 500 kHz'. CCIR Report 265–6; *ibid.*, Volume VI
48 'ELF, VLF and LF propagation in and through the ionosphere'. CCIR Report 262–6; *ibid.*, Volume VI
49 'Ionospheric effects upon Earth–space propagation'. CCIR Report 263–6; *ibid.*, Volume VI
50 'The concept of "worst month"'. CCIR Recommendation 581–1; *ibid.*, Volume V
51 'Worst-month statistics'. CCIR Report 723–2; *ibid.*, Volume V
52 'Statistical distributions in radio-wave propagation'. CCIR Report 1007; *ibid.*, Volume V
53 BEAN, B.R. and DUTTON, E.J.: 'Radio meteorology' (Dover, NY, USA, 1966)
54 'Radiometeorological data'. CCIR Report 563–3; Recommendations and Reports of the CCIR, 1986, Volume V (ITU, Geneva, 1986)
55 'Definitions of terms relating to propagation in the troposphere'. CCIR Recommendation 310–6; *ibid.*, Volume V

56 'Effects of large-scale tropospheric refraction on radiowave propagation'. CCIR Report 718–2; *ibid.*, Volume V

57 CRANE, R.K.: 'Propagation phenomena affecting satellite communication systems operating in the centimetre and millimetre wavelength bands'. *Proc. IEEE*, 1971, **59**, pp. 173–188

58 DOUGHERTY, H.T. and HART, B.A.: 'Recent progress in duct propagation predictions', *IEEE Trans.*, 1979, **AP–27**, pp. 542–548

59 'Effects of propagation on the design and operation of line-of-sight radio relay systems'. CCIR Report 784–1; Recommendations and Reports of the CCIR, 1986, Vol IX–1 (ITU, Geneva, 1986)

60 'Cross-polarisation due to the atmosphere'. CCIR Report 722–2; *ibid.*, Volume V

61 'Reflection from the surface of the Earth'. CCIR Report 1008; *ibid.*, Volume V

62 'Attenuation by hydrometeors, in particular precipitation, and other atmospheric particles'. CCIR Report 721–2; *ibid.*, Volume V

63 'OLSEN, R.L., ROGERS, D.V. and HODGE, D.B.: The aR^{α} relation in the calculation of rain attenuation', *IEEE Trans.*, 1978, **AP–26**, pp. 318–329

64 'Propagation data and prediction methods required for line-of-sight radio-relay systems'. CCIR Report 338–5; Recommendations and Reports of the CCIR, 1986, Volume V (ITU, Geneva, 1986)

65 'Propagation data and prediction methods required for Earth–space telecommunication systems'. CCIR Report 564–3; *ibid.*, Volume V

66 'Use of frequency bands above 10 GHz in the fixed-satellite service'. CCIR Report 552–2; *ibid.*, Volume IV–1

67 MISME, P.: 'Etude experimentale de la propagation des ondes millimetriques dans les bandes de 5 et 3 mm', *Ann. des Telecomm.*, 1966, **21**, pp. 226–234

68 'Attenuation by atmospheric gases'. CCIR Report 719–2; Recommendations and Reports of the CCIR, 1986, Volume V (ITU, Geneva, 1986)

69 'Propagation data and prediction methods required for trans-horizon radio-relay systems'. CCIR Report 238–5; *ibid.*, Volume V

70 'The evaluation of propagation factors in interference problems between stations on the surface of the Earth at frequencies above about 0·5 GHz'. CCIR Report 569–3; *ibid.*, Volume V

71 'Propagation effects on the design and operation of trans-horizon radio relay systems'. CCIR Report 285–6; *ibid.*, Volume IX–1

72 'AWAKA, J.: 'Estimation of rain scatter interference in the case of medium scale broadcasting satellite for experimental purposes (BSE)', *J. Radio Res. Labs. (Japan)*, 1978, **25**, pp. 23–49

73 'Scattering by precipitation'. CCIR Report 882–1; Recommendations and Reports of the CCIR, 1986, Volume V (ITU, Geneva, 1986)

74 Doc. 5/305 (Federal Republic of Germany); CCIR Study Period 1978–82

75 'Propagation by diffraction'. CCIR Report 715–2; Recommendations and Reports of the CCIR, 1986, Volume V (ITU, Geneva, 1986)

76 'Influence of terrain irregularities and vegetation on tropospheric propagation'. CCIR Report 236–6; *ibid.*, Volume V

77 'Propagation statistics required for broadcasting services using the frequency range 30–1000 MHz'. CCIR Report 239–6; *ibid.*, Volume V

78 'VHF and UHF propagation curves for the frequency range from 30 MHz to 1000 MHz, Broadcasting services'. CCIR Recommendation 370–5; *ibid.*, Volume V

79 'Methods and statistics for estimating field strength values in the land mobile services using the frequency range 30 MHz to 1 GHz'. CCIR Report 567–3; *ibid.*, Volume V

80 'Improvement of the reception quality in automobiles for frequency modulation sound broadcasts in band 8 (VHF)'. CCIR Report 1067; *ibid.*, Volume X–1

81 'Short distance radio-wave propagation in special environments, buildings, tunnels, mines etc'. CCIR Report 880–1; *ibid.*, Volume V

82 DELOGNE, P.: 'Radio goes underground', *IEEE Spectrum*, 1980, **17**, pp. 26–29

83 DELOGNE, P.: 'Leaky feeders and subsurface radio communications' (Peter Peregrinus, London, 1982)
84 'Radiating/leaky cable systems in the land mobile service'. CCIR Report 902; Recommendations and Reports of the CCIR, 1986, Volume VIII–1 (ITU, Geneva, 1986)
85 'Measurement of atmospheric radio noise from lightning'. CCIR Report 254–6; *ibid.,* Volume VI
86 'Characteristics and applications of atmospheric radio noise data'. CCIR Report 322–3 (ITU, Geneva, 1988)
87 'Radio emission from natural sources in the frequency range above about 50 MHz'. CCIR Report 720–2; Recommendations and Reports of the CCIR, 1986, Volume V (ITU, Geneva, 1986)
88 'Form of the hypothetical reference circuit and allowable noise standard for frequency division multiplex telephony and television in the fixed-satellite service'. CCIR Report 208–6; *ibid.,* Volume IV–1
89 'Earth station antennas for the fixed-satellite service'. CCIR Report 390–5; *ibid.,* Volume IV–1
90 'Man-made radio noise'. CCIR Report 258–4; *ibid.,* Volume VI
91 'World-wide minimum external noise levels, 0·1 Hz to 100 GHz'. CCIR Report 670; *ibid.,* Volume I
92 'The accuracy of predictions of sky-wave field strength in bands 5 (LF) and 6 (MF)'. CCIR Report 432–2; *ibid.,* Volume VI
93 'Methods for predicting sky-wave field strengths at frequencies between 150 kHz and 1705 kHz'. CCIR Report 575–3; *ibid.,* Volume VI
94 'Analysis of sky-wave propagation measurements for the frequency range 150 to 1600 kHz'. CCIR Report 431–4; *ibid.,* Volume VI
95 'Prediction of sky-wave field strength between 150 and 1600 kHz'. CCIR Recommendation 435–5; *ibid.,* Volume VI
96 'Long-term ionospheric propagation predictions'. CCIR Report 255–6; *ibid.,* Volume VI
97 'Comparison between observed and predicted sky-wave signal intensities at frequencies between 2 and 30 MHz'. CCIR Report 571–2; *ibid.,* Volume VI
98 'Developments in the estimation of sky-wave field strength and transmission loss at frequencies above 1·5 MHz'. CCIR Report 729–2; *ibid.,* Volume VI
99 'Computation of reliability for HF radio systems'. CCIR Report 892–1; *ibid.,* Volume VI
100 'Simple HF propagation prediction method for MUF and field strength'. CCIR Report 894–1; *ibid.,* Volume VI
101 'Propagation data for terrestrial maritime mobile services operating at frequencies about 30 MHz'. CCIR Recommendation 616; *ibid.,* Volume V
102 'Propagation curves for aeronautical services using the VHF, UHF and SHF bands'. CCIR Recommendation 528–2; *ibid.,* Volume V
103 'Propagation data required for terrestrial broadcasting and point-to-multipoint communications systems in the frequency bands above 10 GHz; CCIR Report 562–3; *ibid.,* Volume V
104 'Propagation data for broadcasting from satellites'. CCIR Report 565–3; *ibid.,* Volume V
105 'Propagation data for maritime mobile-satellite systems for frequencies above 100 MHz'. CCIR Report 884–1; *ibid.,* Volume V
106 'Propagation data for land mobile-satellite systems for frequencies above 100 MHz'. CCIR Report 1009, *ibid.,* Volume V
107 'Propagation data required for the evaluation of interference between stations on the surface of the Earth'. CCIR Recommendation 452–4; *ibid.,* Volume V
108 'Determination of the interference potential between earth stations and terrestrial stations'. CCIR Report 448–4; *ibid.,* Volume IV/IX–2
109 'Determination of co-ordination area'. CCIR Report 382–5; *ibid.,* Volume IV/IX–2

110 'Propagation data required for the evaluation of co-ordination distance in the frequency range 1–40 GHz'. CCIR Report 724–2; *ibid.*, Volume V
111 'Propagation data for bi-directional co-ordination of earth stations'. CCIR Report 1010; *ibid.*, Volume V
112 'Determination of the bi-directional co-ordination area'. CCIR Report 999; *ibid.*, Volume IV–1
113 'Propagation data required for evaluating interference between stations in space and those on the surface of the Earth'. CCIR Report 885–1; *ibid.*, Volume V

Appendix C

Abbreviations used in the text

AeFS	aeronautical fixed service
AeM	aeronautical mobile service
AeM (OR)	aeronautical mobile service (off-route)
AeM (R)	aeronautical mobile service (route)
AeMSS	aeronautical mobile-satellite service
AeMSS (OR)	aeronautical mobile-satellite service (off-route)
AeMSS (R)	aeronautical mobile-satellite (route)
AeRN	aeronautical radionavigation service
AeRNSS	aeronautical radionavigation-satellite service
AM	amplitude modulation
AmS	amateur service
AmSS	amateur-satellite service
APT	Asia–Pacific Telecommunity
ARC	administrative radio conference
ARQ	automatic error correction by repetition on request
ATU	Arab Telecommunication Union
BER	bit error ratio
BS	broadcasting service (also used to indicate Broadcasting Satellite in the designation WARC-BS-77)
BSS	broadcasting-satellite service
C	degrees Celsius
CCIR	International Radio Consultative Committee
CCITT	International Telegraph and Telephone Consultative Committee
CEPT	European Conference of Posts and Telecommunications Administrations
CISPR	International Special Committee on Radio Interference
CITEL	Inter-American Telecommunications Conference
CMTT	Joint Study Group for Television and Sound Transmission
CMV	Joint Study Group for Vocabulary
CPM	conference preparatory meeting
CW	continuous wave
dB	decibel
dBd	decibels relative to the maximum gain of a half-wave dipole
dBi	decibels relative to the gain of an isotropic antenna
DBS	direct broadcasting by satellite
dBW	decibels relative to 1 watt

DF	direction-finding
DME	distance measuring equipment
DSB	double sideband
EESS	earth exploration-satellite service
EHF	frequency range 3 to 30 GHz
EIRP	equivalent isotropically radiated power
EMC	electromagnetic compatibility
EMRP	equivalent monopole radiated power
ENG	electronic news gathering
EPIRB	emergency position-indicating radio beacon
ERP	effective radiated power (relative to a half-wave dipole)
FBW	forward band working
FDM	frequency division multiplex
FDMA	frequency division multiple access
FEC	forward error correction
FM	frequency modulation
FS	fixed service
FSK	frequency shift keying
FSS	fixed-satellite service
g	gramme
GESSN	Global Effective Sunspot Number
GHz	gigahertz (= 1000 megahertz)
GMDSS	Global Maritime Distress and Safety System
GPS	Global Positioning System
GSO	geostationary satellite orbit
h	hour
HDTV	high definition television
HF	frequency range 3 to 30 MHz
HFBC	high frequency broadcasting
HRC	hypothetical reference circuit
HRDP	hypothetical reference digital path
HSSWS	high speed solar wind stream
Hz	hertz
IARU	International Amateur Radio Union
ICAO	International Civil Aviation Organisation
IEC	International Electrotechnical Commission
IEV	International Electrotechnical Vocabulary
IF	intermediate frequency
IFL	International Frequency List
IFRB	International Frequency Registration Board
ILS	instrument landing system
IMO	International Maritime Organisation
ISDN	integrated services digital network
ISL	inter-satellite link
ISM	industrial, scientific and medical (applications)
ISS	inter-satellite service
ITU	International Telecommunication Union
IWP	Interim Working Party

J	joule
JIWP	Joint Interim Working Party
K	kelvin
kbit/s	kilobits per second
kHz	kilohertz (= 1000 hertz)
km	kilometre
kW	kilowatt
LF	frequency range 30 to 300 kHz
LM	land mobile service
LMSS	land mobile-satellite service
m	metre
MAC	multiplexed analogue components
Mbit/s	megabits per second
MetA	meteorological aids service
MetS	meteorological-satellite service
MF	frequency range 300 kHz to 3 MHz
MHz	megahertz (= 1000 kilohertz)
MIFR	Master International Frequency Register
MLS	microwave landing system
MM	maritime mobile service
mm	millimetre
MMSS	maritime mobile-satellite service
MOB	mobile (use in the designations of certain WARCs)
MPM	multilateral planning meeting
MRN	maritime radionavigation service
MRNSS	maritime radionavigation-satellite service
MS	mobile service
MSS	mobile-satellite service
mW	milliwatt
MUF	maximum usable frequency
MVDS	multipoint video distribution system
MWARA	Major World Air Route Area
NTSC	a television colour-encoding system
nW	nanowatt
ORB	orbit (used in the designations of certain WARC's)
OWF	optimum working frequency
PAL	a television colour-encoding system
PC	plenipotentiary conference
PCA	polar cap absorption
PCM	pulse code modulation
PDA	pre-determined arc
PEP	peak envelope power
PFD	power flux density
PMR	private mobile radio
p.p.m.	parts per million
PSK	phase shift keying
PSTN	public switched telephone network
pW	picowatt

pW0p	picowatts, psophometrically weighted, measured at a point of zero relative level in public telephone network
RA	radio astronomy service
RARC	regional administrative radio conference
RBW	reverse band working
RD	radiodetermination service
RDARA	Regional and Domestic Air Route Area
RDSS	radiodetermination-satellite service
RF	radio frequency
RL	radiolocation service
r.m.s.	root mean square
RN	radionavigation service
RNSS	radionavigation-satellite service
RPOA	recognised private operating agency
RR	The Radio Regulations
RTCE	real time channel evaluation
SAT	satellite (used in the designation of RARC-SAT-83)
SCPC	single channel per carrier
SECAM	a television colour-encoding system
SETI	search for extra-terrestrial intelligence
SFS	standard frequency and time signal service
SFSS	standard frequency and time signal-satellite service
SHF	the frequency range 3 GHz to 30 GHz
SID	sudden ionospheric disturbance
SIO	scientific or industrial organisation
SNG	satellite news gathering
SpO	space operation service
SRS	space research service
SSB	single sideband
SSR	secondary surveillance radar
TDM	time division multiplex
TDMA	time division multiple access
TEC	total electron content
TTC	telemetry, tracking and command
UHF	frequency range 300 MHz to 3 GHz
UN	The United Nations
URSI	International Union of Radio Science
UT	universal time
V	volt
VHF	frequency range 30 MHz to 300 MHz
VLF	frequency range 3 to 30 kHz
VOR	VHF omnidirectional ranging system
VSAT	very small aperture terminal
W	watt
WARC	World Administrative Radio Conference
WMO	World Meteorological Organisation

Index